Red Hat® Certified
System Administrator
& Engineer

Exams
EX200 and EX300
Red Hat Enterprise Linux 6

Training Guide
&
A Quick Deskside Reference

December 2012

Asghar Ghori

1246 Heil Quaker Blvd., La Vergne, TN USA 37086
Chapter House, Pitfield, Kiln Farm, Milton Keynes, UK MK11-3LW
www.lightningsource.com

HAGAR

Technical Reviewers: Syed R. Ali, Kurt Glasgow, Oleg Waisberg, and Mehmood Khan
Editors: Masood Pehlvi and Joan Weytze
Cover Design: Wajahat Syed
Printers and Distributors: Lightning Source Inc.

Printed in the USA, UK, France, Germany, Italy, Spain, and Australia.

ISBN: 978-1-4675-4940-0
Library of Congress Control Number: 2012955994

Printed and Distributed by: Lightning Source Inc.

To order in bulk at special quantity discounts for sales promotions or for use in training programs, please contact the author at *asghar_ghori2002@yahoo.com*

Preface

Red Hat's EX200 and EX300 exams are performance-based and touch a wide array of topics. The exams present configuration scenarios to be accomplished on live systems within an allotted time frame. This book provides a single, comprehensive resource that equips readers with enough knowledge to comfortably pass both exams. At the same time, this book presents procedures to help configure and administer a Red Hat Enterprise Linux (RHEL) 6 computing environment effectively and efficiently.

Throughout this book, step-by-step implementation procedures have been furnished. My suggestions to you are to study the material thoroughly; perform the full installation, configuration, and administration tasks provided; perform all exercises diligently; answer all questions presented at the end of each chapter; attempt all lab exercises offered at the end of each chapter; and then attempt the sample exam exercises presented in the appendix. While you study the book, have one 64-bit laptop or desktop computer with at least one dual-core processor, 4GB of memory, and built-in support for hardware virtualization available for implementing practice exercises and labs. I would recommend that you purchase a subscription for RHEL6 for one of the systems so as to have access to the software repository and benefits offered by the Red Hat Network. Alternatively, you can download a free copy of either CentOS or Scientific Linux, which are non-commercial versions of RHEL and are 100% compatible with it. While performing exercises and labs, if a command does not produce desired results, see what message it generates and try to resolve it. Minor issues, such as a wrong path, prevent commands from being executed. Sometimes, there are syntax errors in the command construct. You might need to adjust settings presented in exercises to make them work on your systems. If the exercises do not produce the expected results, check settings on your computer, modify what is not working and see if the modifications that you have made work. RHEL manual pages prove helpful and useful in comprehending commands and their syntax. There are numerous commands, command options, service daemons, and configuration, log, and startup/shutdown files in the operating system; discussing all of them is not possible within the scope of this book.

There are four areas where you should be focusing on in order to gain expertise with RHEL (CentOS or Scientific Linux for that matter), as well as preparing yourself for the exams: 1) grasping concepts; 2) mastering implementation procedures, exercises, and labs; 3) learning commands, configuration files, and service daemons; and 4) being able to troubleshoot and resolve problems. An excellent understanding of which command involves which options and updates which files, which daemon provides what services, etc. should also be developed. This way you will have a better overall understanding of what exactly happens in the background when a command is executed. This book provides all that knowledge. Troubleshooting becomes easier when concepts are clear and working knowledge is solid.

I am maintaining the *www.getitcertify.com* website where errors encountered in the book, additional exam information, and links to useful resources are made available. I encourage you to visit this website.

At the end, I would like to ask you to forward your constructive feedback to my personal email address *asghar_ghori2002@yahoo.com* about the content, structure, layout, consistency, clarity, and flow of the book. Please also let me know if you come across any errors or mistakes. Improvement is a continuous process and I am sure your feedback will help me publish a better and improved next edition.

Good luck in your endeavors.

Asghar Ghori
December 2012
Toronto, Canada

Acknowledgments

I am grateful to God who enabled me to write this book successfully.

I would like to thank my friends, colleagues, students, and peers who supported and encouraged me throughout the lifecycle of writing this book. I am grateful to all of them for their invariable support, constant encouragement, constructive feedback, and valuable assistance.

I would like to pay special thanks to my wife, two daughters, and two sons, who endured me while I was away from them hiding in my basement and working on this project. This work could not have been accomplished without the assistance, support, and love that they had demonstrated.

Finally, I would like to pay my best regards to my mother and my deceased father.

Asghar Ghori

About the Author

Currently working as a consultant, Asghar Ghori has been in the IT industry for 22 years. He started his career as a UNIX support engineer working with SCO UNIX and SCO XENIX, visiting customers and providing UNIX technical support services. He has worked with several flavors of the UNIX operating system including HP-UX, Oracle Solaris, and IBM AIX, as well as Red Hat Enterprise Linux and a few other Linux distributions. He has worked in support and administration, and as technical lead, solution design lead, and consultant serving business and government customers. In addition, he has architected and deployed solutions involving clustered servers and tiered storage that span data centers around the world.

Asghar has been involved in planning, developing, and executing disaster recovery procedures for many enterprises.

Asghar holds a BS in Engineering, and has delivered and attended numerous training programs. He has been delivering courses on UNIX & Linux for the past eleven years at local colleges in Toronto, Canada. He is Red Hat Certified System Administrator (RHCSA), HP Certified Systems Administrator, HP Certified Systems Engineer, SUN Certified System Administrator (SCSA), IBM Certified Specialist for AIX, Certified Novell Engineer (CNE), and holds the Project Management Professional (PMP) certification designation.

Conventions Used in this Book

The following typographic and other conventions are used throughout this book:

Book Antiqua Italic is used in text paragraphs for special words or phrases that are emphasized. For example:

> "Linux, like all UNIX operating systems, is a *multi-user, multi-tasking, multi-processing* and *multi-threading* operating system. This means that ….."

Times Roman Italic is used in text paragraphs and Tables to highlight file and directory names, commands, daemons, hostnames, usernames, and URLs. For example:

> "To go directly from */etc* into a sub-directory *dir1* under *aghori's* home ….."

Times Roman Bold is used to highlight commands and command line arguments that the user is expected to type at the command prompt. For example:

> **$ ls –lt**

All headings and sub-headings are in California FB font, and are bolded.

Ctrl+x key sequence implies that you hold down the Ctrl key on the keyboard and then press the other key. Courier New font is used for such sequences, as well as it is used to highlight keyboard keys such as Enter, Esc, and so on.

Andalus 9 is used to segregate command output, shell script contents, and information expected to be entered in configuration files from the surrounding text.

. . . . Dotted lines show the continuation in command output.

 Indicates additional information.

Indicates a task performed using a RHEL graphical tool.

About the RHCSA™ and RHCE™ Exams

The Red Hat Certified System Administrator (RHCSA™) and the Red Hat Certified Engineer (RHCE™) certification exams are performance-based, hands-on practical exams designed for IT professionals. These exams are presented in electronic format. During the exams, the candidates do not have access to the Internet or other printed or electronic documentation except for what comes standard with the RHEL6 operating system software. The official exam objectives are listed at *http://www.redhat.com/training/courses/ex200/examobjective* for RHCSA and that for RHCE at *http://www.redhat.com/training/courses/ex300/examobjective*. Visit the URLs for up-to-date and more in-depth information. The exam objectives are covered in sufficient detail in the chapters throughout this book. A list of the exam objectives is presented below along with chapter numbers where the objectives are discussed.

RHCSA™ Specific Skills:

Understand and Use Essential Tools:

1. Access a shell prompt and issue commands with correct syntax (chapter 2)
2. Use input-output redirection (>, >>, |, 2>, etc) (chapter 4)
3. Use grep and regular expressions to analyze text (chapter 3)
4. Access remote systems using ssh and VNC (chapter 2)
5. Log in and switch users in multiuser run levels (chapter 11)
6. Archive, compress, unpack, and uncompress files using tar, star, gzip, and bzip2 (chapter 2)
7. Create and edit text files (chapter 4)
8. Create, delete, copy, and move files and directories (chapter 3)
9. Create hard and soft links (chapter 3)
10. List, set, and change standard ugo/rwx permissions (chapter 4)
11. Locate, read, and use system documentation including man, info, and files in /usr/share/doc (chapter 2)

Operate Running Systems

12. Boot, reboot, and shut down a system normally (chapter 8)
13. Boot systems into different runlevels manually (chapter 8)
14. Use single-user mode to gain access to a system (chapter 8)
15. Identify CPU/memory intensive processes, adjust process priority with renice, and kill processes (chapter 5)
16. Locate and interpret system log files (chapter 8)
17. Access a virtual machine's console (chapter 7)
18. Start and stop virtual machines (chapter 7)
19. Start, stop, and check the status of network services (chapter 8)

Configure Local Storage

20. List, create, delete, and set partition type for primary, extended, and logical partitions (chapter 9)
21. Create and remove physical volumes, assign physical volumes to volume groups, and create and delete logical volumes (chapter 9)
22. Create and configure LUKS-encrypted partitions and logical volumes to prompt for password and mount a decrypted file system at boot (chapter 10)
23. Configure systems to mount file systems at boot by Universally Unique ID (UUID) or label (chapter 10)
24. Add new partitions and logical volumes, and swap to a system non-destructively (chapters 9 and 10)

Create and Configure File Systems

25. Create, mount, unmount, and use ext2, ext3, and ext4 file systems (chapter 10)
26. Mount, unmount, and use LUKS-encrypted file systems (chapter 10)
27. Mount and unmount CIFS and NFS network file systems (chapters 10, 19, and 20)
28. Configure systems to mount ext4, LUKS-encrypted, and network file systems automatically (chapter 10)
29. Extend existing unencrypted ext4-formatted logical volumes (chapter 9)
30. Create and configure set-GID directories for collaboration (chapter 4)
31. Create and manage Access Control Lists (ACLs) (chapter 10)
32. Diagnose and correct file permission problems (chapter 4)

Deploy, Configure, and Maintain Systems

33. Configure networking and hostname resolution statically or dynamically (chapter 13)
34. Schedule tasks using cron (chapter 5)
35. Configure systems to boot into a specific runlevel automatically (chapter 8)
36. Install Red Hat Enterprise Linux automatically using Kickstart (chapter 7)
37. Configure a physical machine to host virtual guests (chapter 7)
38. Install Red Hat Enterprise Linux systems as virtual guests (chapter 7)
39. Configure systems to launch virtual machines at boot (chapter 7)
40. Configure network services to start automatically at boot (chapter 8)
41. Configure a system to run a default configuration HTTP server (chapter 7)
42. Configure a system to run a default configuration FTP server (chapter 7)
43. Install and update software packages from Red Hat Network, a remote repository, or from the local file system (chapter 6)
44. Update the kernel package appropriately to ensure a bootable system (chapter 8)
45. Modify the system bootloader (chapter 8)

Manage Users and Groups

46. Create, delete, and modify local user accounts (chapter 11)
47. Change passwords and adjust password aging for local user accounts (chapter 11)
48. Create, delete, and modify local groups and group memberships (chapter 11)
49. Configure a system to use an existing LDAP directory service for user and group information (chapter 13)

Manage Security

50. Configure firewall settings using system-config-firewall or iptables (chapter 12)
51. Set enforcing and permissive modes for SELinux (chapter 12)
52. List and identify SELinux file and process context (chapter 12)
53. Restore default file contexts (chapter 12)
54. Use boolean settings to modify system SELinux settings (chapter 12)
55. Diagnose and address routine SELinux policy violations (chapter 12)

RHCE™ Specific Skills:

System Configuration and Management

56. Route IP traffic and create static routes (chapter 16)
57. Use iptables to implement packet filtering & configure network address translation (NAT) (chapter 15)
58. Use /proc/sys and sysctl to modify and set kernel runtime parameters (chapter 17)
59. Configure a system to authenticate using Kerberos (chapter 16)
60. Build a simple RPM that packages a single file (chapter 14)
61. Configure a system as an iSCSI initiator that persistently mounts an iSCSI target (chapter 14)
62. Produce and deliver reports on system utilization (processor, memory, disk, and network) (chapter 15)
63. Use shell scripting to automate system maintenance tasks (chapter 14)
64. Configure a system to log to a remote system (chapter 15)
65. Configure a system to accept logging from a remote system (chapter 15)

Network Services

Network services are an important subset of the exam objectives. RHCE candidates should be capable of meeting the following objectives for each of the network services listed below:

- Install the packages needed to provide the service
- Configure SELinux to support the service
- Configure the service to start when the system is booted
- Configure the service for basic operation
- Configure host-based and user-based security for the service

HTTP/HTTPS (chapter 21)
- Configure a virtual host
- Configure private directories
- Deploy a basic CGI application
- Configure group-managed content

DNS (chapter 16)
- Configure a caching-only name server
- Configure a caching-only name server to forward DNS queries

FTP (chapter 20)
- Configure anonymous-only download

NFS (chapter 19)
- Provide network shares to specific clients
- Provide network shares suitable for group collaboration

SMB (chapter 20)
- Provide network shares to specific clients
- Provide network shares suitable for group collaboration

SMTP (chapter 18)
- Configure a mail transfer agent (MTA) to accept inbound email from other systems
- Configure an MTA to forward (relay) email through a smart host

SSH (chapter 21)
- Configure key-based authentication
- Configure additional options described in documentation

NTP (chapter 17)
- Synchronize time using other NTP peers

Suggestions When Taking the Exams

1. Save time wherever possible as time is of the essence during the exams. Perform tasks using either text or graphical tools, whichever you feel more comfortable with. Install a graphical tool if you need it and if it is not already loaded.
2. Make certain that any changes that you have made must survive across system reboots.
3. Use any text editor that you feel comfortable with to modify text files.
4. Inform the exam proctor right away if you identify a hardware issue with your system.
5. Exams are administered with books closed and no access to the Internet and electronic devices.
6. Read each exam task carefully and understand it thoroughly before attempting it.

Exam Fees and How to Register

The fee for either the RHCSA or the RHCE exam is US$400. To register for an exam, visit *www.redhat.com/training/courses/ex200* or *www.redhat.com/training/courses/ex300* and click ENROLL TODAY to enroll online. Note that the exams are administered on Fridays only. The RHCSA exam begins at 9am and the RHCE exam at 2pm.

About this Book

This book covers three main objectives: 1) to provide a resource to IT administrators who intend to take the new Red Hat Certified System Administrator (EX200) and the Red Hat Certified Engineer (EX300) exams and pass them; 2) to provide a quick and valuable on-the-job resource to Linux/UNIX administrators and programmers and DBAs working in the Linux environment; and 3) to provide an easy-to-understand guide to novice and non-RHEL administrators who intend to learn RHEL.

This book is divided into RHCSA and RHCE sections based on the exam objectives. It includes 21 chapters in total and is structured to facilitate readers to grasp concepts, understand implementation procedures, and learn commands, tools, configuration files, log files, and service daemons involved.

1. **The RHCSA Section** (chapters 1 to 13) covers all the topics required to prepare for the RHCSA exam. Information presented includes general Linux concepts, basic commands, archiving, compression, file manipulation, file security, text editors, shell features, awk programming, hardware management, local installation, package administration, virtualization, network installation, automated installation, boot procedures, kernel management, user & group administration, disk partitioning, file system & swap management, basic firewall, SELinux, TCP Wrappers, basic networking, network interfaces, and DNS, DHCP, and LDAP client configuration.

2. **The RHCE Section** (chapters 14 to 21) covers all the topics required to prepare for the RHCE exam. Information presented includes automating with scripting, building an rpm package, resource utilization reporting, network logging, advanced firewall, iSCSI, DNS server, Kerberos, routing, Internet services, kernel parameters, NTP client, email, NFS, Samba, FTP, Apache web server, and secure shell server.

Each chapter in the book begins with a list of major topics and exam objectives covered in the chapter and ends with a summary followed by review questions and practice lab exercises. Throughout the book, tables, figures, and screen shots have been furnished for ease of understanding. This book includes one sample exam for RHCSA and one sample exam for RHCE in the appendix, and you are expected to accomplish them on your systems using the knowledge and skills gained from reading this book and practicing the exercises and labs presented.

TABLE OF CONTENTS

11. Users & Groups 317

12. Firewall & SELinux 345

Section Two: **RHCE**

List of Figures

List of Tables

Section One
RHCSA

Chapter 01

Local Installation

This chapter covers the following major topics:

- ✓ A brief history of Linux and Open Source
- ✓ An introduction to Linux distributions from Red Hat
- ✓ An overview of the structure and features of Linux
- ✓ How to obtain Red Hat Enterprise Linux 6
- ✓ Plan for the installation
- ✓ An overview of installation logs and virtual consoles
- ✓ Hardware configuration needed for exam preparation
- ✓ Install RHEL6 on a physical computer using local DVD media
- ✓ Perform post-installation configuration

This chapter includes the following RHCSA certification objectives:

None, but sets up the foundation for learning and practicing the exam objectives

This chapter provides a brief overview of Linux and distributions available from Red Hat Inc. It also touches on the structure and features of Linux. The remainder of this chapter explains in detail how to obtain a copy of RHEL6 installation software and perform a fresh installation on a physical computer using a local, physical DVD medium. The install process requires prior planning to identify several system configuration pieces such as locale and networking information, disk drive and disk management software, types and sizes of file systems, software groups and components, etc. Identification of these items makes the RHEL6 installation process smoother. RHEL6 may be installed using either a graphical or text-mode interface. The graphical interface is the default and that is what we are going to use in this chapter for demonstration purposes.

Overview of Linux

Linux is a free computer operating system and is similar to the UNIX operating system in terms of concepts, features, and functionality, and, therefore, is referred to as a UNIX-like operating system.

In 1984, an initiative was undertaken by Richard Stallman whose primary goal was to create a completely free, UNIX-compatible, open source operating system with global collaboration from software developers. The initiative was called the GNU (*GNU's Not Unix*) Project and by 1991, a significant operating system had been developed. The only critical piece missing was a kernel to drive it. This gap was filled by the kernel created by Linus Torvalds in 1991 during his computer science studies. The name "Linux" was given to the new operating system kernel that Torvalds developed, which was very UNIX-like. The Linux operating system was released under the GNU *General Public License* (GPL) and initially written to run on Intel x86 architecture computers. The first version (0.01) of the operating system was released in September 1991 with little more than 10,000 lines of code. In March 1994, the first major release (1.0.0) debuted followed by version 2.0.0 in June 1996, 2.2.0 in January 1999, 2.4.0 in January 2001, and 2.6.0 in December 2003. At the time of this writing, version 3.2, with close to 15 million lines of code, is the latest kernel. The Linux kernel, and the operating system in general, has been enhanced with contributions from thousands of software programmers around the world under the GNU GPL, which provides general public access to the Linux source code free of charge with full consent to amend and redistribute.

Today, Linux runs on a variety of computer hardware platforms, from laptop and desktop computers to massive mainframe systems. Linux also runs as the base operating system on a range of other electronic devices such as routers, switches, RAID arrays, tape libraries, video games, and hand-held devices. Numerous vendors including Red Hat, HP, IBM, Oracle, Novell, and Dell offer support to Linux users worldwide.

The functionality, adaptability, portability, and cost-effectiveness that Linux has to offer has made this operating system the main alternative to proprietary UNIX and Windows operating systems. At present, over a hundred different flavors of Linux are circulating from various vendors, organizations, and individuals; only a few of them are popular and have wide acceptance.

Linux is largely used in government agencies, corporate businesses, academic institutions, and scientific organizations, as well as in home computers. Linux deployment and usage is constantly on the rise.

Linux Distributions from Red Hat

Red Hat, Inc., a company founded in 1993, assembled an operating system called *Red Hat Linux* (RHL) under the GNU GPL and released their first version as Red Hat Linux 1.0 in November 1994. Several versions followed until the last version in the series, called Red Hat Linux 9 (later also referred to as RHEL 3), based on the 2.4.20 kernel, was released in March 2003. Red Hat renamed the Red Hat Linux operating system series to *Red Hat Enterprise Linux* (RHEL) in 2003.

RHL was originally assembled and enhanced within the Red Hat company. In 2003, Red Hat began sponsoring a project called *Fedora* inviting the user community to participate in enhancing the source code. This enabled the company to include the improved code in successive versions of RHEL. The Fedora distribution is completely free, whereas, RHEL is commercial. RHEL 4 (based on the 2.6.9 kernel and released in February 2005), RHEL 5 (based on the 2.6.18 kernel and released in March 2007), and RHEL6 (based on the 2.6.32 kernel and released in November 2010) have been built using Fedora distributions 3, 6, 12, and 13, respectively. The following are RHEL6 editions available for commercial purposes:

- ✓ Red Hat Enterprise Linux for Desktop (targeting desktop and laptop computers)
- ✓ Red Hat Enterprise Linux for Server (targeting small to large deployments)
- ✓ Red Hat Enterprise Linux for Scientific Computing (targeted for high-performance computing users)
- ✓ Red Hat Enterprise Linux for IBM Power (targeting IBM Power series computers)
- ✓ Red Hat Enterprise Linux for IBM System z (targeting IBM mainframe computers)
- ✓ Red Hat Enterprise Linux for SAP Business Applications (including infrastructure software stack to support the operation of SAP applications)

There are two operating systems that are 100% rebuild of Red Hat Enterprise Linux. They are called CentOS (*Community Enterprise Operating System*) and Scientific Linux, and are available for Linux users and learners as free of charge distributions. These rebuilds are not sponsored or supported by Red Hat. CentOS may be downloaded from its official website at *www.centos.org* and Scientific Linux from *www.scientificlinux.org*. For practice and training purposes, you may download and use the latest versions of one of these rebuilds instead of RHEL6.

System Structure

The structure of the Linux system is comprised of three main components: the *kernel*, the *shell*, and the *hierarchical directory structure*. These components are illustrated in Figure 1-1 and explained below.

The Kernel

The kernel controls everything on a system that runs Linux. It controls system hardware including memory, processors, disks, and I/O (Input/Output) devices. It schedules processes, enforces security, manages user access, and so on. The kernel receives instructions from the shell, engages appropriate hardware resources, and acts as instructed.

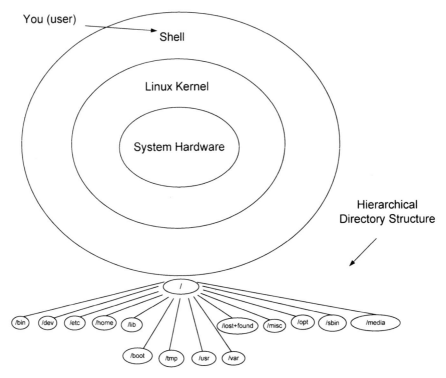

Figure 1-1 Linux System Structure

The Shell

The shell is the interface between a user and the kernel. The user provides instructions (commands) to the shell, which are interpreted and passed to the kernel for processing. The shell handles input and output, keeps track of data stored on disks, and communicates with peripheral devices such as monitors, hard disk drives, tape devices, CD/DVD drives, USB drives, printers, modems, and terminals. Chapter 04 "File Permissions, Text Editors, Text Processors & The Shell" discusses shells in detail.

The Hierarchical Directory Structure

Linux uses the conventional hierarchical directory structure where directories may contain both files and sub-directories. Sub-directories may further hold more files and sub-directories. A sub-directory, also referred to as a *child* directory, is a directory located under a *parent* directory. That parent directory is a sub-directory of some other higher-level directory. In other words, the Linux directory structure is similar to an inverted tree where the top of the tree is the root of the directory, and branches and leaves are sub-directories and files, respectively. The root of the directory is represented by the forward slash (/) character, which is also used as a directory separator in a path as shown below:

/home/user1/dir1/subdir1

In this example, the *home* sub-directory is located under *root* (/), making *root* the parent for *home* (which is the child). *user1* (child) is located under *home* (parent), *dir1* (child) is located under *user1* (parent) and at the very bottom *subdir1* (child) is located under *dir1* (parent).

Each directory has a parent directory and a child directory with the exception of the root (/) and the lowest level directories. The root (/) directory has no parent and the lowest level sub-directory has no child.

 The term sub-directory is used for a directory that has a parent directory.

The hierarchical directory structure keeps related information together in a logical fashion. Compare this concept with a file cabinet that has several drawers with each drawer storing multiple file folders.

Features

Linux, like all UNIX operating systems, is a *multi-user, multi-tasking, multi-processing*, and *multi-threading* operating system. This means that several users can access a Linux system simultaneously and share available resources. The system allows each logged in user to run any number of programs concurrently. The kernel is capable of handling multiple processors (CPUs) installed in the system and breaking large running programs into smaller, more manageable pieces called *threads*, for improved performance.

 A resource may be a hardware device or a software program or service.

The kernel allows *time-sharing* among running programs, and runs them in a round-robin fashion to satisfy processing requirements for each one of them.

Obtaining, Planning & Installing Red Hat Enterprise Linux 6

Red Hat Enterprise Linux 6 may be downloaded from Red Hat's website as explained in the following sub-section. Enough planning needs to happen so that the installation is performed as per the desired requirements. For installing the operating system, there are several options available and we will cover some of them in this chapter and later in the book.

Downloading RHEL6

If you do not already have the RHEL6 installation software, you need to download the ISO image of it and burn it to a DVD. Follow the steps provided below:

1. Go to *www.redhat.com* and click "Customer Portal".
2. Click "Log in" if you already have an account, or click "Register" and fill out the form to open a new account.
3. Click "Download Products and Updates" to get to the "Software & Download Center" page.

4. Click "Download Red Hat Enterprise Linux" if you already have an active subscription, or click "Obtain a free 30-day evaluation subscription" and then "Download a free Red Hat Enterprise Linux 30-day evaluation". Fill out your contact information on the form, and submit. You will receive a confirmation email shortly with instructions on how to download the evaluation copy.
5. Click "Red Hat Enterprise Linux Server (v. 6 for 64-bit x86_64)".
6. Click "Binary DVD" and save the image to your computer.
7. Use a DVD burner tool on your computer and burn the image to an empty, physical DVD.

In addition to allowing you to download the ISO image, the evaluation subscription gives you access to the Red Hat Network for 30 days. The RHEL6 software will continue to function without any issues beyond the 30-day period with the exception that you will be unable to obtain any software updates from the Red Hat Network.

Installation Media Options

There are multiple options available to you with respect to how you can access the installation software. You can have the software located on a DVD, in a Linux partition on the same computer, or on an NFS, HTTP, or FTP server accessible over the network. In this chapter, you will be performing an installation using the RHEL6 DVD that you have just created. (We will assume that you do not have another partition on the same computer to host the installation image. For using an NFS, HTTP, or FTP server as an install source, refer to Chapter 07 "Virtualization & Network Installation").

Hardware Requirements

The first thing you need to do is to understand the requirements for the installation. Critical items such as the type of system and CPUs, amount of physical memory, and types and sizes of hard disks need to be determined. Installation may fail or the system may not function as expected if any of these requirements are not met. The following is a discussion of these requirements.

System Requirements

There is a wide array of hardware platforms on which RHEL6 can be installed and run. The best place to check whether your hardware is certified to run RHEL6 without any glitches is to refer to the Red Hat's *Hardware Compatibility List* (HCL) available at *http://www.redhat.com/rhel/compatibility/hardware*. The HCL lists a number of computer system brands that have been tested and certified to work with RHEL6. However, a large number of computers and laptops (both branded and unbranded) that are not listed on the HCL may also work without issues. For laptop computers, there is a website called Linux on Laptops at *www.linux-laptop.net* which provides a wealth of information on thousands of different laptop models on which people have installed and run Linux and they have shared their experiences. If you are experiencing issues installing or running RHEL6 on your laptop computer, browse this website and search for your laptop model. It is very likely that you find a solution or a hint that would help you fix the issue you are facing.

CPU Requirements

RHEL6 is supported on computers with 32-bit and 64-bit Intel, AMD, and Power processors. However, you must use a 64-bit computer with integrated support for hardware virtualization for the purposes of RHCSA and RHCE exam preparation, or if you plan on hosting virtual machines.

RHEL is generally used as a server operating system, and servers typically have several CPUs. In order for an operating system to be able to use the power of all installed CPUs, the operating system must support them. RHEL6 has the ability to support a large number of installed CPUs and it has the integrated support for virtualization using *Kernel-based Virtual Machine* (KVM) hypervisor software. With KVM, you can create virtual machines, and then install RHEL or Microsoft Windows on them.

Physical Memory Requirements

Enough physical memory should be available in the system to support not only the full RHEL operating system software but also applications that the system is purposed to host. For a typical installation, reserve a minimum of 512MB memory for RHEL; add more for better performance.

Planning the Installation

Installation of RHEL6 requires that you have the following configuration information handy. You will be asked to input this information during the installation process.

- ✓ Language to be used during installation
- ✓ Type of keyboard
- ✓ Local storage or remote on iSCSI or SAN
- ✓ Hostname
- ✓ IP address, subnet mask and default gateway
- ✓ Primary and secondary DNS server names or IP addresses (optional)
- ✓ Time zone
- ✓ Root password
- ✓ Partitioning approach and file system, as well as swap details
- ✓ Location of the bootloader
- ✓ Software group and package selection
- ✓ User creation information (optional)
- ✓ LDAP server name or IP address (optional)
- ✓ NTP server names or IP addresses (optional)

Virtual Console Screens

During the installation, there are five virtual console screens available to monitor the installation process and to diagnose and fix any issues encountered. Additionally, there is a sixth screen, which is where you interact with the anaconda installer program in the graphical mode. You can switch between screens by pressing a combination of keys. The information displayed on the consoles is captured in installation log files.

Console #1 (Ctrl-Alt-F1): This is the screen where you enter keyboard and language information before anaconda begins and switches the default screen to the sixth screen.

Console #2 (Ctrl-Alt-F2): The bash shell interface for running commands.

Console #3 (Ctrl-Alt-F3): Displays installation log, which is also stored in *anaconda.log*. Information on hardware detected is also captured in this file, in addition to other information.

Console #4 (Ctrl-Alt-F4): Displays kernel messages, which are also stored in *install.log.syslog*. The information on hardware detected is captured in greater detail in this file.

Console #5 (Ctrl-Alt-F5): Displays file system creation information.

Console #6 (Ctrl-Alt-F6): This is the default graphical installation console screen.

Installation Logs

There are several log files created and updated during the installation process. These files record configuration and status information. You can view their contents after the installation has been completed to check how the installation progressed. The files are listed and described in Table 1-1. Files in the */var/log* directory are actually created and updated in the */tmp* directory during the installation; however, they are moved to */var/log* once the installation is complete.

File	Description
/root/anaconda-ks.cfg	Records the configuration information entered.
/root/install.log	Lists the packages installed.
/root/install.log.syslog	Stores general messages.
/var/log/anaconda.ifcfg.log	Captures messages related to network interfaces.
/var/log/anaconda.log	Contains informational, debug, and other messages.
/var/log/anaconda.syslog	Records messages related to the kernel.
/var/log/anaconda.xlog	Stores X information.
/var/log/anaconda.program.log	Captures messages generated by external programs.
/var/log/anaconda.storage.log	Records messages generated by storage modules.
/var/log/anaconda.yum.log	Contains messages related to yum packages.

Table 1-1 Installation Logs

Configuration Needed for Practicing RHCSA & RHCE Exam Objectives

Beginning in Chapter 01 "Local Installation" and throughout this book, several administration topics on system, network, and security will be covered along with procedures on how to implement, administer, and troubleshoot them. A number of exercises will be performed and commands executed. The following minimum hardware and virtual machine configuration will be used to explain the procedures and perform the exercises:

PC/laptop architecture: 64-bit Intel dual core (or equivalent AMD processor) with
                                   hardware virtualization support, and a local DVD drive.

Physical memory: 4GB (or more for better performance).

Physical disk size: 100GB (internal disk) or an external USB disk with support in
                                   UEFI / BIOS to boot.

Number of physical network interfaces: 1

Hostname of physical server:	*physical.example.com* with IP 192.168.2.200 on *eth0* and 1x96GB disk (20GB for OS and 77GB for the three virtual machines). This server will be installed locally using the DVD installation media. You are free to use any IP address provided by your home router.
Number of VMs:	3
Memory in each VM:	800MB
CPU in each VM:	1
OS in each VM:	RHEL6
First VM:	*server.example.com* with IP 192.168.2.201 and 1x20GB virtual disk for OS and 4x4GB virtual disks for disk management exercises. All configuration will be implemented on this VM. This VM will be created using the graphical Virtual Machine Manager program and the OS will be installed in it using the installation files located on an HTTP server. You may instead use an IP address on the 192.168.122 network, which is provided on the default virtual network available in the Virtual Machine Manager software.
Second VM:	*insider.example.com* with IP 192.168.2.202 and 1x20GB virtual disk for OS. This will be a test VM and it should be able to access services provided by *server.example.com*. This VM will be built using the graphical Virtual Machine Manager program and the OS will be installed in it via kickstart. You may instead use an IP address on the 192.168.122 network, which is provided on the default virtual network available in the Virtual Machine Manager software.
Third VM:	*outsider.example.net* with IP 192.168.3.200 and 1x20GB virtual disk for OS. This will be a test VM and it should not be able to access services provided by *server.example.com*. This VM will be built using the graphical Virtual Machine Manager program and the OS will be installed in it using the installation files located on an FTP server. This system is intended to be on a network other than the 192.168.2 and 192.168.122 networks.

Exercise 1-1: Install RHEL6 Using Local DVD

In this exercise, you will perform a RHEL6 installation graphically on a physical computer using the local DVD media. This procedure assumes that the physical computer meets the requirements presented earlier in this chapter. You will name this system *physical.example.com* and use the IP, disk, cpu, and memory information provided in the previous sub-section during the installation process. Additional configuration information will be presented as the installation progresses.

Initiating Installation

1. Power on the computer.
2. Insert the RHEL6 installation DVD in the drive.
3. Boot the computer from the DVD.

If you wish to capture screen shots during installation, press the Shift-PrntScrn key combination on each window. The screenshots will be saved in the /root/anaconda-screenshots file.

4. A graphical boot menu will appear displaying five options, as shown in Figure 1-2:

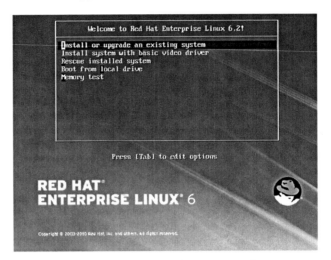

Figure 1-2 Installation – Boot Screen

Initial Installation Steps

5. Press the Enter key to select the first option "Install or upgrade an existing system" to begin the installation in graphical mode.

    i. This is also the default boot option. If Enter is not pressed within 60 seconds, the computer will automatically boot with this option. To begin the installation in text mode, highlight this option, press the TAB key, type text at the end and press Enter.

    ii. If you experience video issues during the installation with the first boot option, reboot the computer and try the second option "Install system with basic video driver".

    iii. The third option "Rescue installed system" is used to recover an already installed but unbootable system.

    iv. The fourth boot option "Boot from local drive" allows you to boot the system from an alternative media such as a USB disk, a boot CD, and so on.

    v. The fifth option "Memory test" is available to run a number of tests on the physical memory of the system.

6. Press "Skip" when asked to test the installation media. It is assumed that the installation DVD is free of errors.

7. The *anaconda* installer program begins and a "Red Hat Enterprise Linux 6" splash screen pops up. Click Next to continue.

8. Choose a language for use during the installation process. The default is English. Make a selection and click Next to continue.

9. Choose a keyboard model attached to the computer. The default is the US keyboard. Make a selection and click Next to continue.

10. The next screen presents you with two storage options on which to install RHEL: i) Basic Storage Devices, and ii) Specialized Storage Devices. The first option is used when you want RHEL installed on a local disk (or an external USB disk) connected to your computer and the second option is used when you want RHEL installed on a disk accessed over an iSCSI or fibre channel link. Select Basic Storage Devices and click Next to continue.

What type of devices will your installation involve?

Basic Storage Devices
Installs or upgrades to typical types of storage devices. If you're not sure which option is right for you, this is probably it.

Specialized Storage Devices
Installs or upgrades to enterprise devices such as Storage Area Networks (SANs). This option will allow you to add FCoE / iSCSI / zFCP disks and to filter out devices the installer should ignore.

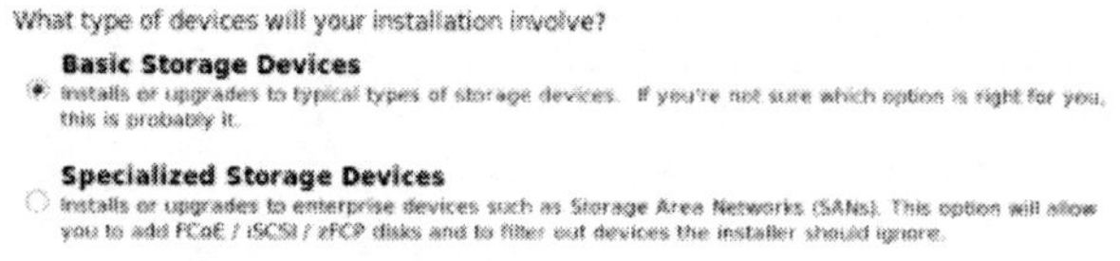

Figure 1-3 Installation – Storage Device Selection

11. The installation program then asks if you wish to perform a fresh installation or an upgrade. Choose Fresh Installation and click Next to continue.

Configuring Network Interfaces

12. The next screen allows you to choose a hostname for your system and configure one or more network interfaces, as shown in Figure 1-4. Specify the hostname (*physical.example.com*) and click Configure Network to bring up the Network Connections dialog box. By default, all network interfaces detected by the installation program appears under their respective tabs.

   i. Click the Wired tab, highlight "System eth0" and click Edit to edit the first Ethernet network interface detected. Click Add to add an interface if none was found. Leave all options as defaulted on the Wired tab.

   ii. Click the IPv4 Settings tab. If you wish the system to obtain IP information for the selected interface from a configured, active DHCP server, choose Automatic (DHCP) and leave the rest of the fields in this tab to their default values. Alternatively, select Manual from the drop down menu and click Add to supply the information manually. For this demonstration, choose Manual and enter IP (192168.2.200), netmask (255.255.255.0), and gateway (192.168.2.1) values. Do not forget to check the Connect Automatically box to activate the interface at each system reboot, and the "Available to all users" options at the top and bottom of the dialog box, respectively.

   iii. Leave all other settings intact. Click Apply and then Close to close the Network Connections dialog box. Click Next to continue.

Selecting a Time Zone

13. Choose an appropriate time zone for the system. For this demonstration, select America/New_York, and select "System clock uses UTC". Click Next to continue.

Choosing a root Password

14. Enter a password for the *root* user twice and click Next to continue.

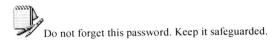 Do not forget this password. Keep it safeguarded.

Figure 1-4 Installation – Network Interface Configuration

Disk Partitioning

15. The next screen lists five options where you can place the RHEL operating system: i) Use All Space; ii) Replace Existing Linux System(s); iii) Shrink Current System; iv) Use Free Space; and v) Create Custom Layout.

    i. The Use All Space option wipes out any and all partitions including any non-Linux partitions that exist on the selected disk
    ii. The Replace Existing Linux System(s) option overwrites partitions created by a previous installation of Linux and use the reclaimed space for this RHEL installation, however, it does not remove any non-Linux partitions.
    iii. The Shrink Current System option shrinks a current partition to a specified size to free up space for this RHEL installation.
    iv. The Use Free Space option uses any unpartitioned, free space available on the disk without touching the current partitioning.
    v. The Create Custom Layout option lets you fully customize partitioning on the selected disk using a Red Hat disk partitioning tool known as *Disk Druid*. This tool is only available during the installation process.

    For the first four options you can check the "Review and modify partitioning layout" checkbox, otherwise an automated installation kicks in without providing you with an opportunity to partition your storage devices manually. You can leave the "Encrypt system" option unchecked. For our installation, choose Create Custom Layout option and then click Next to continue.

Figure 1-5 Installation – Select Layout

16. The next screen displays the current disk layout of all the detected disks as well as their associated device names (*sda* for the first disk, *sdb* for the second, and so on) and their sizes. Since our installation is being performed on a brand new disk, no partitioning information currently exists.

Figure 1-6 Installation – Current Partitioning Layout

There are four menu options at the bottom of this screen, as depicted in Figure 1-6 above: Create, Edit, Delete, and Reset.

Create allows you to create a standard, software RAID or an LVM partition, Edit and Delete modify or delete any of these, and Reset undoes any changes made on this screen.

Click Create to open up the Create Storage dialog box. The following two sub-sections explain how to create standard and LVM partitions.

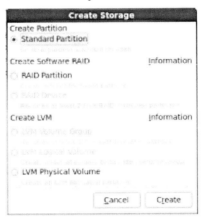

Figure 1-7 Installation – Standard Partition Selection

Standard Partitioning Example

17. Select Standard Partition and click Create to bring up the Add Partition dialog box.

Figure 1-8 Installation – Add Standard Partition

Create partitions as per Table 1-2 below. Figure 1-8 above shows how to fill in information for the */boot* partition. Select Fixed size for each partition. Two other options – Fill all space up to (MB) and Fill to maximum allowable size – are also available under Additional Size Options. The former limits the usage of the file system space to the specified maximum and the latter allows it to use the maximum size allocated. Select "Force to be a primary partition" for *boot*. Click OK when done. Repeat this procedure for the remainder of the partitions but leave "Force to be a primary partition" unchecked for them.

Mount Point	File System Type	Size (MB)
/boot	ext4	200
/	ext4	4,000
	swap	4,000
/home	ext4	800
/tmp	ext4	1,000
/usr	ext4	10,000
/var	ext4	80,000

Table 1-2 Installation – File System Information for Standard Partitioning

 By default, only three partitions – /boot, /, and swap – are created.

For the swap partition, choose "swap" under "File System type". A swap partition does not have a mount point. A detailed discussion on file systems, swap and mount points is covered in Chapter 10 "File Systems & Swap".

18. A summary of all partitions created will be displayed including the available free space at the bottom, as shown in Figure 1-9.

Device	Size (MB)	Mount Point/ RAID/Volume	Type	Format
▽ Hard Drives				
▽ sda				
sda1	200	/boot	ext4	✓
sda2	80000	/var	ext4	✓
sda3	4000		swap	✓
▽ sda4	18199		Extended	
sda5	4000	/	ext4	✓
sda6	1000	/tmp	ext4	✓
sda7	800	/home	ext4	✓
Free	12395			

Figure 1-9 Installation – Standard Partitioning Summary

Click Next to continue.

LVM Partitioning Example

19. To practice LVM partitioning, remove all the partitions created in the previous sub-section, except for */boot* as it must exist outside of LVM configuration otherwise the system will not boot up.
20. The first step is to create a physical volume. Click Create to bring up the Create Storage dialog box, select LVM Physical Volume and click Create.

Figure 1-10 Installation – LVM Partitioning Selection

21. Specify a size for the physical volume. For this demonstration, specify 100,000MB and leave the rest of the options to their default values. Click OK when done.

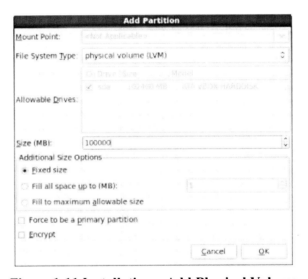

Figure 1-11 Installation – Add Physical Volume

22. If no other partitions besides */boot* exist, the physical volume partition will be created as *sda2*.
23. From the Create Storage dialog box, choose LVM Volume Group under Create LVM and click Create to bring up the Make LVM Volume Group dialog box. Specify a name for the volume group, such as *vg00*, and select *sda2* as the physical volume to use in this volume group.

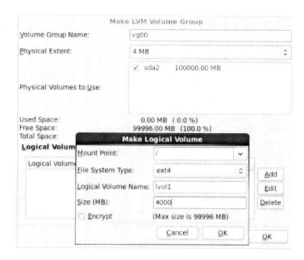

Figure 1-12 Installation – Add Volume Group & Logical Volumes

Then click Add to bring up the Make Logical Volume dialog box and create logical volumes in *vg00* as demonstrated in Figure 1-12 above with information provided in Table 1-3 below. For the swap logical volume, choose "swap" under File System Type. A detailed discussion on file systems, swap and mount points is covered in Chapter 10 "File Systems & Swap".

Mount Point	File System Type	Size (MB)	Logical Volume
/	ext4	4,000	*lvol1*
	swap	4,000	*lvol2*
/home	ext4	800	*lvol3*
/tmp	ext4	1,000	*lvol4*
/usr	ext4	10,000	*lvol5*
/var	ext4	80,000	*lvol6*

Table 1-3 Installation – File System Information for LVM Partitioning

Click OK after adding all the logical volume information. If you make an error, use Edit or Delete as appropriate to make adjustments.

24. Figure 1-13 summarizes the LVM partitions created.

    LVM is explained at length in Chapter 09 "Disk Partitioning". Click Next to continue with the installation process.

After the disk has been carved up using one or both disk management solutions, follow the steps below to continue with the installation process.

25. The next screen displays a warning message saying that the disk you have selected will be formatted and will destroy any existing data on it. Click Format to continue.
26. The next screen displays a warning message saying that the partitioning you have selected will now be written to the disk causing any existing data to be lost. Click "Write changes to

disk" to continue with initializing the disk, creating and formatting partitions and logical volumes, and mounting file systems.

Device	Size (MB)	Mount Point/ RAID/Volume	Type	Format
▽ LVM Volume Groups				
▽ vg00	99996			
lvol1	4000	/	ext4	✓
lvol2	4000	swap	swap	✓
lvol3	800	/home	ext4	✓
lvol4	1000	/tmp	ext4	✓
lvol5	10000	/usr	ext4	✓
lvol6	80000	/var	ext4	✓
Free	196			
▽ Hard Drives				
▽ sda				
sda1	200	/boot	ext4	✓
sda2	100000	vg00	physical volume (LVM)	✓
Free	2199			

Figure 1-13 Installation – LVM Partitions

Configuring Bootloader

27. The next screen brings up the GRUB bootloader menu. GRUB stands for *Grand Unified Bootloader*. By default, the "Install bootloader on /dev/sda" is selected, which points to the *Master Boot Record* (MBR) of the first detected disk on the system (MBR is a region for storing partition table information). Choose this option if RHEL is going to be the only operating system running on this computer. Click "Change device" to choose the alternative location "First sector of boot partition – /dev/sda1". This option is recommended if you already have an existing operating system such as Microsoft Windows running on your system. You may want to expand "BIOS Drive Order" if you think the order of disks presented is not right or you would like to reorganize the drive order. Leave the label "Red Hat Enterprise Linux 6" unchanged. Check the box "Choose a bootloader password" if you wish to. For this installation, leave the box unchecked.

Figure 1-14 Installation – Bootloader Options

Choosing a System Role & Package Groups

28. The next screen, Figure 1-15, allows you to choose a role that this server is going to play. The default is the Basic Server installation. Additional choices are Database Server, Web Server, Identity Management Server, Virtualization Host, Desktop, Software Development

Workstation, and Minimal. Details for roles are provided in Table 1-4. Leave the role to the default and click "Customize now" at the bottom of the screen. Click Next to continue.

Figure 1-15 Installation – Server Role Selection

Server Role	Description
Basic Server	Installs basic tools and services.
Database Server	Installs MySQL and PostgreSQL database software.
Web Server	Installs Apache web server software.
Identity Management Server	Installs OpenLDAP and System Security Services Daemon (SSSD) services for identity and authentication management.
Virtualization Host	Installs KVM virtualization software.
Desktop	Installs graphical support, and other desktop productivity tools.
Software Development Workstation	Installs software compilation tools.
Minimal	Installs basic RHEL.

Table 1-4 Installation – Server Role Selection

29. The next screen, Figure 1-16, displays a list of all software package groups available on the RHEL DVD along with individual packages within each group. Here, you have an opportunity to view and select or unselect specific packages from different package groups. The default packages are already checked. Highlighting a package group in the right pane shows a brief description of what it contains and lists associated packages. For this installation, select all four packages under the Virtualization package group and select Desktop, Graphical Administration Tools, and X Window System from the Desktops package group. Click Next when the selection has been made.

Figure 1-16 Installation – Software Package Groups

30. The installation program now checks for package dependencies and adds any required (dependent) packages to the list of software for installation. It then begins loading the selected packages on the system and updates the bootloader. As the installation continues, informational messages appear on the screen apprising you of the progress.
31. After the installation has completed, the installation media will be ejected and you will be asked to press the Reboot button to restart the system.

Performing Post-Installation Tasks – The First Boot Process

32. Following the reboot, the system will automatically initiate a program called *firstboot* (a.k.a. *Setup Agent*) for you to perform certain post-installation configuration tasks.
33. A Welcome screen appears along with a list of configuration tasks for you to perform. Click Forward to continue.
34. The license agreement window now appears. You must agree to the terms and conditions of use.
35. If you have an active subscription, you should register the system with Red Hat Network to get benefits such as automatic software updates for the system. Choose "No, I prefer to register at a later time" if you do not wish to do so at this time. Refer to Chapter 06 "Package Management" for details on how to register.

Figure 1-17 Firstboot – Setup Software Updates

Click Forward to continue and then choose "No thanks, I'll connect later." on the pop up screen that provides benefits of registration. The next screen highlights how to register your system with RHN later. Click Forward to continue.

36. Create a user account in addition to the default *root* user, which was already created during installation. Refer to Figures 1-18 through 1-21. Supply username and password information. The password must not be less than six characters in length. If you have an LDAP server in place for user identity and authentication, click "Use Network Login" and configure client functionality for the available service. To bring the User Manager interface up for creating user and group accounts, click Advanced. For details on how to create user and group accounts and how to manage them, consult Chapter 11 "Users & Groups". For this installation, enter *aghori* as the username and supply a password twice to create a user account. Also create an additional user account *user1* for use in later chapters by clicking on Advanced and entering username and password. Leave other settings to their default values.

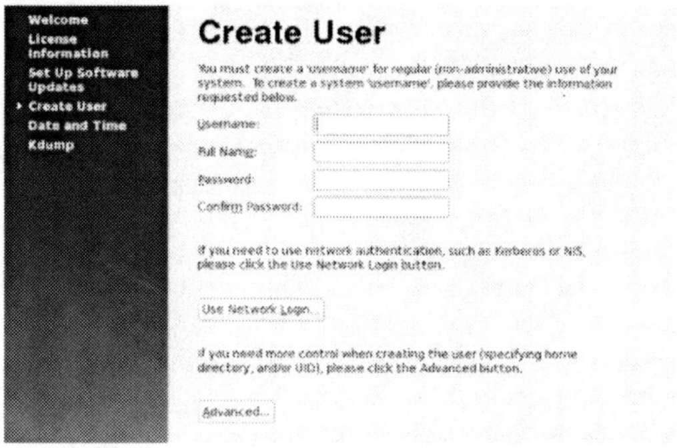

Figure 1-18 Firstboot – Create User

Clicking Use Network Login brings the Authentication Configuration interface up, as shown in Figure 1-18. Expand on User Authentication Database to see the choices available. There are four choices – local accounts only, LDAP, NIS, and Winbind. For this demonstration, we are interested only in LDAP configuration.

Figure 1-19 Firstboot – Authentication Configuration Interface

Choose LDAP and enter the required information. Select either Kerberos or LDAP as an authentication method.

Figure 1-20 Firstboot – LDAP Configuration

Go to the Advanced Options tab and choose options appropriately.

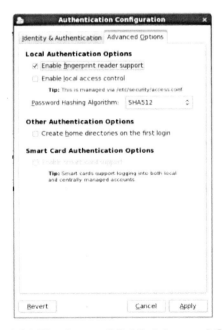

Figure 1-21 Firstboot – LDAP Advanced Options

37. Set the correct date and time for the system. If you wish to connect to an NTP server to obtain time, enable NTP by clicking "Synchronize date and time over the network", which will use default NTP servers listed there. You may specify hostname(s) or IP address(es) of your own or other Internet-based public NTP server(s). Figure 1-22 shows non-NTP date/time settings and Figure 1-23 displays NTP settings. If you do not wish to set date/time at this point, you can do so later using one of the methods described in Chapter 17 "Internet Services, Kernel Parameters & NTP".

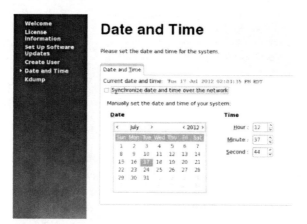

Figure 1-22 Firstboot – Date and Time Settings

Figure 1-23 Firstboot – NTP Settings

38. The next window allows you to configure kernel dump settings. You should reserve some memory, the default is usually sufficient, for kernel crash dump purposes in the work environment. The dump can be helpful in determining the root cause of a system crash. By default, it is disabled. If you make any modifications here, the system must be rebooted for the changes to take effect. If you do not wish to enable Kdump at this point, you can do so later using the Kdump Configuration tool *system-config-kdump*.

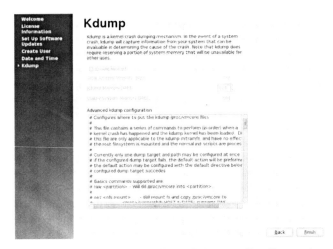

Figure 1-24 Firstboot – Kernel Dump Configuration

39. Click Finish to complete the firstboot process. The default graphical login interface screen called GNOME will appear (Figure 1-25) and you should now be able to log in as either *root* or one of the user accounts created earlier.

Figure 1-25 RHEL Login Screen

On successful login, you will see a window similar to the one displayed in Figure 1-26.

Figure 1-26 RHEL Desktop

After you are done, click System at the top and choose "Log Out root" to log off.

Figure 1-27 Log Out

This completes the step-by-step procedure for installing RHEL6 on a physical computer via local DVD media using the graphical user interface.

Chapter Summary

In this chapter, you studied the basics of Linux and learned about the available distributions from Red Hat. The chapter provided some basic pre-installation information on planning, available installation methods, installation log files, and virtual console screen that are available during the installation phase. It demonstrated how to download the RHEL6 software and load it on a physical computer using the local DVD media in graphical mode. It covered several post-installation steps for additional configuration and key system settings.

Chapter Review Questions

1. Can RHEL6 be installed in text mode?
2. What is the description of the first detected network interface?
3. How many console screens are available during the installation process?
4. The */boot* partition within LVM can be used to boot RHEL successfully. True or False?
5. Which kernel version is RHEL6 based on?
6. What are the two 100% rebuilds of Red Hat Enterprise Linux?
7. Are large processes divided into threads?
8. RHEL6 is available on DVDs as well as CDs from Red Hat's website. True or False?
9. Several log files are created and updated in the */tmp* directory during the installation process. Where are these files moved to after the installation has been completed?
10. What is the first step to create a volume group?
11. What are the two key hardware requirements for a computer to support KVM?
12. What is the name of the RHEL installer program?
13. What is the default location to store the bootloader program?
14. Support for iSCSI and fibre channel disks is available by choosing Specialized Storage Devices during RHEL installation. True or False?

15. What happens if you do not select "Review and modify partitioning layout" with any of the first four disk layout options during installation?
16. What is the default name of the first local disk?
17. Can you run the User Manager during the installation process to create user and group accounts?
18. What are the three disk partitioning techniques that RHEL supports?
19. Which is the default console screen for anaconda?
20. What server role would you choose if you want to install support for KVM?
21. Red Hat Network provides access to automatic software updates. True or False?
22. What is the disk partitioning tool that is only available during installation?
23. NTP does not provide automatic time updates with a remote time server. True or False?

Answers to Chapter Review Questions

1. Yes, RHEL6 can be installed in text mode.
2. The description of the first detected network interface is "System eth0"
3. There are six screens available during the installation process.
4. False. */boot* must not be defined in the LVM.
5. RHEL6 is based on kernel version 2.6.32.
6. Two 100% rebuilds of Red Hat Enterprise Linux are CentOS and Scientific Linux.
7. Yes, large processes are divided into smaller threads by the kernel.
8. False. RHEL6 software is available only on a single DVD.
9. These files are moved into the */var/log* directory.
10. The first step to create a volume group is to initialize a disk or partition with the *pvcreate* command.
11. A computer has to be 64-bit and support hardware virtualization.
12. The name of the RHEL installer program is Anaconda.
13. The Master Boot Record is the default location to store the bootloader program.
14. True.
15. Anaconda will create the default partitioning on the disk.
16. The name of the first local disk is */dev/sda*.
17. Yes, you can run the User Manager during the installation process.
18. The three disk partitioning techniques that RHEL support are the standard partitioning, the software RAID partitioning, and the LVM.
19. The default console screen for anaconda is the sixth screen.
20. KVM support is available in the Virtualization server role.
21. True.
22. The disk partitioning tool that is only available during installation is Disk Druid.
23. False.

Labs

Lab 1-1: Perform a Local Installation

Perform a local installation of RHEL6 on your physical computer. Use standard partitioning to create */boot* 200MB, */* 2GB, */usr* 5GB, */var* 2GB, */opt* 2GB, */tmp* 1GB, */home* 1GB, and swap 2GB. Select packages to support X Window and GNOME desktop. After the installation, run the firstboot process and create a local user account called *testuser*. Use information at will for additional configuration.

Lab 1-2: Perform a More Advanced Local Installation

Overwrite the installation implemented in Lab 1-1. Perform another local installation of RHEL6 on your physical computer but use standard partitioning for */boot* file system and LVM-based partitioning for the rest of the file systems. Create */boot* 200MB, *lvol1* for / 4GB, *lvol2* for swap 4GB, *lvol3* for */usr* 8GB, *lvol4* for */var* 80GB, *lvol5* for */opt* 2GB, *lvol6* for */tmp* 1GB, and *lvol7* for */home* 1GB in that order. Select packages to support virtualization, X Window, GNOME desktop, KDE desktop, graphical administration tools, CIFS, NFS, Apache web server, and vsFTP. Create a local user account called *testuser* with all the defaults and sync the system time with NTP. Use information at will for additional configuration.

Chapter 02

Basic Linux Commands

This chapter covers the following major topics:

- ✓ Access the system remotely using ssh and VNC
- ✓ Access the command prompt
- ✓ Command line components
- ✓ General Linux commands and how to execute them
- ✓ Linux online help

This chapter includes the following RHCSA objectives:

01. Access a shell prompt and issue commands with correct syntax

04. Access remote systems using ssh and VNC

06. Archive, compress, unpack, and uncompress files using tar, star, gzip, and bzip2

11. Locate, read, and use system documentation including man, info, and files in /usr/share/doc

A RHEL system may be accessed over the network using either the secure shell protocol or the Virtual Network Computing client/server program. These programs may be run on a Windows system, a Unix system, another RHEL system, or systems running other Linux distributions.

A RHEL system comes with a variety of commands categorized as privileged and non-privileged. The privileged commands are for system management purposes and can be executed by the *root* user or regular users with *root* privileges. On the other hand, the non-privileged commands do not require extra rights for execution and can be run by regular users. In this chapter and throughout the book, you will interact with scores of user and administrative commands. Each command may or may not take arguments. Some commands do require that you specify one, more than one, or a fixed number of arguments. Arguments are supplied with commands for better, restricted, or enhanced results, or a combination thereof. An argument may be an option, a sub-command, text, or the name of a file or directory. You may be able to memorize essential commands as well as key arguments they accept depending on how much you practice and interact with your system. This chapter provides an introduction to a number of user and administrative commands.

Accessing a RHEL System Remotely

A user must log in to the Linux system to use it. The login process identifies the user to the system. Towards the end of Chapter 01 "Local Installation", logging in and out at the RHEL console was demonstrated. For accessing a RHEL system remotely, the most common tool is the *ssh* (secure shell) command. Secure shell software (both client and server packages) is loaded by default during RHEL6 installation, and is covered in detail in Chapter 21 "Apache Web Server & Secure Shell Server". A graphical equivalent of the *ssh* client may be downloaded and run on MS Windows to access a RHEL system. The following sub-section provides basic information on how to use the *ssh* client from another Linux system and from a Windows system to access *physical.example.com* and how to use *Virtual Network Computing* (VNC) to access the desktop of *physical.example.com*.

Remotely Accessing from Another Linux System

The *ssh* command can be executed from another Linux system using the hostname or IP address of the remote RHEL system to be accessed. For example, if you are logged in as *aghori* on a Linux system and wish to access *physical.example.com*, run the following:

```
$ ssh physical.example.com
The authenticity of host 'physical.example.com (192.168.2.200)' can't be established.
RSA key fingerprint is 12:78:84:2a:38:47:12:0a:4b:e0:ec:67:b1:00:f6:85.
Are you sure you want to continue connecting (yes/no)? yes
Warning: Permanently added 'physical.example.com,192.168.2.200' (RSA) to the list of known hosts.
aghori@physical.example.com's password:
[aghori@physical ~]$
```

Answer yes to the question and press Enter. This adds the system's hostname to *aghori*'s *~/.ssh/known_hosts* file. This message will no longer appear for this user on subsequent login attempts from this client. Enter *aghori*'s password to enter *physical.example.com*. After you are done, use either the *exit* or the *logout* command to log out. You may alternatively press Ctrl+d to log off.

If you wish to log on as a different user such as *user1* (assuming *user1* exists on *physical.example.com*), use any of the following three commands:

$ **ssh physical.example.com –l user1**
$ **ssh –l user1 physical.example.com**
$ **ssh user1@physical.example.com**

For running graphical tools on *physical.example.com* and having the output redirected to your local Linux system, run the *ssh* command with –X option. Once you are logged in to *physical.example.com*, you can execute any graphical application such as *system-config-network* there. Enter the *root* user's password when prompted.

$ **ssh –X physical.example.com**

Remotely Accessing from Windows

On the Windows side, several ssh client programs such as PuTTY, are available. PuTTY may be downloaded free of charge from the Internet. Figure 2-1 shows the PuTTY interface.

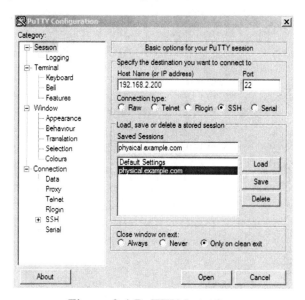

Figure 2-1 PuTTY Interface

Enter the IP address of *physical.example.com* and select the radio button next to SSH under the Connection Type heading. The ssh protocol uses port 22, which is exhibited in the Port field. Assign a name to this session (typically the hostname) in the Saved Sessions field and click Save to store this information so as to avoid retyping it in the future. Now click Open to attempt a connection.

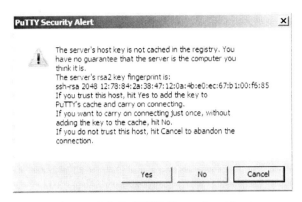

Figure 2-2 PuTTY Security Alert

Click yes to the security alert question to get a login prompt. This alert will not reappear on subsequent attempts. Enter a username and password to log in.

Remotely Accessing Using VNC

The desktop of *physical.example.com* can be accessed from remotely by using Virtual Network Computing (VNC). VNC is a platform-independent client/server graphical desktop sharing application for viewing and controlling networked systems. The server piece allows users from remote systems to be able to see and control their systems. VNC uses port 5900 by default for the service, 5901 for the first connection, and 5902 to 5909 for subsequent connections. The VNC server software called *vino* is installed as part of RHEL6, and is available in GNOME desktop. To allow access to remote users for viewing and controlling your RHEL system, run the *vino-preferences* command or go to System → Preferences → Remote Desktop and make appropriate selections. Figure 2-3 shows the Remote Desktop Preferences screen that pops up.

Figure 2-3 Remote Desktop Preferences

There are options grouped in Sharing, Security, and Notification Area categories. Checking the first two options under Sharing let remote users view and control the desktop. Checking the first option under Security forces confirmation for each access to this computer, the second option forces the user to enter the specified password, and the third option permits automatic access from remote networks. The choices under the Notification Area control if a notification icon is to be exhibited or concealed in the upper right hand corner on the GNOME desktop.

Once *physical.example.com* has been configured for remote user access, log on to a Windows or another Linux system and download a VNC client program such as VNC Viewer. Run this program and type in the IP address of *physical.example.com* and click Connect. See Figure 2-4.

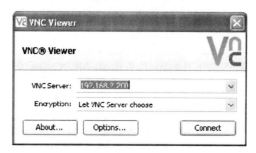

Figure 2-4 Remote Desktop Preferences

This will connect to the desktop of the system and you should be able to view and work on the system as if you are sitting right in front of it.

Common Linux Commands

There are hundreds of commands available in RHEL. This section first describes how to access the command prompt and then provides an understanding of how commands are formed, describing in detail some of them that are commonly used in routine system administration.

Access the Command Prompt

You need access to the command prompt in order to run commands. It was demonstrated at the end of Chapter 01 "Local Installation" how to log on at the console of a RHEL system. And, the previous section discussed how to access a RHEL system via *ssh*. To practice the commands provided in this chapter, you will need to log in as *root* either at the console or using ssh, run the commands, and observe their output. You should also log in as user *aghori* on a different terminal window.

The basic syntax of a command is:

    command argument1 argument2 argument3 ……..

The following examples of commands demonstrate the use of arguments. The text on the right in brackets state how many arguments are supplied with the command. (You may have noticed that you've already seen the ssh command with arguments in the previous sections of this chapter: i.e. "ssh –X physical.example.com").

```
$ cal 2012                  (one argument)
$ cal 10 2012               (two arguments)
$ ls                        (no arguments)
$ ls –l                     (one argument, which is an option)
$ ls directory_name         (one argument)
$ ls –l directory_name      (two arguments of which the first one is an option)
```

The ls Command and its Variations

The *ls* (list) command displays a list of files and directories. It supports several options some of which are listed in Table 2-1 along with a short description of each.

Option	Description
–a	Lists hidden files also. If a file or directory name starts with a dot, it is referred to as hidden.
–F	Displays file types. Shows / for directories, * for executable files, @ for symbolic links and nothing for text files.
–lh	Displays long listing with file sizes in human readable format.
–l	Displays long listing with detailed file information including file type, permissions, link count, owner, group, size, date and time of last modification, and name of the file.
–ld	Displays long listing of the specified directory, but hides its contents.
–R	Lists contents of the specified directory and all its sub-directories (recursive listing).
–lt	Lists all files sorted by date and time with the newest file first.
–ltr	Lists all files sorted by date and time with the oldest file first.

Table 2-1 ls Command Options

The following seven examples will help you understand the impact of options used with the *ls* command:

To list files in the current directory with an assumption that you are in the / directory:

```
# ls
bin  cgroup  etc    lib    lost+found misc  net  proc  sbin   srv  tmp  var
boot dev     home   lib64  media      mnt   opt  root  selinux sys  usr
```

To list files in the current directory with detailed information, use one of the following:

```
# ll
# ls –l
total 106
dr-xr-xr-x.   2 root root  4096 Jul 17 13:10 bin
dr-xr-xr-x.   5 root root  1024 Jul 17 12:31 boot
drwxr-xr-x.  10 root root  4096 Jul 17 12:36 cgroup
drwxr-xr-x.  19 root root  3700 Jul 20 07:28 dev
drwxr-xr-x. 110 root root 12288 Jul 20 07:35 etc
drwxr-xr-x.   6 root root  4096 Jul 19 10:04 home
dr-xr-xr-x.  11 root root  4096 Jul 17 12:26 lib

. . . . . . . .
```

To display all files in the current directory with their file types:

```
# ls -F
bin/      dev/    lib/         media/   net/    root/     srv/   usr/
boot/     etc/    lib64/       misc/    opt/    sbin/     sys/   var/
cgroup/   home/   lost+found/  mnt/     proc/   selinux/  tmp/
```

To list all files in the current directory with detailed information and sorted by date and time with the newest file first:

```
# ls -lt
total 106
drwxr-xr-x. 110 root root  12288 Jul 20 07:35 etc
dr-xr-x---.  22 root root   4096 Jul 20 07:32 root
drwxrwxrwt.  12 root root   4096 Jul 20 07:30 tmp
drwxr-xr-x.  19 root root   3700 Jul 20 07:28 dev
drwxr-xr-x.   2 root root      0 Jul 20 07:27 net
drwxr-xr-x.   2 root root      0 Jul 20 07:27 misc
drwxr-xr-x.   7 root root      0 Jul 20 07:27 selinux
. . . . . . . .
```

To display all files in the current directory with their sizes in human readable format:

```
# ls -lh
total 106K
dr-xr-xr-x.    2 root root 4.0K Jul 17 13:10 bin
dr-xr-xr-x.    5 root root 1.0K Jul 17 12:31 boot
drwxr-xr-x.   10 root root 4.0K Jul 17 12:36 cgroup
drwxr-xr-x.   19 root root 3.7K Jul 20 07:28 dev
drwxr-xr-x. 110 root root  12K Jul 20 07:35 etc
drwxr-xr-x.    6 root root 4.0K Jul 19 10:04 home
. . . . . . . .
```

To list all files, including the hidden files, in the current directory with detailed information:

```
# ls -la
total 118
dr-xr-xr-x.  26 root root  4096 Jul 20 07:27 .
dr-xr-xr-x.  26 root root  4096 Jul 20 07:27 ..
dr-xr-xr-x.   2 root root  4096 Jul 17 13:10 bin
dr-xr-xr-x.   5 root root  1024 Jul 17 12:31 boot
drwxr-xr-x.  10 root root  4096 Jul 17 12:36 cgroup
drwx------.   3 root root  4096 Jul 17 12:37 .dbus
drwxr-xr-x.  19 root root  3700 Jul 20 07:28 dev
drwxr-xr-x. 110 root root 12288 Jul 20 07:35 etc
drwxr-xr-x.   6 root root  4096 Jul 19 10:04 home
. . . . . . . .
```

To list contents of the *etc* directory recursively:

```
# ls –R /etc
    < a very lengthy output will be generated >
```

The pwd Command

The *pwd* (present working directory) command displays a user's current location in the directory tree. The following example shows that *aghori* is presently in the */home/aghori* directory:

```
$ pwd
/home/aghori
```

The cd Command

The *cd* (change directory) command is used to navigate the directory tree. Run the following commands as user *aghori*.

To change directory to */usr/bin*:

```
$ cd /usr/bin
```

To go back to the home directory, issue either of the following two commands:

```
$ cd
$ cd ~
```

To go directly from */etc* into a sub-directory *dir1* under *aghori's* home directory:

```
$ cd ~/dir1
```

tilde (~) is used as an abbreviation for the absolute pathname to a user's home directory. Refer to Chapter 03 "Files & Directories" to understand what an absolute path is.

To go to the home directory of user *user1* from anywhere in the directory structure, use the ~ character and specify the login name. Note that there is no space between ~ and *user1*. This command is only successful if user *user1* has an execute permission bit set on his home directory at the public level. Refer to Chapter 04 "File Permissions, Text Editors, Text Processors" for details on file permissions.

```
$ cd ~user1
```

Usage of the ~ character as demonstrated, is called *tilde substitution*. Refer to Chapter 04 "File Permissions, Text Editors, Text Processors & The Shell" for more information.

To go to the root directory, use the forward slash character:

```
$ cd /
```

To go one directory up to the parent directory, use period twice:

$ **cd ..**

To switch between current and previous directories, repeat the *cd* command with the dash character:

$ **cd –**

The tty Command

This command displays the pseudo terminal where you are currently logged in:

$ **tty**
/dev/pts/1

The who Command

The *who* command consults the */var/run/utmp* file and displays the information about all currently logged on users.

$ **who**
root   tty7   2012–07–20 07:28 (:0)
aghori   pts/0   2012–07–20 10:28 (192.168.2.16)

The 1st column displays the username; the 2nd column shows the terminal session. tty1 represents the first virtual console session, pts/0 represents the first remote pseudo terminal session, and so on; the 3rd and 4th column show the date and time the user logged in; and the 5th column indicates if the terminal session is graphical (:0) or remote.

The *who* command shows information only about the user that runs it if executed with "am i" arguments:

$ **who am i**
aghori   pts/0   2012–07–20 10:28 (192.168.2.16)

The w Command

The *w* (what) command displays information similar to the *who* command, but in more detail. It also tells how long the user has been idle, his CPU utilization, and current activity. On the first line, it shows the current system time, the length of time the system has been up for, number of users currently logged in, and the current average load on the system over the past 1, 5, and 15 minutes.

$ **w**
06:08:49 up 3 min, 2 users, load average: 1.90, 0.80, 0.30

USER	TTY	FROM	LOGIN@	IDLE	JCPU	PCPU	WHAT
aghori	pts/0	192.168.2.16	06:08	0.00s	0.29s	0.23s	w
root	tty7	:0	06:08	2:50	3.31s	3.31s	/usr/bin/Xorg :

The whoami Command

The *whoami* (who am i) command displays the username of the user that executes this command. The output may either be the current or the effective username. The current username is the name of the user that logs in and runs this command. Note that if the user issues the *su* command to switch identity, he becomes the effective user.

> **$ whoami**
> aghori

The logname Command

The *logname* (login name) command shows the name of the real user that logs in initially. If that user uses the *su* command to switch identity, the *logname* command, unlike the *whoami* command, still shows the real username.

> **$ logname**
> aghori

The id Command

The *id* (identification) command displays a user's UID (*user identification*), username, GID (*group identification*), group name, all secondary groups that the user is a member of, and SELinux security context:

> **$ id**
> uid=500(aghori) gid=500(aghori) groups=500(aghori) context=unconfined_u:unconfined_r:unconfined_t:s0-
> s0:c0.c1023

> Each user and group has a corresponding number (called UID and GID) for identification purposes. See Chapter 11 "Users & Groups" for more information. For SELinux, see Chapter 12 "Firewall & SELinux".

The groups Command

The *groups* command lists all groups that a user is a member of:

> **$ groups**
> aghori

The first group listed is the primary group for the user, all others are secondary (or supplementary) groups. Consult Chapter 11 "Users & Groups" for further details.

The last Command

The *last* command reports the history of successful user login attempts and system reboots by reading the */var/log/wtmp* file. This file keeps a record of all login and logout activities including login time, duration a user stayed logged in, and tty where the user session took place. Consider the following three examples.

To list all user login, logout, and system reboot activities, type the *last* command without any arguments:

```
# last
root     tty7         :0                Tue Jul 24 06:08   still logged in
aghori   tty1         :0                Tue Jul 24 06:07 - 06:08 (00:00)
root     pts/0        192.168.2.16      Tue Jul 24 06:07 - 06:08 (00:01)
reboot   system boot  2.6.32-220.el6.x  Tue Jul 24 06:06 - 07:10 (01:04)
aghori   pts/0        192.168.2.16      Sun Jul 22 05:37 - down  (00:02)
aghori   pts/1        192.168.2.16      Fri Jul 20 10:28 - 19:39 (09:11)
. . . . . . . .
```

To list only system reboot details:

```
# last reboot
reboot   system boot  2.6.32-220.el6.x  Tue Jul 24 06:06 - 07:12 (01:06)
reboot   system boot  2.6.32-220.el6.x  Fri  Jul 20 07:27 - 05:40 (1+22:12)
reboot   system boot  2.6.32-220.el6.x  Thu Jul 19 07:52 - 07:26 (23:34)
reboot   system boot  2.6.32-220.el6.x  Tue Jul 17 15:40 - 07:26 (2+15:46)
reboot   system boot  2.6.32-220.el6.x  Tue Jul 17 12:36 - 15:38 (03:01)
wtmp begins Tue Jul 17 12:36:53 2012
```

There is another command that lists more detailed information on recent logins and reboots. This command is *utmpdump* and is executed the following way:

```
# utmpdump /var/log/wtmp
Utmp dump of /var/log/wtmp
. . . . . . . .

[7] [01892] [3 ] [root   ]  [tty3 ]  [    ]  [0.0.0.0] [Tue Jul 24 06:17:50 2012 EDT]
[8] [01892] [3 ] [       ]  [tty3 ]  [    ]  [0.0.0.0] [Tue Jul 24 06:18:49 2012 EDT]
[6] [02768] [3 ] [LOGIN ]  [tty3 ]  [    ]  [0.0.0.0] [Tue Jul 24 06:18:49 2012 EDT]
```

The lastb Command

The *lastb* command reports the history of unsuccessful user login attempts by reading the */var/log/btmp* file. This file keeps a record of all unsuccessful login attempt activities including login name, time, and the tty where the attempt was made. Consider the following two examples.

To list all unsuccessful login attempts, type the *lastb* command without any arguments:

```
# lastb
. . . . . . . .

aghori   ssh:notty  192.168.2.16      Tue Jul 24 06:04 - 06:04 (00:00)
root               :0                 Mon Jul 23 12:57 - 12:57 (00:00)

btmp begins Tue Jul 17 12:33:44 2012
```

Alternatively, you can use the following command to display the above information:

```
# utmpdump /var/log/btmp
```
. . . .
```
[6] [19482] [ ] [root ]   [ssh:notty ]  [192.168.2.16 ]  [192.168.2.16]  [Mon Jul 23 12:57:19 2012 EST]
[6] [21767] [ ] [aghori] [ssh:notty ]  [192.168.2.16]   [192.168.2.16]  [Tue Jul 24 06:04:15 2012 EST]
```

The lastlog Command

The *lastlog* command displays the recent user logins by reading the */var/log/lastlog* file:

```
# lastlog
Username     Port     From              Latest
root         tty3                       Tue Jul 24 06:17:50 -0400 2012
bin                   **Never logged in**
```
. . . .
```
aghori       pts/0    192.168.2.16      Tue Jul 24 06:08:48 -0400 2012
user1        pts/2    192.168.2.16      Fri Jul 20 10:09:54 -0400 2012
```

The uname Command

The *uname* command produces basic information about the system. Without any options, this command displays the operating system name only. You can use the –a option to get more information.

```
# uname
Linux
# uname –a
Linux physical.example.com 2.6.32-220.el6.x86_64 #1 SMP Wed Nov 9 08:03:13 EST 2011 x86_64 x86_64
x86_64 GNU/Linux
```

The information returned by the second command is explained as follows:

Linux	Kernel name.
physical.example.com	Hostname of this system.
2.6.32-220.el6.x86_64 #1 SMP	Kernel version and release.
Wed Nov 9 08:03:13 EST 2011	Date and time of this kernel built.
x86_64	Machine hardware name.
x86_64	Processor type.
x86_64	Hardware platform.
GNU/Linux	Operating system name.

Try running the *uname* command with –s, –n, –r, –v, –m, –p, –i, and –o options to view specific information.

The hostname Command

The *hostname* command displays the system name:

```
# hostname
physical.example.com
```

The clear Command

The *clear* command clears the terminal screen and places the cursor at the beginning of the screen:

clear

The date Command

The *date* command displays current system date and time.

date
Tue Jul 24 07:37:49 EDT 2012

This command can also be used to modify system date and time. For example, to change the system date and time to July 24, 2012 07:46, run the *date* command as follows:

date --set "Tue Jul 24 07:46:00 EDT 2012"
Tue Jul 24 07:48:00 EDT 2012

The hwclock Command

The *hwclock* command displays the date and time based on the hardware clock.

hwclock
Tue 24 Jul 2012 07:38:03 AM EDT −0.493312 seconds

This command can also be used to modify system date and time. For example, to change the system date and time to July 24, 2012 07:48, run the *hwclock* command as follows:

hwclock --set --date "Tue Jul 24 07:48:00 EDT 2012"

The cal Command

The *cal* (calendar) command displays the calendar for the current month:

cal
```
      July 2012
Su Mo Tu We Th Fr Sa
 1  2  3   4  5  6  7
 8  9 10  11 12 13 14
15 16 17  18 19 20 21
22 23 24  25 26 27 28
29 30 31
```

The uptime Command

The *uptime* command shows the system's current time, how long it has been up for, number of users currently logged in, and average number of processes over the past 1, 5, and 15 minutes.

```
# uptime
07:42:23 up 1:36, 3 users, load average: 0.00, 0.00, 0.00
```

The output above shows that the current system time is 7:42 am, the system has been up for 1 hour and 36 minutes, there are currently three users logged in, and the system load averages over the past 1, 5, and 15 minutes are 0.00, 0.00, and 0.00, respectively.

The which Command

The *which* command shows the absolute path of the command that will execute if run without using the absolute path:

```
# which cat
/bin/cat
```

The output means that the *cat* command will execute from */bin* directory if you run it without specifying its full path.

The whereis Command

The *whereis* command displays the binary name and the full pathname of the command along with the location of its manual pages:

```
# whereis cat
cat: /bin/cat /usr/share/man/man1p/cat.1p.gz /usr/share/man/man1/cat.1.gz
```

The wc Command

The *wc* (word count) command displays number of lines, words, and characters (or bytes) contained in a text file or input supplied. For example, when you run this command on the */etc/profile* file, you will see output similar to the following:

```
# wc /etc/profile
78  257 1793 /etc/profile
```

The 1st column indicates the number of lines (78); the 2nd column indicates the number of words (257); the 3rd column indicates the number of characters (or bytes) (1793); and the 4th column indicates the file name (*/etc/profile*).

You can use the options listed in Table 2-2 to obtain the desired output.

Option	Action
–l	Prints line count.
–w	Prints word count.
–c	Prints byte count.
–m	Prints character count.

Table 2-2 wc Command Options

The following example displays only the number of lines in *etc/profile*:

wc –l /etc/profile
78 /etc/profile

Try running *wc* with the other options and review the results.

The wall Command

The *wall* command is used to broadcast a message to all logged in users on the system.

Type the *wall* command and press Enter. Start typing a message and press Enter when finished. Press Ctrl+d to broadcast it.

wall
Hello Users!
[root@physical ~]#
Broadcast message from root@physical.example.com (pts/0) (Tue Jul 24 07:49:47 2012):

Hello Users!

Archiving Tools

There are many native tools in RHEL that can be utilized to archive files. These tools include *tar*, *star*, and *cpio*. The following sub-sections discuss these tools in detail.

Using tar

The *tar* (tape archive) command creates, appends, updates, lists, and extracts files to and from a single file, which is called a *tar* file (also called a *tarball*).

tar supports several options, some of which are summarized in Table 2-3.

Switch	Definition
–c	Creates a tarball.
–f	Specifies a tarball name.
–j	Compresses a tarball with *bzip2* command.
–r	Appends files to the end of an existing tarball. Does not append to compressed tarballs.
–t	Lists contents of a tarball.
–u	Appends files to the end of an existing tarball if the specified files are newer. Does not append to compressed tarballs.
–v	Verbose mode.
–x	Extracts from a tarball.
–z	Compresses a tarball with *gzip* command.

Table 2-3 tar Command Options

A few examples have been provided below to explain how *tar* works. Note that the use of the −
character is optional.

To create a tarball called */tmp/home.tar* of the */home* directory:

> # **tar cvf /tmp/home.tar /home**
> /home/
> /home/aghori/
> /home/aghori/Pictures/
> /home/aghori/Desktop/
> /home/aghori/.bash_profile
> . . . .

To create a tarball called */tmp/files.tar* containing multiple files located in the */etc* directory:

> # **tar cvf /tmp/files.tar /etc/host.conf /etc/ntp.conf /etc/yum.conf**

To append files located in the */etc/xinetd.d* directory to *home.tar*:

> # **tar rvf /tmp/home.tar /etc/xinetd.d**

To list the contents of *home.tar*:

> # **tar tvf /tmp/home.tar**
> drwxr-xr-x root/root   0 2012-07-24 08:37 home/
> drwx------ aghori/aghori   0 2012-07-24 06:08 home/aghori/
> drwxr-xr-x aghori/aghori   0 2012-07-24 06:07 home/aghori/Pictures/
> drwxr-xr-x aghori/aghori   0 2012-07-24 06:07 home/aghori/Desktop/
> -rw-r--r-- aghori/aghori 176 2011-01-27 08:41 home/aghori/.bash_profile
> . . . .
> drwxr-xr-x root/root   0 2012-07-17 12:29 etc/xinetd.d/
> -rw-r--r-- root/root   523 2010-10-22 05:30 etc/xinetd.d/cvs
> -rw-r--r-- root/root   332 2009-05-20 06:49 etc/xinetd.d/rsync

To list *files.tar* contents:

> # **tar tvf /tmp/files.tar**
> -rw-r--r-- root/root   9 2011-06-30 09:35 host.conf
> -rw-r--r-- root/root   1917 2009-07-22 01:37 ntp.conf
> -rw-r--r-- root/root   813 2011-09-21 10:30 yum.conf

To restore */home* from *home.tar*:

> # **tar xvf /tmp/home.tar**

To extract files from *files.tar*:

> # **tar xvf /tmp/files.tar**

To create a tarball called *tmp/home.tar.gz* of the */home* directory and compress it with *gzip*:

tar cvzf /tmp/home.tar.gz /home

To create a tarball called *tmp/home.tar.bz* of the */home* directory and compress it with *bzip2*:

tar cvjf /tmp/home.tar.bz /home

Using star

The *star* (standard tar) command is an enhanced, POSIX-compliant version of *tar*. It supports SELinux security contexts as well. Options for creating, listing, appending, updating, and extracting tarballs are the same as the *tar* command's. This utility is not installed by default, use **yum install star-1*** to install it.

To create a tarball */tmp/home.tar* of the */home* directory and save all extended file attributes including the SELinux file contexts, use the xattr and exustar options with *star*:

star cvf /tmp/home.tar –xattr –H=exustar /home

To list and extract, run the following respectively:

star tvf /tmp/home.tar
star xvf /tmp/home.tar

Using cpio

The *cpio* (copy in/out) command copies, lists, and extracts files to and from a single file.

Some options available with *cpio* are summarized in Table 2-4. The *cpio* command requires that one of the o, i, or p options must be specified.

Option	Description
–o	Copies data.
–i	Extracts from a copy.
–t	Lists copy contents.
–v	Verbose mode.
–p	Reads from a copy to get pathnames.
–a	Resets access times on files after they are copied.

Table 2-4 cpio Command Options

Here are a few examples to understand the usage of the *cpio* command.

To copy the contents of */home*, run the *find* command as demonstrated and redirect the output to */tmp/home.cpio*:

```
# find /home | cpio –ov > /tmp/home.cpio
/home/aghori
/home/aghori/Pictures
/home/aghori/Desktop
/home/aghori/.bash_profile
. . . . . . . .
```

To list the contents of *home.cpio*:

```
# cpio –itv < /tmp/home.cpio
```

To restore files from *home.cpio*:

```
# cpio –iv < /tmp/home.cpio
```

To copy files directly from */home* into a new directory called */tmp/home.bkp*:

```
# find /home | cpio –pvd /tmp/home.bkp
```

Compression Tools

Compression tools are used to compress the size of an archive to save space. Once a compressed archive is created, it can be copied to a remote system faster than a non-compressed archive. These tools may be used with the archive commands, such as *tar*, to create a single compressed archive of hundreds of files and directories. RHEL provides a number of tools such as *zip* (*unzip*), *bzip2* (*bunzip2*), and *gzip* (*gunzip*) that can be used for compression purposes.

Using zip and unzip

The *zip* command is a popular compression utility available on a number of operating system platforms. This command adds the *.zip* extension to a zipped file.

The following example compresses three files into one called */tmp/files.zip*:

```
# zip /tmp/files.zip /etc/host.conf /etc/ntp.conf /etc/yum.conf
  adding: etc/host.conf (stored 0%)
  adding: etc/ntp.conf (deflated 55%)
  adding: etc/yum.conf (deflated 39%)
# ll /tmp/files.zip
-rw-r--r--. 1 root root 1839 Jul 24 11:42 /tmp/files.zip
```

The *unzip* command performs the opposite of what *zip* does. It uncompresses a zipped file and restores the files to their original state.

To uncompress the three files, issue the *unzip* command:

```
# unzip /tmp/files.zip
```

Using gzip and gunzip

The *gzip* command creates a compressed file of each of the files specified at the command line and adds the *.gz* extension to each one of them.

To compress the two files */root/install.log* and */root/install.log.syslog*, issue the following *gzip* command:

```
# gzip /root/install.log /root/install.log.syslog
# ll /root | grep install.log
-rw-r--r--. 1 root  root  9719 Jul 17 12:32 install.log.gz
-rw-r--r--. 1 root  root  1669 Jul 17 12:29 install.log.syslog.gz
```

To uncompress the files, run either of the following commands on each file:

```
# gunzip /root/install.log.gz
# gzip –d /root/install.log.gz
```

Using bzip2 and bunzip2

The *bzip2* command creates a compressed file of each of the files specified at the command line and adds the *.bz2* extension to each one of them.

To compress the two files */root/install.log* and */root/install.log.syslog*, issue the following *bzip2* command:

```
# bzip2 /root/install.log /root/install.log.syslog
# ll /root | grep install.log
-rw-r--r--. 1 root  root  8216 Jul 17 12:32 install.log.bz2
-rw-r--r--. 1 root  root  1550 Jul 17 12:29 install.log.syslog.bz2
```

To uncompress the files, run either of the following commands on each file:

```
# bunzip2 /root/install.log.bz2
# bzip2 –d /root/install.log.bz2
```

Online Help

While working on the system, you may require help to obtain information about a command, its usage, and options available with it. RHEL offers online help via *man* (manual) pages. Manual pages are installed as part of the package installation, and provide detailed information on commands and configuration files including short and long description, usage, options, bugs, additional references, and the author. In addition to the man pages, *apropos*, *whatis*, and *info* commands, as well as documentation located in the */usr/share/doc* directory, are also available. These are also discussed in this section.

Using man

Use the *man* command to view help on a command. The following example shows how to check man pages of the *passwd* command:

 # man passwd

PASSWD(1)   User utilities   PASSWD(1)

NAME
   passwd – update user's authentication tokens

SYNOPSIS
   passwd [–k] [–l] [–u [–f]] [–d] [–n mindays] [–x maxdays] [–w warndays]
   [–i inactivedays] [–S] [--stdin] [username]

DESCRIPTION
   The passwd utility is used to update user's authentication token(s).
. . . .
      # a proposed password before updating it.

:

While you are in man pages, some common keys listed in Table 2-5 help you navigate efficiently.

Key	Action
Enter / Down arrow	Moves forward one line.
Up arrow	Moves backward one line.
f / Spacebar / Page down	Moves forward one page.
b / Page up	Moves backward one page.
d / u	Moves down / up half a page.
g / G	Moves to the beginning / end of the man pages.
:f	Displays line number and bytes being viewed.
q	Quits the man pages.
/pattern	Searches forward for the specified pattern.
?pattern	Searches backward for the specified pattern.
n / N	Finds the next / previous occurrence of a pattern.
h	Gives help on navigational keys.

Table 2-5 Navigating within man Pages

man Sections

There are several sections within man pages. For example, section 1 refers to user commands, section 4 contains special files, section 5 describes system configuration files, section 8 includes system administration commands, and so on.

To look for information on a configuration file *(etc/passwd*, issue the following:

 # man 5 passwd

PASSWD(5)   Linux Programmer's Manual   PASSWD(5)

NAME
>   passwd – password file

DESCRIPTION
>   Passwd is a text file, that contains a list of the systemâs accounts,
>   giving for each account some useful information like user ID, group ID,
>   home directory, shell, etc. Often, it also contains the encrypted
>   passwords for each account. It should have general read permission
>   (many utilities, like ls(1) use it to map user IDs to usernames), but
>   write access only for the superuser.
>   . . . .
>   If the encrypted password, whether in /etc/passwd or in /etc/shadow, is
>
>   ;

Searching by Keyword

Sometimes you need to use a command but do not know its name. Linux allows you to perform a keyword search on all available man pages using the *man* command with the *–k* option or the *apropos* command. The search is performed in the *whatis* database that contains details about commands and files, and lists names of all man pages that include the specified keyword. The whatis database is updated once a day by the */etc/cron.daily/makewhatis.cron* script. You can manually update the whatis database as well using the *makewhatis* command.

 # man –k password
 # apropos password
chage   (1) – change user password expiry information
chpasswd   (8) – update passwords in batch mode
cracklib-check   (8) – Check passwords using libcrack2
. . . .

Another way of quickly determining the options available with a command and a short description associated with it is to supply the --help or -? option with the command. For example, to get quick help on the *passwd* command, run either of the following:

 # passwd --help
 # passwd -?
Usage: passwd [OPTION...]   <accountName>
 -k, --keep-tokens   keep non-expired authentication tokens
 -d, --delete   delete the password for the named account (root only)
. . . .
Help options:
 -?, --help   Show this help message
 --usage   Display brief usage message

Displaying Short Description

The *whatis* command provides a quick method for searching the specified command or file in the whatis database for a short description. For example, the following shows outputs of the command when run on *yum.conf* and *passwd* files:

```
# whatis yum.conf
yum.conf [yum]        (5)    – Configuration file for yum(8)
# whatis passwd
Passwd                (1)    – update a user's authentication tokens(s)
Passwd                (5)    – password file
passwd [sslpasswd]    (1ssl) – compute password hashes
```

The first output indicates that the specified file is a configuration file associated with the *yum* command and the second output points to three entries for the *passwd* file.

Alternative to the *whatis* command, you can use the *man* command with –f option to produce identical results:

```
# man –f yum.conf
# man –f passwd
```

The info Command

The *info* command is available in RHEL as part of the info package to allow users to be able to read command documentation as distributed by the GNU Project. It provides more detailed information than the *man* command does. Documentation displayed is divided into sections called *nodes*. The header is at the top of the screen and shows the name of the file being displayed, names of the current, next, and previous nodes, and the name of the node prior to the current node. The following example shows the first screen exhibited when the *info* command is executed on the command *passwd*:

```
# info passwd
File: *manpages*.  Node: passwd,  Up: (dir)
PASSWD(1)              User utilities              PASSWD(1)
NAME
      passwd – update userâs authentication tokens
SYNOPSIS
      passwd [-k] [-l] [-u [-f]] [-d] [-n mindays] [-x maxdays] [-w warndays]
      [-i inactivedays] [-S] [--stdin] [username]
DESCRIPTION
      The passwd utility is used to update user's authentication token(s).
. . . . . . . .
      -----Info: (*manpages*)passwd, 265 lines --Top----------------------------------
Welcome to Info version 4.13. Type h for help, m for menu item.
```

While you practice the info command, some common keys listed in Table 2-6 will help you navigate efficiently.

Key	Action
Down arrow	Moves forward one line.
Up arrow	Moves backward one line.
Spacebar	Moves forward one page.
Del	Moves backward one page.
Home	Goes to the beginning of current node.
End	Goes to the end of current node.
Tab	Goes to the next hypertext link.
q	Quits the info tutorial.
[	Goes to the previous node in the document.
]	Goes to the next node in the document.
t	Goes to the top node of this document.
s	Searches forward for a specified string.
{	Searches for the previous occurrence of the string.
}	Searches for the next occurrence of the string.

Table 2-6 Navigating within info Documentation

Documentation in the /usr/share/doc Directory

The */usr/share/doc* directory stores documentation of all installed packages under sub-directories that match package names following by the -<version> of the package. For example, the entry for the gzip package looks like:

 # ll –d /usr/share/doc/gzip*
drwxr-xr-x. 2 root root 4096 Jul 17 12:18 gzip-1.3.12

In this example, gzip is the name of the package followed by dash and the package version -1.3.12. Now run the *ll* command again but without the –d option to see what files are in there:

 # ll /usr/share/doc/gzip-1.3.12
total 116
–rw–r--r--. 1 root root   98 Oct 8 1999 AUTHORS
–rw–r--r--. 1 root root   69384 Apr 13 2007 ChangeLog
–rw–r--r--. 1 root root   15479 Apr 13 2007 NEWS
–rw–r--r--. 1 root root   7056 Apr 13 2007 README
–rw–r--r--. 1 root root   13037 Nov 20 2006 THANKS
–rw–r--r--. 1 root root   3592 Nov 20 2006 TODO

These files contain a huge amount of useful information on the gzip package. Read the contents of these files for more information.

Red Hat Enterprise Linux 6 Documentation

The Red Hat's documentation website at *docs.redhat.com* contains product documentation on RHEL6 in HTML, PDF, and EPUB formats. This set of documentation includes release and technical notes, and guides on installation, deployment, virtualization, Logical Volume Manager, storage administration, security, and SELinux. You should be able to download any of these guides for reference.

Chapter Summary

In this chapter, the basic commands of Linux were covered. The chapter provided an overview of how to log in to a RHEL system using ssh and virtual network computing clients.

You saw how to construct and then execute a number of basic Linux commands.

Finally, you learned how to access online help on commands and configuration files. You performed a keyword search on all available man pages. You saw the command that performed a search operation on the specified command or file.

Chapter Review Questions

1. The *wall* command can be used to broadcast an announcement to all logged in and logged off users. True or False?
2. Which command can be used to check if the installed RHEL6 is 32 bit or 64?
3. The *ssh* tool provides a non-secure tunnel over the network for accessing a RHEL system. True or False?
4. What file does the *who* command consult to provide a list of logged in users?
5. A file compressed with *bzip2* can be uncompressed using *gunzip* command. True or False?
6. For what purpose can we use the *apropos* command?
7. What is the purpose of the *lastlog* command?
8. Which option would you use with the *ssh* command to bring the graphical user interface of a remote system to your desktop?
9. What are the –R and –a options used for with the *ls* command?
10. Linux commands may be divided into user and administrative commands. True or False?
11. Which other command besides *uptime* can be used to display system load averages?
12. Which command displays the line number and word count in a given file?
13. The *logname* command does not show the real user name after you switch user. True or False?
14. Can you see the calendar of 2030 with the *cal* command?
15. The *id* and *groups* commands are useful for listing a user's identification. True or False.
16. Which file does the *last* command consult to display reports?
17. What type of information does section 5 in the manual pages contain?
18. What information does the *lastb* command provide?
19. What is the function of the *pwd* command?
20. Which command provides more accurate system time: *date* or *hwclock*?
21. Which option would you use with the *cd* command to switch between the current and previous working directories?
22. The *who* command is used to view currently logged out users. True or False?
23. The *tar* command archives SELinux file contexts while archiving a file. True or False?
24. Which command can be used to display the tutorial of a command?
25. The whatis database is generated and updated automatically by the system. True or False?

Answers to Chapter Review Questions

1. False. The shutdown command broadcasts message only to the logged in users.
2. The *uname* command can be used for this purpose.
3. False.

4. The *who* command consults the */var/run/utmp* file to list logged in users.
5. False.
6. The *apropos* command can be used to perform a keyword search in the manual pages.
7. The *lastlog* command provides information about recent user logins.
8. The –X option is used with the *ssh* command for this purpose.
9. The –R option is used for recursive directory listing and the –a option for listing hidden files.
10. True.
11. The *w* command may alternatively be used to display system load averages.
12. The *wc* command is used to display the line number and word count.
13. False.
14. Yes, you can see the calendar of any year with the *cal* command.
15. False.
16. The *last* command consults the */var/log/wtmp* file to display reports.
17. Section 5 of the manual pages contain information on configuration files.
18. The *lastb* command reports the history of unsuccessful user login attempts.
19. The *pwd* command shows the absolute path of the current working directory.
20. Both commands provide accurate system time.
21. The dash character is used with the *cd* command to toggle between the current and previous working directories.
22. True.
23. False.
24. The *info* command can be used to display the tutorial of a command.
25. True.

Labs

Lab 2-1: Navigate the Directory Structure

Log on to *physical.example.com* as *user1* and execute the *pwd* command to check your location in the directory tree. Run the *ls* command with appropriate switches to show files in the current directory along with all the hidden files. Change directory into */etc* and run the *pwd* command again to confirm the directory change. Switch back to the directory where you were before and run *pwd* again to verify.

Lab 2-2: Verify User and Group Identity

Log on to *physical.example.com* as *user1* and execute the *logname, whoami, who* and *w* commands one at a time. Verify *user1*'s identity by comparing the output from these commands. The output should show *user1* as the logged on user running these commands. Now execute the *id* and *groups* commands and verify *user1*'s identity by comparing the output from these commands. What additioinal information does the *groups* command provide that the *id* command does not.

Lab 2-3: Check User Log In Attempts

Log on to *physical.example.com* as *root* and execute the *last* and *lastb* commands to check which users have recently logged in to and out of the system successfully and unsuccessfully. Also list the timestamp when the system was last rebooted.

Lab 2-4: Identify the System and Kernel

Log on to *physical.example.com* as *root* and execute the *hostname* and *uname* commands. Identify the system's name, the kernel version, the RHEL release, and the hardware architecture of this system.

Lab 2-5: Check and Alter System Date and Time

Log on to *physical.example.com* as *root* and execute the *date* and *hwclock* commands to check the current system date and time. Identify the differences between the command outputs. Run the *cal* command to view the calendar of the current year. Use the *hwclock* command and change the system date to a date in January of next year. Issue the *date* command and change the system time to one hour ahead of the current time. Observe the new date and time with both commands. Reverse the date and time to the current time using either the *date* or the *hwclock* command.

Lab 2-6: Check System Uptime and Count Lines

Log on to *physical.example.com* as *root* and execute the *uptime* command. Identify how long the system has been up for and how many users are currently logged on to the system. Run the *wc* command and show how many lines, words, and bytes are in the */etc/profile* file.

Lab 2-7: Locate Command and File Paths

Log on to *physical.example.com* as *root* and execute the *which* command and supply the argument passwd to it. Note down the path that the command has returned. Now execute the *whereis* command on the same argument. Compare the output of both commands. The *which* command has only shown the location of the *passwd* command whereas the *whereis* command has provided the locations of the *passwd* command, *passwd* file, and their manual pages.

Lab 2-8: Archive, List, and Restore Files

Log on to *physical.example.com* as *root* and execute the *tar* command to archive the contents of the entire */etc* directory structure. Run the *star* and *cpio* commands and archive the */etc* directory contents again. Compare the archive file sizes created by the three commands. Run the *zip*, *gzip*, and *bzip2* commands on these archives to produce **.zip*, **.gz*, and **.bz2* compressed files. Compare the compressed archive file sizes created by the three compression commands. Run the commands to uncompress (*unzip*, *gunzip*, and *bunzip2*) the files and then use the archiving tools to restore the contents of the */etc* directory.

Chapter 03

Files & Directories

This chapter covers the following major topics:

- ✓ The Red Hat Enterprise Linux directory structure
- ✓ Static and dynamic directories
- ✓ Access files using absolute and relative pathnames
- ✓ Types of files
- ✓ Naming convention for files and directories
- ✓ Manage and manipulate files and directories including creating, listing, displaying, copying, moving, renaming, and removing them
- ✓ Search for text within files
- ✓ Search for files in the directory system
- ✓ Sort contents of text files
- ✓ Create file and directory links

This chapter includes the following RHCSA objectives:

03. Use grep and regular expressions to analyze text
08. Create, delete, copy, and move files and directories
09. Create hard and soft links

$Linux$ files are organized in a logical fashion for ease of administration. This logical division of files is maintained in hundreds of directories that are located in larger containers called *file systems*. Red Hat Enterprise Linux follows the *File system Hierarchy Standard* (FHS) for file organization. There are two types of file systems – *disk-based* and *memory-based*. Disk-based file systems are created on physical media such as a hard disk, and memory-based file systems, also called *virtual file systems*, are created at system boot up and destroyed at shut down.

Files are stored in static and dynamic directories, and may be referenced using absolute and relative paths. There are certain rules that should be followed when naming a new file or directory. There are a number of operations that can be performed during file and directory management. Linux includes thousands of files that may or may not be readable, but it allows the viewing of contents of many of them. There are tools available that prove to be very helpful in searching for text within files, searching for files in the specified location, sorting on input, and forming links for files and directories.

File System Tree

The Linux file system structure is like an inverted tree with the root of the tree at the top and branches and leaves at the bottom. The top-level is referred to as the *root* and represented by the forward slash (/) character. This is the point where the entire file system structure is ultimately connected.

Two file systems, / and */boot*, are created by default when RHEL is installed. The custom installation procedure covered in Chapter 01 "Local Installation" also allows you to create */var*, */usr*, */tmp*, */opt*, and */home* file systems in addition to / and */boot*. The main directories under the / and other file systems are shown in Figure 3-1. Some of these directories hold *static* data while others contain *dynamic* (or *variable*) information. The static data refers to file contents that are usually not modified. The dynamic or variable data refers to file contents that are modified as required. Static directories normally contain commands, library routines, kernel files, device files, etc. and dynamic directories hold log files, status files, configuration files, temporary files, etc.

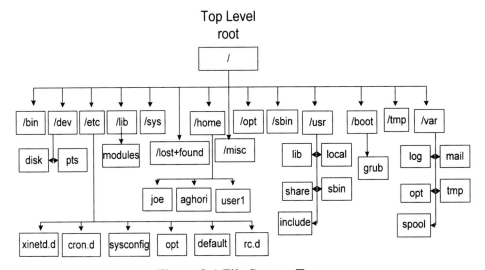

Figure 3-1 File System Tree

A brief description of disk-based and virtual file systems is provided in the following sub-sections.

The Root File System (/) – Disk-Based

The *root* file system is the top-level file system in the FHS and contains many higher-level directories holding specific information. Some of the key directories are:

The Binary Directory (/bin)

The *binary* directory contains crucial user executable commands. This directory holds static data files.

The Library Directory (/lib)

The *library* directory contains shared library files required by the kernel and other programs. It contains sub-directories that hold library routines.

In Figure 3-1, *modules* under */lib* refers to library files used by kernel modules.

The */lib* directory holds static data files.

The System Binary Directory (/sbin)

Most commands required at system boot up are located in the *system binary* directory. In addition, most commands requiring root privileges to run are also located here. In other words, this directory contains crucial system administration commands that are not intended for regular users (although they can still run a few of them). This directory is not included in the default search path for normal users because of the nature of the commands it contains.

The */sbin* directory holds static data files.

The Etcetera Directory (/etc)

The *etcetera* directory holds most system configuration files. Some of the more common sub-directories under */etc* are: *sysconfig, default, opt, lvm, xinetd.d, rc.d*, and *skel*. These sub-directories contain, in that sequence, configuration files for various system services, user account defaults, additional software installed on the system, the Logical Volume Manager, Internet services, system startup and shutdown scripts, and user profile templates.

The */etc* directory contains dynamic data files.

The lost+found Directory (/lost+found)

This directory holds files that become orphan after a system crash. An *orphan* file is a file that has lost its name. A detailed discussion on this is covered in Chapter 10 "File Systems & Swap".

If the *lost+found* directory is deleted, it should be re-created with the *mklost+found* command:

```
# mklost+found
mklost+found 1.41.12 (17-May-2010)
```

This directory is automatically created when the file system is created and holds dynamic information.

The /root Directory

This is the default home directory location for the *root* user.

The /srv Directory

This directory holds server data associated with databases, web sites, etc.

The /net Directory

If AutoFS is used to mount NFS file systems using a special map, all available NFS file systems on the network get mounted under their corresponding hostnames beneath */net*.

The Media Directory (/media)

This directory is used to automatically mount removable media such as floppy, CD, DVD, USB, and Zip drives.

The Miscellaneous Directory (/misc)

This location is used by AutoFS to mount local resources.

The Boot File System (/boot) – Disk-Based

The */boot* file system contains the Linux kernel, bootloader, and boot configuration files in addition to other files required to boot RHEL. The default size of this file system is 200MB, and may need to be altered only when an update to the kernel is required. More information on the */boot* file system is provided in Chapter 08 "Linux Boot Process & Kernel Management".

The */boot* file system contains static data files.

The Variable File System (/var) – Disk-Based

/var contains data that frequently changes while the system is operational. Files holding log, status, spool, and other dynamic data are typically located in this file system.

Some common sub-directories under */var* are briefly discussed here:

The /var/log Directory

Most system log files are located here. This directory contains system logs, boot logs, failed user logs, user logs, installation logs, cron logs, mail logs, etc.

The /var/spool/mail Directory

This is the location for user mailboxes.

The /var/opt Directory

For additional software installed in */opt*, this directory contains log, status, and other variable data files for that software.

The /var/spool Directory

Directories that hold print jobs, cron jobs, email messages, and other queued items before being sent out, are located here.

The /var/tmp Directory

Large temporary files or temporary files that need to exist for extended periods of time than what is allowed in */tmp,* are stored here. These files survive system reboots and are not automatically deleted.

The UNIX System Resources File System (/usr) – Disk-Based

This file system contains general files related to the system.

Some of the more important sub-directories under */usr* are briefly discussed here:

The /usr/bin Directory

The */usr/bin* directory contains additional user executable commands.

The /usr/sbin Directory

The */usr/sbin* directory contains additional system administration commands.

The /usr/local Directory

This directory is the system administrator repository for storing commands and tools that system administrators download from the web, develop in-house, or obtain elsewhere. These commands and tools are not generally included with the original Linux distribution. In particular, */usr/local/bin* holds executable files, */usr/local/etc* contains their configuration files, and */usr/local/man* holds associated man pages.

The /usr/include Directory

The */usr/include* directory contains header files for the *C* language.

The /usr/share Directory

This is the directory location for man pages, documentation, sample templates, configuration files, etc. that may be shared on multi-vendor Linux platforms with heterogeneous hardware architectures.

The /usr/lib Directory

The */usr/lib* directory contains library files pertaining to programming sub-routines.

The Temporary File System (/tmp) – Disk-Based

This file system is a repository for temporary files. Many programs create temporary files as they run or as they are being installed. Some programs delete the temporary files that they create after they are finished, while others do not.

The Optional File System (/opt) – Disk-Based

This file system holds additional software installed on the system. A sub-directory is created for each installed software.

The Home File System (/home) – Disk-Based

The */home* file system is designed to hold user *home* directories. Each user account is assigned a home directory for storing personal files. Each home directory is owned by the user the directory is assigned to. No other user has access to other users' home directories by default.

In Figure 3-1, users *joe, aghori,* and *user1* have their home directories located under */home*.

The directories discussed thus far are RHEL system related. It is highly recommended that you create separate file systems for data and applications. A detailed discussion on file systems and how to create and access them is covered in Chapter 10 "File Systems & Swap".

The Devices File System (/dev) – Virtual

The */dev* file system contains device files for hardware and virtual devices. The Linux kernel communicates with system hardware and virtual devices through corresponding device files located here.

There are two types of device files: *character* device files (a.k.a. *raw* device files) and *block* device files. The kernel accesses devices using one or both types of device files.

Character devices are accessed in a serial manner, meaning that streams of bits are transferred during kernel and device communication. Examples of such devices are serial printers, mice, keyboards, terminals, floppy disks, hard disk devices, tape drives, etc.

Block devices are accessed in a parallel fashion, meaning that data is transferred between the kernel and the device in blocks (parallel) when communication between the two takes place. Examples of block devices are hard disk devices, CD/DVD drives, floppy disks, parallel printers, etc.

 Some utilities access hard disk devices as block devices while others access them as character devices.

Some key sub-directories under the */dev* file system are *disk, pts,* and *vg00,* which contain device files for hard disks, pseudo terminals, and the root volume group, respectively.

The */dev* file system holds static data files.

The Process File System (/proc) – Virtual

The */proc* file system maintains information about the current state of the running kernel including details on CPU, memory, partitioning, interrupts, I/O addresses, DMA channels, and running processes. This file system is represented by various files which do not actually store the information, they point to the information in the memory. The */proc* file system is maintained automatically by the system.

The */proc* file system contains dynamic data files.

The System File System (/sys) – Virtual

Information about the currently configured hardware is stored and maintained in the */sys* file system. This file system is maintained automatically by the system.

The SELinux File System (/selinux) – Virtual

If SELinux packages are installed, this file system stores all current settings for SELinux. A detailed discussion on SELinux is covered in Chapter 12 "Firewall & SELinux".

Absolute and Relative Paths

A *path* is like a road map which shows how to get from one place in the directory tree to another. It uniquely identifies a particular file or directory by its absolute or relative location in the directory structure.

At any given time, the directory you are located in within the directory tree is referred to as your *present* (or *current*) working directory. When you log in to the system, you are placed in your home directory by default. Use the *pwd* command to display your current location.

Absolute Path

An *absolute path* (a.k.a. a *full path* or a *fully qualified path*) points to a file or directory in relation to the root (/). It always starts with the forward slash (/) character. The *pwd* command displays your current location in the tree.

> **$ pwd**
> /home/aghori/dir1/scripts

This example shows that you are in */home/aghori/dir1/scripts* directory, which represents the full path with respect to the root (/).

Relative Path

A *relative path* points to a file or directory in relation to your current location in the directory tree. A relative path never begins with the forward slash (/) character, rather it always begins in one of the following three ways:

<u>With a period:</u> A period represents the current working directory. For example, if you are located in the */home/aghori/dir1/scripts* directory and wish to run a script *script1* from there, you would type:

> **./script1**

<u>With a pair of periods:</u> A pair of period represents a parent directory in relation to your current working directory. A parent directory is one level higher than the current working directory. For example, to go back one level up to the parent directory, type:

> **$ cd ..**

<u>With a sub-directory name:</u> Say you are currently in the */home/aghori* directory and want to go to the *scripts* sub-directory under *dir1*, you will need to run the following:

   $ **cd dir1/scripts**

File Types

RHEL supports several different types of files. Some of the more common file types are regular files, directory files, executable files, symbolic link files, and device files, and are described in the following sub-sections.

Regular Files

Regular files may contain text or binary data. These files may be shell scripts or commands. When you execute *ll* on a directory, all line entries for files in the output that begin with – represent regular files:

   $ **ll /bin**
   . . . .
   –rwxr–xr–x. 1 root root 391360 Oct 25 2011 tcsh
   –rwxr–xr–x. 1 root root   52776 Oct   5 2011 touch
   –rwxr–xr–x. 1 root root   11392 Mar 22 2011 tracepath
   . . . .

You can use the *file* command to determine a file type. For example, the following shows that *.bash_profile* contains ascii text:

   $ **file .bash_profile**
   .bash_profile: ASCII English text

Directory Files

Directories are logical containers that hold files and sub-directories. Run *ll* on the root directory and you should see an output similar to the following:

   $ **ll /**
   total 102
   dr–xr–xr–x.   2 root root   4096 Jul 17 13:10 bin
   dr–xr–xr–x.   5 root root   1024 Jul 17 12:31 boot
   drwxr–xr–x. 10 root root   4096 Jul 17 12:36 cgroup
   drwxr–xr–x. 19 root root   3700 Jul 27 20:34 dev
   . . . .

The letter d at the beginning of each line entry indicates a directory. Use the *file* command to determine a type. For example, the following shows that */root* is a directory file:

   $ **file /root**
   /root: directory

Executable Files

Executable files could be commands or shell scripts. In other words, any file that can be run is an executable file. A file that has an x in the 4th, 7th, or the 10th field in the output of the *ll* command is executable.

$ ll /usr/sbin

```
. . . . . . . .
-rwxr-xr-x. 1  root  root      53792 Jun 24  2010 ypbind
-rwxr-xr-x. 1  root  root      16744 Aug 22  2011 yppoll
-rwxr-xr-x. 1  root  root      19840 Aug 22  2011 ypserv_test
-rwxr-xr-x. 1  root  root      14616 Aug 22  2011 ypset
. . . . . . . .
```

The *file* command can be used to identify the type of the executable file. For example:

$ file /usr/bin/who
/usr/bin/who: ELF 64-bit LSB executable, x86-64, version 1 (SYSV), dynamically linked (uses shared libs), for GNU/Linux 2.6.18, stripped

Symbolic Link Files

A *symbolic link* (a.k.a. a *soft link* or a *symlink*) may be considered as a shortcut to another file or directory. When you issue *ll* on a symbolically linked file or directory, you will notice two things. One, the line entry begins with the letter l; and two, there is an arrow pointing to the linked file or directory. For example:

$ ll /usr/sbin/sendmail
lrwxrwxrwx. 1 root root 21 Jul 17 12:22 /usr/sbin/sendmail -> /etc/alternatives/mta

The *file* command tells if the specified file or directory is linked, as follows:

$ file /usr/sbin/sendmail
/usr/sbin/sendmail: symbolic link to `/etc/alternatives/mta'

Device Files

Each piece of hardware in the system has an associated file used by the kernel to communicate with it. This type of file is called a *device file*. There are two types of device files: a *character* (or *raw*) device file and a *block* device file. The following example uses the *ll* command to display device files:

$ ll /dev/sd*
```
brw-rw----. 1 root disk 8, 0 Jul 27 20:33 /dev/sda
brw-rw----. 1 root disk 8, 1 Jul 27 20:33 /dev/sda1
```
$ ll /dev/tty*
```
crw-rw-rw-. 1 root tty   5, 0 Jul 27 20:33 /dev/tty
crw--w----. 1 root tty   4, 0 Jul 27 20:33 /dev/tty0
crw--w----. 1 root tty   4, 1 Jul 27 20:34 /dev/tty1
```

The first character in each line entry tells if the file type is block or character. A b denotes a block device file and a c represents a character device file. The *file* command shows their type:

$ file /dev/sda
/dev/sda: block special
$ file /dev/tty0
/dev/tty0: character special

Named Pipe Files

A *named pipe* allows two unrelated processes running on the same system or on two different systems to communicate with each other and exchange data. Named pipes are uni-directional. They are also referred to as FIFO because they use *First In First Out* mechanism. Named pipes make *Inter Process Communication* (IPC) possible. The output of the *ll* command shows a "p" as the first character in the line entry to represent a named pipe file:

 # ll /var/spool/postfix/public/pickup
prw--w--w-. 1 postfix postfix 0 Dec 12 09:57 /var/spool/postfix/public/pickup

IPC allows processes to communicate directly with each other by sharing parts of their virtual memory address space, and then reading and writing data stored in that shared virtual memory.

The *file* command shows the type of a named pipe file as fifo. See the following example:

 # file /var/spool/postfix/public/pickup
/var/spool/postfix/public/pickup: fifo (named pipe)

Socket Files

A *socket* is a named pipe that works in both directions. In other words, a socket is a two-way named pipe. It is also a type of IPC. Sockets are used by client/server programs. Notice an "s" as the first character in the output of the *ll* command below that points to a socket file:

$ ll /dev/log
srw-rw-rw-. 1 root root 0 Dec 9 04:33 /dev/log

The *file* command shows the type of socket file as follows:

$ file /dev/log
/dev/log: socket

File and Directory Operations

This section elaborates on file and directory naming rules and describes various operations on files and directories that you can perform. These operations include creating, listing, displaying contents of, copying, moving, renaming, and deleting files and directories.

File and Directory Naming Convention

Files and directories are assigned names when they are created. There are certain rules that you should remember and follow while assigning names. A file or directory name:

- ✓ Can contain a maximum of 255 alphanumeric characters (letters and numbers).
- ✓ Can contain non-alphanumeric characters such as an underscore, a hyphen, a space, and a period.
- ✓ Should not include special characters such as an asterisk, a question mark, a tilde, an ampersand, a pipe, double quotes, a single back quote, a single forward quote, a semi-colon, redirection symbols, and a dollar sign. These characters hold special meaning to the shell.
- ✓ May or may not have an extension. Some users prefer using extensions, others do not.

Creating Files and Directories

Files can be created in multiple ways; however, there is only one command to create directories.

Creating Files Using the touch Command

The *touch* command creates an empty file. If the file already exists, the time stamp is updated on it with the current system date and time. Execute the following as *aghori* to create *file1*:

```
$ cd
$ touch file1
$ ll file1
-rw-rw-r--. 1 aghori aghori 0 Jul 28 07:14 file1
```

As expected, the fifth field in the output is zero meaning that *file1* is created with zero bytes. Now, if you re-run the command on *file1* you will notice that the time stamp is updated.

```
$ touch file1
$ ll file1
-rw-rw-r--. 1 aghori aghori 0 Jul 28 07:15 file1
```

Creating Files Using the cat Command

The *cat* command allows you to create short text files:

```
$ cat > newfile
```

Nothing will be displayed when you execute this command because the system is waiting on you to input something; it expects you to input something that it will capture into the output file. Press Ctrl+d when done to save what you have typed to a file called *newfile*.

Creating Files Using the vi, vim, or nano Editors

These are Linux text editing tools of which the most common is the vi editor, but any of these may be used to create and modify text files.

Refer to Chapter 04 "File Permissions, Text Editors, Text Processors & The Shell" for a detailed understanding of the *vi* editor. Chapter 04 also touches on the *vim* and *nano* editors.

Creating Directories Using the mkdir Command

The *mkdir* command is used to create directories. The following example shows how to create a new directory called *scripts.dir1* in *aghori* user's home directory (*/home/aghori*):

```
$ cd
$ pwd
/home/aghori
$ mkdir scripts.dir1
```

You must have appropriate permissions to create a directory; otherwise, you will get an error message complaining about the lack of permissions.

You can create a hierarchy of sub-directories using *mkdir* with the –p option. In the following example, *mkdir* creates a directory *scripts.dir2* in *aghori*'s home directory. At the same time, it creates a directory *perl* as a sub-directory of *scripts.dir2* and a sub-directory *perl5* under *perl*.

```
$ mkdir –p scripts.dir2/perl/perl5
```

Listing Files and Directories

To list files and directories, use the *ls* or *ll* command. The following example runs *ll* as *aghori* in *aghori's* home directory:

```
$ ll
total 12
-rw-rw-r--. 1 aghori  aghori    48  Jul 28 07.21 file1
-rw-rw-r--. 1 aghori  aghori     0  Jul 28 07.30 newfile
drwxrwxr-x. 3 aghori  aghori  4096  Jul 28 07.28 scripts.dir1
drwxrwxr-x. 3 aghori  aghori  4096  Jul 28 07.30 scripts.dir2
```

Notice that there are nine columns in the output; these columns are described as follows:

Column 1: The 1st character tells the type of file. The next 9 characters indicate permissions. File permissions are explained at length in Chapter 04 "File Permissions, Text Editors, Text Processors & The Shell".
Column 2: Displays the number of links that the file or the directory has.
Column 3: Shows the owner name of the file or directory.
Column 4: Displays the group name that the owner of the file or directory belongs to.
Column 5: Shows the file size in bytes. For directories, this number reflects the number of blocks being used by the directory to hold information about its contents.
Columns 6, 7, and 8: Display the month, day of the month, and the time at which the file or directory was created or last accessed/modified.
Column 9: Shows the name of the file or directory.

Displaying File Contents

There are several commands that Linux offers to display file contents. Directory contents are simply the files and sub-directories within it. Use the *ll* or the *ls* command as explained earlier.

The *cat, more, less, head, tail, view, vi, vim, nano,* and *strings* commands are available to display file contents. The following is an explanation of each of these commands.

Using the cat Command

The *cat* command displays the contents of a text file. In the example below, *.bash_profile* in *aghori*'s home directory is displayed using the *cat* command:

```
$ cat /home/user1/.bash_profile
# .bash_profile
# Get the aliases and functions
. . . . . . . .
PATH=$PATH:$HOME/bin
export PATH
```

Using the more Command

The *more* command displays the contents of a long text file one page at a time starting at the beginning. In the example below, */etc/profile* is shown with *more*. It shows the percentage of the file being viewed in the last line.

```
$ more /etc/profile
# /etc/profile
# System wide environment and startup programs, for login setup
# Functions and aliases go in /etc/bashrc
. . . . . . . .
    esac
}
--More--(34%)
```

Navigation keys listed in Table 3-1 are helpful when viewing a large file with *more*.

Key	Purpose
Spacebar / f	Scrolls forward one screen.
Enter	Scrolls forward one line.
b	Scrolls backward one screen.
d	Scrolls forward half a screen.
h	Displays help.
q	Quits and returns to the command prompt.
/string	Searches forward for a string.
?string	Searches backward for a string.
n / N	Finds the next / previous occurrence of a string.

Table 3-1 Navigating with more

Using the less Command

The *less* command is similar to the *more* command but with extended capabilities. It does not need to read the entire file before it starts to display it, thus its output appears faster. In the example below, */etc/profile* is shown with *less*:

```
$ less /etc/profile
# /etc/profile
........
    esac
}
/etc/profile
```

The same navigation keys listed in Table 3-1 are helpful when viewing a large file with *less*.

Using the head Command

The *head* command displays the first few lines of a text file. By default, the first 10 lines are displayed. The following example shows the first 10 lines from the */etc/profile* file:

```
$ head /etc/profile
# /etc/profile

# System wide environment and startup programs, for login setup
# Functions and aliases go in /etc/bashrc

# It's NOT a good idea to change this file unless you know what you
# are doing. It's much better to create a custom.sh shell script in
# /etc/profile.d/ to make custom changes to your environment, as this
# will prevent the need for merging in future updates.
```

Supply a number with the command as an argument to view a different number of lines. The following example displays the first three lines from */etc/profile*:

```
$ head -3 /etc/profile
# /etc/profile

# System wide environment and startup programs, for login setup
```

Using the tail Command

The *tail* command displays the last few lines of a file. By default, the last 10 lines are displayed. The following example shows the last 10 lines from */etc/profile*:

```
$ tail /etc/profile
    if [ "${-#*i}" != "$-" ]; then
      . "$i"
    else
      . "$i" >/dev/null 2>&1
    fi
```

```
        fi
    done

    unset i
    unset pathmunge
```

You can specify a numerical value as an argument with the command to view a different set of lines. The following example displays the last 8 lines from /etc/profile:

$ tail –8 /etc/profile
```
        else
            . "$i" >/dev/null 2>&1
        fi
    fi
done

unset i
unset pathmunge
```

The *tail* command proves to be very useful when you wish to view a log file while it is being updated. The –f option enables this function. The following example shows how to view the RHEL system log file /var/log/messages in this manner. Try running this command on your system and notice the behavior.

 # tail –f /var/log/messages

Using the view, vi, and vim Commands

These commands open up the specified text file in the *vi* editor for modification. Here is how to run them to open .bash_profile in aghori's home directory:

$ view .bash_profile
$ vi .bash_profile
$ vim .bash_profile

See Chapter 04 "File Permissions, Text Editors, Text Processors & The Shell" for details on *vi* and *vim*.

Using the nano Command

Many system administrators and users prefer to use a menu-driven text editor call *nano* (nano's another). Here is how you would run it to open .bash_profile in user aghori's home directory:

$ nano .bash_profile

See Chapter 04 "File Permissions, Text Editors, Text Processors & The Shell" for details on *nano*.

Using the uniq Command

The *uniq* command identifies any duplicate line entries in a file or input provided. Without any options, this command prints unique lines. With the –d option, it prints only the duplicate lines. You can specify the –i option to instruct the command to ignore the case of the text. Here is an example:

$ uniq .bash_profile
```
# .bash_profile

# Get the aliases and functions
if [ –f ~/.bashrc ]; then
    . ~/.bashrc
fi

# User specific environment and startup programs

PATH=$PATH:$HOME/bin

export PATH
```

Using the strings Command

The *strings* command finds and displays legible information embedded within a non-text file. For example, when you run the *strings* command on the */bin/cat* file, you will observe output similar to the following. Although */bin/cat* is a non-text file, it does contain some legible information.

$ strings /bin/cat
```
/lib64/ld–linux–x86–64.so.2
fff.
AUATU
ffff.
AWAVAUATUH
D$pH
|$hH
. . . . . . . .
```

Copying Files and Directories

The copy operation duplicates a file or directory. There is a single command called *cp* which is used for this purpose.

Copying Files

The *cp* command copies one or more files to either the current directory or to another. If you want to duplicate a file in the same directory, you must give a different name to the target file. If you want to copy a file to a different directory, you can use the same file name or you can assign it a different name. Consider the following examples.

To copy *file1* as *newfile1* in the same directory:

$ cp file1 newfile1

To copy *file1* by the same name into another directory called *scripts.dir1*:

$ cp file1 scripts.dir1

By default, when you copy a file, the destination is overwritten and a warning message is not generated. In order to avoid such a situation use the –i option with *cp*, which prompts for confirmation before overwriting.

$ cp –i file1 file2
cp: overwrite `scripts.dir1/file1'?

Copying Directories

The *cp* command with the –r (recursive) option copies a directory along with its contents to another location. In the following example, *scripts.dir1* directory is copied under *scripts.dir2*:

$ cp –r scripts.dir1 scripts.dir2

You may wish to use the –i option if needed.

Moving and Renaming Files and Directories

The move operation copies a file or directory to an alternative location and deletes the original file or directory. The rename operation simply changes the name of a file or directory.

Moving and Renaming Files

The *mv* command is used to move or rename files. The –i option can be specified for user confirmation if the destination file exists. The following example moves *file1* to *scripts.dir1* and prompts for confirmation if a file by the same name exists in *scripts.dir1*:

$ mv –i file1 scripts.dir1
mv overwrite: scripts.dir1/file1? (y/n)

To rename *newfile* as *newfile1*:

$ mv newfile newfile1

You may want to use the –i option with *mv*, if needed.

Moving and Renaming Directories

To move a directory along with its contents to some other directory location or simply change the name of the directory, use the *mv* command. For example, to move *scripts.dir1* into *scripts.dir2* (*scripts.dir2* must exist), issue the following:

$ mv scripts.dir1 scripts.dir2

To rename *scripts.dir2* as *scripts.dir20* (*scripts.dir20* must not exist):

  $ mv scripts.dir2 scripts.dir20

Removing Files and Directories

The remove operation deletes a file or directory.

Removing Files

You can remove a file using the *rm* command, which deletes one or more specified files at once. The following example deletes *newfile*:

  $ rm newfile

The –i option can be used to prevent accidental file removal. The option instructs the command to prompt for confirmation before removing. The following example prompts for confirmation before deleting *newfile*:

  $ rm –i newfile
  rm: remove regular empty file `newfile'?

Removing Directories

There are two commands available to remove directories, which are demonstrated in the following examples.

To remove an empty directory, use the *rmdir* command:

  $ rmdir scripts.dir30

To remove a non-empty directory, use *rm* with the –r option:

  $ rm –r scripts.dir10

Use *rm* with the –i option to remove interactively:

  $ rm –ri scripts.dir2
  rm: descend into directory `scripts.dir2'?

Summary of File and Directory Operations

Table 3-2 lists commands for file and directory operations that you have learned in this chapter.

Command to	File	Directory
Create	*cat, touch, vi, vim, nano*	*mkdir*
List	*ll, ls*	*ll, ls*
Display contents	*cat, more, less, head, tail, view, vi, vim, nano, uniq, strings*	*ll, ls*
Copy	*cp*	*cp*
Move	*mv*	*mv*
Rename	*mv*	*mv*
Remove	*rm*	*rm −r, rmdir*

Table 3-2 Summary of File & Directory Operations

File and Directory Control Attributes

There are certain control attributes that may be set on a file or directory in order to allow data to be appended, to prevent it from being changed or deleted, etc. For example, you can enable attributes on a critical system file or directory so that no users, including *root*, can delete or change it, disallow a backup utility such as the *dump* command to back up a specific file or directory, and so on. These attributes can only be set on files and directories located in an ext2, ext3, or an ext4 file system. See Chapter 10 "File Systems & Swap" for details on ext2, ext3, and ext4 file systems.

Table 3-3 lists common control attributes.

Attribute	Effect on File or Directory
a (append)	File can only be appended.
A	Prevents updating access time.
c (compressed)	File is automatically compressed on the disk.
d (dump)	File cannot be backed up with the *dump* command.
D	Changes on a directory are written synchronously to the disk.
e (extent format)	File uses extents for mapping the blocks on disk. Set on all files by default.
i (immutable)	File cannot be changed, renamed, or deleted.
j (journaling)	File has its data written to the journal before being written to the file itself. See Chapter 10 "File Systems & Swap" for details on journaling.
S (synchronous)	Changes in a file are written synchronously to the disk.

Table 3-3 File and Directory Control Attributes

There are two commands *lsattr* and *chattr* that are used for attribute management. The following examples demonstrate the usage of these commands with an assumption that you are in the */root* directory and the *install.log* file exists.

To list current attributes on *install.log*:

lsattr install.log
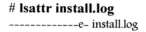
-------------e- install.log

The output indicates that presently there are no attributes, except for the default extent format attribute, set on the file.

To prevent *install.log* from deletion and any modifications:

chattr +i install.log
lsattr install.log
----i--------e- install.log

Now, try deleting this file as *root*:

rm install.log
rm: remove regular file `install.log'? y
rm: cannot remove `install.log': Operation not permitted

To allow only the append operation on the file:

chattr +a install.log
cat /etc/fstab >> install.log

To unset both attributes:

chattr –ia install.log

Pattern Matching

RHEL provides a powerful tool to search the contents of one or more text files, or to search the contents of input provided, for matching a pattern. This is referred to as *pattern matching* (a.k.a. *regular expression* or *globbing*). A pattern can be a single character, a series of characters, a word, or a sentence. You must enclose the pattern in double quotes if it contains one or more white spaces.

The command used for pattern matching is called *grep* (*global regular expression print*), and it searches contents of one or more specified files for a regular expression. If the expression is found, *grep* prints every line containing that expression on the screen without changing the contents of the original file. Consider the following examples.

To search for the pattern aghori in the */etc/passwd* file:

$ **grep aghori /etc/passwd**
aghori:x:500:500::/home/aghori:/bin/bash

To search for all occurrences of the pattern aghori in both the */etc/passwd* and */etc/group* files:

$ **grep aghori /etc/passwd /etc/group**
/etc/passwd:aghori:x:500:500::/home/aghori:/bin/bash
/etc/group:aghori:x:500:

To display only the names of those files that contain the pattern aghori from the specified file list, use the –l option:

$ grep –l aghori /etc/group /etc/passwd /etc/hosts
/etc/group
/etc/passwd

To look for the pattern root in the */etc/group* file along with associated line number(s), use the –n option:

$ grep –n root /etc/group
1:root:x:0:root
2:bin:x:1:root,bin,daemon
3:daemon:x:2:root,bin,daemon
4:sys:x:3:root,bin,adm
5:adm:x:4:root,adm,daemon
7:disk:x:6:root
11:wheel:x:10:root

To search for the pattern root in */etc/group* and exclude the lines in the output that contain this pattern, use the –v option:

$ grep –v root /etc/group
tty:x:5:
lp:x:7:daemon,lp
mem:x:8:
. . . .

To search for all lines in the */etc/passwd* file that begin with the pattern root. The bash shell treats the caret ^ sign as a special character which marks the beginning of a line or word. This is useful, for instance, if you wish to know whether there is more than one user with that name.

$ grep ^root /etc/passwd
root:x:0:0:root:/root:/bin/bash

To list all lines from the */etc/passwd* file that end with the pattern bash. The bash shell treats the dollar sign ($) as a special character which marks the end of a line or word. This is useful, for example, to determine which users have their shells set to the bash shell.

$ grep bash$ /etc/passwd
root:x:0:0:root:/root:/bin/bash
aghori:x:500:500::/home/aghori:/bin/bash
user1:x:501:501::/home/user1:/bin/bash

To search for all empty lines in the */etc/passwd* file:

$ grep ^$ /etc/passwd

To search for all lines in the */etc/passwd* file that contain only the pattern root:

$ grep ^root$ /etc/passwd

To search for all lines in the */etc/passwd* file that contain the pattern root. The –i option used with *grep* here ignores the letter case. This is useful to determine if there are *root* user accounts with a combination of lowercase and uppercase letters.

> **$ grep –i root /etc/passwd**
> root:x:0:0:root:/root:/bin/bash
> operator:x:11:0:operator:/root:/sbin/nologin

To print all lines from the output of the *ll* command that contain either the drwx or the xin pattern, run either of the following:

> **$ ll /etc | grep –E 'drwx|xin'**
> **$ ll /etc | egrep 'drwx|xin'**
> drwxr-xr-x. 3 root root 4096 Jul 17 12:20 abrt
> drwxr-xr-x. 4 root root 4096 Jul 17 12:28 acpi

Comparing File Contents

The *diff* (difference) command finds differences between contents of text files and prints them line-by-line on the screen. Two options –i and –c are commonly used to ignore letter case and produce a list of differences in three sections, respectively. Assume that you have two text files *testfile1* and *testfile2* with the following contents:

testfile1	*testfile2*
apple	Apple
pear	tomato
mango	guava
tomato	mango
guava	banana

Run the *diff* command to display the results in three sections:

> **$ diff –c testfile1 testfile2**
> *** testfile1   2012-07-28 13:27:39.778219980 -0400
> --- testfile2   2012-07-28 13:27:55.221232218 -0400
> ***************
> *** 1,5 ****
>   apple
> - pear
> - mango
>   tomato
>   guava
> --- 1,6 ----
> + testfile2
>   apple
>   tomato
>   guava
> + mango
> + banana

The first section shows the file names being compared along with time stamps on them and some fifteen * characters to mark the end of this section.

The second section tells the number of lines in *testfile1* that differs from *testfile2* and the total number of lines *testfile1* contains. Then it prints the actual line entries from *testfile1*. Each line that differs from *testfile2* is preceded by the – symbol.

The third section tells the number of lines in *testfile2* that differs from *testfile1* and the total number of lines *testfile2* contains. Then it prints the actual line entries from *testfile2*. Each line that differs from *testfile1* is preceded by the + symbol.

Now, if you wish to make the contents of the two files identical, you will need to remove entries for pear and mango from *testfile1* and append entries for mango and banana to *testfile1*.

You can also use the *diff* command to find differences in directory contents. The syntax is the same.

Finding Files

Sometimes you need to find one or more files or directories in the file system structure based on a criterion. RHEL offers two tools to help you in that situation. These tools are called *find* and *locate*, and are explained in the following sub-sections.

Using the find Command

The *find* command recursively searches the directory tree, finds files that match the specified criteria, and optionally performs an action. This powerful tool can be customized to look for files in a number of ways. The search criteria may include searching for files by name, size, ownership, group membership, last access or modification time, permissions, file type, and inode number. Here is the command syntax:

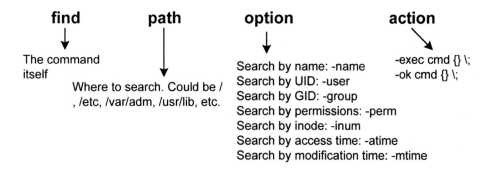

With the *find* command, files that match the specified criteria are located and the full path to each file is displayed on the screen. Let us look at a few examples.

To search for *newfile* in *aghori*'s home directory */home/aghori* (assuming *newfile* exists):

```
$ cd
$ find . –name newfile –print
```

–print is optional. The *find* command, by default, displays results on the screen. You do not have to specify this option.

To search for files and directories in */dev* that begin with vol followed by any characters. The –i option forces the command to perform a case insensitive search.

 # find /dev –iname vg00*

To find files smaller than 1MB in size in *root*'s home directory:

 # find ~ –size –1M

The tilde ~ character represents a user's home directory. See Chapter 04 "File Permissions, Text Editors, Text Processors & The Shell" for details.

To find files larger than 10MB in size in */usr* directory:

 # find /usr –size +10M

To find files in */home* with ownership set to *aghori* and group membership set to any group but *aghori*:

 # find /home –user aghori –not –group aghori

To find files in the */etc/rc.d* directory that were modified more than 120 days ago:

 # find /etc/rc.d –mtime +120

To find files in the */etc/rc.d* directory that have not been accessed in the last 90 days:

 # find /etc/rc.d –atime –90

To find files in the */etc/rc.d* directory that were modified exactly 10 days ago:

 # find /etc/rc.d –mtime 10

To search for character device files in the */dev* directory with permissions set to 666. In this example, two criteria – character device files with read and write permissions for all users – are defined and files that match both the criteria are displayed.

 # find /dev –type c –perm 666

To search for character device files in the */dev* directory that are world writeable:

 # find /dev –type c –perm –222

To search for *core* files in the entire directory tree and delete them as found without prompting for confirmation:

find / –name core –exec rm {} \;

 The pattern {} \; is part of the syntax and must be defined that way.

There are numerous other options available with the *find* command. Try them to build a better understanding. Refer to the command's man pages.

Using the locate Command

The *locate* command is used for locating all occurrences of the specified string as it appears in file pathnames. It can also be used to locate files with certain extensions. The command searches the */var/lib/mlocate/mlocate.db* database file and displays matching occurrences. This file is updated automatically every day when the */etc/cron.daily/mlocate.cron* script is executed by the *cron* daemon. Alternatively, it can be updated manually with the *updatedb* command. The output is the absolute path of files and directories for which the user has access permissions. The security enhanced version of the *locate* command is the *slocate* command whose database file is */var/lib/slocate/slocate.db*. Here are a few examples to explain how the *locate* command works.

To locate all occurrences of the string passwd:

$ locate passwd
/etc/passwd
/etc/passwd–
. . . .

Use the –n option to specify how many occurrences you wish to display:

$ locate –n 3 passwd
/etc/passwd
/etc/passwd–
/etc/pam.d/passwd

To locate all files with .sh extension and list the first two of them:

$ locate –n 2 .sh
/etc/X11/xinit/xinitrc.d/00–start–message–bus.sh
/etc/X11/xinit/xinitrc.d/localuser.sh
. . . .

Sorting Input

Sorting allows you to arrange columns of text in a specified order. The *sort* command is used for this purpose. This command sorts contents of one or more files and prints the result on the screen. You can sort file contents in either alphabetic (default) or numeric order.

Let us look at a few examples to understand the usage of *sort*.

Consider a file *file10* in *aghori*'s home directory with the following text in two columns. The first column contains alphabetic names and the second contains numbers.

```
Maryland 667
Mississippi 662
Pennsylvania 445
Missouri 975
Florida 772
Montana 406
Massachusetts 339
```

To sort this file alphabetically:

$ sort file10
```
Florida 772
Maryland 667
Massachusetts 339
Mississippi 662
Missouri 975
Montana 406
Pennsylvania 445
```

To sort this file numerically (–n option) on the second column (–k option):

$ sort –k 2 –n file10
```
Massachusetts 339
Montana 406
Pennsylvania 445
Mississippi 662
Maryland 667
Florida 772
Missouri 975
```

To sort *file10* numerically but in reverse order (–r option):

$ sort –k 2 –nr file10
```
Missouri 975
Florida 772
Maryland 667
Mississippi 662
Pennsylvania 445
Montana 406
Massachusetts 339
```

To sort the output of the *ll* command:

```
# ll / | sort
drwx------.    2 root root   16384 Jul 17 08:09  lost+found
drwxrwxrwt. 13 root root    4096 Jul 28 14:03  tmp
drwxr-xr-x.  10 root root    4096 Jul 17 12:36  cgroup
drwxr-xr-x. 110 root root   12288 Jul 28 13:25  etc
. . . . . . . .
```

To sort on the 6th column (month) of the *ll* command output:

```
$ ll –a /etc/skel | sort –k 6M
total 36
-rw-r--r--.    1  root  root       124 Jan 27  2011  .bashrc
-rw-r--r--.    1  root  root       176 Jan 27  2011  .bash_profile
-rw-r--r--.    1  root  root        18 Jan 27  2011  .bash_logout
drwxr-xr-x. 110 root  root     12288 Jul 28 13:25  ..
drwxr-xr-x.    2  root  root     4096 Jul 14  2010  .gnome2
drwxr-xr-x.    4  root  root     4096 Jul 17 12:15  .mozilla
drwxr-xr-x.    4  root  root     4096 Jul 17 12:24  .
```

By default, the output of *sort* is displayed on the screen but if you want it redirected to a file, you can use the –o option. The example below saves the output in */tmp/sort.out* and does not display it on the screen:

$ ll /etc/skel | sort –k 6M –o /tmp/sort.out

To sort on the 6th and then on the 7th column, run the following. This is an example of multi-level sorting.

```
$ ll –a /etc/skel | sort –k 6M –k 7
total 36
-rw-r--r--.    1  root  root        18 Jan 27  2011 .bash_logout
-rw-r--r--.    1  root  root       176 Jan 27  2011 .bash_profile
-rw-r--r--.    1  root  root       124 Jan 27  2011 .bashrc
drwxr-xr-x.    2  root  root     4096 Jul 14  2010 .gnome2
drwxr-xr-x.    4  root  root     4096 Jul 17 12:15 .mozilla
drwxr-xr-x.    4  root  root     4096 Jul 17 12:24 .
drwxr-xr-x. 110 root  root    12288 Jul 28 13:25 ..
```

There are numerous other options available with the *sort* command. Try them to build a better understanding. Refer to the command's man pages.

Linking Files and Directories

Each file in the system has a unique number assigned to it at the time of its creation. This number is referred to as its *inode* (index node) number. All file attributes such as the name, type, size, permissions, ownership, group membership, and last access/modification time are maintained in that inode. Moreover, the inode points to the exact location in the file system where the file data is actually stored. See Chapter 10 "File Systems & Swap" for details on inodes.

Linking files or directories means that you have more than one instance of them pointing to the same physical data location in the directory tree.

There are two types of links: *soft* links and *hard* links.

Soft Link

A *soft* link (a.k.a. a *symbolic* link or a *symlink*) makes it possible to associate one file with another. It is similar to a shortcut in MS Windows where the actual file is resident somewhere in the directory structure but you may have multiple shortcuts or pointers with different names pointing to it. This means accessing the file via the actual file name or any of the shortcuts would yield an identical result. Each soft link has a unique inode number.

A soft link can cross file system boundaries and can be used to link directories.

To create a soft link for *newfile* as *newfile10* in the same directory, use the *ln* command with the –s option:

```
$ cd /home/aghori
$ ln –s newfile newfile10
```

where:

> *newfile* is an existing file
> *newfile10* is soft linked to *newfile*

After you have created the link, issue *ll* with –i option. Notice the letter l as the first character in the second column of the output. Also notice an arrow pointing from the linked file to the original file. This indicates that *newfile10* is merely a pointer to *newfile*. The –i option displays associated inode numbers in the first column.

```
$ ll –i
36 –rw–rw–r––. 1 aghori aghori   0 Jul 28 07:30 newfile
32 lrwxrwxrwx. 1 aghori aghori   7 Jul 28 14:23 newfile10 -> newfile
```

If you remove the original file (*newfile* in this example), the link *newfile10* will stay but points to something that does not exist.

Hard Link

A *hard* link associates two or more files with a single inode number. This allows the files to have identical permissions, ownership, time stamp, and file contents. Changes made to any of the files are reflected on the other linked files. All files actually contain identical data.

A hard link cannot cross file system boundaries and cannot be used to link directories.

The following example uses the *ln* command and creates a hard link for *newfile2* located under */home/aghori* to *newfile20* in the same directory. Note that *newfile20* does not currently exist, it will be created.

```
$ cd /home/aghori
$ ln newfile2 newfile20
```

After creating the link, run *ll* with the –i option:

$ ll –i
32 –rw–rw–r––. 2 aghori aghori 0 Jul 28 14:29 newfile2
32 –rw–rw–r––. 2 aghori aghori 0 Jul 28 14:29 newfile20

Look at the first and the third columns. The first column indicates that both files have identical inode numbers and the third column tells that each file has two hard links. *newfile2* points to *newfile20* and vice versa. If you remove the original file (*newfile2* in this example), you will still have access to the data through the linked file *newfile20*.

Chapter Summary

This chapter presented an overview of the RHEL file system structure and significant higher level sub-directories that consisted of either static or variable files and grouped logically into lower level sub-directories. You looked at how files and sub-directories were accessed using a path relative to either the top-most directory of the file system structure or your current location in the tree.

You learned about different types of files and the set of rules to adhere to when creating files and directories. You looked at several file and directory manipulation tools for creating, listing, displaying, copying, moving, renaming, and removing them.

Searching for text within files and searching for files within the directory structure using specified criteria provided you with an understanding and explanation of tools required to perform such tasks.

Finally, you studied how to sort contents of a text file or output generated by executing a command in ascending and descending orders. The last topic discussed creating soft and hard links for files and directories.

Chapter Review Questions

1. An executable file in Linux must have an extension in its name otherwise you cannot execute it. True or False?
2. What is another name for a fully qualified path?
3. */proc* is a disk-based file system. True or False?
4. Name the two types of paths.
5. The *egrep* command and the *grep* command with the –E option would produce identical results. True or False?
6. Which command can be used to determine a file type?
7. What are the two distinctions in the output of the *ll* command to determine if the file listed is a symbolically linked file?
8. Which option can be specified with the *mkdir* command to create a hierarchy of directories?
9. The *rmdir* command without any options can be used to remove an empty directory. True or False?
10. What are the two types of file systems?
11. Which option should be used with the *tail* command to view the contents of a log file in realtime?
12. What would the command *find / -name –mtime 20* do?
13. The –m option with the *chattr* command makes a file immutable. True or False?
14. What would the command *egrep 'root|qemu' /etc/passwd* produce?

15. What would the command *ll /etc/sysconfig | sort –k 6M –k 7* produce?
16. Hard linked directories can cross file system boundaries but soft linked directories cannot. True or False?
17. What is the difference between moving and renaming a file?
18. Which command refers to the */var/lib/slocate/slocate.db* file?
19. The *diff* command may be used to identify differences between two files. True or False?
20. What are the two commands for managing file and directory attributes?
21. What is the purpose of the *strings* command?
22. What are the purposes of the –i and –r options with the *cp, rm,* and *mv* commands?
23. What would the *find / -name core –exec rm {} \;* command do?
24. The *ll* command produces 9 columns in the output by default. True or False?

Answers to Chapter Review Questions

1. False. File extensions are not necessary in Linux.
2. Another name for a fully qualified path is the absolute path.
3. False. */proc* is a virtual file system.
4. The two types of paths are absolute and relative.
5. True.
6. The *file* command can be used to determine a file type.
7. The line entry for a symbolically linked file begins with an *l* and has an arrow with the file name.
8. The –p option can be specified with the *mkdir* command to create a hierarchy of directories.
9. True.
10. The two type of file systems are disk-based and virtual.
11. The –f option is used with the tail command for this purpose.
12. True.
13. The *find* command provided will search for all files in the entire directory tree that were modified exactly 20 days ago.
14. False.
15. The *egrep* command provided will display all lines from the */etc/passwd* file containing the words root and qemu.
16. The *ll* command provided will display the listing of the */etc/sysconfig* directory sorted by date and time.
17. False.
18. The move operation makes a copy of the original file and then deletes it, whereas, the rename operation simply change the file's name.
19. The *slocate* command refers to this file.
20. True.
21. The *lsattr* and *chattr* commands are used to manage file and directory attributes.
22. The *strings* command is used to extract and display the legible contents of a non-text file.
23. The –i option is used to run the commands interactively and the –r option to run them recursively.
24. The *find* command provided will remove all files with the name core from the entire directory hierarchy.
25. True.

Labs

Lab 3-1: Find Files and Determine File Types

Log on to *physical.example.com* as *root* and execute the *find* command to search for all files in the entire directory structure that have been modified in the past 15 days. Run the *file* command as part of the *find* command to display the type of each file as well. Run the *ll* command on the *root* home directory and identify directories and regular files. Check the type of some of the files and directories. Locate one block special and one character special file in the system and run the *file* command on them to confirm their types. Run the *find* command again to search for three named pipe and three socket files in the directory structure, and identify their types with the *file* command.

Lab 3-2: File and Directory Operations

Log on to *physical.example.com* as *user1* and create one file and one directory in the home directory with any names of your choice. List the file and directory and observe the permissions, ownership, and group membership. Try to move the file and the directory to the */var/log* directory and observe the output. Now try again to move them to the */tmp* directory. Duplicate the file with the *cp* command and then rename the duplicated file to any name. Finally, remove the file and directory created for this lab.

Lab 3-3: Extracting Text

Log on to *physical.example.com* as *root*. Execute the *ll* command recursively on the */usr* directory and *grep* for directories that have read/write/execute permissions for the owner. Run the *grep* command on the */etc/profile* file and show, along with the line number, if the shell scripts located in the */etc/profile.d* directory are executed when a user attempts to log on to the system. Issue the combination of the *grep* and the *wc* commands and provide a count of empty lines in the */etc/profile* file. List which users do not have the bash shell as their default login shell (hint: *grep* on the */etc/passwd* file).

Lab 3-4: Find Files Using Different Criteria

Log on to *physical.example.com* as *root* and execute the *find* command to search for regular files in the entire directory structure that were accessed more than 10 days ago, are not bigger than 5MB in size, and are owned by the user *root*.

Lab 3-5: Sort the Input

Log on to *physical.example.com* as *root* and execute the *ll* command on the */etc/init.d* directory sorted by the date and time.

Chapter 04

File Permissions, Text Editors, Text Processors & The Shell

This chapter covers the following major topics:

✓ Permissions assigned to owners, members of owner's group, and others
✓ Permission types – read, write, and execute
✓ Permission modes – adding, revoking, and assigning
✓ Modify permissions using symbolic and octal notations
✓ Set default permissions on new files and directories
✓ Modify ownership and group membership
✓ Configure special permissions with setuid, setgid, and sticky bits
✓ Configure a shared directory for group collaboration
✓ The vi and the nano text editors
✓ Manipulate columns and rows of text using awk and sed
✓ The bash shell and its features – local and environment variables, input, output, and error redirection, tab completion, command line editing, command history, tilde substitution, special characters, pipe, and filters

This chapter includes the following RHCSA objectives:

02. Use input-output redirection (>, >>, |, 2>, etc)
07. Create and edit text files
10. List, set, and change standard ugo/rwx permissions
30. Create and configure set-GID directories for collaboration
32. Diagnose and correct file permission problems

Permissions

Permissions are set on files and directories to prevent access from unauthorized users. Users on the system are grouped into three distinct categories. Each user category is then assigned required permissions. Permissions may be modified using one of two available methods. The user mask may be defined for individual users so new files and directories they create always get preset permissions. Every file in Linux has an owner and a group. Special permissions may be defined for specific users within a group.

As a system administrator you need to edit text files as required. RHEL provides several text editors for this purpose. The vi editor is very popular in the Linux administrator community, however there are additional text editors included in RHEL. These include an enhanced and improved version of vi called vim (vi improved) and the nano editor.

The vim editor is basically the vi editor with several enhanced features such as multi-level undo, multi-window support, syntax highlighting, command line editing, tab completion, online help, and visual selection. Nano, on the other hand, is an enhanced editor of yet another text editor called *pico*. Nano provides simple editing functions similar to what notepad provides in MS Windows.

Text processors are tools for manipulating columns and rows of text, and prove to be invaluable for system administrators when specific data needs to be extracted from the input provided.

Shells interface users with the kernel by enabling them to submit their requests for processing. RHEL supports several shells of which the bash shell is the most popular. It is also the default shell in RHEL6. The bash shell offers a variety of features that help administrators perform their job with greater ease and convenience.

Determining Access Permissions

Access permissions on files and directories allow administrative control over which users can access them and to what level. Permissions discussed in this section are referred to as standard ugo/rwx file permissions. The following sub-sections elaborate on this topic.

Permission Classes

Users on the system are categorized into three distinct classes for the purpose of maintaining file security through permissions. These classes are described in Table 4-1.

Permission Class	Description
User (u)	Owner of file or directory. Usually, the creator of a file or directory is the owner of it.
Group (g)	A set of users that need identical access on files and directories that they share. Group information is maintained in the */etc/group* file and users are assigned to groups according to shared file access needs.
Others (o)	All other users that have access to the system except for the owner and group members. Also called *public*.

Table 4-1 Permission Classes

Permission Types

Permissions control what actions can be performed on a file or directory and by whom. There are four types of permissions as defined in Table 4-2.

Perm Type	Symbol	File	Directory
Read	r	Displays file contents or copies contents to another file.	Displays contents with the *ll* command.
Write	w	Modifies file contents.	Creates, removes, or renames files and sub-directories.
Execute	x	Executes a file.	*cd* into the directory.
Access Denied	-	None.	None.

Table 4-2 Permission Types

Permission Modes

A permission mode is used to add, revoke, or assign a permission type to a permission class. Table 4-3 shows various permission modes.

Permission Mode	Description
Add (+)	Gives specified permission(s).
Revoke (-)	Removes specified permission(s).
Assign (=)	Gives specified permission(s) to owner, group members, and public in one go.

Table 4-3 Permission Modes

The output of the *ll* command lists files and directories along with their type and permission settings. This information is shown in the first column of the output where 10 characters are displayed. The first character indicates the type of file: d for directory, – for regular file, l for symbolic link, c for character device file, b for block device file, n for named pipe, s for socket, and so on. The next nine characters – three groups of three characters – show read (r), write (w), execute (x), or none (-) permission for the three user classes: user, group, and others, respectively.

Figure 4-1 illustrates the *ll* command output and its various components.

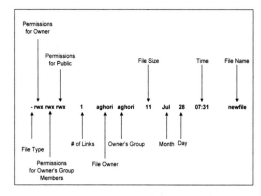

Figure 4-1 Permission Settings

From Figure 4-1, it is obvious who the owner (3rd column) of the file is and which group (4th column) the owner belongs to.

Changing Access Permissions

The *chmod* command is used to modify access permissions on files and directories. It works identically on both file and directories. *chmod* can be used by *root* or the file owner, and can modify permissions specified in one of two methods: *symbolic* or *octal*. Symbolic notation uses a combination of letters and symbols to add, revoke, or assign permissions to each class of users. The octal notation (a.k.a. absolute notation), on the other hand, uses a three-digit numbering system ranging from 0 to 7 to specify permissions for the three user classes. Octal values are given in Table 4-4.

Octal Value	Binary Notation	Symbolic Notation	Explanation
0	000	---	No permissions.
1	001	--x	Execute permission only.
2	010	-w-	Write permission only.
3	011	-wx	Write and execute permissions.
4	100	r--	Read permission only.
5	101	r-x	Read and execute permissions.
6	110	rw-	Read and write permissions.
7	111	rwx	Read, write, and execute permissions.

Table 4-4 Octal Permission Notation

From Table 4-4, it is obvious that each 1 corresponds to an r, a w, or an x, and each 0 corresponds to the – character for no permission at that level. Figure 4-2 shows weights associated with each digit location in the 3-digit octal numbering model. The right-most location has weight 1, the middle location carries weight 2, and the left-most location has 4. When you assign a permission of 6, for example, it would correspond to the two left-most digit locations. Similarly, a permission of 2 would mean only the middle digit location.

Figure 4-2 Permission Weights

Exercise 4-1: Modify File Permissions Using the Symbolic Notation

For this exercise, it is assumed that the following file *file1* with read permission for the owner, group members, and others, exists. The assumption also includes that this file is owned by *user1* and you are logged in as *user1*.

```
-r--r--r-- 1 user1 user1 0 Nov 26 20.45 file1
```

In this exercise, you will add the execute permission for the owner in the first step, add the write permission for the owner in the second step, and add the write permission for the group and public in the third step. You will then revoke the write permission from the public and assign the read, write, and execute permissions to the three user categories in one shot.

1. Add the execute permission for the owner and run the *ll* command to verify:

    $ chmod u+x file1
    -r-xr--r-- 1 user1 user1 0 Nov 26 20:45 file1

2. Add the write permission for the owner and run the *ll* command to verify:

    $ chmod u+w file1
    -rwxr--r-- 1 user1 user1 0 Nov 26 20:45 file1

3. Add the write permission for group members and public and run the *ll* command to verify:

    $ chmod go+w file1
    -rwxrw-rw- 1 user1 user1 0 Nov 26 20:45 file1

4. Remove the write permission for public and run the *ll* command to verify:

    $ chmod o-w file1
    -rwxrw-r-- 1 user1 user1 0 Nov 26 20:45 file1

5. Assign read, write, and execute permissions to all three user categories and run the *ll* command to verify:

    $ chmod a=rwx file1
    -rwxrwxrwx 1 user1 user1 0 Nov 26 20:45 file1

Exercise 4-2: Modify File Permissions Using the Octal Notation

For this exercise, it is assumed that the following file *file2* with permissions 444 exists. The assumption also includes that this file is owned by *user1* and you are logged in as *user1*.

    -r--r--r-- 1 user1 user1 0 Nov 26 21:19 file2

In this exercise, you will add the execute permission for the owner in the first step, add the write permission for the owner in the second step, and add the write permission for the group and public in the third step. You will then revoke the write permission from the public and assign the read, write, and execute permissions to the three user categories in one shot.

1. Add the execute permission for the file owner and run the *ll* command to verify:

    $ chmod 544 file2
    -r-xr--r-- 1 user1 user1 0 Nov 26 21:19 file2

2. Add the write permission for the owner and run the *ll* command to verify:

   $ chmod 744 file2
   -rwxr--r-- 1 user1 user1 0 Nov 26 21:19 file2

3. Add the write permission for group members and others, and run the *ll* command to verify:

   $ chmod 766 file2
   -rwxrw-rw- 1 user1 user1 0 Nov 26 21:19 file2

4. Remove the write permission for public and run the *ll* command to verify:

   $ chmod 764 file2
   -rwxrw-r-- 1 user1 user1 0 Nov 26 21:19 file2

5. Assign read, write, and execute permissions to all three user categories and run the *ll* command to verify:

   $ chmod 777 file2
   -rwxrwxrwx 1 user1 user1 0 Nov 26 21:19 file2

Default Permissions

The system assigns *default permissions* to a file or directory at the time of its creation. Default permissions are calculated based on the *umask* (user mask) permission value subtracted from a preset value called *initial* permissions.

The umask is a three-digit value that refers to read/write/execute permissions for owner, group, and others. Its purpose is to set default permissions on new files and directories created without touching the existing files and directories. In RHEL, the default umask value is set to 0022 for the *root* and other system users and 0002 for all regular users who have the bash shell assigned. Note that the left-most 0 has no significance. Run the *umask* command without any options and it will display the current umask value:

$ umask
0002

Run the *umask* command with the –S option to display the current umask in symbolic notation:

$ umask –S
u=rwx,g=rwx,o=rx

Different shells have different default umask values for regular users. For instance, the Korn shell has 0022 and the C shell 22.

The pre-defined initial permission values are 666 (rw-rw-rw-) for files and 777 (rwxrwxrwx) for directories. Even if the umask is set to 777, the new files always get a maximum of 666 permissions. This is part of Linux security. You have to use the *chmod* command to add executable bits explicitly.

Calculating Default Permissions

Consider the following example to calculate default permission values on files for regular users:

Initial Permissions	666	
umask	– 002	(subtract)
=======================		
Default Permissions	664	

This indicates that every new file will have read and write permissions assigned to the owner and owner's group members, and read-only permission to others.

To calculate default permission values on directories for regular users:

Initial Permissions	777	
umask	– 002	(subtract)
=======================		
Default Permissions	775	

This indicates that every new directory will have read, write, and execute permissions assigned to the owner and owner's group members, and read and execute permissions to others.

If you wish to have different default permissions set on new files and directories, you need to modify the umask. First determine what default values you need. For example, if you want all your new files and directories to have 640 and 750 permissions, respectively, run *umask* and set the value to 027 as follows:

$ umask 027

The new umask value becomes effective right away, and it will be applicable only on files and directories created after this value change. The existing files and directories will remain intact. Create a file *file10* and a directory *dir10* as *user1* under */home/user1* and test the effect of the change.

$ touch file10
$ ll file10
–rw–r–––––. 1 user1 user1 0 Jul 29 07:00 file10
$ mkdir dir10
$ ll –d dir10
drwxr-x–––. 2 user1 user1 4096 Jul 29 07:00 dir10

The above examples show that new files and directories are created with different permissions. The files have (666 – 027 = 640) and directories have (777 – 027 = 750) permissions.

The umask value set at the command line will be lost as soon as you log off. In order to retain the new setting, place it in one of the shell initialization files. Customizing shell initialization files is covered later in this chapter.

File Ownership and Group Membership

In RHEL, every file and directory has an owner associated with it. By default, the creator becomes the owner. The ownership may be altered and allocated to some other user, if required.

Similarly, every user is a member of one or more groups. A group is a collection of users with the exact same privileges. By default, the owner's group is assigned to a file or directory.

For example, below is an output from the *ll* command ran on *file10*:

```
$ ll file10
-rw-r-----. 1 user1 user1  0 Jul 29 07:00 file10
```

The output indicates that the owner of *file10* is *user1* who belongs to group *user1*. If you wish to view the corresponding UID and GID instead, run the *ll* command with –n option:

```
$ ll –n file10
-rw-r-----. 1 501 501  0 Jul 29 07:00 file10
```

The *chown* and *chgrp* commands are used to alter ownership and group membership on files and directories. You must be *root* to make this modification. Consider the following examples.

To change ownership from *user1* to *user2*:

```
# chown user2 file10
# ll file10
-rw-r-----. 1 user2 user1  0 Jul 29 07:00 file10
```

To change group membership from *user1* to *user2*:

```
# chgrp user2 file10
# ll file10
-rw-r-----. 1 501 user2  0 Jul 29 07:00 file10
```

To modify both ownership and group membership in one go:

```
# chown user1:user1 file10
# ll file10
-rw-r-----. 1 user1 user1  0 Jul 29 07:00 file10
```

To modify recursively (–R option) all files and sub-directories under *dir1* to be owned by *user2* and group *user2*:

```
# chown –R user2:user2 dir1
```

Special Permissions

There are three types of special permission bits that may be set up on executable files or directories. These permission bits are:

- ✓ *setuid* (set user identification) bit
- ✓ *setgid* (set group identification) bit
- ✓ *sticky* bit

The setuid Bit

The setuid bit is set on executable files at the file owner level. With this bit set, the file is executed by other users with the exact same privileges as that of the file owner. For example, the *su* command is owned by *root* with group membership set to *root*. This command has setuid bit enabled on it by default. See the highlighted s in the owner's permission class below:

$ ll /bin/su
-rwsr-xr-x. 1 root root 34904 Oct 5 2011 /bin/su

When a normal user executes this command, it will run as if *root* (the owner) is running it and, therefore, the user is able to run it successfully and gets the desired result.

The *su* (switch user) command allows a user to switch to some other user's account provided the switching user knows the password of the user he is trying to switch to.

Now, let us remove the bit from *su* and replace it with an x. You must be *root* in order to make this change.

 # chmod 755 /bin/su
-rwxr-xr-x. 1 root root 34904 Oct 5 2011 /bin/su

The file is still executable, but when a normal user runs it, he will be running it as himself and not as *root*. Here is what will happen when *user1* tries to *su* into *aghori* and enters a valid password:

$ su – aghori
Password:
su: incorrect password

user1 gets the "incorrect password" message even though he entered the correct login credentials.

To set the setuid bit back on *su* (or on some other file):

 # chmod 4755 /bin/su

When digit 4 is used with the *chmod* command in this manner, it sets the setuid bit on the file. Alternatively, you can use the symbolic notation to get the exact same results:

 # chmod u+s /bin/su

To unset the setuid bit, use either of the following:

 # chmod 755 /bin/su
 # chmod u-s /bin/su

To search for all files in the system with this bit set, use the *find* command:

    # **find / –perm –4000**

The setgid Bit

The setgid bit is set on executable files at the group level and on shared directories for a group collaboration purpose. With this bit set, the file is executed by other users with the exact same privileges that the group members have on it. For example, the *wall* command is owned by *root* with group membership set to *tty*. This command has the setgid bit enabled on it. See the highlighted s in the group's permission class below:

    $ **ll /usr/bin/wall**
     –r–xr–**s**r–x. 1 root tty 15224 Jan 18 2011 /usr/bin/wall

When a normal user executes this command, it will run as if one of the members of the *tty* group is running it and, therefore, the user will be able to run it successfully and the message will be printed on the screens of all logged in users whether they are regular users or *root*.

Now, let us remove the bit from *wall* and replace it with an x. You must be *root* in order to make this change.

    # **chmod 555 /usr/bin/wall**
     –r–xr–**x**r–x. 1 root tty 15224 Jan 18 2011 /usr/bin/wall

The file is still executable, but when a normal user runs it, he will be running it as himself and not as *root*. When a normal user now tries to *wall* a message, the message will be displayed successfully on the screens of all logged in regular users, but not on *root*'s screen (assuming *root* is logged on).

To set the setgid bit on */home/user1/file10* you must either be the owner of the file or *root*:

    $ **chmod 2555 file10**
    $ **ll**
     –r–xr–**s**r–x. 1 user1 user1 0 Jul 29 07:00 file10

When digit 2 is used with the *chmod* command in this manner, it sets the setgid bit on the file. Alternatively, you can use the symbolic notation to get the exact same results:

    $ **chmod g+s file10**

To unset the setgid bit, use either of the following:

    # **chmod 555 /bin/su**
    # **chmod g-s /bin/su**

To search for all files in the system with this bit set, use the *find* command:

    # **find / –perm –2000**

Exercise 4-3: Use setgid for Group Collaboration

You may wish to set up the setgid bit on a directory shared amongst members of a group for collaboration. In this exercise, you will create a directory called */sdata*, create a group called *sdatagrp*, add members *aghori* and *user1* to *sdatagrp*, set ownership to user *nobody*, set group membership to *sdatagrp*, and set the setgid bit on */sdata*. For details on *groupadd* and *usermod* commands, consult Chapter 11 "Users & Groups".

1. Create */sdata* directory:

   # mkdir /sdata

2. Run the *groupadd* command to add group *sdatagrp* with GID 9999:

   # groupadd –g 9999 sdatagrp

3. Execute the *usermod* command to add *aghori* and *user1* as members to *sdatagrp*:

   # usermod –G aghori sdatagrp
   # usermod –G user1 sdatagrp

4. Issue the *chown* command to set ownership on */sdata* to user *nobody* and group membership to *sdatagrp*:

   # chown nobody:sdatagrp /sdata

5. Use the *chmod* command to set the setgid bit on */sdata*:

   # chmod 2770 /sdata

6. Run the *ll* command on the directory to verify the attributes set in the above steps:

   # ll –d /sdata
   drwxrws---. 2 nobody sdatagrp 4096 Sep 27 13:31 /sdata

If users *aghori* and *user1* are already logged on, they need to log off and log back in. They should now be able to store their data in the */sdata* directory and share it.

The Sticky Bit

The sticky bit is typically set on public writable directories (or other directories with rw permissions for everyone) to protect files and sub-directories owned by regular users from being deleted or moved by other regular users. This bit is set on */tmp* and */var/tmp* directories by default as depicted below:

 # ll –d /tmp /var/tmp
drwxrwxrwt. 13 root root 4096 Jul 29 15:01 /tmp
drwxrwxrwt. 2 root root 4096 Jul 17 12:32 /var/tmp

Notice a t in other's permissions. This indicates that the sticky bit is enabled on the two directories.

To set the sticky bit on */var* (for example):

```
# ll –d /var
drwxr–xr–x. 22 root  root    4096 Jul 17 12:29 var
# chmod 1755 /var
```

When digit 1 is used with the *chmod* command in this manner, it sets the sticky bit on the specified directory. Alternatively, you can use the symbolic notation to do exactly the same:

```
# chmod o+t /var
```

To unset the sticky bit, use either of the following:

```
# chmod 755 /var
# chmod o-t /var
```

To search for all directories in the system with this bit set, use the *find* command:

```
# find / –type d –perm –1000
```

The vi (vim) Editor

The vi editor is an interactive *visual* text editing tool that allows you to create and modify text files. All text editing within vi takes place in a buffer (a small chunk of memory used to hold updates being done to the file). Changes can either be written to the disk or discarded. A graphical version of the vim editor, called *gvim*, is available and can be downloaded and installed.

It is essential for system administrators to master the vi editor skills. The following sub-sections provide details on how to use and interact with vi.

Modes of Operation

The vi editor has three modes of operation:

1. Command mode
2. Edit mode
3. Last line mode

Command Mode

The *command* mode is the default mode of vi. The vi editor places you into this mode when you start it. While in the command mode, you can carry out tasks such as copy, cut, paste, move, remove, replace, change, and search on text, in addition to performing navigational tasks. This mode is also known as the *escape* mode because the Esc key is pressed to enter it.

Input Mode

In the *input* mode, anything you type at the keyboard is entered into the file as text. Commands cannot be run in this mode. The input mode is also called the *edit* mode or the *insert* mode. To return to the command mode, press the Esc key.

Last Line Mode

While in the command mode, you may carry out advanced editing tasks on text by pressing the colon (:) character, which places the cursor at the beginning of the last line of the screen, and hence referred to as the *last line* mode. This mode is considered a special type of command mode.

Starting the vi Editor

The vi editor may be started in one of the ways described in Table 4-5. Use the *vimtutor* command to view the man pages of vi.

Method	Description
vi	Starts vi and opens up an empty screen for you to enter text. You can save or discard the text entered at a later time.
vi existing_file	Starts vi and loads the specified file for editing or viewing.
vi new_file	Starts vi and creates the specified file when saved.

Table 4-5 Starting The vi Editor

Inserting text

To enter text, issue one of the commands described in Table 4-6 from the command mode to switch into the edit mode.

Command	Action
i	Inserts text before the current cursor position.
I	Inserts text at the beginning of the current line.
a	Appends text after the current cursor position.
A	Appends text at the end of the current line.
o	Opens up a new line below the current line.
O	Opens up a new line above the current line.

Table 4-6 Inserting Text

Press the Esc key when done to return to the command mode.

Navigating within vi

Table 4-7 elaborates key sequences that control the cursor movement while you are in vi.

Command	Action
h / left arrow / Backspace / Ctrl+h	Moves left (backward) one character.
j / down arrow	Moves down one line.
k / up arrow	Moves up one line.
l / right arrow / Spacebar	Moves right (forward) one character.
5 right arrow / Spacebar	Moves right (forward) five characters. Change the number to move that many characters.
5 left arrow	Moves left (backward) five characters. Change the number to move that many characters.
5 up / down arrow	Moves up / down five lines. Change the number to move that many lines.
W or w	Moves forward one word.
B or b	Moves backward one word.
E or e	Moves forward to the last character of the next word.
M	Moves to the line in the middle of the page.
$	Moves to the end of the current line.
0 (zero) or ^	Moves to the beginning of the current line.
Enter	Moves down to the beginning of the next line.
Ctrl+f / Page Down	Moves forward (scrolls down) to the next page.
Ctrl+d	Moves forward (scrolls down) one-half page.
Ctrl+b / Page Up	Moves backward (scrolls up) to the previous page.
Ctrl+u	Moves backward (scrolls up) one-half page.
G or]]	Moves to the last line of the file.
(	Moves backward to the beginning of the current sentence.
)	Moves forward to the beginning of the next sentence.
{	Moves backward to the beginning of the preceding paragraph.
}	Moves forward to the beginning of the next paragraph.
1G or [[or :1	Moves to the first line of the file.
:11 or 11G	Moves to the specified line number (such as line number 11).
Ctrl+g	Tells you what line number you are at.

Table 4-7 Navigating Within vi

Deleting Text

Commands listed in Table 4-8 are available to carry out delete operations.

Command	Action
x	Deletes the character at the current cursor position. You may type a digit before this command to delete that many characters. For example, 2x would remove two characters, 3x would remove 3 characters, and so on.
X	Deletes the character before the current cursor location. You may type a digit before this command to delete that many characters. For example, 2X would remove two characters, 3X would remove 3 characters, and so on.

Command	Action
dw	Deletes the word or part of the word to the right of the current cursor location. You may type a digit before this command to delete that many words. For example, 2w would remove two words, 3w would remove 3 words, and so on.
dd	Deletes the current line. You may type a digit before this command to delete that many lines. For example, 2dd would remove two lines (current line plus next line), 3dd would remove 3 lines (current line plus next two lines), and so on.
d)	Deletes at the current cursor position to the end of the current sentence.
d(	Deletes at the current cursor position to the beginning of the last sentence.
d}	Deletes at the current cursor position to the end of the current paragraph.
d{	Deletes at the current cursor position to the beginning of the last paragraph.
D	Deletes at the current cursor position to the end of the current line.
:6,12d	Deletes lines 6 through 12.

Table 4-8 Deleting Text

Undoing and Repeating

Table 4-9 explains the commands available to undo the last change you did and repeat the last command you ran.

Command	Action
u	Undoes the last command.
U	Undoes all the changes done on the current line.
:u	Undoes the previous last line mode command.
. (dot)	Repeats the last command run.
Ctrl+r	Repeats the last undone command.

Table 4-9 Undoing and Repeating

Searching and Replacing Text

Search and replace text functions are performed using the commands mentioned in Table 4-10.

Command	Action
/string	Searches forward for a string.
?string	Searches backward for a string.
n	Finds the next occurrence of a string. This would only work if you have run either a forward or a backward string search.
N	Finds the previous occurrence of a string. This would only work if you have run either a forward or a backward string search.
:%s/old/new	Searches and replaces the first occurrence of *old* with *new*. For example, to replace the first occurrence of profile with Profile, you would use *:%s/profile/Profile*.

Command	Action
:%s/old/new/g	Searches and replaces all occurrences of *old* with *new*. For example, to replace all the occurrences of profile with Profile in a file, you would use *:%s/profile/Profile/g*.

Table 4-10 Searching and Replacing Text

Copying, Moving, and Pasting Text

Table 4-11 describes the vi commands available to perform copy, move, and paste functions.

Command	Action
yl	Yanks the current letter into the buffer. You may specify a digit before this command to yank that many letters. For example, 2yl would yank two characters, 3yl would yank three characters, and so on.
yw	Yanks the current word into the buffer. You may specify a digit before this command to yank that many words. For example, 2yw would yank two words, 3yw would yank three words, and so on.
yy	Yanks the current line into the buffer. You may specify a digit before this command to yank that many lines. For example, 2yy would yank two lines, 3yy would yank three lines, and so on.
y)	Yanks the current sentence into the buffer. You may specify a digit before this command to yank that many sentences. For example, 2y) would yank two sentences, 3y) would yank three sentences, and so on.
y(	Yanks the previous sentence into the buffer. You may specify a digit before this command to yank that many sentences. For example, 2y(would yank two previous sentences, 3y(would yank three previous sentences, and so on.
y}	Yanks the current paragraph into the buffer. You may specify a digit before this command to yank that many paragraphs. For example, 2y} would yank two paragraphs, 3y} would yank three paragraphs, and so on.
y{	Yanks the previous paragraphs into the buffer. You may specify a digit before this command to yank that many paragraphs. For example, 2y{ would yank two previous paragraphs, 3y{ would yank three previous paragraphs, and so on.
p	Pastes yanked data below the current line.
P	Pastes yanked data above the current line.
:1,3co5	Copies lines 1 through 3 and pastes them after line 5.
:4,6m8	Moves lines 4 through 6 after line 8.

Table 4-11 Copying, Moving, and Pasting Text

Changing Text

Use the commands given in Table 4-12 to change text. Some of these commands will switch you into the edit mode. To return to the command mode, press the Esc key.

Command	Action
cl	Changes the letter at the cursor location.
cw	Changes the word (or part of the word) at the current cursor location to the end of the current word.
C	Changes at the current cursor position to the end of the current line.
r	Replaces the character at the current cursor location with the character entered following this command.
R	Overwrites or replaces the text on the current line.
s	Substitutes a string for character(s).
S or cc	Substitutes an entire line.
c)	Changes at the current cursor location to the end of the current sentence.
c(	Changes at the current cursor location to the beginning of the last sentence.
c}	Changes at the current cursor location to the end of the current paragraph.
c{	Changes at the current cursor location to the beginning of the last paragraph.
J	Joins the current line and the line below it.
xp	Switches the position of the character at the current cursor position with the character to the right of it.
~	Changes the letter case (uppercase to lowercase and vice versa) at the current cursor location.

Table 4-12 Changing Text

Importing Contents of Another File

While working in vi you may want to import contents of another file. The vi editor allows you to do that. You must be at the last line mode to accomplish this task. See Table 4-13.

Command	Action
:r file2	Reads *file2* and imports its contents below the current line.

Table 4-13 Importing Contents of Another File

Customizing vi Edit Sessions

The vi editor supports settings to customize edit sessions to display the line numbers, invisible characters, and so on. Use the *set* command to control these options. Consult Table 4-14.

Command	Action
:set nu (nonu)	Shows (hides) line numbering.
:set ic (noic)	Ignores (does not ignore) the letter case when carrying out searches.
:set list (nolist)	Displays (hides) invisible characters such as Tab and End Of Line (EOL).
:set showmode (noshowmode)	Displays (hides) the current mode of operation.
:set	Displays the current vi variable settings.
:set all	Displays all available vi variables and their current settings.

Table 4-14 Customizing vi Settings

Saving and Quitting vi

When you are done with modifications, you will want to save or discard them. Commands listed in Table 4-15 will help.

Command	Action
:w	Writes changes into the file without quitting vi.
:w file3	Writes changes into a new file called *file3*.
:w!	Writes changes into the file even if the file owner does not have write permission on the file.
:wq or :x or ZZ	Writes changes to the file and quits vi.
:wq! or :x!	Writes changes into the file and quits vi even if the file owner does not have write permission on the file.
:q	Quits vi if no modifications were made.
:q!	Quits vi if modifications were made, but you do not wish to save them.

Table 4-15 Saving and Quitting vi

Miscellaneous vi Commands

Table 4-16 describes additional commands available to perform specific tasks within vi.

Command	Action
Ctrl+l	Refreshes the vi screen (if proper terminal type is set).
:sh	Exits vi temporarily. Type *exit* or Ctrl+d to return.
:!cmd	Executes the specified command without quitting vi.

Table 4-16 Miscellaneous vi Commands

The nano Editor

Many users prefer to use the nano text editor. This editor displays a menu of commands at the bottom of the screen for help with editing a file efficiently. Figure 4-3 shows the interface that pops up when *nano* is executed on the */home/user1/.bash_profile* file:

$ nano /home/user1/.bash_profile

Figure 4-3 Nano Text Editor

Several key combinations are available as commands within the editor to perform navigation and other tasks. Some of these are listed and described in Table 4-17.

Key (main screen)	Action
Ctrl+g	Displays help.
Ctrl+x	Exits nano.
Ctrl+o	Saves file.
Ctrl+j	Justifies the current paragraph.
Ctrl+r	Reads a text file for insertion.
Ctrl+w	Searches for a pattern.
Ctrl+y	Moves to the previous page.
Ctrl+v	Moves to the next page.
Ctrl+k	Deletes the current line.
Ctrl+u	Brings back the last deleted line.
Ctrl+c	Displays the current cursor location.
Ctrl+t	Checks spelling.

Table 4-17 nano Commands

The Text Processors

RHEL supports two famous Linux/Unix text processors for performing various operations on columns and rows of text. These processors are *awk* and *sed*, and both work on input taken either from a specified file or the output of a command such as *ll*. Neither text processor makes any modification to the input provided; they merely read the input and display results on the screen. If you wish to save the result, you need to use the output redirection which is explained later in this chapter. Let us take a look at both the processors and try to comprehend them with the help of examples.

The awk Processor

The name *awk* was derived from the first initial of the last names of those who developed it: Alfred Aho, Peter Weinberger, and Brian Kenigham.

awk works on columns of text to generate desired reports. It scans a file (or input provided) one line at a time. It starts from the first line, searches for lines matching the specified pattern enclosed in quotes and curly braces, and performs a selected action on those lines.

In order to understand the behavior of the *awk* utility, create a file by running the *ll* command and redirect its output to a file called *ll.out*. Then use this file as input to *awk* and examine results displayed on the screen. Before doing that, let us see how *awk* interprets columns (in other words, how *awk* differentiates between columns). Let us run *ll* on *user1*'s home directory */home/user1* and assume that four files exist in it:

```
$ ll
total 0
-rw-rw-r--. 1 user1 user1 0 Jul 31 05:38 file1
-rw-rw-r--. 1 user1 user1 0 Jul 31 05:38 file2
-rw-rw-r--. 1 user1 user1 0 Jul 31 05:38 file3
-rw-rw-r--. 1 user1 user1 0 Jul 31 05:38 file4
```

awk automatically breaks a line into columns and assigns a variable to each column. A white space such as a tab is used as the default delimiter between columns for separation.

Each line from the *ll* command output contains nine columns of text. Figure 4-4 shows how *awk* represents each column with respect to its position. The figure shows that $1 represents the first column, $2 represents the second column, $3 represents the third column, and so on. All columns are collectively represented by $0.

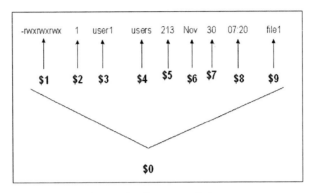

Figure 4-4 awk Command Arguments

Let us now save the output of *ll* into *ll.out*:

$ ll > ll.out

The following examples help you build an understanding on the usage of *awk*.

To display only the file name (column 9), file size (column 5), and file owner (column 3) in that sequence, execute the following. Notice a single white space between variables in the *awk* statement. The output will not contain any spaces between columns and will print each row as one single string of characters.

$ awk '{print $9 $5 $3}' ll.out

```
file10user1
file20user1
file30user1
file40user1
ll.out0user1
```

awk requires that you enclose the pattern in quotes and curly braces. The print function within the curly braces is part of the syntax. The above output does not contain any spaces between columns.

To place a single white space between columns in the output, use a comma between variables:

```
$ awk '{print $9, $5, $3}' ll.out
```

file1 0 user1
file2 0 user1
file3 0 user1
file4 0 user1
ll.out 0 user1

To provide for exact alignment between columns, insert a tab or two:

```
$ awk '{print $9 "      " $5 "      " $3}' ll.out
```

file1 0   user1
file2 0   user1
file3 0   user1
file4 0   user1
ll.out 0   user1

To re-arrange columns to display file owner, file size, and file name in that order:

```
$ awk '{print $3,$5,$9}' ll.out
```

user1 0 file1
user1 0 file2
user1 0 file3
user1 0 file4
user1 0 ll.out

To add text between columns:

```
$ awk '{print $9,"was last modified/accessed on",$6,$7,"at",$8}' ll.out
```
file1 was last modified/accessed on Jul 31 at 05:38
file2 was last modified/accessed on Jul 31 at 05:38
file3 was last modified/accessed on Jul 31 at 05:38
file4 was last modified/accessed on Jul 31 at 05:38
ll.out was last modified/accessed on Jul 31 at 05:46

Note that when inserting text between columns, each text insert must be enclosed in double quotes and all but the last text string inserted must be followed by a comma.

The sed Processor

Unlike *awk*, which works on columns of text, the *sed* (stream editor) processor works on rows of text. Following examples illustrate how *sed* works.

To search the */etc/group* file for all lines containing the word root and hide those lines in the output. Note again that *sed* does not remove anything from the specified file. Let us first see what */etc/group* file contains:

$ cat /etc/group
root:x:0:root
bin:x:1:root,bin,daemon
daemon:x:2:root,bin,daemon
. . . .

Now run *sed* to get the desired result:

$ sed '/root/d' /etc/group
tty:x:5:
lp:x:7:daemon,lp
mem:x:8:
. . . .

/root/d is enclosed in single quotes. The /root portion of the command tells *sed* to search for the pattern root and the /d portion instructs *sed* to delete that pattern from the output.

To remove all lines from the output of the *ll* command containing digit 4:

$ ll | sed '/4/d'
–rw–rw–r––. 1 user1 user1   0 Jul 31 05:38 file1
–rw–rw–r––. 1 user1 user1   0 Jul 31 05:38 file2
–rw–rw–r––. 1 user1 user1   0 Jul 31 05:38 file3

To print all lines in duplicate that contain the pattern root from the */etc/group* file. All other lines to be printed one time:

$ sed '/root/p' /etc/group
root:x:0:root
root:x:0:root
bin:x:1:root,bin,daemon
. . . .

To print only those lines from the */etc/group* file that contain the pattern root:

$ sed –n '/root/p' /etc/group
root:x:0:root
bin:x:1:root,bin,daemon
daemon:x:2:root,bin,daemon
sys:x:3:root,bin,adm
adm:x:4:root,adm,daemon
disk:x:6:root
wheel:x:10:root

To append the character string RHEL to the end of every line in the output of the *ll* command, run the command as *user1* on *user1*'s home directory */home/user1*:

```
$ ll | sed 's/$/ RHEL/'
total 4 RHEL
-rw-rw-r--. 1 user1  user1    0 Jul 31 05:38 file1 RHEL
-rw-rw-r--. 1 user1  user1    0 Jul 31 05:38 file2 RHEL
-rw-rw-r--. 1 user1  user1    0 Jul 31 05:38 file3 RHEL
-rw-rw-r--. 1 user1  user1    0 Jul 31 05:38 file4 RHEL
-rw-rw-r--. 1 user1  user1  244 Jul 31 05:46 ll.out RHEL
```

Note in the preceding command, s is used for substitution and $ represents the end of line.

To perform two edits on the */etc/group* file where the first edit replaces all occurrences of root with ROOT and the second edit replaces daemon with USERS:

```
$ sed -e 's/root/ROOT/g' -e 's/daemon/USERS/g' /etc/group
ROOT:x:0:ROOT
bin:x:1:ROOT,bin,USERS
USERS:x:2:ROOT,bin,USERS
. . . . . . . .
```

You can perform multiple edits using this method.

The BASH Shell

In the Linux world, the *shell* is referred to as the command interpreter, and is an interface between the user and the kernel. It accepts instructions from users (or scripts), interprets them, and passes them on to the kernel for processing. The kernel utilizes all hardware and software components required to process the instructions. When finished, the results are returned to the shell and displayed on the screen. The shell also shows appropriate error messages, if generated.

A widely used shell by Linux administrators is the *bash* (bourne again shell) shell. bash is a free replacement for the older Bourne shell with a number of enhancements. It is the default shell in RHEL6 as well, and offers several features such as input, output, and error redirection; tab completion; pattern matching; environment variables; job control; command line editing; command aliasing; command history; tilde substitution; quoting mechanisms; wildcarding; piping; conditional execution; flow control; and writing shell scripts. These features are discussed in detail in this section.

The bash shell is identified by the $ sign for regular users and the # sign for *root*. This shell is resident in the */bin/bash* file.

Other Linux Shells

Many of the features listed above are also supported by additional shells available in RHEL6. These shells include the *dash* (Debian Almquist Shell), the *Korn* shell, and the *TC* shell.

 RHEL6 does not include the Korn shell in the base RHEL6 installation. You need to install it if you require it.

The dash shell is faster than the bash since it relies on fewer library routines for execution. Much smaller in size than the bash, it lacks many bash shell features. The prompt for the dash shell is $ and it resides in the */bin/dash* file.

The Korn shell is similar to the bash shell in terms of features. The prompt for this shell is $ and it resides in the */bin/ksh* file.

The TC shell is mainly used by developers. It provides a programming interface similar to the C language and offers many bash and Korn shell functions in addition to numerous other features. It is backward compatible with the older C shell. The prompt for the TC shell is $ and it resides in the */bin/tcsh* file.

The shell is changeable, providing users with the flexibility of choosing an alternative command interpreter.

Variables

A *variable* is a temporary storage of data in memory. It stores information that is used for customizing the shell environment and by many system and application processes to function properly. The shell allows you to store a value in a variable.

There are two types of variables: *local* and *environment*.

Local Variable

A local variable is private to the shell in which it is created and its value cannot be used by processes that are not started in that shell. This introduces the concept of *current* shell and *sub*-shell (or *child* shell). The current shell is where you execute a program, whereas, a sub-shell is created by a running program. The value of a local variable is available only in the current shell, and not in the child shell.

Environment Variable

The value of an environment variable, a.k.a. a *global* variable, is passed from the current shell to the sub-shell during the execution of a script. In other words, the value stored in an environment variable is passed from the parent process to the child process. Any local or environment variable set in a sub-shell loses its value when the sub-shell terminates. Some environment variables are defined automatically through system and user initialization files at log in. A variable may also be set in shell scripts or at the command line as required.

Setting, Unsetting, and Viewing Variables

To set, unset, or view shell variables, consult Table 4-18. Two variables – V1 and V2 – have been used as examples. It is recommended that you use uppercase letters for variable names to distinguish them from the names of any commands or programs that you might have on the system.

Action	Local Variable	Environment Variable
Setting a variable	Syntax: VR1=value	Syntax: VR1=value; export VR1 OR export VR1=value
	Examples: **$ VR1=college** **$ VR2="I use RHEL6"**	Examples: **$ VR1=college; export VR1** **$ VR2="I use RHEL6"; export VR2** or **$ export VR1=college** **$ export VR2="I use RHEL"**
Listing variables and displaying their values	**$ echo $VR1** **$ echo $VR2** **$ set | grep VR**	**$ echo $VR1** **$ echo $VR2** **$ set | grep VR** **$ env | grep VR** **$ export | grep VR**
Unsetting a variable	**$ unset VR1 VR2**	**$ unset VR1 VR2**

Table 4-18 Setting, Unsetting, and Viewing Variables

To set a local variable VR1 to contain the value 'college', simply define it as shown in Table 4-18 (Setting a variable). Note that there must not be any white spaces before or after the = sign. To make VR1 a global variable, use the *export* command. The same example sets a variable VR2 containing a series of white space-separated words. Make sure to enclose them in double quotes.

To display the value of a variable, use the *echo*, *set*, *env*, or the *export* command.

The *set* command lists current values for all shell variables including local and environment variables. However, the *export* and *env* commands list only environment variables.

```
$ set
BASH=/bin/bash
BASHOPTS=checkwinsize:cmdhist:expand_aliases:extquote:force_fignore:hostcomplete
:interactive_comments:login_shell:progcomp:promptvars:sourcepath
BASH_ALIASES=()
. . . . . . . .
$ env
HOSTNAME=physical.example.com
SHELL=/bin/bash
TERM=xterm
. . . . . . . .
$ export
declare -x CVS_RSH="ssh"
declare -x G_BROKEN_FILENAMES="1"
declare -x HISTCONTROL="ignoredups"
. . . . . . . .
```

Finally, to unset (remove) a variable, use the *unset* command and specify the variable name to be removed.

Pre-Defined Environment Variables

Some environment variables such as those listed and described in Table 4-19 are defined by the shell at user log in. You may also set custom environment variables in scripts as required.

Variable	Description
DISPLAY	Stores the hostname or IP address for X terminal sessions.
HISTFILE	Defines the file for storing the history of executed commands.
HISTSIZE	Defines the maximum size for the HISTFILE.
HOME	Contains the home directory path.
LOGNAME	Stores the login name.
MAIL	Contains the path to the user mail directory.
PATH	Defines a colon-separated list of directories to be searched when executing a command.
PS1	Defines the primary command prompt.
PS2	Defines the secondary command prompt.
PWD	Stores the current directory location.
SHELL	Holds an absolute path for the primary shell.
TERM	Holds the terminal type value.

Table 4-19 Pre-Defined Environment Variables

Command and Variable Substitution

The primary command prompt for the *root* user is the # sign and that for the regular users is the $ sign. Customizing the primary command prompt to display useful information such as who you are, the system you are currently logged on to, and your current location in the directory tree is a good practice. The examples below illustrate how to modify *user1*'s primary prompt using any of the following:

```
$ export PS1="< $LOGNAME@`hostname`:\$PWD > "
$ export PS1="< $LOGNAME@$(hostname):\$PWD > "
```

user1's command prompt will now look like:

```
< user1@physical.example.com:/home/user1 >
```

The value of the PWD variable will reflect the directory location in *user1*'s prompt as he navigates the directory tree. This is called *variable substitution*. For example, if *user1* moves to */usr/bin*, the prompt will change to:

```
< user1@physical.example.com:/usr/bin >
```

Also, the value of LOGNAME is used to display *user1*'s login name in this example.

Running the command *hostname* and assigning its output to a variable is an example of a shell feature called *command substitution*. Note that the command whose output you want to assign to a variable must be either enclosed within single forward quotes or within parentheses preceded by the $ sign.

Input, Output, and Error Redirection

Many programs read input from the keyboard and write output to the terminal window where they are initiated. Any errors, if encountered, are displayed on the terminal window too. This is the default behavior. What if you do not wish to take input from the keyboard or write output to the terminal window? The bash shell allows you to redirect input, output, and error messages to enable programs and commands to read input from something other than the keyboard and send output and errors to something other than the terminal window.

The default (or the standard) locations for input, output, and error are referred to as *stdin*, *stdout*, and *stderr*, respectively. Table 4-20 demonstrates that each of the three has an association with a symbol and a digit. These symbols and digits may be used at the command line and in shell scripts for redirection purposes.

File Descriptor	Symbol	Associated Digit	Description
stdin	<	0	Standard input
stdout	>	1	Standard output
stderr	>	2	Standard error

Table 4-20 Input, Output, and Error Redirection

The following text explains how redirection of input, output, and error takes place.

Redirecting Standard Input

Input redirection instructs a command to read the required information from an alternative source such as a file, instead of the keyboard. The < character is used for input redirection. For example, run the following to have the *mailx* command mail the contents of */etc/group* to *user1*:

$ mailx user1 < /etc/group

In order for *mailx* to work properly, a mail server must be configured and running. See Chapter 18 "Electronic Mail" on how to configure a mail server and use it.

Redirecting Standard Output

Output redirection sends the output generated by a command to an alternative destination such as a file, instead of sending it to the terminal window. The > character is used for output redirection. For example, execute the following to direct the *sort* command to send the sorted output of *ll.out* to a file called *sort.out*. This will overwrite any *sort.out* file contents. If *sort.out* does not exist, it will be created.

$ sort ll.out > sort.out

To direct the *sort* command to append the output to *sort.out* file instead, use the >> characters:

    $ sort ll.out >> sort.out

Redirecting Standard Error

Error redirection sends any error messages generated to an alternative destination such as a file, rather than sending them to the terminal window. For example, issue the following to direct the *find* command to search for all occurrences of files by the name core in the entire root directory tree and send any error messages produced to */dev/null*. The */dev/null* file is a special system file used to discard data.

    $ find / –name core –print 2> /dev/null

Redirecting both Standard Output and Error

Issue the following to redirect both stdout and stderr to a file called *testfile1*:

    $ ls /usr /cdr 1> testfile1 2>&1

This example will produce a listing of the */usr* directory and save the result in *testfile1*. At the same time, it will generate an error message complaining about the non-existence of the */cdr* directory. This error message will be sent to the same file as well and saved.

Tab Completion, Command Line Editing, and Command History

Tab completion (a.k.a. *command line completion*) is a bash shell feature whereby typing a partial filename at the command line and then hitting the Tab key twice completes the filename if there are no other possibilities. If multiple possibilities exist, it will complete up to the point they have in common.

Command line editing allows you to edit a line of text right at the command prompt. Pressing the Esc+k key combination or the up arrow key brings the last command you executed at the prompt. Pressing the letter k or the up arrow key repeatedly scrolls backward to previous commands in reverse chronological order. The letter j or the down arrow key scrolls forward through the command history in chronological order. When you get the desired command, you may edit it right at the command prompt using the vi editor commands. If you do not wish to edit the command or are done with editing, simply press the Enter key to execute it.

Ensure that you run the following command to set the editor to vi so that you are able to use the command line editing feature:

    $ set –o vi

Command history, or simply *history*, keeps a log of all commands that you run at the command prompt. The shell stores command history in a file located in the user's home directory. You may retrieve these commands, modify them at the command line, and re-run them using the command line editing feature.

There are three variables that enable the features just discussed. These variables are listed below along with sample values:

HISTFILE=~/.bash_history
HISTSIZE=1000
EDITOR=vi

The HISTFILE variable tells the bash shell to store all commands run by a user in the *.bash_history* file in that user's home directory. This file is created automatically if it does not already exist. Each command, along with options and arguments, is stored on a separate line.

The HISTSIZE variable controls the maximum number of commands that can be stored in HISTFILE.

The EDITOR variable defines what text editor to use. If this variable is not set, issue the following to be able to use the tab completion, command line editing, and command history features:

$ set –o vi

The HISTSIZE variable is pre-defined in the */etc/system* initialization file. The EDITOR variable and the **set –o vi** command may be defined in the user or system initialization file. Refer to Chapter 11 "Users & Groups" on how to define variables and commands in initialization files to customize the shell behavior.

RHEL provides the *history* command to display previously executed commands. It gets the information from the *~/.bash_history* file. By default, the last 500 entries are displayed.

$ history
  1 logout
  2 exit
  3 ssh test
. . . .

Let us use some of the *history* command options to alter its behavior.

To display this command and the 17 commands preceding it:

$ history 17

To re-execute a command by its line number (line 38 for example) in the history file:

$!38

To re-execute the most recent occurrence of a command that started with a particular letter or series of letters (ch for example), use any of the following:

$!ch
$!?ch

To repeat the last command executed, run any of the following:

```
$ !!
$ !$
```

Tilde Substitution

Tilde substitution (or *tilde expansion*) is performed on words that begin with the tilde ~ sign. The rules are:

1. If ~ is used as a standalone character, the shell refers to the $HOME directory of the user running the command. The following example displays the $HOME directory of *user1*:

   ```
   $ echo ~
   /home/user1
   ```

2. If ~ is used prior to the + sign, the shell refers to the current directory. For example, if *user1* is in the */etc/init.d* directory and does ~+, the output displays the user's current directory location:

   ```
   $ echo ~+
   /etc/init.d
   ```

3. If ~ is used prior to the – sign, the shell refers to the last working directory. For example, if *user1* switches into the */usr/share/man* directory from */etc/init.d* and does ~–, the output displays the user's last working directory location:

   ```
   $ echo ~–
   /etc/init.d
   ```

4. If ~ is used prior to a username, the shell refers to the $HOME directory of that user:

   ```
   $ echo ~user2
   /home/user2
   ```

You can use tilde substitution with any commands such as *cd*, *ls*, and *echo* that refer to a location in the directory structure. Additional examples of tilde substitution were provided in Chapter 02 "Basic Linux Commands".

Special Characters

Special characters are the symbols on the keyboard that possess special meaning to the shell. These characters are also referred to as *metacharacters* or *wildcard* characters, and are used for globbing purposes (see Pattern Matching in Chapter 03 "Files & Directories"). Some of these special characters such as –, ~, and < > have already been covered earlier in this chapter. Four additional special characters: * (asterisk), ? (question mark), [ ] (square brackets), and ; (semicolon) are discussed next.

The * Character

The * matches zero to an unlimited number of characters (except for the leading period in a hidden file). See the following examples.

To list names of all files that begin with letters fi followed by any characters in *user1*'s home directory:

```
$ ls fi*
file1  file2  file3  file4
```

To list names of all files that begin with letter d. You will notice in the output that the contents of *dir1* and *dir2* sub-directories are also listed.

```
$ ls d*
dir1:
scripts1  scripts2

dir2:
newfile1  newfile2  newfile3  newfile4
```

To list names of all files that end with digit 4:

```
$ ls *4
file4
```

To list names of all files that have a period followed by letters "out":

```
$ ls *.out
ll.out  sort.out
```

The ? Character

The ? character matches exactly one character, except for the leading period in a hidden file. See the following example to understand its usage.

To list all files that begin with "file" followed by one character only:

```
$ ls file?
file1  file2  file3  file4
```

The [ ] Characters

The [ ] can be used to match either a set of characters or a range of characters for a single character position.

When you specify a set of characters, order is unimportant therefore, [xyz], [yxz], [xzy], and [zyx] are treated alike. In the following example, two characters are enclosed within the square brackets. The output will include all files and directories that begin with either of the two characters followed by any number of characters.

```
$ ls [cf]*
car car1 car2 file1 file2 file3 file4
```

A range of characters must be specified in a proper order such as [a-z] or [0-9]. The following example matches all file and directory names that begin with any letter between a and f:

```
$ ls [a–f]*
car car1 car2 file1 file2 file3 file4

dir1:
scripts1 scripts2

dir2:
newfile1 newfile2 newfile3 newfile4
```

The ; Character

The ; allows you to enter multiple commands on a single command line by acting as a separator between them. The following example shows three commands: *cd*, *ls*, and *date* separated by semicolon. The three commands will be executed in the order they are specified.

```
$ cd; ls fil*; date
file1 file2 file3 file4
Tue Jul 31 20:47:19 EDT 2012
```

Pipe and Filters

This section discusses pipe and filters which are often used at the command line and in shell scripts.

Pipe

A *pipe*, represented by the | character and resides with the \ on the keyboard, is a special character that sends output of one command as input to another command.

The following example uses the *ll* command to display contents of the */etc* directory. The output is piped to the *more* command, which displays the listing one screen at a time.

```
$ ll /etc | more
total 1920
drwxr-xr-x. 3 root root  4096 Jul 17 12:20 abrt
drwxr-xr-x. 4 root root  4096 Jul 17 12:28 acpi
-rw-r--r--. 1 root root    45 Jul 22 05:40 adjtime
. . . . . . . .
--More--
```

In another example, the *who* command is run and its output is piped to the *nl* command so associated line numbers are also visible:

```
$ who | nl
     1  aghori  pts/0    2012-07-31 13:19 (192.168.2.16)
     2  root    pts/1    2012-07-31 05:32 (192.168.2.16)
     3  user1   pts/2    2012-07-31 20:46 (192.168.2.16)
```

The following example creates a pipeline whereby the output of *ll* is sent (piped) to the first *grep* command, which filters out all lines that do not contain the pattern root. The new output is then sent (piped) to the second *grep* command that filters out all lines that do not contain the pattern apr. Finally, the output is sent (piped) to the *nl* command which numbers and displays the result on the screen. A construct like this with multiple pipes is referred to as a *pipeline*.

```
$ ll /etc | grep root | grep –i apr | nl
     1  –rw–r––r––.  1 root  root  4439 Apr 28  2010 DIR_COLORS
     2  –rw–r––r––.  1 root  root  5139 Apr 28  2010 DIR_COLORS.256color
     3  –rw–r––r––.  1 root  root  4113 Apr 28  2010 DIR_COLORS.lightbgcolor
     4  –rw–r––r––.  1 root  root   478 Apr  6  2010 ksmtuned.conf
     5  –rw–r––r––.  1 root  root  1962 Apr 19  2010 vimrc
     6  –rw–r––r––.  1 root  root  1962 Apr 19  2010 virc
```

The tee Filter

The *tee* filter is used to send an output to more than one destination. It can send one copy of the output to a file and another to the screen (or some other program) if used with pipe.

In the following example, the output from *ll* is numbered and captured in */tmp/ll.out* file. The output is also displayed on the screen.

```
$ ll /etc | nl | tee /tmp/ll.out
```

Issue the *cat* command on the */tmp/ll.out* file and notice that the file contains the exact same information displayed on the screen when you executed the command.

By using –a with the *tee* command, the output may be appended to the file rather than overwriting the existing contents.

```
$ date | tee –a /tmp/ll.out
```

The cut Filter

The *cut* filter is used to extract selected columns from a line. The default column separator used is a white space such as a tab. The following example command cuts out columns 1 and 4 from the */etc/group* file as specified with the –f option. The colon character is used as a field separator.

```
$ cut –d: –f 1,4 /etc/group
root:root
bin:root,bin,daemon
daemon:root,bin,daemon
. . . . . . . .
```

Chapter Summary

This chapter started off with a study of file and directory permissions. It covered classes, types, and modes of permissions and went on to demonstrate how to modify permissions using symbolic and octal notations. You studied how default permissions could be set up for new files and directories, and the role of the umask value in determining the new default permissions.

The next topic explained how a user could alter his primary group membership temporarily and how a user or *root* could modify ownership and group membership on files and directories.

You learned about setting special permission bits on executable files and directories to gain privileged access and prevent individual files and directories from being deleted by other users. This topic also explained how to set up a shared directory for group collaboration.

A detailed study on the usage and available commands within the vi editor was covered in the chapter. You learned how to navigate and manipulate text in vi. Following the coverage of the vi editing tool, the nano editor's benefits and usage were described.

The next topic in the chapter covered manipulating columns and rows of text using the *awk* and *sed* text processors along with demonstrating several examples on their use.

The bash shell has many features such as variable settings, command prompt customization, redirection, tab completion, command line editing, command history, tilde substitution, and special characters. These features were explained at length.

Finally, you looked at the usage of pipe and filters.

Chapter Review Questions

1. The output generated by the *umask* command shows the current user mask in four digits. What is the significance of the left-most digit?
2. Name the four types of permissions.
3. The *chgrp* command may be used to modify both ownership and group membership on a file. True or False?
4. What would the command *export PS1="$LOGNAME@`hostname`:\$PWD:* do?
5. Name the two notations for specifying file permissions.
6. Which command is used to modify file permissions?
7. Default permissions are calculated by subtracting the initial permissions from the umask value. True or False?
8. Define Linux pipeline.
9. Name the three permission classes.
10. What would the command *ls /cdr /var 1> out.out 2>&1* do?
11. The default umask for a regular user in bash shell is 0027. True or False?
12. What would the command *sed –e 's/root/ROOT/g' /etc/group* do?
13. Name the three modes of permissions.
14. What digit represents the setuid bit in the *chmod* command?
15. What would the command *find / -perm -2000* do?
16. What would the dot command do in vi?
17. What would the command *chmod g+s file1* do?
18. Sticky bit can be set on any directory. True or False?
19. Name the three modes of the vi editor operation?

20. What would the command *set noic* do in vi?
21. What would the command *ll | awk '{print $3,"owns",$9}'* do?
22. The setuid bit enables a regular user to run a command at a lower priority. True or False?
23. The dash shell has many more features than the bash shell. True or False?
24. Which command is used to make a local variable a global variable?
25. The *chown* command may be used to modify both ownership and group membership on a file. True or False?
26. What would the command *set –o vi* do?
27. The square brackets may be used for defining a range or set of characters. True or False?
28. What is the primary purpose of the semicolon character?
29. The *tee* command may be used to direct output to two different locations. True or False?
30. What would the command *cut –d : –f 1,3 /etc/passwd* do?
31. What permissions would the owner of the file get if the *chmod* command is executed with 731?
32. What is the equivalent octal value for permissions rwxrw-r--?

Answers to Chapter Review Questions

1. The left-most digit has no significance in the *umask* value.
2. The four types of permissions are read, write, execute, and none.
3. False.
4. The *export* command provided will change the login prompt to display username, hostname, and the current working directory.
5. The two notations for specifying file permissions are symbolic and octal.
6. The *chmod* command is used to modify file permissions.
7. True.
8. Output of one command is sent as input to the other command, and this feature is used more than once in the same command.
9. The three permission classes are user, group, and other.
10. The *ls* command provided will redirect both the standard output and the standard error to the *out.out* file.
11. False. It is 0002.
12. The *sed* command provided will display file contents of the */etc/group* file with all occurrences of root replaced with ROOT.
13. The three modes of permissions are dd, revoke, and assign.
14. The digit 4 represents the setuid bit in the *chmod* command.
15. The *find* command provided will search for all files with the setgid bit set.
16. The dot command will repeat the last command executed within vi.
17. The *chmod* command provided will enable the setgid bit on *file1*.
18. True, but it is typically set on shared directories.
19. The three modes of the vi editor operation are the command mode, the edit mode, and the last line mode.
20. The *set noic* command will not ignore the letter case when searching for text within vi.
21. The *ll* command provided will display file names and their owners.
22. False. The setuid bit has nothing to do with file processing priority.
23. False. The bash shell is more powerful and feature-rich than the dash shell.
24. The *export* command is used to make a local variable a global variable.
25. True.

26. The *set* command provided will set vi as the default editor at the command prompt.
27. True.
28. The primary purpose of the semicolon character is to separate commands specified on the same command line.
29. True.
30. The *cut* command provided will display the username and associated UID from the */etc/passwd* file.
31. The owner will get read, write, and execute permissions.
32. The equivalent octal value for permissions rwxrw-r-- will be 764.

Labs

Lab 4-1: Manipulate File Permissions

Log on to *physical.example.com* as *user1* and create file *file1* and directory *dir1* in the user's home directory. Make a note of the permissions on them. Run the *umask* command and determine the current umask. Change the umask value to 0035 using the symbolic notation and then create file *file2* and directory *dir2* in the user's home directory. Observe the permissions on *file2* and *dir2* and compare them with the permissions on *file1* and *dir1*. Using the *chmod* command, modify the permissions on *file1* to match that on *file2*. Using the *chmod* command, modify the permissions on *dir2* to match that on *dir1*. Do not remove *file1*, *file2*, *dir1*, and *dir2* yet.

Lab 4-2: Manipulate File Ownership & Membership

Continue from Lab 4-1. Who do the ownership and group membership belong to on both files and directories that were created in Lab 4-1? As *user1*, try changing the ownership on *file1* to *user2* and observe the result. Now log in as *user2* and repeat changing the ownership on *file1* to *user2*. As either user, you will not be permitted to make that modification. Now log in as *root* and attempt to change both the ownership and the group membership to *user2* on *file1*. It should work.

Lab 4-3: Configure Group Collaboration

Log on to *physical.example.com* as *root* and create directory */shared_dir1*. Create a group called *shared_grp* and assign *user2* and *user3* to it (create these users if they do not already exist). Set up appropriate ownership, group membership, and permissions on the directory to support group collaboration.

Lab 4-4: Practice the vi Editor

Log on to *physical.example.com* as *user1* and create a file called *vipractice* in the user's home directory. Type (do not copy and paste) the first four sentences from under Lab 4-2 in such a way that each sentence occupies one line (do not worry about the line wrapping). Save the file and quit the *vi* editor. Open *vipractice* in the *vi* editor again and enable line numbering. Copy the 2nd and 3rd lines to the end of the file to make the total number of lines to 6 in the file. Move the 3rd line to make it the very 1st line in the file. Go to the last line and append the contents of the *.bash_profile*. Substitute all the occurrences of the word "Profile" with "Pro File" and all the occurrences of the word "profile" with "pro file". Remove lines 5 to 8. Save the file and quit the *vi* editor. Provide a count of lines, words, and characters in the *vipractice* file using the *wc* command.

Lab 4-5: Use the awk Processor

Log on to *physical.example.com* as *user1*. Execute the *ps* command with the –efl options and pipe it to the *awk* processor. Display the output in a manner that it reads "PID <PID> is owned by user <UID > with PPID <PPID> and nice value <NI>. This PID belongs to the <CMD> process.".

Lab 4-6: Customize the Shell Prompt

Log on to *physical.example.com* as *user1* and customize the primary shell prompt to display the information enclosed within the quotes "<username@hostname in directory_location >: ".

Lab 4-7: Redirect the Standard Input, Output, and Error

Log on to *physical.example.com* as *user1* and run the *ll* command on the */etc*, */dvd*, and the */var* dirctories. Have the output printed on the screen as well as redirected to the */tmp/ioutput* file and the errors forwarded to the */tmp/ioerror* file. Check both files after the execution of the command to validate.

Chapter 05

Processes, Scheduling & Basic Hardware

This chapter covers the following major topics:

- ✓ Understand system and user executed processes
- ✓ Display processes
- ✓ View process states
- ✓ Process priority
- ✓ Modify nice value of a running process
- ✓ Signals and their use
- ✓ Run a command immune to hangup signals
- ✓ Understand job scheduling
- ✓ Schedule and manage jobs using at and cron
- ✓ Basic hardware information
- ✓ Device files
- ✓ Major and minor numbers

This chapter includes the following RHCSA objectives:

15. Identify CPU/memory intensive processes, adjust process priority with renice, and kill processes
34. Schedule tasks using cron

A process is any program or command running on the system. It has a unique identification number and is managed by the kernel. It may be viewed, listed, niced, and reniced. It is in one of five states at any given time during its lifecycle. One or several files are opened during a program or command execution and their content may be listed. There are several pre-defined signals that may be passed to a running process to kill or terminate it, among other actions. A program or command may be run in a way that no hangup signals can terminate it (such as the disconnection of the terminal session where the program or command is running).

Job scheduling allows a user to schedule a command for one-time execution or recurring execution in a pre-determined future timeframe. A job may be submitted and managed by any user who is authorized to do so. All executed jobs are logged. Anacron is a service that automatically runs jobs missed while the system was down.

Basic knowledge of hardware-related RHEL features such as interrupt request, memory addresses, direct memory access, hardware abstraction layer, plug and play, device files, and major and minor numbers is essential for a Linux administrator. The last section provides that information.

Understanding Processes

A *process* is created in memory when a program or command is initiated. A unique identification number, known as the *process identification* (PID), is assigned to it and is used by the kernel to manage the process until the program or command it is associated with terminates. For example, when a user logs on to the system, the shell (which is a process) is started. Similarly, when a user executes a command or opens up an application, a process is created. Thus, a process is any program, application, or command that runs on the system.

Several processes are spawned at system boot up, many of which sit in the memory and wait for an event to trigger a request to use their service. These background system processes are called *daemons* and are critical to system functionality.

Viewing Processes

There are two commands commonly used to view currently running processes. These are *ps* (process status) and *top*.

The *ps* command without any options or arguments, lists processes specific to the terminal where this command is run:

```
$ ps
  PID TTY    TIME    CMD
 3679 pts/0  00:00:00 bash
 3703 pts/0  00:00:00 ps
```

The output has four columns that show the PID of the process in the first column, the terminal the process belongs to in the second column, the cumulative time the process is given by the system CPU in the third column, and the name of the actual command or program being executed in the last column.

Two options –e (every) and –f (full) are popularly used to generate detailed information on every process running on the system. There are a number of additional options available with the *ps* command. Check the man pages for details.

```
# ps –ef
1 UID    PID PPID C  STIME TTY   TIME      CMD
2 root    1   0   0  07:14  ?     00:00:01  /sbin/init
3 root    2   0   0  07:14  ?     00:00:00  [kthreadd]
4 root    3   2   0  07:14  ?     00:00:00  [migration/0]
5 root    4   2   0  07:14  ?     00:00:00  [ksoftirqd/0]
. . . . . . . .
```

The output shows more details about the running processes. Table 5-1 describes the content type of each column.

Heading	Description
UID	User ID of the process owner.
PID	Process ID of the process.
PPID	Process ID of the parent process.
C	The process priority.
STIME	The process start time.
TTY	The terminal on which the process was started. Console represents the system console and ? indicates that the process is a daemon.
TIME	Cumulative execution time for the process.
CMD	The name of the command or the program.

Table 5-1 ps Command Output Explanation

Notice that in the *ps* output above there are scores of daemon processes running in the background that have no association with any terminals. Also notice the PID and PPID numbers. The smaller the number, the earlier it is started. The process with PID 0 is started first at system boot, followed by the process with PID 1, and so on. Each PID has an associated PPID in the 3rd column. The owner of each process is also shown along with the name of the command or program.

Information about each running process is kept and maintained in a process table, which the *ps* and other commands read to display information.

The second method for viewing the process information is the *top* command, which also displays the CPU, memory, and swap utilization. A sample output from a running *top* session is shown in Figure 5-1:

$ top

```
top - 20:20:33 up 22:30,  1 user,  load average: 0.00, 0.00, 0.00
Tasks: 136 total,   1 running, 135 sleeping,   0 stopped,   0 zombie
Cpu(s):  0.3%us,   0.3%sy,  0.0%ni, 99.0%id,   0.3%wa,  0.0%hi,  0.0%si,  0.0%st
Mem:    791428k total,    427820k used,    363608k free,    89612k buffers
Swap:  4095992k total,        0k used,   4095992k free,   137536k cached

  PID USER      PR  NI  VIRT  RES  SHR S %CPU %MEM    TIME+  COMMAND
23007 user1     20   0 15088 1192  896 R  0.7  0.2   0:00.05 top
 2107 root      20   0 97820 4108 3148 S  0.3  0.5   0:01.68 sshd
    1 root      20   0 19396 1500 1192 S  0.0  0.2   0:01.25 init
    2 root      20   0     0    0    0 S  0.0  0.0   0:00.04 kthreadd
    3 root      RT   0     0    0    0 S  0.0  0.0   0:00.00 migration/0
    4 root      20   0     0    0    0 S  0.0  0.0   0:00.03 ksoftirqd/0
    5 root      RT   0     0    0    0 S  0.0  0.0   0:00.00 migration/0
    6 root      RT   0     0    0    0 S  0.0  0.0   0:00.00 watchdog/0
    7 root      20   0     0    0    0 S  0.0  0.0   0:02.22 events/0
    8 root      20   0     0    0    0 S  0.0  0.0   0:00.00 cpuset
    9 root      20   0     0    0    0 S  0.0  0.0   0:00.01 khelper
   10 root      20   0     0    0    0 S  0.0  0.0   0:00.00 netns
   11 root      20   0     0    0    0 S  0.0  0.0   0:00.00 async/mgr
   12 root      20   0     0    0    0 S  0.0  0.0   0:00.00 pm
   13 root      20   0     0    0    0 S  0.0  0.0   0:00.03 sync_supers
   14 root      20   0     0    0    0 S  0.0  0.0   0:00.00 bdi-default
   15 root      20   0     0    0    0 S  0.0  0.0   0:00.00 kintegrityd/0
```

Figure 5-1 top in Text Mode

Press q or Ctrl+c to quit *top*.

If *top* is run in an X terminal window, the screen will look similar to the one shown in Figure 5-2.

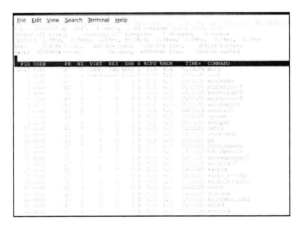

Figure 5-2 top in Graphical Mode

The third method to list, monitor, and manage processes is a graphical tool called *system monitor* which can be started with the *gnome-system-monitor* command or by choosing System → Administration → System Monitor. A sample is displayed in Figure 5-3.

Listing a Specific Process

The *pidof* command can be used to list the PID of a specific process if you know the name of the process. For example, to list the PID of the *crond* daemon, run the command as follows:

$ **pidof crond**
1816

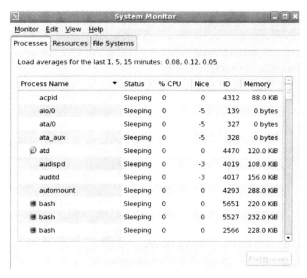

Figure 5-3 GNOME System Monitor

Determining Processes by Ownership

Processes can be listed by their ownership or group membership. The *pgrep* command is used for this purpose. For example, to list all the processes owned by *root*, use any of the following:

ps –U root
pgrep –U root

The first command lists PID, TTY, Time, and process name for all the running processes owned by the *root* user, while the *pgrep* command only lists the PIDs. With the –G option, both commands list all the processes owned by the specified group.

Process States

After a process is spawned, it does not run continuously. It may be in a non-running condition for a while or waiting for some other process to feed it with information so that it can continue to run. There are five process states and each process is in one state at any given time. These states are *running*, *sleeping*, *waiting*, *stopped*, and *zombie*, and are explained below:

✓ The running state determines that the process is currently being executed by the system CPU.
✓ The sleeping state shows that the process is currently waiting for input from a user or another process.
✓ The waiting state means that the process has received the input it has been waiting for and it is now ready to run as soon as its turn arrives.
✓ The stopped state indicates that the process is currently halted and will not run even when its turn comes, unless it is sent a signal.
✓ The zombie state designates that the process is dead. A zombie process exists in the process table just as any other process entry, but takes up no resources. The entry for a zombie process is retained until the parent process permits it to die. A zombie process is also called a *defunct* process.

Process Niceness & How to Set it

The *priority* of a process (niceness) is determined using the *nice* value. The system assigns a nice value to a process at initiation to establish a priority.

There are 40 nice values with –20 being the highest and +19 the lowest. Most system-started processes use the default nice value of 0. A child process inherits the nice value of its parent process.

Use the *ps* command and specify the –l option to determine the niceness of running processes. See the associated nicenesses for each process under the NI column:

```
# ps -efl
F S UID     PID PPID C PRI NI ADDR  SZ   WCHAN STIME TTY     TIME      CMD
4 S root     1   0   0 80  0  -    4851  poll_s 07.14 ?      00.00.01  /sbin
1 S root     2   0   0 80  0  -       0  kthrea 07.14 ?      00.00.00  [kth]
1 S root     3   2   0 -40 -  -       0  migrat 07.14 ?      00.00.00  [mig]
1 S root     4   2   0 80  0  -       0  ksofti 07.14 ?      00.00.00  [kso]
1 S root     5   2   0 -40 -  -       0  cpu_st 07.14 ?      00.00.00  [mig]
. . . . . . . .
```

To determine the default niceness, use the *nice* command without any options or arguments:

```
# nice
0
```

A different priority may be assigned to a program or command at its startup. For example, to run the *top* command at a lower priority of +2:

```
# nice -2 top
```

Use the *ps* command with the –l option, or view the output of the *top* command, and validate the niceness of the process. It should be +2.

To run the same program at a higher priority with the niceness of –2, specify the value with a pair of dashes:

```
# nice --2 top
```

Verify the new value with the *ps* command or view the output of the *top* command. It should be –2.

Altering Niceness of a Running Process

The niceness of a running program may be altered using the *renice* command. For example, to change the nice value of *top* while it is running from –2 to –5, specify the PID (5366) with the *renice* command:

```
# renice -5 5366
5366, old priority -2, new priority -5
```

To alter the nice values of all the processes owned by members of a particular group, use the –g option with *renice*. Similarly, to alter the nice values of all the processes owned by a particular user, use the –u option with it. Run the *renice* command without any options to view its usage.

Listing Open Files

A file is opened when the process or program stored in it is executed and closed when it is no longer required or the associated process or program has terminated. To determine the information such as which files are open, which processes are using them, and who the owners are, the *lsof* (list open files) command is used. Without any options, this command displays a list of all open files.

```
# lsof
COMMAND   PID     USER    FD      TYPE    DEVICE   SIZE          NODE    NAME
init      1       root    cwd     DIR     253,0    4096          2       /
init      1       root    rtd     DIR     253,0    4096          2       /
init      1       root    txt     REG     253,0    38652         95475   /sbin/init
. . . . . . . .
```

The command generated nine columns in the output; these are listed and explained in Table 5-2.

Column	Description
COMMAND	Displays the first nine characters of the command or process name.
PID	Displays the PID of the process.
USER	Displays the owner of the process.
FD	Displays the file descriptor of the file. Some of the values in this field would be: cwd = current working directory; rtd = root directory; txt = text file; mem = memory-mapped file; pd = parent directory
TYPE	Displays the node type of the file.
DEVICE	Displays the major and minor numbers of the device on which the file is located.
SIZE	Displays the file size or offset in bytes.
NODE	Displays the inode number of the file.
NAME	Displays the file name or the file system name where the file resides.

Table 5-2 lsof Output Description

Signals and Their Use

A system runs several processes simultaneously. Sometimes it becomes necessary to pass a notification to a process alerting it of an event. A user or the system uses a signal to pass that notification to the process. A signal contains a signal number and is used to control processes.

There are a number of signals available for use but most of the time you deal with only a few of them. Each signal is associated with a unique number, a name, and an action. A list of the available signals can be displayed with the *kill* command using the –l option:

kill –l

1) SIGHUP	2) SIGINT	3) SIGQUIT	4) SIGILL
5) SIGTRAP	6) SIGABRT	7) SIGBUS	8) SIGFPE
9) SIGKILL	10) SIGUSR1	11) SIGSEGV	12) SIGUSR2
13) SIGPIPE	14) SIGALRM	15) SIGTERM	16) SIGSTKFLT
17) SIGCHLD	18) SIGCONT	19) SIGSTOP	20) SIGTSTP
21) SIGTTIN	22) SIGTTOU	23) SIGURG	24) SIGXCPU
25) SIGXFSZ	26) SIGVTALRM	27) SIGPROF	28) SIGWINCH
29) SIGIO	30) SIGPWR	31) SIGSYS	34) SIGRTMIN
35) SIGRTMIN+1	36) SIGRTMIN+2	37) SIGRTMIN+3	38) SIGRTMIN+4
39) SIGRTMIN+5	40) SIGRTMIN+6	41) SIGRTMIN+7	42) SIGRTMIN+8
43) SIGRTMIN+9	44) SIGRTMIN+10	45) SIGRTMIN+11	46) SIGRTMIN+12
47) SIGRTMIN+13	48) SIGRTMIN+14	49) SIGRTMIN+15	50) SIGRTMAX-14
51) SIGRTMAX-13	52) SIGRTMAX-12	53) SIGRTMAX-11	54) SIGRTMAX-10
55) SIGRTMAX-9	56) SIGRTMAX-8	57) SIGRTMAX-7	58) SIGRTMAX-6
59) SIGRTMAX-5	60) SIGRTMAX-4	61) SIGRTMAX-3	62) SIGRTMAX-2
63) SIGRTMAX-1	64) SIGRTMAX		

Table 5-3 describes the signals that are most often used.

Number	Name	Action	Response
1	SIGHUP	Hang up signal causes a phone line or terminal connection to be dropped. Also used to force a running daemon to re-read its configuration file.	Exit
2	SIGINT	Interrupt signal issued from keyboard, usually by ^c.	Exit
9	SIGKILL	Kills a process abruptly by force.	Exit
15	SIGTERM	Sends a process a soft TERMination signal to stop it in an orderly fashion. This signal is default.	Exit

Table 5-3 Key Signals

The commands used to pass a signal to a process are *kill* and *pkill*. These commands are usually used to terminate a process. Ordinary users can kill processes they own, while the *root* user can kill any process.

The syntax of the *kill* command to kill a process is:

kill PID
kill –s <signal name or number> PID

Specify multiple PIDs if you wish to kill all of them in one go.

The syntax of the *pkill* command to kill a process is:

pkill process_name
pkill –s <signal name or number> process_name

Specify multiple process names if you wish to kill all of them in one go.

Let us look at a few examples for a better understanding.

To pass the soft terminate signal to the Cron daemon *crond*, use one of the following to determine its PID:

```
# ps –ef | grep crond
root    1816    1 0 07:15 ?    00:00:01 crond
# pgrep smbd
1816
# pidof smbd
1816
```

Now pass signal 15 to the process using any of the following:

```
# kill 1816                      # pkill crond
# kill –s 15 1816                # pkill –s 15 crond
# kill –15 1816                  # pkill –15 crond
# kill –SIGTERM 1816             # pkill –SIGTERM crond
# kill –s SIGTERM 1816
```

Using the *kill* or *pkill* command without specifying a signal name or number sends the default signal of 15 to the process. This signal usually causes the process to terminate.

Some processes ignore signal 15 as they might be waiting for an input to continue processing. Such processes may be terminated forcefully using signal 9:

```
$ kill –s 9 1816
$ pkill –9 crond
```

You may wish to run the *killall* command to kill all running processes that match a specified criteria. Here is how you would kill all *crond* processes (assuming there are more than one running):

```
# killall crond
```

Executing a Command Immune to Hangup Signals

A command or program ties itself to the terminal session where it is initiated and does not release the control of the terminal session to the shell until it finishes execution. During this time, if the terminal session is closed or terminated, the running command or program is also terminated. Imagine a large file transfer of several GBs taking place which is about to finish when this unwanted termination occurred.

To avoid such a situation, use the *nohup* (no hang up) command to execute commands or programs that need to run for extended periods of time without being interrupted. For example, to copy */sdata1* directory recursively containing 20GB of data to */sdata2*, issue the *cp* command as follows:

```
# nohup cp –av /sdata1 /sdata2 &
```

Understanding Job Scheduling

Job scheduling is a feature that allows a user to submit a command for execution at a specified time in the future. The execution of the command could be one time or periodically based on a pre-determined time schedule. All this is taken care of by two daemons – *atd* and *crond*. While the *atd* daemon manages the jobs scheduled to run one time in the future, the *crond* daemon takes care of jobs scheduled to run at pre-specified times in the future.

Usually, a one-time execution is scheduled for an activity that needs to be performed at times of low system usage. One example of such an activity is running a lengthy shell program. In contrast, recurring activities could include performing backups, trimming log files, and removing unwanted files from the system.

There are no SELinux requirements for the *atd* daemon, however, the *crond* daemon has two booleans – cron_can_relabel and fcron_crond – associated with it. Neither of the two applies here. Also, there is no need to alter the SELinux file context as long as the crontab files are placed in the */var/spool/cron* and */etc/cron.d* directories.

Scheduler Daemons

The *atd* daemon is started when the system enters run level 3 and runs the */etc/rc.d/rc3.d/S95atd* script. It is terminated when the run level changes to 2, 1, 0, or 6, or shuts down, and it calls the */etc/rc.d/rc#.d/K05atd* script. Likewise, the *crond* daemon is started when the system enters run level 2 and runs the */etc/rc.d/rc2.d/S90crond* script. It is terminated when the run level changes to 1, 0, or 6, or shuts down, and it calls the */etc/rc.d/rc#.d/K60crond* script. These scripts are symbolically linked to their corresponding */etc/rc.d/init.d/atd* and */etc/rc.d/init.d/crond* files, which include the start and stop functions coinciding with service startup and termination.

Both daemons can be started, restarted, reloaded, and stopped manually, and can also be configured to start automatically at specific run levels. The following examples demonstrate performing these operations on *crond*. These commands and procedures can be identically applied to *atd*.

To start *crond*:

```
# service crond start
Starting crond:                    [ OK ]
```

To restart *crond*:

```
# service crond restart
Stopping crond:                    [ OK ]
Starting crond:                    [ OK ]
```

To force *crond* to re-read its configuration:

```
# service crond reload
Reloading crond:                   [ OK ]
```

To stop *crond*:

> # **service crond stop**
> Stopping crond: [OK]

To enable *crond* to start at each system reboot, and validate:

> # **chkconfig crond on**
> # **chkconfig --list crond**
> crond   0:off   1:off   2:on   3:on   4:on   5:on   6:off

To check the status of *crond*:

> # **service crond status**
> crond (pid 4571) is running...

The *crond* daemon checks every minute for any changes made in the */etc/crontab* or the */etc/anacrontab* files, or any files in the */etc/cron.d* or the */var/spool/cron* directories, and loads any modifications found in the memory. Hence, there is no need to restart *crond* after an alteration has been made to any of these files.

Controlling User Access

Which users can or cannot submit an *at* or *cron* job is controlled through files located in the */etc* directory. For *at* job control, the *at.allow* and *at.deny* files are used. For *cron*, the *cron.allow* and *cron.deny* files are used.

The syntax of all four files is identical. You only need to list usernames that require allow or deny access to these tools. Each file takes one username per line. The *root* user is always permitted to use these tools and is neither impacted by the existence or non-existence of these files nor by the presence or absence of an entry for it in these files.

Table 5-4 shows various combinations and the impact on user access.

at.allow / cron.allow	at.deny / cron.deny	Impact
Exists, and contains user entries	Existence does not matter	All users listed in *.allow files are permitted.
Exists, but is empty	Existence does not matter	No users are permitted.
Does not exist	Exists, and contains user entries	All users, other than those listed in *.deny files, are permitted.
Does not exist	Exists, but is empty	All users are permitted.
Does not exist	Does not exist	No users are permitted.

Table 5-4 Controlling User Access

By default, the *.deny* files exist and are empty and the *.allow* files do not exist. This means that, by default, no users (other than *root*) are authorized to submit an *at* or *cron* job.

The following message will appear on the screen if you attempt to execute the *at* command, but are not authorized:

You do not have permission to use at.

Similarly, the following message will appear on the screen if you attempt to execute the *crontab* command, but are not authorized:

You (user1) are not allowed to use this program (crontab)
See crontab(1) for more information

Scheduler Log File

All activities involving the *atd* and *crond* daemons are logged to the */var/log/cron* file. Information such as owner and start time for each invocation of *at* and *crontab* is captured. The file also keeps track of when the *crond* daemon was started, the PID associated with it, spooled *cron* jobs, etc. Sample entries from the log file are shown below:

cat /var/log/cron

. . . .

Aug 3 07:27:02 physical CROND[7925]: (user1) CMD (ls > /dev/pts/0)
Aug 3 07:28:01 physical CROND[7952]: (user1) CMD (ls > /dev/pts/0)
Aug 3 07:29:01 physical CROND[7983]: (user1) CMD (ls > /dev/pts/0)
Aug 3 07:30:01 physical CROND[7998]: (root)   CMD (/usr/lib64/sa/sa1 –S DISK 1 1)
Aug 3 07:30:01 physical CROND[7997]: (user1) CMD (ls > /dev/pts/0)

. . . .

What is anacron?

Anacron is a service that runs after every system reboot, checking for any *cron* and *at* scheduled jobs that were to run while the system was down and hence, have not yet run. It scans the */etc/cron.hourly/0anacron* file for three factors to determine whether to run these missed jobs. The three factors are the presence of the */var/spool/anacron/cron.daily* file, the elapsed time of 24 hours since anacron last ran, and the presence of the AC power to the system. If all of the three factors are affirmative, anacron goes ahead and automatically executes the scripts located in the */etc/cron.daily*, */etc/cron.weekly*, and */etc/cron.monthly* directories, based on the settings and conditions defined in anacron's main configuration file */etc/anacrontab*. The default contents of the */etc/anacrontab* file are displayed below:

cat /etc/anacrontab

. . . .
SHELL=/bin/sh
PATH=/sbin:/bin:/usr/sbin:/usr/bin
MAILTO=root
RANDOM_DELAY=45
START_HOURS_RANGE=3-22
1   5   cron.daily   nice run–parts /etc/cron.daily
7   25   cron.weekly   nice run–parts /etc/cron.weekly
@monthly 45   cron.monthly   nice run–parts /etc/cron.monthly

This file has five environment variables defined: the SHELL and PATH variables set the shell and path to be used for executing the scripts (defined at the bottom of this file); MAILTO defines the username or an email which is sent any output and error messages; RANDOM_DELAY expresses the maximum random delay in minutes (added to the base delay of the jobs as defined in the second column of the last three lines); and START_HOURS_RANGE states the range of hours when the jobs could begin. The last three lines, in the above sample output, define the schedule and the scripts to be executed. The first column represents the period in days (or @daily, @weekly, @monthly, or @yearly) which anacron uses to check whether the specified job has been executed in this many days or period, the second specifies the delay in minutes for anacron to wait before executing the job, the third identifies a job identifier, and the fourth column specifies the command to be used to execute the contents of the */etc/cron.daily*, */etc/cron.weekly*, and */etc/cron.monthly* files. Here the *run-parts* command is used to execute all files under the three directory locations at the default niceness.

For each job, anacron checks whether the job was run previously in the specified days or period (column 1) and executes it after waiting for the number of minutes (column 2) if it was not.

Anacron may be run manually at the command prompt. For example, to run all the jobs that are scheduled in the */etc/anacrontab* file but were missed, you can issue the following command:

> # **anacron**

Anacron stores its execution date in the files located in the */var/spool/anacron* directory for each defined schedule.

By default, anacron service is already installed on the system as part of RHEL6 installation. Check the presence of it using the *rpm* command:

> # **rpm –qa | grep anacron**
> cronie-anacron-1.4.4-7.el6.x86_64

If this package does not already exist, run the *yum* command to install it. See Chapter 06 "Package Management" for details on package management.

> # **yum –y install cronie-anacron**

Using at

The *at* command is used to schedule a one-time execution of a program in the future. All submitted jobs are spooled in the */var/spool/at* directory and executed by the *atd* daemon when the scheduled time arrives. Each job submitted will have a file created containing all variable settings for establishing the user's shell environment so that the job is properly carried out. This file also includes the name of the command or script to be executed at the bottom.

There are multiple ways of specifying an execution time with the *at* command. Some examples are:

at 11am	(executes the task at the next 11am)
at noon	(executes the task at 12pm)
at 23:00	(executes the task at 11pm)
at midnight	(executes the task at 12am)
at 17:30 tomorrow	(executes the task at 5:30pm on the next day)

at now + 2 minutes	(executes the task after 2 minutes. You can specify hours, days, or weeks also)
at 3:00 8/13/12	(executes the task at 3am on August 13, 2012)

 at assumes the current year if no year is mentioned. Similarly, it assumes today's date if no date is mentioned.

You may supply a filename with the *at* command using the –f option to execute it at the specified time. For example, the following will run *script1.sh* (create this file, add the *ls* command to it, and ensure it is executable by the user) from the user's home directory at the current time but after five days:

$ at –f ~/script1.sh now + 5 days

By default, at is already installed on the system as part of RHEL6 installation. Check the presence of it using the *rpm* command:

 # rpm –q at
at-3.1.10-43.el6.x86_64

If this package does not already exist, run the *yum* command to install it. See Chapter 06 "Package Management" for details on package management.

 # yum –y install at

Using crontab

Using the *crontab* command is the other method for scheduling tasks for execution in the future. Unlike *atd*, *crond* executes cron jobs on a regular basis if the cron jobs comply with the format defined in the */etc/crontab* file. Crontab files for users are located in the */var/spool/cron* directory. Each user, who is allowed to run crontab and has a job scheduled, has a file matching his login name in this directory. The other location where system crontab files are stored is the */etc/cron.d* directory. The *crond* daemon scans entries in the files at the two directory locations to determine a job execution schedule. The daemon runs the commands or scripts at the specified time and adds a log entry to the */var/log/cron* file.

By default, cron is already installed on the system as part of RHEL6 installation. Check the presence of it using the *rpm* command:

 # rpm –qa | grep cron | grep –v anacron
crontabs-1.10-33.el6.noarch
cronie-1.4.4-7.el6.x86_64

If these packages do not already exist, run the *yum* command to install them. See Chapter 06 "Package Management" for details on package management.

 # yum –y install cronie-1.* crontabs

The *crontab* command is used to edit (–e), list (–l), and remove (–r) crontab files. Option –u is available if you wish to modify a different user's crontab file provided you have permissions to do so and that user is listed in the *cron.allow* or *cron.deny* file. By default, a crontab file is opened in the vi editor. The *root* user can add or modify the contents of any other user's crontab file even if the user is not allowed to schedule a cron job.

Syntax of User Crontab Files

Each line in a user crontab file that contains an entry for a scheduled job is comprised of six fields. These fields must be in a precise sequence in order for the *crond* daemon to interpret them correctly. See Figure 5-4 for the syntax.

Compared to user crontab files, the system crontab files have seven fields. The first five and the last fields are identical in both user and system files. However, the sixth field specifies the user name who will be executing the specified commands or scripts.

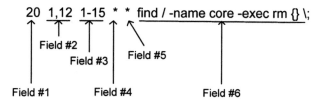

Figure 5-4 Scheduling Syntax of the Crontab Files

A description of each field is given in Table 5-5.

Field	Field Content	Description
1	Minute of the hour	Valid values are 0 (representing an exact hour) to 59. This field can have one specific value (see #1), multiple comma-separated values (see #2), a range of values (see #3), a list of comma-separated or range of values ("1,4,6" and "1-5,6-19"), or an * that represents every minute of the hour (see #4 and #5).
2	Hour of the day	Valid values are 0 (representing midnight) to 23. Values are defined similarly to the way they are defined for the minute of the hour.
3	Date of the month	Valid values are 1 to 31. Values are defined similarly to the way they are defined for the minute of the hour.
4	Month of the year	Valid values are 1 to 12 or jan to dec. Values are defined similarly to the way they are defined for the minute of the hour.
5	Day of the week	Valid values are 0 to 6 or sun to sat, with 0 or sun representing Sunday, 1 or mon representing Monday, and so on. Values are defined similarly to the way they are defined for the minute of the hour.
6	Command or script to execute	Specifies the full path name of the command or script to be executed.

Table 5-5 Crontab File Syntax Description

Moreover, step values may be specified with * and ranges in the crontab files. Step values allow you to specify the number of skips to be done for a given value. For example, */2 in the minute field would mean every 2nd minute, */3 would mean every 3rd minute, 0-59/4 would mean every 4th minute, and so on. Step values are also supported in the same manner in the hour, date of month, month of year, and day of week fields.

Exercise 5-1: Submit, View, List, and Remove an at Job

In this exercise, you will submit an at job as the *root* user to run the *find* command at 11pm on December 31, 2012 to search for all the core files in the entire directory structure and remove them as they are found. You will have the output and any error messages generated redirected to the */tmp/find.out* file. You will list the submitted job and show its contents for verification. Finally, you will remove the job.

1. Run the *at* command and specify the correct time and date for the job execution. Press Ctrl+d at the at> prompt when done.

    # at 11pm 12/31/12
    at> find / –name core –exec rm {} \; > /tmp/find.out 2>&1
    at> Ctrl+d
    job 13 at 2012-12-31 23:00

2. Issue *ll* on */var/spool/at* to list the file created for the submitted job:

    # ll /var/spool/at
    -rwx------. 1 root   root   2875 Aug 5 09:07 a0000d01591b30

3. Display the contents of this file with both the *cat* and the *at* command:

    # cat /var/spool/at/a0000d01591b30
    # at –c 13
    umask 22
    HOSTNAME=physical.example.com; export HOSTNAME
    SELINUX_ROLE_REQUESTED=; export SELINUX_ROLE_REQUESTED
    . . . .
    find / –name core –exec rm {} \;
    marcinDELIMITER0b0f25d5

4. Use both the *at* and the *atq* command to list the spooled job:

    # at –l
    # atq
    13   2012-12-31 23:00 a root

5. Use either the *at* or the *atrm* command to remove the spooled job:

    # at –d 13
    # atrm 13

Exercise 5-2: Add, List, and Remove a Cron Job

In this exercise, you will submit a cron job as the *root* user for *user1* to run the *script1.sh* (ensure this script is executable by *user1*) located in *user1*'s home directory at five minutes past the hour from 1am to 5am only on the first and the fifteenth days of alternate months if the date falls on a Sunday, Tuesday, or Friday. You will have the output and any error messages generated redirected to the */tmp/script1.out* file. You will list the cron entry and then remove it.

1. Open *user1*'s crontab file as *root* and append an entry with the scheduling information provided:

   # crontab –e –u user1
   */5 1-5 1,15 */2 sun,tue,fri /home/user1/script1.sh > /tmp/script1.out 2>&1

2. Go into the */var/spool/cron* directory and see if a new file by the name *user1* exists there:

   # cd /var/spool/cron
   # ll user1
   –rw–––––––. 1 root root 13 Aug 5 16:19 user1

3. Edit */etc/cron.allow* file and add *user1* to it:

   # vi /etc/cron.allow
   user1

4. Run the following as *user1* to list the contents of his crontab file:

   $ crontab –l
   */5 1-5 1,15 */2 sun,tue,fri /home/user1/script1.sh > /tmp/script1.out 2>&1

5. Run the following as *user1* to remove the entire file if you wish to:

   $ crontab –r

6. Confirm the file deletion:

   $ crontab –l
   no crontab for user1

Basic Hardware Information

This section briefly examines the concepts around IRQ, I/O address, DMA, processor, hardware abstraction layer (HAL), plug and play (PnP), hardware devices, and major and minor numbers.

Computer communication takes place via three key channels – *Interrupt Request* (IRQ), *Input/Output address* (I/O address), and *Direct Memory Address* (DMA). Modern computers use the *plug and play* feature which allows operating systems such as RHEL to manage these communication channels automatically without human interaction.

IRQ

An IRQ is a signal sent by a device to the CPU to request processing time. The requesting device may be a network interface, graphics adapter, mouse, modem, printer, keyboard, USB device, or a serial port. Each device attached to the computer may require a dedicated IRQ. Newer devices such as the USB devices can share IRQs. To check which IRQs are in use, view the /proc/interrupts file. The assigned IRQ values are shown in the first column.

```
# cat /proc/interrupts
        CPU0
   0:     163  IO–APIC–edge    timer
   1:       7  IO–APIC–edge    i8042
   8:       0  IO–APIC–edge    rtc0
. . . . . . . .
```

I/O Address

An I/O address is a memory storage location used for communication between different parts of the computer and the CPU. To check for a list of assigned I/O addresses, view the /proc/ioports file. The assigned address values are shown in the first column in hexadecimal notation.

```
# cat /proc/ioports
0000-001f : dma1
0020-0021 : pic1
0040-0043 : timer0
. . . . . . . .
```

DMA

A DMA channel is used when a device has its own processor and can bypass the system CPU when exchanging data with other devices. Examples of such devices include sound and fibre channel cards. There are eight (0 to 7) DMA channels. To check assigned DMA addresses, view the /proc/dma file. The assigned DMA values are shown in the first column.

```
# cat /proc/dma
 2: floppy
 4: cascade
```

Hardware Abstraction Layer and Plug and Play

The *hardware abstraction layer* (HAL) is a piece of software implemented to function between the Linux kernel and the underlying system hardware. It hides differences in hardware from the kernel to enable it to run on a variety of hardware platforms.

The implementation of HAL in RHEL has enabled the operating system to work on a variety of computer system hardware. The *hald* daemon runs on the system and maintains a list of devices. It automatically detects any new hardware added to the system on the fly. If the added hardware is a CD/DVD or a USB device, the daemon mounts the device on the pre-defined mount point.

The *plug and play* (PnP) functionality uses HAL to allow a removable device added to the system to work with RHEL without any manual intervention. HAL autoconfigures the necessary IRQs, I/O addresses, and DMA settings for the PnP device.

The *lshal* command displays a list of all detected hardware in the system:

```
# lshal
Dumping 52 device(s) from the Global Device List:
---------------------------------------------------
udi = '/org/freedesktop/Hal/devices/computer'
  info.addons = {'hald-addon-acpi'} (string list)
  info.callouts.add = {'hal-storage-cleanup-all-mountpoints'} (string list)
  info.interfaces = {'org.freedesktop.Hal.Device.SystemPowerManagement'} (string list)
  info.product = 'Computer' (string)
  info.subsystem = 'unknown' (string)
  info.udi = '/org/freedesktop/Hal/devices/computer' (string)
. . . . . . . .
```

See the man pages of this command for details.

Listing Device Information

Information about PCI, USB, and PC card devices can be gathered and displayed using commands such as *lspci*, *lsusb*, and *pccardctl* in addition to the *lshal* command.

The *lspci* command displays information about PCI buses and the devices attached to them. Specify –v, –vv, or –vvv for detailed output. With the –m option, the command produces more legible output.

```
# lspci –m
00.00.0 "Host bridge" "Intel Corporation" "440FX - 82441FX PMC [Natoma]" -r02 "" ""
00.01.0 "ISA bridge" "Intel Corporation" "82371SB PIIX3 ISA [Natoma/Triton II]" "" ""
00.01.1 "IDE interface" "Intel Corporation" "82371AB/EB/MB PIIX4 IDE" -r01 -p8a "" ""
. . . . . . . .
```

The *lsusb* command displays information about USB buses and the devices connected to them. Specify the –v option for verbosity.

```
# lsusb
Bus 001 Device 001: ID 1d6b:0001 Linux Foundation 1.1 root hub
```

The *pccardctl* command monitors and controls the state of PCMCIA sockets on a laptop computer. See its man pages for details.

Processor/Core

RHEL runs on several types of hardware architectures – Intel, AMD, PowerPC, and mainframe. The Intel, AMD, and PowerPC processors are available in dual-, quad-, hexa-, and octa-core packaging meaning that a single physical processor chip has two, four, six, or eight independent

---

processors on it that share the chip connections to the system board. To know what kind of CPUs your system has, check the /proc/cpuinfo file:

```
# cat /proc/cpuinfo
processor     : 0
vendor_id     : AuthenticAMD
cpu family    : 16
model         : 6
model name    : AMD Athlon(tm) II P320 Dual-Core Processor
stepping      : 3
cpu MHz       : 2011.093
........
```

Alternatively, you may use the *lscpu* command to view cpu details:

```
# lscpu
Architecture:        x86_64
CPU op-mode(s):      32-bit, 64-bit
Byte Order:          Little Endian
CPU(s):              1
On-line CPU(s) list: 0
........
```

Major and Minor Numbers

Every hardware device in the system has an associated device driver loaded in the kernel. Some of the hardware device types are disks, CD/DVD, tape, printers, terminals, and modems. The kernel talks to hardware devices through their respective device drivers. Each device driver has a unique number called a *major* number assigned to it by which the kernel recognizes its type.

Furthermore, there is a possibility that more than one devices of the same type are installed in the system. In this case the same driver is used to control all of them. For example, SATA device driver controls all SATA hard disks and CD/DVD drives. The kernel in this situation allots another unique number called a *minor* number to each individual device within that device driver category to identify it as a separate device. This applies to disk partitions as well. In summary, a major number points to the device driver and a minor number points to an individual device or partition controlled by that device driver.

The major and minor numbers for disk devices and partitions, for instance, can be viewed using the *ll* command on the */dev/sd** files:

```
# ll /dev/sd*
brw-rw----. 1 root disk 8,  0 Aug  6 21:50 /dev/sda
brw-rw----. 1 root disk 8,  1 Aug  6 21:50 /dev/sda1
brw-rw----. 1 root disk 8,  2 Aug  6 21:50 /dev/sda2
```

Column 5 in the output shows the major number and column 6 shows the minor numbers. Major number 8 represents the block device driver for SATA disks. All minor numbers are unique.

Chapter Summary

You studied about processes in the beginning of this chapter. A good understanding of what user and system processes are running, what resources they are consuming, who is running them, what their execution priorities are, etc. is vital for overall system performance and health, as well as for general system administration. You learned how to list processes in different ways. You looked at the five process states, niceness and reniceness for increasing and decreasing a process priority, signals and how they are passed to running processes, and running commands immune to hangup signals.

The next topic covered submitting and managing tasks to run in the future one time or on a recurring basis. You looked at the daemons that control the task execution and the control files where you list users who may or may not be able to submit jobs. You looked at the log file where all executed jobs log their information. You studied anacron service and how it is configured and used to run jobs missed while the system was down. You reviewed the syntax of the crontab file and looked at a variety of date/time formats for use with both at and cron job submission.

Lastly, you reviewed some basic hardware-related information on IRQ, memory addresses, DMA, processor and core, hardware abstraction layer, plug and play, major and minor numbers, and how to list and view device files associated with various hardware devices.

Chapter Review Questions

1. What command would you use to alter the priority of a running process?
2. When would the *cron* daemon execute a job that is submitted as */10 * 2-8 */6 1 /home/user1/script1.sh
3. What is the other command besides the *ps* command to view processes running on the system?
4. What is the command to list the PID of a specific process?
5. What are the background processes normally referred to in Linux?
6. Which command is used to run a process immune to hangup signals?
7. What is the default nice value?
8. What are the four *ls** commands to view pci, usb, cpu, and hal information?
9. The parent process gets the nice value of its child process. True or False?
10. What would the *nice* command display without any options or arguments?
11. Every process running on the system has a unique identication number called UID. True or False?
12. Why would you use the *renice* command?
13. Which user does not have to be explicitly defined in either *.allow* or *.deny* file to be able to run the *at* and *cron* jobs?
14. What command would you use to list open files?
15. What does the *run-parts* command do?
16. When would the *at* command execute a job that is submitted as *at 01:00 12/12/12*
17. What are the two commands that you can use to kill a process?
18. What is the directory location where user crontab files are stored?
19. By default the *.allow* files exist. True or False?
20. Where does the scheduling daemons store log information of executed jobs?
21. How often does anacron checks and runs missed jobs?
22. Which command would you use to edit a crontab file?

---

23. You must restart the *crond* daemon after making any changes in either the */etc/crontab* or the */etc/anacrontab* file. True or False?
24. What are the five process states?
25. Which virtual file system contains information on IRQs, I/O addresses, and DMAs being used?
26. Signal 15 is used for soft termination of a process. True or False?
27. The major number represents the specific type of device driver being used whereas the minor number points to the individual devices or partitions within it. True or False?

Answers to Chapter Review Questions

1. The *renice* command.
2. The *cron* daemon will run the script every tenth minute of the hour on the 2nd, 3rd, 4th, 5th, 6th, 7th, and 8th of every 6th month provided the day falls on a Monday.
3. The *top* command.
4. The *pidof* command can be used to list the PID of a specific process.
5. The background processes are referred to as daemons.
6. The *nohup* command with an ampersand sign at the end of the command line.
7. The default nice value is zero.
8. The *lspci, lsusb, lscpu,* and *lshal* commands.
9. True.
10. The *nice* command displays the default nice value when executed without any options.
11. False. It is called the PID.
12. The *renice* command can be used to change the niceness of a running process.
13. The *root* user.
14. The *lsof* command.
15. The *run-parts* command is used to run scripts listed in the specified directory.
16. The *at* command will run it at 1am on December 12, 2012.
17. The *kill* and *pkill* commands.
18. The user crontab files are stored in the */var/spool/cron* directory.
19. False.
20. The scheduling daemons store log information of executed jobs in the */var/log/cron* file.
21. Anacron checks and runs missed jobs on a daily basis.
22. The *crontab* command.
23. False. It is not required to restart the *crond* daemon.
24. The five process states are running, sleeping, waiting, stopped, and zombie.
25. The */proc* file system.
26. True.
27. True.

Labs

Lab 5-1: Set and Change Nice Values

Open two terminal sessions on *physical.example.com* as *root*. Run the *system-config-users* command on one of the terminals. Run a command on the other terminal to determine the PID and the nice value of the *system-config-users* command. Stop *system-config-users* on the first terminal and re-run it at a lower priority of +8. Confirm the new nice value of the process by running the appropriate command on the second terminal. Execute the *renice* command on the second terminal and increase the priority of the *system-config-users* process to −10, and validate.

Lab 5-2: Configure a User Crontab File

Log on to *physical.example.com* as *user1* and create a cron entry to run the *find* command to search for the core files in the entire directory structure and delete them as they are found. Schedule this command in cron in such a way that it runs every other day of every other month at 15 minutes past 6am. As *root*, create an entry for *user1* if he is not authorized to schedule cron jobs.

Chapter 06

Package Management

This chapter covers the following major topics:

- ✓ Software package concepts including naming convention, dependency, database, and repository
- ✓ Red Hat Network and its benefits
- ✓ Administer software using Red Hat Network
- ✓ List, install, upgrade, freshen, query, remove, extract, validate, and verify packages using the rpm command
- ✓ Overview of yum repository and how to create one
- ✓ List, install, update, search, remove, and check availability of packages as well as synchronize package header information using yum
- ✓ The PackageKit

This chapter includes the following RHCSA objectives:

RHCSA:
43. Install and update software packages from Red Hat Network, a remote repository, or from the local file system

Red Hat software packaging is based on a special format called *Redhat Package Manager* (RPM). All packages available in and for RHEL are in this format. Packages have meaningful names and contain necessary files, as well as metadata structures such as ownership, permissions, and directory location for each included file. Packages may be downloaded and saved locally or on a network share for quick access. Packages may have dependencies over other packages. In other words, a package may require the presence of another package or a group of packages in order to be installed and operate properly. Once a package is installed, its metadata information is stored in a directory; each time the package is updated or upgraded, this information is updated as well.

The Red Hat Network is a web-based secure environment for Red Hat's clients with a subscription to manage package updates, schedule tasks for future execution, and monitor health of their systems remotely. It requires customer registration for access and use.

RHEL provides powerful tools for installing and managing software packages on the system. The *rpm* command has scores of options for flexibility. *Yum* is superior to *rpm* in the sense that it resolves package dependencies automatically. PackageKit provides graphical tools for adding, updating, and removing packages.

Package Overview

As you know RHEL is essentially a set of packages grouped together to form an operating system. It is built around the Linux kernel and includes thousands of packages that are digitally signed, tested, and certified. There are several concepts associated with software package management that are touched upon in the following sub-sections.

Packages and Packaging

A software *package* is a group of files organized in a directory structure and makes up a Red Hat software application. Files contained in a package include installable scripts, configuration files, commands, and related documentation. The documentation includes detailed instructions on how to install and uninstall the package, man pages of the included files and commands, and any other necessary information pertaining to the installation and usage of the package.

All data related to packages is stored at a central location and includes information such as package versioning, the location it is installed at, and checksum values. This allows package management tools to efficiently handle package administration tasks by referencing this data.

Package Naming Convention

Red Hat software packages follow a standard naming convention. Typically, there are five parts to naming a package: the first part contains the package name, the second part lists the package version, the third part shows the package release (revision or build), the fourth part shows the Enterprise Linux for which the package is built, and the last (fifth) part shows the processor architecture for which the package is built. An installable package name always has the .rpm extension. The extension is removed after the package has been installed. For example:

     openssl-1.0.0-20.el6.x86_64.rpm   (package name before it is installed)
     openssl-1.0.0-20.el6.x86_64   (package name after it has been installed)

The following is a description of each part of the package:

- ✓ **openssl** – package name
- ✓ **1.0.0** – package version
- ✓ **20** – package release
- ✓ **el6** – stands for Enterprise Linux 6. Some packages have it, some do not.
- ✓ **x86_64** – processor architecture the package is built for. If you see "noarch" instead, the package will be platform-independent and can be installed on any hardware architecture. If you see "src", it will contain source code for the package.
- ✓ **.rpm** – the extension

Package Dependency

A package to be loaded may require the presence of certain files or other packages in order for a successful installation. Similarly, many software packages require certain files or other packages to be present in order to run and operate smoothly. This is referred to as *package dependency* where one package depends on one or more other packages for installation or execution.

Package Database

Metadata information of installed package files is stored in the */var/lib/rpm* directory. This directory location is referred to as the *package database*, and is referenced by package management tools to obtain information such as ownership, permissions, timestamp, and file size. The package database also contains information on package dependencies. The information contained herein aids package management commands in verifying dependencies and file attributes, upgrading and uninstalling existing packages, and adding new packages.

Package Repository

A *package repository* is a storage location from where one or several packages may be downloaded at cost or cost-free for installation. Red Hat maintains its own package repositories. In addition to accessing Internet-based repositories, you can create your own and add packages to it for later installation. It is highly advisable that you obtain packages from authentic and reliable sources such as the Red Hat Network, which is described next, to prevent damage to your system or to prevent software corruption.

Red Hat Network (RHN)

The *Red Hat Network* (RHN), found online at *rhn.redhat.com* is Red Hat's web interface for customers to manage software updates, download RHEL installation images, perform custom kickstart installations, schedule tasks for future execution, and monitor and report the health of their systems remotely, efficiently, and conveniently. Access to the RHN is available to subscribers only. To check for available software updates and schedule tasks, the *rhnsd* daemon must be running on your system; this daemon polls the RHN every 4 hours, as defined in the */etc/sysconfig/rhn/rhnsd* file, and performs any required actions without human intervention. Alternatively, you can execute the *rhn_check* command to manually perform the check.

RHN Benefits

Several benefits come with the RHN subscription. Some of them are:

- ✓ Schedule commands for later execution on one or more systems.
- ✓ Install, update, and remove packages on one or more systems.
- ✓ Group systems based on requirements.
- ✓ Download RHEL installation images.
- ✓ Add and edit custom configuration files.
- ✓ Create custom kickstart installations.
- ✓ Create system snapshots.

Registering with RHN

In order to obtain the benefits that the RHN offers, you must register your system using an active subscription. This registration opportunity is presented after the system reboots following a graphical installation of RHEL. This process does not automatically begin after the text-based and network-based installations.

Exercise 6-1: Register with the Red Hat Network

In this exercise, you will register your system with the RHN. It is assumed that a valid, active subscription is available for this purpose.

1. Run *rhn_register* at the command prompt in an X terminal window or choose Applications → System Tools → Software Updater. The screen shown in Figure 6-1 will appear. Click Forward.

Figure 6-1 RHN Registration – Intro Screen

2. The next screen will allow you to choose either the Red Hat Network as a source to receive software updates or the Red Hat Network Satellite or Proxy for this purpose. Choose the first option for this demonstration. If there is an internal HTTP proxy server in the environment,

click Advanced Network Configuration and specify its location and authentication information, if applicable. Click Forward when done.

Figure 6-2 RHN Registration – Choose an Update Location

3. On the subsequent screen, enter the login information if you are already registered. If you do not have an account, go to *rhn.redhat.com* and create an account for yourself.

Figure 6-3 RHN Registration – Login Screen

4. The next screen shows the system name. It will also show the system's hardware information and the packages installed. Click on the buttons to view the information. Ensure that both Send Hardware Profile and Send Package Profile are checked. Click Forward and the system will send the profiles to RHN.

Figure 6-4 RHN Registration – Subscription Activation

5. Review the system subscription details on the next screen that pops up. Click Forward.

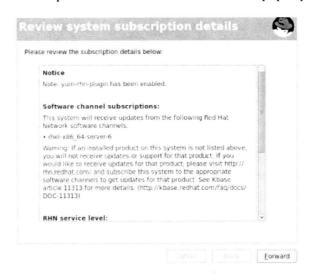

Figure 6-5 RHN Registration – Review Subscription Details

6. The last screen, Figure 6-6, in the registration process indicates that the software update setup has been completed for this system. Click Finish to end the setup.

Figure 6-6 RHN Registration – Finishing the Setup

Administering Software via RHN

After you have completed the registration process for your system, you can log in to the Red Hat Network at *rhn.redhat.com* and manage the system remotely, which includes installing, updating, and removing packages on the system automatically or manually, and viewing pending actions in addition to performing other management tasks. Ensure that the *rhnsd* daemon is running on the system by checking it with the following command:

> # **service rhnsd status**
> rhnsd (pid 20338) is running...

If the daemon is not running, perform the below steps to start it manually and make certain that it gets started at each system reboot:

> # **service rhnsd start**
> # **chkconfig rhnsd on**
> rhnsd   0:off 1:off 2:on   3:on   4:on   5:on   6:off

Managing Packages with rpm

This section discusses the package management tasks including installing, upgrading, freshening, overwriting, querying, removing, extracting, validating, and verifying packages using the *rpm* command. One caveat with this command is that it does not automatically satisfy package dependencies, which can become a big headache in the software installation process.

Before getting into the details, let us take a look at Table 6-1, which provides a list of options commonly used with the *rpm* command.

Option	Description
–a (--all)	Queries all installed packages if used with the –q option.
–c (--configfiles)	Displays configuration files only if used with the –q option.
–d (--docfiles)	Displays documentation files only if used with the –q option.
–e (--erase)	Removes a package.
–f (--file)	Displays information about a file if used with the –q option.
–F (--freshen)	Upgrades an existing package. An older version of the package must exist in order to upgrade it.
--force	Installs a package even if the same version already exists.
–h (--hash)	Displays progress of package installation / upgrade.
–i (--info)	Displays information about a package if used with the –q option.
--import	Imports the specified public key.
–i (--install)	Installs a package.
–K (--checksig)	Validates the signature and also the package integrity.
–l (--list)	Lists files in a package if used with the –q option.
–p (--package)	Queries an installed package against an installable package.
–q (--query)	Queries and displays packages. You may use the *rpmquery* command instead of using rpm –q.
--replacepkgs	Overwrites existing packages.
–R (--requires)	If used with the –q option, it lists dependencies without which a package cannot be installed.
–U (--upgrade)	Upgrades an existing package or installs it if not already installed.
–v or –vv	Displays detailed information.
–V (--verify)	Verifies the integrity of package files. You may use the *rpmverify* command instead of using *rpm –V*.

Table 6-1 rpm Command Options

Before you look at various package management tasks, use the *mount* command and mount the RHEL6 DVD on the */mnt* mount point. Insert the DVD medium in the drive and run the following:

mount /dev/cdrom /mnt

If you have the RHEL6 ISO image copied to a directory on this system, you can mount it instead of the actual DVD medium. The following assumes that the ISO image is located in the */var/downloads* directory and it is called *rhel-server-6.2-x86_64-dvd.iso*:

mount –o loop /var/downloads/rhel-server-6.2-x86_64-dvd.iso /mnt

Installing a Package

Installing a package creates the necessary directory structure for the package and installs the required files. The following example command will attempt to install a package called zsh-4.3.10-4.1.el6.x86_64.rpm:

rpm –ivh /mnt/Packages/zsh-4.3.10-4.1.el6.x86_64.rpm
```
Preparing...        ############################################# [100%]
   1:zsh             ############################################# [100%]
```

If this package requires the presence of other missing packages, you will see an error message that would say "failed dependencies". You must first install the missing packages in order for this package to be loaded successfully.

Alternatively, you may specify the --force option to forcibly install the package without satisfying the dependencies. Using the --force option is not recommended. This option will forcibly install a package and if the same version of the package is already installed, it will overwrite existing files.

To install the zsh package located in a remote repository such as *ftp.example.com*:

> # **rpm –ivh ftp://ftp.example.com/pub/outgoing/zsh-4.3.10-4.1.el6.x86_64.rpm**
> # **rpm –ivh ftp://aghori:Welcome01@ftp.example.com/pub/outgoing/zsh-4.3.10-4.1.el6.x86_64.rpm**

The first command logs in as *anonymous* user and the second uses username *aghori* and password Welcome01 to log in and install the package.

Upgrading a Package

Upgrading a package will upgrade the specified package if an older version of the package is already installed. If an older version is not already installed, the upgrade will go ahead and install it.

To upgrade the package zsh-4.3.10-4.1.el6.x86_64.rpm, use the –U option:

> # **rpm –Uvh /mnt/Packages/zsh-4.3.10-4.1.el6.x86_64.rpm**
> Preparing...   ### [100%]
>   package zsh-4.3.10-4.1.el6.x86_64 is already installed

The command makes a backup of all the affected configuration files during the upgrade process and adds the extension *.rpmsave* to them.

Freshening a Package

Freshening a package requires that an older version of the package must already exist.

To freshen the package zsh-4.3.10-4.1.el6.x86_64.rpm, use the –F option:

> # **rpm –Fvh /mnt/Packages/zsh-4.3.10-4.1.el6.x86_64.rpm**

To freshen all installed packages from the packages on the DVD, execute the following:

> # **rpm –Fvh /mnt/Packages/*.rpm**

Overwriting a Package

Overwriting a package replaces existing files associated with the package with the same version.

To overwrite the package zsh-4.3.10-4.1.el6.x86_64.rpm, use the --replacepkgs option:

> # **rpm –ivh --replacepkgs /mnt/Packages/zsh-4.3.10-4.1.el6.x86_64.rpm**

Querying One or More Packages

Querying one or all packages searches for package information and displays it on the screen. You can use the *rpm* command with the –q or --query option, or the *rpmquery* command.

To query whether a package is installed, run any of the following:

> # **rpm –q zsh**
> # **rpmquery zsh**
> zsh-4.3.10-4.1.el6.x86_64

To identify which package the specified file is associated with, run any of the following:

> # **rpm –qf /bin/zsh**
> # **rpmquery –f /zsh**
> zsh-4.3.10-4.1.el6.x86_64

To list all files in a package, run any of the following:

> # **rpm –ql zsh**
> # **rpmquery –l zsh**
> /bin/zsh
> /etc/skel/.zshrc
> . . . .

To list all configuration files in a package, run any of the following:

> # **rpm –qc zsh**
> # **rpmquery –c zsh**
> /etc/skel/.zshrc
> /etc/zlogin
> . . . .

To list all documentation files in a package, run any of the following:

> # **rpm –qd zsh**
> # **rpmquery –d zsh**
> /usr/share/doc/zsh-4.3.10/BUGS
> /usr/share/doc/zsh-4.3.10/CONTRIBUTORS
> . . . .

To display basic information about a package, run any of the following:

> # **rpm –qi zsh**
> # **rpmquery –i zsh**

Name	: zsh	Relocations: (not relocatable)
Version	: 4.3.10	Vendor: Red Hat, Inc.
Release	: 4.1.el6	Build Date: Tue 08 Dec 2009 03:35:26 PM EST
Install Date	: Fri 10 Aug 2012 08:46:01 AM EDT	Build Host: ls20-bc2-13.build.redhat.com

---

Group	: System Environment/Shells	Source RPM: zsh-4.3.10-4.1.el6.src.rpm
Size	: 5011302	License: BSD
Signature	: RSA/8, Mon 16 Aug 2010 04:35:42 PM EDT, Key ID 199e2f91fd431d51	
Packager	: Red Hat, Inc. <http://bugzilla.redhat.com/bugzilla>	
URL	: http://zsh.sunsite.dk/	
Summary	: A powerful interactive shell	

. . . .

To list all dependencies without which the specified package cannot be installed, run any of the following:

rpm –qR zsh
rpmquery –R zsh
/bin/sh
/bin/zsh

. . . .

To query all installed packages, run any of the following:

rpm –qa
rpmquery –a
xorg-x11-drv-elographics-1.3.0-2.el6.x86_64
hicolor-icon-theme-0.11-1.1.el6.noarch

. . . .

Removing a Package

Removing a package uninstalls the package and all the associated files and directory structure.

To remove a package, use the –e option:

rpm –e zsh

This command performs a dependency check to see if there are any packages that require the existence of the package being removed, and will fail if it determines that another package has dependencies on it.

Extracting Files from a Package

RPM packages are created and packaged for distribution using the *cpio* command. Each package contains several associated directories and files, and may be retrieved to replace a corrupted or lost command or a critical configuration file. For example, to extract a corrupted */etc/inittab* file, use the below command to determine what package contains it:

rpm –qf /etc/inittab
initscripts-9.03.27-1.el6.x86_64

Assuming that the package is located in the */mnt/Packages* directory, use the *rpm2cpio* command to extract (–i) all files from the initscripts package and create (–d) directory structure as required:

```
# cd /tmp
# rpm2cpio /mnt/Packages/ initscripts-9.03.27-1.el6.x86_64.rpm | cpio –id
11226 blocks
```

Search for the *inittab* file under *//tmp/etc* and copy it to the */etc* directory.

Validating Package Signature

A package may be checked for integrity (completeness and an error-free state) and originality after it has been copied to another location, downloaded from the web, or obtained elsewhere, and before it is installed. You can use the MD5 checksum for package integrity and the *GNU Privacy Guard* (GPG) public key signatures to ensure that the package is in fact an official Red Hat package. This will ensure that you are using an authentic piece of software.

Each package contains the GPG signature on the installation media, in the appropriate RPM-GPG-KEY-* file, or in the */etc/pki/rpm-gpg* directory. RHEL6 supports six GPG keys based on whether the package was released before or after November 2006. The RPM-GPG-KEY-redhat-former is used for packages released prior to November 2006 and the RPM-GPG-KEY-redhat-release and RPM-GPG-KEY-redhat-auxilliary are used for packages released after November 2006.

Here is how you would check the integrity and originality of a package such as zsh-4.3.10-4.1.el6.x86_64.rpm located in the */mnt/Packages* directory. The first command will import the specified GPG key(s), the second command will verify that the import is successful, the third command will validate the signature and also the package integrity, and the last command will display the detailed information about the key(s):

```
# rpm --import /etc/pki/rpm-gpg/RPM-GPG-KEY-redhat-release
# rpm –qa gpg-pubkey
gpg-pubkey-2fa658e0-45700c69
gpg-pubkey-fd431d51-4ae0493b
# rpm –K /mnt/Packages/zsh-4.3.10-4.1.el6.x86_64.rpm
/mnt/Packages/zsh-4.3.10-4.1.el6.x86_64.rpm: rsa sha1 (md5) pgp md5 OK
# rpm –qi gpg-pubkey
```

Name	: gpg-pubkey	Relocations: (not relocatable)
Version	: fd431d51	Vendor: (none)
Release	: 4ae0493b	Build Date: Fri 10 Aug 2012 08:20:35 AM EDT
Install Date	: Fri 10 Aug 2012 08:20:35 AM EDT	Build Host: localhost
Group	: Public Keys	Source RPM: (none)
Size	: 0	License: pubkey
Signature	: (none)	
Summary	: gpg(Red Hat, Inc. (release key 2) <security@redhat.com>)	
Description :		

```
-----BEGIN PGP PUBLIC KEY BLOCK-----
Version: rpm-4.8.0 (NSS-3)

mQINBErgSTsBEACh2A4b0O9t+vzC9VrVtL1AKvUWi9OPCjkvR7Xd8DtJxeeMZ5eF
0HtzIG58qDRybwUe89FZprB1ffuUKzdE+HcL3FbNWSSOXVjZIersdXyH3NvnLLLF
0DNRB2ix3bXG9Rh/RXpFsNxDp2CEMdUvbYCzE79K1EnUTVh1L0Of023FtPSZXX0c

. . . . . . . .

-----END PGP PUBLIC KEY BLOCK-----
```

Verifying Attributes of an Installed Package

Verifying an installed package compares several package attributes with the original attributes saved and stored in the package database at */var/lib/rpm* at the time the package was installed. The *rpm* command with –V or --verify option can be used for package attribute verification purposes. You may use the *rpmverify* command instead.

To verify attributes of all installed packages, run any of the following:

```
# rpm –Va
# rpm --verify –a
# rpmverify –a
..5....T.   c /etc/yum/pluginconf.d/rhnplugin.conf
SM5....T.  c /etc/sysconfig/rhn/up2date
.M....G..     /var/log/gdm
. . . . . . . .
```

The commands perform a total of eight checks on each package as illustrated by the eight character codes in the first column of the output and displays what, if any, changes have occurred since the file was installed or created. Each of these codes have an associated meaning. Table 6-2 lists the codes as they appear from left to right, and describes them. A dot character appears for an attribute that has not been modified.

Code	Description
S	Appears if the file size is different.
M	Appears if the permission or file type has altered.
5	Appears if MD5 checksum does not match.
D	Appears if the file is a device file and its major or minor number has changed.
L	Appears if the file is a symlink and its path has altered.
U	Appears if the ownership has been modified.
G	Appears if the group membership has modified.
T	Appears if timestamp has changed.
.	Appears if no modifications have been detected.

Table 6-2 Package Verification Codes

The second column in the output above indicates a code that represents the type of file. Table 6-3 lists them.

File Type	Description
c	Configuration file.
d	Documentation file.
g	Ghost file.
l	License file.
r	Readme file.

Table 6-3 File Type Codes

To verify a single file, use any of the following. There will be no output if no modifications are detected.

```
# rpm –Vf /bin/cp
# rpm –V --file /bin/cp
# rpmverify –f /bin/cp
```

To verify an installed package against an installable package, use any of the following:

```
# rpm –Vp /mnt/Packages/zsh-4.3.10-4.1.el6.x86_64.rpm
# rpm –V --package /mnt/Packages/zsh-4.3.10-4.1.el6.x86_64.rpm
# rpmverify --package /mnt/Packages/zsh-4.3.10-4.1.el6.x86_64.rpm
```

Exercise 6-2: Perform Package Management Tasks Using rpm

In this exercise, you will install a package called dcraw located in the */mnt/Packages* directory on the RHEL installation DVD. You will verify the package's integrity before the installation. After the installation, you will display the package's attributes, show basic information about it, show files it contains, list only the configuration and documentation files it contains, and remove it.

1. Make sure the RHEL6 installation media is in the DVD drive and is mounted on */mnt*. If not, run the following to mount it:

   # **mount /dev/cdrom /mnt**
   mount: block device /dev/sr0 is write-protected, mounting read-only

2. Run the *ll* command on the */mnt/Packages* directory and *grep* for dcraw to ensure that the package is available.

   # **ll /mnt/Packages | grep dcraw**
   –r--r--r--. 124 root root   212784 Aug 16 2010 dcraw-8.96-1.1.el6.x86_64.rpm

3. Run the rpm command and verify the package's integrity:

   # **rpm –K /mnt/Packages/dcraw-8.96-1.1.el6.x86_64.rpm**
   /mnt/Packages/dcraw-8.96-1.1.el6.x86_64.rpm: rsa sha1 (md5) pgp md5 OK

4. Install the package:

   # **rpm –ivh /mnt/Packages/dcraw-8.96-1.1.el6.x86_64.rpm**
   Preparing...   ### [100%]
      1:dcraw   ### [100%]

5. Display the package's attributes:

   # **rpm –V dcraw**

6. Show basic information about the package:

```
# rpm –qi dcraw
Name        : dcraw                              Relocations: (not relocatable)
Version     : 8.96                               Vendor: Red Hat, Inc.
Release     : 1.1.el6                            Build Date: Mon 07 Dec 2009 01:18:51 PM EST
Install Date: Sat 11 Aug 2012 08:33:07 AM EDT    Build Host: hs20-bc1-5.build.redhat.com
Group       : Applications/Multimedia            Source RPM: dcraw-8.96-1.1.el6.src.rpm
Size        : 437155                             License: GPLv2+
Signature   : RSA/8, Mon 16 Aug 2010 11:50:37 AM EDT, Key ID 199e2f91fd431d51
. . . . . . . .
```

7. Show all the files the package contains:

```
# rpm –ql dcraw
/usr/bin/dcraw
/usr/share/locale/ca/LC_MESSAGES/dcraw.mo
. . . . . . . .
```

8. List only the configuration files the package contains:

```
# rpm –qc dcraw
```

9. List only the documentation files the package contains:

```
# rpm –qd dcraw
/usr/share/man/ca/man1/dcraw.1.gz
. . . . . . . .
```

10. Remove the package:

```
# rpm –e dcraw
```

Managing Packages with yum

The *yum* command (*yellowdog updater, modified*) is the front-end to the *rpm* command and is the preferred tool for package management. This tool requires that your system has access to one or more configured software repositories such as the RHN with a valid user account. Alternatively, packages to be installed may be downloaded and stored in a local yum repository. The location of the repository is then defined in the */etc/yum.repos.d* directory. The primary benefit of using this tool is that it performs dependency checks and automatically resolves dependencies by downloading any additional required packages in order to successfully install the specified package. With multiple repositories defined, *yum* can extract the specified software package from wherever it finds it. The default yum repository is the RHN. When the *yum* command is executed for the first time on a system to connect to the RHN repository, it downloads the header information associated with software packages and keeps them in cache. The next time you access the RHN via *yum*, it will download only the updated headers into cache.

Before getting into the details, let's take a look at Table 6-4, which provides a list of sub-commands commonly used with *yum*.

Option	Description
check-update	Checks if any updates are available for the installed packages.
clean	Synchronizes the package header information.
groupinstall	Installs or updates a group of packages.
groupinfo	Provides information on the specified package group.
grouplist	Lists available package groups.
groupremove	Removes the specified package group.
info	Displays the package header information.
install	Installs or updates the specified package(s).
list	Lists packages that are installed or available for installation or update.
localinstall	Installs or updates packages located locally on the system.
provides (or whatprovides)	Searches for packages that contain the specified file.
remove / erase	Removes the specified package(s).
search	Searches for packages that contain the specified string.
update	Updates already installed package(s).

Table 6-4 yum Sub-Commands

Yum Repository

Although several yum repositories are available on the Internet such as the RHN, *www.rpmforge.net*, and *atrpms.net*, you can configure one (or more) for your network. This is a good practice if you have a large number of RHEL systems and you want packages managed with dependencies to be satisfied automatically. This also aids in maintaining software consistency across the board. If you have developed a new package or built one, it can be kept in that repository as well. You may create separate sub-directories within a yum repository to store dissimilar package versions.

Exercise 6-3: Create a Local Yum Repository

In this exercise, you will create a local yum repository for your own system. You will install a package from RHN that will allow you to set up the repository. You will create a directory to store packages and then copy a single package to that directory. Finally, you will create a definition file for the repository.

1. Download and install a package called createrepo:

   # **yum –y install createrepo**
   . . . .
   Installed:
     createrepo.noarch 0.0.9.8-5.el6
   Dependency Installed:
     deltarpm.x86_64 0.3.5-0.5.20090913git.el6
     python-deltarpm.x86_64 0.3.5-0.5.20090913git.el6

2. Create a directory */var/yum/repos.d/local* and *cd* into it:

   # **mkdir –p /var/yum/repos.d/local && cd /var/yum/repos.d/local**

3. Copy the package dcraw-8.96-1.1.el6.x86_64.rpm from */mnt/Packages* into this directory:

   # **cp /mnt/Packages/dcraw* .**

4. Execute the *createrepo* command on the */var/yum/repos.d/local* directory to remove the header information from the package, create a sub-directory called *repodata*, generate XML files to describe the repository, gzip them, and place them into *repodata*:

   # **createrepo –v .**
   1/1 - dcraw-8.96-1.1.el6.x86_64.rpm
   Saving Primary metadata
   Saving file lists metadata
   Saving other metadata
   # **ll repodata**
   total 16
   –rw–r--r--. 1 root root   448 Aug 11 09:16 filelists.xml.gz
   –rw–r--r--. 1 root root   685 Aug 11 09:16 other.xml.gz
   –rw–r--r--. 1 root root   894 Aug 11 09:16 primary.xml.gz
   –rw–r--r--. 1 root root 1353 Aug 11 09:16 repomd.xml

5. Create a definition file */etc/yum.repos.d/local.repo* for the repository. Enter the information as shown:

   # **vi /etc/yum.repos.d/local.repo**
   [local]
   name=local yum repository
   baseurl=file:///var/yum/repos.d/local/
   enabled=1
   gpgcheck=0

6. Execute the following command to clean up the yum cache directory:

   # **yum clean all**

7. Run the following to confirm that the repository has been created and is available for use:

   # **yum repolist**

Exercise 6-4: Create a DVD Yum Repository

In this exercise, you will create a DVD yum repository. You will mount the RHEL installation DVD on */var/yum/repos.d/dvd* and create a definition file for the repository.

1. Complete step #1 from the previous expercise if you have not already done.
2. Create a directory */var/yum/repos.d/dvd* and *cd* into it:

> # mkdir –p /var/yum/repos.d/dvd && cd /var/yum/repos.d/dvd

3. Mount the RHEL6 installation DVD on this directory:

> # mount /dev/cdrom /var/yum/repos.d/dvd

4. Create a definition file */etc/yum.repos.d/dvd.repo* for the repository. Enter the information as shown:

> # vi /etc/yum.repos.d/dvd.repo
> [dvd]
> name=DVD yum repository
> baseurl=file:///var/yum/repos.d/dvd/
> enabled=1
> gpgcheck=0

5. Execute steps 6 and 7 from Exercise 6-3 above.

Exercise 6-5: Create a Remote Yum Repository

In this exercise, you will create a remote yum repository for use on your network. You will configure an NFS server on *physical.example.com* that will host the RHEL6 ISO installation image which is assumed to be located in the */var/downloads* directory. You will perform additional tasks as outlined in the previous exercise.

1. Complete step #1 from Exercise 6-3 if you have not already done.
2. Create a directory */var/yum/repos.d/remote*:

> # mkdir –p /var/yum/repos.d/remote

3. Mount the installation image located in the */var/downloads* directory on to */var/yum/repos.d/remote*:

> # mount –o loop /var/downloads/rhel-server-6.2-x86_64-dvd.iso \
> /var/yum/repos.d/remote

4. Create a definition file */etc/yum.repos.d/remote.repo* for the repository. Enter the information as shown:

> # vi /etc/yum.repos.d/remote.repo
> [remote]
> name=remote yum repository
> baseurl=file://192.168.2.200/var/yum/repos.d/remote/
> enabled=1
> gpgcheck=0

5. Edit the */etc/exports* file and insert the following entry:

 # vi /etc/exports

/var/yum/repos.d/remote   *(ro)

6. Start the NFS service:

 # service nfs start

Starting NFS services:	[OK]
Starting NFS quotas:	[OK]
Starting NFS daemon:	[OK]
Starting NFS mountd:	[OK]

7. Run the following to ensure that the */var/yum/repos.d/remote* has been exported:

 # exportfs

/var/yum/repos.d/remote
                        <world>

8. Execute steps 6 and 7 from Exercise 6-3 above.

Yum Configuration File

The key configuration file for *yum* is */etc/yum.conf*. The default contents are listed below:

 # cat /etc/yum.conf

```
[main]
cachedir=/var/cache/yum/$basearch/$releasever
keepcache=0
debuglevel=2
logfile=/var/log/yum.log
exactarch=1
obsoletes=1
gpgcheck=1
plugins=1
installonly_limit=3
. . . . . . . .
# in /etc/yum.repos.d
```

Table 6-5 explains various directives defined in the file.

Directive	Description
cachedir	Specifies the location to store *yum* downloads. Default is */var/cache/yum/x86_64/6Server*.
Keepcache	Specifies whether to store the package and header cache following a successful installation. Default is 0 (disabled).
debuglevel	Specifies the level at which the debug is to be recorded in the logfile. Default is 2.
logfile	Specifies the name and location of the log file for *yum* activities. Default is */var/log/yum.log*.

Directive	Description
exactarch	Specifies the CPU architecture for the packages to be downloaded. Default is 1 (enabled).
obsoletes	Checks and removes any obsolete packages. Default is 1 (enabled).
gpgcheck	Specifies whether to check the GPG signature for package authenticity. Default is 1 (enabled).
plugins	Specifies to include plug-ins with the packages to be downloaded. Default is 1 (enabled).
installonly_limit	Specifies the maximum number of versions of a single package, as defined by the installonlypkgs directive, to be kept installed simultaneously. Default is 3.

Table 6-5 Directives as Defined in /etc/yum.conf File

To view additional directives that you may define in the */etc/yum.conf* file, run the *yum-config-manager* command. This command will show you the directives that may be defined in the main section of the file as well as any configured repository.

Yum Plugin Directory

The yum plugin directory at */etc/yum/pluginconf.d* holds connection information between yum and the RHN. One important file in this directory is *rhnplugin.conf* that has only two directives defined by default:

cat /etc/yum/pluginconf.d/rhnplugin.conf
[main]
enabled=1
gpgcheck=1

Another file in this directory is *refresh-packagekit-conf* which holds connection information between yum and the PackageKit. It has only one directive defined by default:

cat /etc/yum/pluginconf.d/refresh-packagekit-conf
[main]
enabled=1

Listing Packages and Package Groups

Listing packages allows you to search for installed and available packages in a number of ways and display them on the screen.

To list packages available for installation from all configured yum repositories:

yum list available
. . . .
Available Packages
389-ds-base.x86_64   1.2.10.2-20.el6_3   rhel-x86_64-server-6
389-ds-base-libs.i686   1.2.10.2-20.el6_3   rhel-x86_64-server-6
. . . .

To list all packages available for installation from all configured yum repositories as well as those that are already installed, use any of the following:

 # yum list
 # yum list all

. . . .

Installed Packages
ConsoleKit.x86_64   0.4.1–3.el6   @anaconda-RedHatEnterpriseLinux-201111171049.x86_64/6.2
ConsoleKit-libs.x86_64   0.4.1–3.el6   @anaconda-RedHatEnterpriseLinux-201111171049.x86_64/6.2

. . . .

To list all packages available from all configured yum repositories that you should be able to update:

 # yum list updates

. . . .

Updated Packages
NetworkManager.x86_64   1.0.8.1–33.el6   rhel-x86_64-server-6
NetworkManager-glib.x86_64   1.0.8.1–33.el6   rhel-x86_64-server-6

. . . .

To list if a package (bc for instance) is installed or available for installation from any configured repository:

 # yum list bc

. . . .

Installed Packages
bc.x86_64 1.06.95–1.el6 @anaconda-RedHatEnterpriseLinux-201111171049.x86_64/6.2

To list all installed packages that contain "gnome" in their names:

 # yum list installed *gnome*

. . . .

Installed Packages
NetworkManager-gnome.x86_64 1.0.8.1–15.el6 @anaconda-RedHatEnterpriseLinux201111171049.x86_64/6.2

. . . .

To list any recently added packages from any configured repository:

 # yum list recent

. . . .

Recently Added Packages
dcraw.x86_64   8.96–1.1.el6   physical

To list all installed groups as well as those that are available from all configured repositories:

yum grouplist

. . . .

   Additional Development
   Base
   Client management tools

. . . .

To list all packages a specific group contains:

yum groupinfo Base

. . . .

Group: Base
 Description: The basic installation of Red Hat Enterprise Linux.
 Mandatory Packages:
  alsa-utils
  at

. . . .

Installing and Updating Packages and Package Groups

Installing a package creates the necessary directory structure for the package and installs the required files. If the package is already installed, the command updates it to the latest available version. The following example command will attempt to install / update a package called zsh:

yum –y install zsh

. . . .

Updated:
 zsh.x86_64 0:4.3.10-5.el6
Complete!

Indicate the version information with the package name if you want to obtain and install a specific version of the package.

To install or update several packages:

yum –y install system-config-services system-config-keyboard

. . . .

Updated:
 system-config-keyboard.x86_64 0:1.3.1-4.el6
Complete!

To install or update a package such as dcraw located locally on the system:

yum –y localinstall /var/yum/repos.d/local/dcraw-8.96-1.1.el6.x86_64.rpm

. . . .

Installed:
 dcraw.x86_64 0:8.96-1.1.el6
Complete!

To install or update a group of packages such as Backup Client:

yum –y groupinstall "Backup Client"

. . . .
Installed:
  amanda-client.x86_64 0:2.6.1p2-7.el6
Dependency Installed:
  amanda.x86_64 0:2.6.1p2-7.el6   xinetd.x86_64 2:2.3.14-34.el6

To update a package to the latest available version, issue the following command. Note that *yum* will only update the package to the latest version. The command will fail if the specified package is not already installed.

yum –y update autofs

. . . .
Updated:
  autofs.x86_64 1:5.0.5-54.el6
Complete!

To update all installed packages to the latest version:

yum –y update

. . . .
Setting up Update Process
Resolving Dependencies
--> Running transaction check
---> Package NetworkManager.x86_64 1:0.8.1-15.el6 will be updated
---> Package NetworkManager.x86_64 1:0.8.1-33.el6 will be an update
. . . .

To update autofs group to the latest version:

yum –y groupupdate virtualization

. . . .
Dependency Installed:
  usbredir.x86_64 0:0.4.3-1.el6
Updated:
  qemu-kvm.x86_64 2:0.12.1.2-2.295.el6_3.1
Dependency Updated:
  qemu-img.x86_64 2:0.12.1.2-2.295.el6_3.1
Complete!
. . . .

Displaying Package and Package Group Header Information

To display header information for the autofs package:

---

yum info autofs

```
. . . . . . . .
Name        : autofs
Arch        : x86_64
Epoch       : 1
Version     : 5.0.5
Release     : 54.el6
Size        : 3.1 M
Repo        : installed
From repo   : rhel-x86_64-server-6
Summary     : A tool for automatically mounting and unmounting filesystems
URL         : http://wiki.autofs.net/
License     : GPLv2+
. . . . . . . .
```

To display header information for the virtualization group:

yum groupinfo virtualization

```
. . . . . . . .
Setting up Group Process
Group: Virtualization
 Description: Provides an environment for hosting virtualized guests.
 Mandatory Packages:
   qemu-kvm
 Optional Packages:
   qemu-kvm-tools
   vios-proxy
```

Searching Packages

To search for all the packages that contain a specific file such as /bin/bash, use either provides or whatprovides sub-command with *yum*:

yum provides /bin/bash

```
. . . . . . . .
bash-4.1.2-9.el6_2.x86_64 : The GNU Bourne Again shell
Repo                      : rhel-x86_64-server-6
Matched from:
Filename                  : /bin/bash
. . . . . . . .
```

Use the wildcard character to match all filenames:

yum whatprovides /usr/bin/system-config*

```
. . . . . . . .
authconfig-gtk-6.1.12-5.el6.x86_64 : Graphical tool for setting up
                                   : authentication from network services
Repo                               : rhel-x86_64-server-6
. . . . . . . .
```

To search for all packages that contain the specified string in their name, description, or summary:

yum search system-config
. . . .
system-config-firewall-base.noarch   : system-config-firewall base components and command line tool
system-config-keyboard-base.x86_64 : system-config-keyboard base components
. . . .

Removing Packages and Package Groups

To remove the dcraw package and any packages that depend on it:

yum –y remove dcraw
. . . .
Removed:
  dcraw.x86_64 0:8.96–1.1.el6
Complete!

To remove a specific installed group:

yum –y groupremove "Tajik Support"
. . . .
Removed:
  dejavu–sans–fonts.noarch 0:2.30–2.el6
  dejavu–sans–mono–fonts.noarch 0:2.30–2.el6
  dejavu–serif–fonts.noarch 0:2.30–2.el6
Complete!

Checking Availability of Updated Packages

To check whether any updates are available for packages installed on your system:

yum check-update
. . . .
NetworkManager.x86_64   1.0.8.1–33.el6   rhel–x86_64–server–6
NetworkManager-glib.x86_64   1.0.8.1–33.el6   rhel–x86_64–server–6
. . . .

Synchronizing Package Header Information

By default, the header information associated with packages in the */var/cache/yum* directory is automatically synchronized with that in the Red Hat repositories every 90 minutes by the *yum* command if the system is registered with the RHN. This default expiry period is defined in the */etc/yum.conf* file with the *metadata_expire* directive. If you wish to synchronize it instantly, run the following:

yum clean all
. . . .
Cleaning up Everything

Downloading Packages

The *yumdownloader* command is used to download individual packages from a configured repository.

To download a package, specify the name of the package with the command. Make sure to change into the directory where you want the package downloaded.

> # **cd /var/yum/repos.d/local**
> # **yumdownloaer dhclient**
>
> . . . .
> dhclient-4.1.1–31.P1.el6_3.1.x86_64.rpm | 317 kB   00.00

Exercise 6-6: Perform Package Management Tasks Using yum

In this exercise, you will remove an installed package called wireless-tools. After removing it, you will download and reinstall it, and display its header information. You will also install a package group called Backup server. After the installation, you will display the package group's header information and then remove it.

1. Remove the wireless-tools package:

    > # **yum –y erase wireless-tools**
    >
    > . . . .
    > Removed:
    >   wireless-tools.x86_64 1:29–5.1.1.el6
    > Complete!

2. Change into the /var/yum/repos.d/local directory:

    > # **cd /var/yum/repos.d/local**

3. Download the wireless-tools package using the *yumdownloader* command:

    > # **yumdownloader wireless-tools**

4. Run the *ll* command on the wireless-tools file to ensure the package has been downloaded:

    > # **ll wireless-tools***
    > –rw–r--r--. 1 root root 95892 Aug 25 2010 wireless-tools–29–5.1.1.el6.x86_64.rp

5. Install the package:

    > # **yum –y localinstall /var/yum/repos.d/local/wireless-tools-29-5.1.1.el6.x86_64.rpm**
    >
    > . . . .
    > Installed:
    >   wireless-tools.x86_64 1:29–5.1.1.el6
    > Complete!

6. Display the package header information:

 # yum info wireless-tools

```
. . . . . . . .
Installed Packages
Name        : wireless-tools
Arch        : x86_64
Epoch       : 1
Version     : 29
Release     : 5.1.1.el6
Size        : 207 k
Repo        : installed
From repo   : /wireless-tools-29-5.1.1.el6.x86_64
Summary     : Wireless ethernet configuration tools
URL         : http://www.hpl.hp.com/personal/Jean_Tourrilhes/Linux/Tools.html
License     : GPL+
. . . . . . . .
```

7. Install the Backup server package group:

 # yum –y groupinstall "Backup server"

```
. . . . . . . .
Installed:
  amanda-server.x86_64 0:2.6.1p2-7.el6
Dependency Installed:
  amanda.x86_64 0:2.6.1p2-7.el6        xinetd.x86_64 2:2.3.14-34.el6
Complete!
```

8. Display the Backup server package group information:

 # yum groupinfo "Backup server"

```
. . . . . . . .
Setting up Group Process
Group: Backup Server
 Description: Software to centralize your infrastructure's backups.
 Mandatory Packages:
   amanda-server
 Optional Packages:
   mt-st
   mtx
```

9. Remove the Backup server package group:

 # yum –y groupremove "Backup server"

```
. . . . . . . .
Removed:
  amanda-server.x86_64 0:2.6.1p2-7.el6
Complete!
```

PackageKit

RHEL6 provides graphical package management tools in a single package called the PackageKit for those who prefer to use the graphical interface. This package is installed automatically during the RHEL6 installation when you choose Graphical Administration Tools under Desktop. If this package is not already installed for whatever reasons, you may run the following command to install it:

> # **yum –y install gnome-packagekit**

The following sub-sections discuss the tools available within PackageKit.

Package Updater

The Package Updater is the graphical front-end to the *yum* command with the update option. This tool allows you to view and select the updates available to apply on your system. Follow the steps below to start and use it:

☞Execute *gpk-update-viewer* at the command line, or click System → Administration → Software Update. The interface will appear as shown in Figure 6-7. It shows that there are 13 updates currently available for this system. Click Install Updates if you wish to install them.

Figure 6-7 The Package Updater

Automatic Updates

Configuring to download and install RHEL updates may be scheduled using the graphical Software Updates Preferences tool. Follow the steps below to start and interact with it:

☞Execute *gpk-prefs* at the command line, or click System → Preferences → Software Updates. The interface will appear as shown in Figure 6-8. There are two primary settings: Check for updates and Automatically install. The first setting enables you to choose how often you would like your system to check for available updates. Choices available are hourly, daily, weekly, and never. The second setting allows you to choose how you would like the available updates to be handled. Choices are to automatically install all available updates, only the security patches, or do nothing.

Figure 6-8 Software Update Preferences

Add / Remove Software

The Add/Remove Software tool allows you to add, update, or remove one or more packages or package groups in one shot. Follow the steps below to start and use it:

☞Execute *gpk-application* at the command line, or click System → Administration → Add/Remove Software. The interface will appear as shown in Figure 6-9. Here you have an opportunity to add, update, or remove an individual package, a collection of packages, or any new packages available. The Selected Packages option displays a list of the packages or package groups that have been marked for addition, updating, or deletion. Down in the left hand pane, software package categories are shown. These categories are the same as you saw during the installation process. Consult Chapter 01 "Local Installation" for details. You may expand these categories to list available packages. You can obtain a list of files and dependency information for a selected package. Context sensitive help is also available when you highlight a package. You can filter the list using a string of characters. Click Apply after the desired choices have been made. A list of all selected packages including any dependents will be displayed for confirmation to proceed.

Figure 6-9 Add/Remove Software

Chapter Summary

This chapter discussed software package management. You learned concepts around packages, packaging, naming convention, dependency, and patch database. You looked at the benefits of RHN and how to register a system to administer software. You studied and performed a number of package management tasks using the *rpm* command. You looked at the concepts and benefits of having a yum repository, and then performed scores of package management tasks using the *yum* command.

Finally, you reviewed three graphical package administration tools that are part of the PackageKit package.

Chapter Review Questions

1. What would the *rpm –ql dcraw* command do?
2. What is the purpose of the *rpm2cpio* command?
3. What is the difference between freshing and upgrading a package?
4. What is the command that you would run to register your system with the RHN?
5. What are the names of the three graphical administration tools included in the PackageKit?
6. What would the *yum groupinfo Base* command do?
7. What is the use of the –y option with the *yum install* and *yum remove* commands?
8. The Red Hat Network provides free access to package repositories. True or False?
9. What is the biggest advantage of using the *yum* command over the *rpm* command?
10. What is the difference between installing and upgrading a package?
11. Package database is located in the */var/lib/rpm* directory. True or False?
12. What sub-command would you use with the *yum* command to check for the availability of the updates for the installed packages?
13. What is the equivalent of the *rpmquery* command?
14. What would the *rpm –qf /bin/bash* command do?
15. Which directory on an installed system does RHEL6 store GPG signatures in?
16. What should be the extension of a yum repository configuration file?
17. What would the *yum list dcraw* command do?

18. What is the name of the RHN daemon that must be running on the system in order for the system to be able to communicate with the RHN?
19. What would the options ivh cause the *rpm* command to do?
20. You can use the *downloadyum* command to download a package. True or False?
21. What would the *yum list installed *gnome** command do?
22. What option would you use with the *rpmquery* command if you wish to list dependent packages without which the specified package cannot be installed?
23. What would the *rpm –qa* command do?
24. How many package names can be specified at a time with the *yum install* command?
25. You can update all the packages within a package group using the groupupdate sub-command with *yum*. True or False?
26. What would the *yum info dcraw* command do?
27. Automatic software updates may be set up using the *gpk-prefs* command. True or False?
28. Which graphical tool may be used for adding and removing software packages?
29. Which package needs to be installed in order to set up a private yum repository on the system?
30. The Package Updater is the front-end to the *rpm* command. True or False?

Answers to Chapter Review Questions

1. The *rpm* command provided will list files in the dcraw package.
2. The purpose of the *rpm2cpio* command is to extract files from the specified package.
3. Both are used to upgrade an existing package, but freshing requires an older version of the package to exist.
4. The *rnh_register* command.
5. The three graphical administration tools included in the PackageKit are the Package Updater, the Software Updates Preferences, and the Add/Remove Software.
6. The *yum* command provided will list all packages in the Base package group.
7. The *yum* command will not prompt for user confirmation if the –y option is used with it.
8. False. RHN requires a subscription.
9. The *yum* command resolves and installs dependent packages automatically.
10. Installing will install a new package whereas upgrading will upgrade an exitsing package or install it if it does not already exist.
11. True.
12. The *check-update* sub-command.
13. The equivalent for the *rpmquery* command is the *rpm* command with the –q option.
14. The *rpm* command provided will display information about the */bin/bash* file.
15. The */etc/pki/rpm-gpg* directory.
16. The extension of a yum repository configuration file should be .repo.
17. The *yum* command provided will display if the dcraw package is already installed or available for installation.
18. The *rhnsd* daemon.
19. It will install the specified package and show installation details and hash signs for progress.
20. False. There is no such command.
21. The *yum* command provided will display all installed packages that contain gnome in their names.
22. The –R (--require) option.
23. The *rpm* command provided will display all installed packages.

24. There is no limit.
25. True.
26. The *yum* command provided will display the header information for the dcraw package.
27. True.
28. The Add/Remove Software program.
29. The createrepo package.
30. True.

Labs

Lab 6-1: Set up a Third Party Yum Repository

Set up a third party yum repository to access packages located at *atrpms.net* for RHEL6. Install a package called aalib from this repository. Consider using the --disablerepo option when installing the package to ensure that the *yum install* command does not search for this program in other configured repositories.

Lab 6-2: Configure a Local Yum Repository

Configure a local yum repository to access packages located locally in a directory. Create a directory called */var/yum.repos.d/local* and copy all the contents of the RHEL6 installation DVD to it. Create a repo file for this repository and install packages policycoreutils* from this repository to validate it.

Lab 6-3: Install Package Groups

Install package groups Backup Server, Remote Desktop Clients, and Security Tools from the local yum repository configured in Lab 6-2. Review the *yum.log* file for confirmation after the installation is complete. Display the information for the three package groups.

## Chapter 07

Virtualization & Network Installation

This chapter covers the following major topics:

- ✓ Overview of virtualization
- ✓ Overview of network installation servers
- ✓ Configure FTP and HTTP installation servers
- ✓ Interact with the Virtual Machine Manager program
- ✓ Create a virtual machine and installing RHEL using network installation servers
- ✓ Benefits of using kickstart
- ✓ Create a kickstart configuration file
- ✓ Install RHEL using kickstart

This chapter includes the following RHCSA objectives:

37. Configure a physical machine to host virtual guests
41. Configure a system to run a default configuration HTTP server
42. Configure a system to run a default configuration FTP server
38. Install Red Hat Enterprise Linux systems as virtual guests
39. Configure systems to launch virtual machines at boot
17. Access a virtual machine's console
18. Start and stop virtual machines
36. Install Red Hat Enterprise Linux automatically using kickstart

Virtualization is a feature that enables a single physical computer system to host

several virtual machines with each virtual machine acting as a standalone computer running a unique instance of RHEL, Solaris, Windows, or some other Linux distribution.

RHEL may be installed over the network using a configured FTP, HTTP, or NFS server hosting the installation files. Installing over the network is much faster than a local DVD based installation. A client system, where RHEL needs to be installed, can be booted locally and then redirected to one of these network installation servers for loading the operating system software. The client system can be configured during the installation or supplied with a file that contains all the configuration information including disk partitioning. This way there is no need to go through the configuration process during installation, which makes the installation faster and fully unattended.

Virtualization & KVM

Virtualization enables a single physical computer to run multiple instances of various operating systems concurrently with complete isolation from one another. For this purpose, virtualization software such as the *Kernel-based Virtual Machine* (KVM) is needed on the physical computer to control the physical hardware of the system and to allow for the creation of one or many *virtual machines* on that physical computer. Each virtual machine will operate as a separate, standalone computer. Each virtual machine will share the physical computer's processor, memory, DVD drive, and network interfaces. KVM is a native, bare-metal *hypervisor* software that comes standard with RHEL6 and is loaded directly on the physical computer to virtualize the hardware to host any number of *guest* operating systems. A hypervisor is a software layer between the physical hardware and the guest operating systems (guests), and presents a virtual operating platform to the guests and manages their execution.

Features and Benefits

Some of the features of virtualization and key benefits of its implementation are listed below:

- ✓ Allows creating multiple virtual machines on a single physical computer.
- ✓ Supports running instances of a variety of operating systems as guests.
- ✓ Each virtual machine and its guest is completely isolated and independent of other virtual machines and their guests.
- ✓ Provides the ability to consolidate older systems onto newer hardware platforms.
- ✓ Decreases the overall cost associated with computer hardware, and network and storage switches.
- ✓ Reduces the overall cost associated with power and cooling, floor and rack space, and network and fibre cabling.
- ✓ Enhanced scalability.
- ✓ Rapid virtual server deployment.
- ✓ Better utilization of computer hardware resources.
- ✓ Improved security, reliability, availability, and fault tolerance.
- ✓ Increased operating system uptime.
- ✓ Better return on investment.
- ✓ Scalability of running virtual machines, i.e. you can add more disk space on the fly.

KVM Packages

In order for a physical computer to be able to host virtual machines, RHEL must be directly installed on it. During the RHEL6 installation, you have the opportunity to select all packages within the virtualization package group. To check if the virtualization packages are already installed, run the *rpm* command as follows:

 # rpm –qa | egrep 'virt|kvm'
libvirt-python-0.9.10–21.el6_3.4.x86_64
virt-viewer-0.5.2–9.el6.x86_64
virt-who-0.6–6.el6.noarch
libvirt-0.9.10–21.el6_3.4.x86_64
virt-manager-0.9.0–14.el6.x86_64
libvirt-client-0.9.10–21.el6_3.4.x86_64
qemu-kvm-0.12.1.2–2.295.el6_3.1.x86_64
virt-top-1.0.4–3.13.el6.x86_64
virt-what-1.11–1.1.el6.x86_64
python-virtinst-0.600.0–8.el6.noarch

If the packages are not already installed, use the following command to check if they are available for installation. Refer to Chapter 06 "Package Management" for details.

 # yum grouplist | grep –i virtualization
Virtualization
Virtualization Client
Virtualization Platform
Virtualization Tools

The output indicates that there are four package groups available for virtual machine management. A short description of each of these package groups is provided in Table 7-1 below.

Package Group	Description
Virtualization	Provides the foundation to host virtual machines. Includes the qemu-kvm required package.
Virtualization Client	Provides the support to install and manage virtual machines. Includes the python-virtinst, virt-manager, virt-top, and virt-viewer required and default packages.
Virtualization Platform	Provides an interface to access and control virtual machines. Includes virt-who, libvirt, and libvirt-client required packages.
Virtualization Tools	Provides tools for offline management of virtual machines. Includes the libguestfs required package group.

Table 7-1 KVM Package Groups

If the package groups are available for installation in a yum repository, use the following command to install all of them. Refer to Chapter 06 "Package Management" for details.

> # **yum –y groupinstall virtualization "virtualization client" "virtualization platform" \\
> "virtualization tools"**

. . . .

Installing:

libguestfs	x86_64	1.1.16.19-1.el6	rhel-x86_64-server-6	1.5 M
libvirt	x86_64	0.9.10-21.el6_3.4	rhel-x86_64-server-6	1.9 M
libvirt-client	x86_64	0.9.10-21.el6_3.4	rhel-x86_64-server-6	3.2 M
python-virtinst	noarch	0.600.0-8.el6	rhel-x86_64-server-6	490 k
qemu-kvm	x86_64	2.0.12.1.2-2.295.el6_3.1	rhel-x86_64-server-6	1.2 M
virt-manager	x86_64	0.9.0-14.el6	rhel-x86_64-server-6	1.0 M

. . . .

Configuring Network Installation Servers

Before you are able to perform an over the network RHEL installation, you must have the installation files available on a configured network server using the FTP, HTTP, or NFS protocol. The following exercises provide step by step procedures on how to configure FTP and HTTP servers. Note that configuring an NFS installation server is beyond the scope of the RHCSA exam.

Exercise 7-1: Configure an FTP Installation Server

In this exercise, you will configure an FTP installation server on *physical.example.com* using the *very secure FTP* program available in RHEL6. You will install the necessary packages associated with vsFTP, copy the files from the installation DVD to the */var/ftp/pub/rhel6* directory, set proper SELinux context, enable port 21 to allow the traffic to pass through, and start the FTP service. Finally, you will open a browser window and test access to the files.

1. Install the vsftpd software:

    > # **yum –y install vsftpd**

2. Create the directory */var/ftp/pub/rhel6* for storing the RHEL installation files. Ensure that there is at least 4GB of free space available in the */var* file system.

    > # **mkdir –p /var/ftp/pub/rhel6**

3. Load the installation DVD in the drive. The DVD should automatically mount on */media/"RHEL_6.2 x86_64 Disc 1"*. Unmount the DVD using the following command and remount it on */mnt*:

    > # **umount /dev/cdrom**
    > # **mount /dev/cdrom /mnt**

4. Change directory into /mnt and opy the entire directory structure from */mnt* to */var/ftp/pub/rhel6*:

cd /mnt && find . | cpio –pmd /var/ftp/pub/rhel6

5. Unmount and eject the DVD after the copy has been finished:

   # umount /mnt
   # eject

6. Copy the SELinux context set on the */var/ftp/pub* directory to the */var/ftp/pub/rhel6* directory:

   # chcon –Rv --reference=/var/ftp/pub /var/ftp/pub/rhel6

7. Issue the *semanage* command and modify the contexts on the directory to ensure that the new contexts survive a SELinux relabeling:

   # semanage fcontext –a –s system_u –t public_content_t /var/ftp/pub/rhel6

8. Configure host-based access by allowing vsFTP traffic on port 21 to pass through the firewall:

   # iptables –I INPUT –s 192.168.2.0/24 –p tcp --dport 21 –j ACCEPT

9. Save the rule in the */etc/sysconfig/iptables* file and restart the firewall to activate the new rule:

   # service iptables save ; service iptables restart

10. Restart the vsFTP service and check the running status:

    # service vsftpd restart
    # service vsftpd status
    vsftpd (pid 5677) is running...

11. Set the vsFTP service to autostart at each system reboot, and validate:

    # chkconfig vsftpd on
    # chkconfig --list vsftpd
    rhnsd   0:off 1:off 2:on   3:on   4:on   5:on   6:off

12. Open up a browser window and type the following command to test access:

    ftp://192.168.2.200/pub/rhel6

Exercise 7-2: Configure an HTTP Installation Server

In this exercise, you will configure an HTTP installation server on *physical.example.com* using the *Apache* program available in RHEL6. You will install the necessary packages associated with Apache, copy the files from the installation DVD to the */var/www/html/rhel6* directory, set proper SELinux context, enable port 80 to allow the traffic to pass through, and start the HTTP service. Finally, you will open a browser window and test access to the files.

---

1. Install the httpd software:

   # **yum –y install httpd**

2. Create the directory */var/www/html/rhel6* for storing the RHEL installation files. Ensure that there is at least 4GB of free space available in the */var* file system.

   # **mkdir –p /var/www/html/rhel6**

3. Perform steps 3 to 5 from the previous exercise to copy installation files, but specify the target copy location as */var/www/html/rhel6*.

4. Copy the SELinux context set on the */var/www/html* directory to the */var/www/html/rhel6* directory:

   # **chcon –Rv --reference=/var/www/html /var/www/html/rhel6**

5. Issue the *semanage* command and modify the contexts on the directory to ensure that the new contexts survive a SELinux relabeling:

   # **semanage fcontext –a –s system_u –t httpd_sys_content_t /var/www/html/rhel6**

6. Configure host-based access by allowing http traffic on port 80 to pass through the firewall:

   # **iptables –I INPUT –s 192.168.2.0/24 –p tcp --dport 80 –j ACCEPT**

7. Save the rule in the */etc/sysconfig/iptables* file and restart the firewall to activate the new rule:

   # **service iptables save ; service iptables restart**

8. Restart the http service and check the running status:

   # **service httpd restart**
   httpd (pid 5921) is running...

9. Set the Apache service to autostart at each system reboot, and validate:

   # **chkconfig httpd on**
   # **chkconfig --list httpd**
   httpd   0:off 1:off 2:on 3:on 4:on 5:on 6:off

10. Open up a browser window and type the following command to test access:

    http://192.168.2.200/rhel6

The Virtual Machine Manager

The Virtual Machine Manager is a graphical application for creating and managing virtual machines on the KVM hypervisor. This desktop program includes a wizard that makes it easy for you to supply information as you set up a new virtual machine. It allows you to modify, clone, or delete virtual machines. It provides you with the ability to add, modify, or delete virtual storage pools, virtual networks, and network interfaces.

Interacting with the Virtual Machine Manager

Start the Virtual Machine Manager by using the *virt-manager* command in an X terminal or by clicking Applications → System Tools → Virtual Machine Manager. See Figure 7-1 for the main screen.

Figure 7-1 Virtual Machine Manager Interface

The first time you start the Virtual Machine Manager, you will need to connect to the KVM's QEMU (Quick Emulator) hypervisor. Highlight "localhost (QEMU) – Not Connected", right click and select Connect to connect to the hypervisor. If you are still unable to connect, try restarting the *libvirtd* daemon and checking/loading the kvm module in memory, as follows:

To restart the *libvirtd* daemon:

> # **service libvirtd restart**

Ensure the KVM module is loaded in the memory:

> # **lsmod | grep kvm**

If the module is not already loaded in the memory, issue the *modprobe* command to load it:

> # **modprobe kvm**

Now you should be able to connect to the hypervisor.

To view the details of the hypervisor, right click on it and select Details. You will see four tabs in the details window, as shown in Figure 7-2. These tabs are Overview, Virtual Networks, Storage, and Network Interfaces.

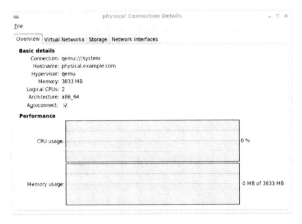

Figure 7-2 QEMU Hypervisor Details

The Overview tab shows the basic information about the hypervisor including the hypervisor's URI (Universal Resource Identifier), hostname of the server that it is running on, name of the hypervisor, memory and number of CPUs available for virtual machines, CPU architecture, and CPU/memory utilization information at the bottom. There is also a checkmark whether to connect to the hypervisor each time the server is started.

The Virtual Networks tab displays the default virtual network configuration (see Figure 7-3).

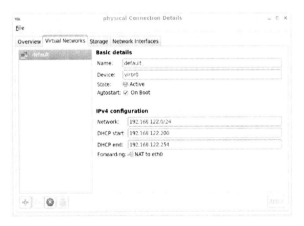

Figure 7-3 QEMU Hypervisor Virtual Networks Tab

This tab provides basic details and displays IP configuration. The basic details include the virtual network name, the device associated with it, the state of the device, and whether to autostart it when a virtual machine using this virtual network is started. The IP configuration includes the subnet IP, the netmask, and a range of IP addresses that it may be able to supply to virtual machines if they are configured to use DHCP. The Forwarding button indicates that NATting is enabled on the physical network interface. The four little square buttons at the bottom left may be used if you wish to create, start, stop, or delete a virtual network.

The Storage tab displays the default virtual storage configuration (see Figure 7-4).

Figure 7-4 QEMU Hypervisor Storage Tab

This tab provides details on the configured storage. It indicates that the default location for storing virtual machine files is the */var/lib/libvirt/images* directory. It also tells the current usage of the selected storage pool. It is recommended to create a separate partition or logical volume large enough to satisfy the storage requirements for all additional VMs that you plan to have on this server. The New Volume button provides you with an opportunity to create an additional volume for storing VM files. The Delete Volume button allows you to remove an existing volume. The State and Autostart buttons in the middle of the window as well as the four square buttons at the bottom left are self-explanatory.

The Network Interfaces tab displays the virtual network interfaces that are available for VMs to connect to (see Figure 7-5).

Figure 7-5 QEMU Hypervisor Network Interfaces Tab

This tab provides a list of all currently configured network interfaces. Highlighting one of them in the left window pane provides associated details on the right side of the window. *eth0* represents the first physical Ethernet interface on the system, *lo0* denotes the loopback interface, and *wlan0* the wireless interface. Each interface shows the MAC address, state, whether to autostart it, and IP information if available. The + button at the bottom left allows you to set up a new network interface. KVM supports four types of network interfaces for the use of virtual machines. These are Bridge (binds a physical interface with a virtual interface), Bond (forms teaming of two or more interfaces), Ethernet (creates a bridged interface), and VLAN (hooks up a physical or virtual interface to VMs).

Exercise 7-3: Create a Virtual Network

In this exercise, you will create a virtual network called rhnet with subnet 192.168.3.0/24 for use later by *outsider.example.net* and any other new or existing virtual machines.

1. Run the *virt-manager* command or click Applications → System Tools → Virtual Machine Manager to start the VM manager (if it is not already running).
2. Highlight "localhost (QEMU)", right-click, and select Details.
3. Click Virtual Networks to go to the physical connection details tab.
4. Click the + sign at the bottom of the screen to start the "Create a new virtual network" wizard.
5. Enter rhnet as the name of the virtual network and click Forward.

Figure 7-6 Add a Virtual Network – Assign a Name

6. Enter 192.168.3.0/24 as the IPv4 address space and click Forward.

Figure 7-7 Add a Virtual Network – Choose an Address Space

7. Unselect the "Enable DHCP" checkbox and click Forward.

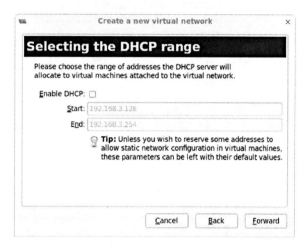

Figure 7-8 Add a Virtual Network – Disable DHCP

8. Select "Forwarding to physical network", destination "Any physical device", and mode "NAT". Click Forward.

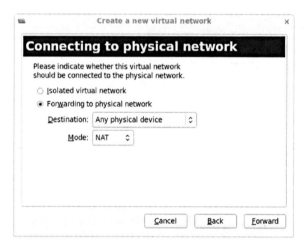

Figure 7-9 Add a Virtual Network – Connection to Physical Network

9. View the summary of the selections (Figure 7-10) and click Finish to create the virtual network.

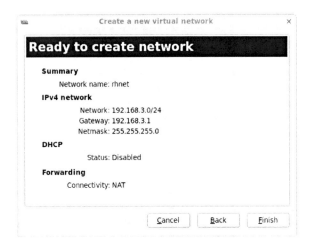

Figure 7-10 Add a Virtual Network – Summary of Selections

Exercise 7-4: Create a Storage Pool

In this exercise, you will create a storage pool called rhstr and add a new volume to it for use by new or existing virtual machines. You will use a file system directory as the pool type and */var/lib/libvirt/images/rhstr* as its location.

1. Run the *virt-manager* command or click Applications → System Tools → Virtual Machine Manager to start the VM manager (if it is not already running).
2. Highlight "localhost (QEMU)", right-click, and select Details.
3. Click Storage to go to the storage details tab. The information on this tab depicts that there is currently one "default" pool of size 48.2GB exist and is located in the */var/lib/libvirt/images* directory. There is 28.7GB of space in use, it is currently inactive, and has no volumes defined.

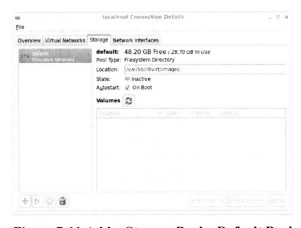

Figure 7-11 Add a Storage Pool – Default Pool

4. Click the + sign at the bottom of the screen to start the "Add Storage Pool" wizard.
5. Enter rhstr as the name of the storage pool and filesystem directory as the type. Click Forward to continue.

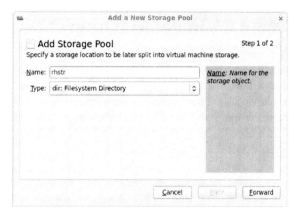

Figure 7-12 Add a Storage Pool – Assign a Name and Type

6. Specify */var/lib/libvirt/images/rhstr* as the target path and click Finish to complete adding a storage pool and go back to the storage tab.

Figure 7-13 Add a Storage Pool – Specify a Target Path

7. The storage tab now shows the rhstr pool in addition to the default pool. You should now be able to add volumes by clicking on New Volume and allocate them to virtual machines.

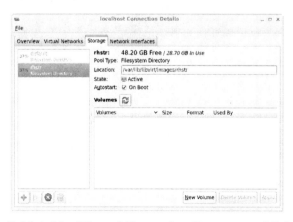

Figure 7-14 Add a Virtual Network – Summary of Selections

Exercise 7-5: Create a Virtual Machine and Install RHEL6

In this exercise, you will create and configure a virtual machine on the localhost hypervisor using the Virtual Machine Manager application. The name of the virtual machine will be *server.example.com* and it will use the HTTP server, configured earlier in this chapter, as the installation source. It will have 800MB of memory, one cpu, and one 20gb virtual hard disk for the OS. During the RHEL installation, you will provide hostname *server.example.com*, IP 192.168.2.201, netmask 255.255.255.0, gateway 192.168.2.1, DNS 192.168.2.1, and will use the default file system layout and software package groups.

Use an IP address on the 192.168.122 network if the above does not work. Review the virtual network configuration in the Virtual Machine Manager. Create a virtual network if you prefer to put this VM on a network of your choice.

1. Run the *virt-manager* command or click Applications → System Tools → Virtual Machine Manager to start the VM manager.
2. Right click on localhost and select Connect to connect to the localhost hypervisor.
3. Right click on localhost again and select New to create a virtual machine.
4. Enter *server.example.com* for the new virtual machine and select Network Install as the installation source. Click Forward to continue.

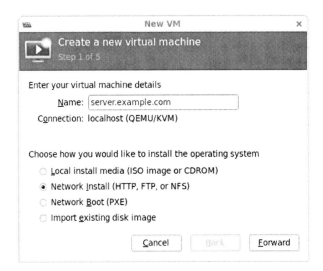

Figure 7-15 Create a Virtual Machine – Step 1 of 5

5. Specify the URL for the network install. Type http://192.168.2.200/rhel6. Uncheck the "Automatically detect operating system based on install media" and select OS type Linux and Version Red Hat Enterprise Linux 6. Click Forward to continue.

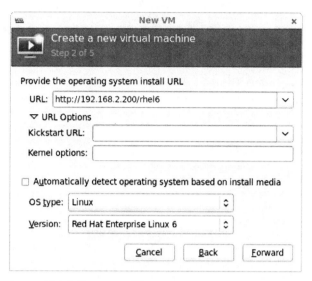

Figure 7-16 Create a Virtual Machine – Step 2 of 5

6. Enter 800MB for memory and 1 for cpu. Click Forward to continue.

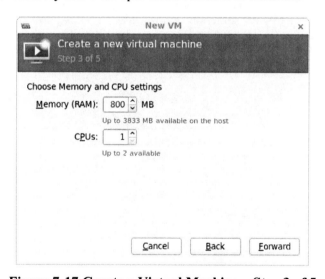

Figure 7-17 Create a Virtual Machine – Step 3 of 5

7. Check the "Enable storage for this virtual machine" and the "Allocate entire disk now" boxes. Select "Create a disk image on the computer's hard drive" and specify 20gb disk size. This will use the */var/lib/libvirt/images* directory to hold files for this virtual machine. Click Forward to continue.

Figure 7-18 Create a Virtual Machine – Step 4 of 5

The other option "Select managed or other existing storage" will work if you already have alternative storage pools defined.

8. Finally, a summary of your selections is displayed. You can still choose "Customize configuration before install" to make additional modifications prior to starting the actual installation of the operating system. Click Finish to begin RHEL installation.

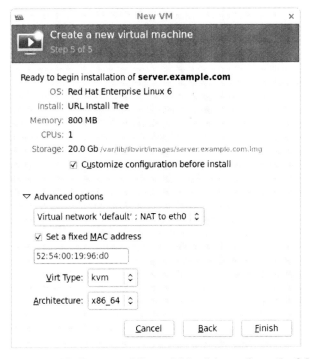

Figure 7-19 Create a Virtual Machine – Step 5 of 5

All configuration for the VM is stored in XML format in the */etc/libvirt/qemu* directory with file name matching the VM name. This file may be viewed for reference.

9. Supply information for the OS. Refer to Chapter 01 "Local Installation" for details. Install the default software packages, and also choose X Window and Desktop.
10. Reboot the VM after the installation has been completed and follow the firstboot process to complete the installation.

Exercise 7-6: Adding Virtual Disks to a Virtual Machine

In order to perform disk management exercises in Chapter 09 "Disk Partitioning" and in Chapter 10 "File Systems & Swap", you will need to allocate additional disks to *server.example.com*. As indicated in Chapter 01 "Local Installation", *server.example.com* will have four 4gb virtual disks available on it to facilitate the exercises.

In this exercise, you will allocate three 4gb virtual disks to *server.example.com* using the Virtual Machine Manager.

1. Run the *virt-manager* command or click Applications → System Tools → Virtual Machine Manager to start the VM manager (if it is not already running).
2. Highlight *server.example.com* under "localhost (QEMU)" and click Open.

Figure 7-20 Open Virtual Machine

3. Click View from the main menu and choose Details.

Figure 7-21 View Virtual Machine Details

---

4. Click Add Hardware at the bottom of the screen to open up the Add New Virtual Hardware dialog box.
5. Click Storage in the left window pane to bring up the Storage window.
6. Click "Create a disk image on the computer's hard drive" and specify 4GB. Select "Allocate entire disk now" and choose Virtio Disk as the Device type. Leave other selections to their default values. Click Finish to complete the disk allocation procedure.

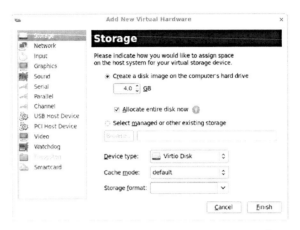

Figure 7-22 Add Storage to Virtual Machine

7. Repeat steps 4 to 6 and add the remaining virtual disks to *server.example.com*.

Installing RHEL via Kickstart

Kickstart allows you to perform a hands-off, fully-customized, and faster installation of RHEL. What you need is a single configuration file supplied to the installation program, which will use the information defined in the file and perform a custom installation on the system. This eliminates the lengthy question and answer session that you otherwise have to go through. The following are major advantages associated with using the kickstart method of installation:

✓ Fully automated, unattended installation.
✓ No configuration questions asked.
✓ Simultaneous installation on a number of systems.
✓ Identical configuration on a number of systems.
✓ Rapid deployment of any number of servers.

There are four major steps that need to be completed before you are able to perform a successful kickstart installation. These steps are:

1. Making the installation files available.
2. Creating a kickstart configuration file.
3. Preparing a bootable USB flash drive.
4. Beginning the installation process.

The following sub-sections explain each step.

Making the Installation Files Available

This step is to copy the installation files to a directory and make that directory available over the network via the FTP or HTTP protocol. Refer to the section earlier in this chapter on how to do this.

Creating a Kickstart Configuration File

Before you begin an automated kickstart installation, you will need to create a file and capture all the required configuration information in it. This information will be referenced and used by the kickstart process to install the system and configure it accordingly. You can create this file by either customizing the */root/anaconda-ks.cfg* file or running the Red Hat Kickstart Configurator tool *system-config-kickstart*. Both methods are explained below.

Using the /root/anaconda-ks.cfg File

During the RHEL installation in Chapter 01 "Local Installation" and earlier in this chapter, you entered several pieces of configuration information. The installation program stored that information in the */root/anaconda-ks.cfg* file. You can customize this file as per the requirements and use it for other client deployments.

The following displays the contents of the *anaconda-ks.cfg* file, which was generated by the anaconda program during the installation performed in Chapter 01 "Local Installation". I have made a copy of it and have modified it for use in our next installation on a system to be called *insider.example.com* with IP 192.168.2.202 as follows:

> # **cp /root/anaconda-ks.cfg /tmp/ks.cfg**
> # **nl /tmp/ks.cfg**
> 1 install
> 2 url --url-"http://192.168.2.200/rhel6"
> 3 lang en_US
> 4 keyboard us
> 5 network --onboot yes --device eth0 --bootproto static --ip 192.168.2.202 --netmask 255.255.255.0 --gateway 192.168.2.1 --nameserver 192.168.2.1 --hostname insider.example.com --noipv6

Use an IP address on the 192.168.122 network if the above does not work. Review the virtual network configuration in the Virtual Machine Manager. Create a virtual network if you prefer to put this VM on a network of your choice.

> 6 rootpw --iscrypted
> 6Bg5Nrmc9UKuXbsgL$kCvPTz.ECm37NKaVQhXYQpEbMmwX5zAg01fcUI55s9Xs0D725igYj3pPGF40LD4hYD
> spB6H.1ICbFe2qtcAsX.
> 7 firewall --enabled --ssh --ftp --http
> 8 auth --useshadow --passalgo-sha512
> 9 graphical
> 10 logging --level-info
> 11 selinux --enforcing
> 12 timezone --isUtc America/Toronto
> 13 firstboot --disable
> 14 bootloader --location-mbr --driveorder-vda --append-" rhgb crashkernel-auto quiet"

```
15  clearpart –all --drives=vda --initlabel
16  part /boot --asprimary --fstype=ext4 --onpart=vda1 --size=200
17  part pv.008002 --onpart=vda2
18  volgroup vg00 --pesize=16384 pv.008002
19  logvol / --fstype=ext4 --name=lvol1 --vgname=vg00 --size=4000
20  logvol swap --name=lvol2 --vgname=vg00 --size=1000
21  logvol /home --fstype=ext4 --name=lvol3 --vgname=vg00 --size=500
22  logvol /tmp --fstype=ext4 --name=lvol4 --vgname=vg00 --size=500
23  logvol /usr --fstype=ext4 --name=lvol5 --vgname=vg00 --size=5000
24  logvol /var --fstype=ext4 --name=lvol6 --vgname=vg00 --size=4000
25  %packages
26  @base
27  @client-mgmt-tools
28  @console-internet
29  @core
30  @debugging
31  @basic-desktop
32  @directory-client
33  @graphical-admin-tools
34  @hardware-monitoring
35  @java-platform
36  @large-systems
37  @legacy-unix
38  @network-file-system-client
39  @network-tools
40  @performance
41  @perl-runtime
42  @server-platform
43  @server-policy
44  @system-admin-tools
45  @x11
46  mtools
47  pax
48  python-dmidecode
49  oddjob
50  sgpio
51  certmonger
52  pam_krb5
53  krb5-workstation
54  tcp_wrappers
55  perl-DBD-SQLite
56  %post
57  chkconfig sendmail off
```

Let us analyze the file contents line by line so that you have a good understanding of each entry. The empty lines and the comments have been omitted.

Line #1: Instructs anaconda to begin a fresh installation process. The other option is upgrade, which upgrades an existing version of RHEL.

Line #2: Instructs anaconda to use this source for installation. Other options are cdrom, hard drive, and URL for ftp, and these may be specified as follows:

    a. cdrom
    b. harddrive --partition=/dev/sda3 --dir=/var/html/pub/rhel6
    c. url --url="ftp://192.168.2.200/pub/rhel6"

cdrom tells anaconda to install RHEL using the installation DVD. The second option will use the ISO image located on the same computer but in a different partition in the specified directory, and the last option will use a configured FTP server at 192.168.2.200.

Line #3: Instructs anaconda to use US English during installation.

Line #4: Instructs anaconda to use the US keyboard type.

Line #5: Instructs anaconda to configure the first network interface *eth0*, set it to autostart at system boot, assign it IP address 192.168.2.202, netmask 255.255.255.0, gateway 192.168.2.1, name server IP 192.168.2.1, hostname *insider.example.com*, and keep IPv6 support disabled. If you wish to obtain networking information from a configured DHCP server, use "network --device eth0 --bootproto dhcp" instead.

Line #6: Instructs anaconda to assign the specified password to the *root* user. You can copy and paste the *root* password from the */etc/shadow* file from an existing RHEL system.

Line #7: Instructs anaconda to enable the firewall and allow ssh, ftp, and http traffic to pass through it.

Line #8: Instructs anaconda to enable password shadowing for user authentication and encrypt all user passwords using sha512.

Line #9: Instructs anaconda to perform the installation in graphical mode. This is default.

Line #10: Instructs anaconda to log informational messages during the installation process. This is default.

Line #11: Instructs anaconda to activate SELinux in enforcing mode. This is default. Other options are disabled and permissive.

Line #12: Instructs anaconda to set the hardware clock to America/Toronto timezone using the offset from *Universal Time Coordinated* (UTC) (previously called GMT).

Line #13: Instructs anaconda not to run the firstboot program after the installation has been completed and the system has been rebooted. This is default.

Line #14: Instructs anaconda to install the GRUB bootloader program on the master boot record on the *vda* disk drive and append "rhgb crashkernel=auto quiet" to the boot string.

Line #15: Instructs anaconda to wipe off partition table information from the *vda* disk and initialize the disk.

Line #16: Instructs anaconda to create a primary partition */dev/vda1* of size 200MB, format it to the ext4 type file system, and mount it on the */boot* mount point.

Line #17: Instructs anaconda to create an LVM physical volume */dev/vda2* on the remaining portion of the disk.

Line #18: Instructs anaconda to create a volume group *vg00* with PE size 16MB on the physical volume created in the previous step.

Line #19: Instructs anaconda to create a logical volume *lvol1* of size 4000MB in the *vg00* volume group, format it to the ext4 type file system, and mount it on the / mount point.

Line #20: Instructs anaconda to create a logical volume *lvol2* of size 1000MB in the *vg00* volume group to be used for swap purposes.

Lines #21-24: Instructs anaconda to create logical volumes *lvol3* to *lvol6* of the specified sizes in the *vg00* volume group, format them to the ext4 type file system, and mount them on the specified mount points.

Lines #25 to #45: Instructs anaconda to install all the listed package groups as part of the installation process. %packages marks the start of the software package list.

Lines #46 to #55: Instructs anaconda to install all the listed packages as part of the installation process. Replace the lines 26 to 55 with the asterisk character if you wish to install all available packages and package groups.

Lines #56 to #57: Instructs anaconda to run the listed command after the installation has been completed. You can specify additional post-installation commands in this section. Also available is the %pre section which is not defined in the file here. Commands listed under the %pre section are executed before the *ks.cfg* file is parsed.

There are several other keywords and options available that you may want to use in the kickstart configuration file, but not all of them are mandatory. If you do not define a mandatory option, anaconda will prompt you to enter that piece of information during the installation process, which will defeat the purpose of this automated installation. Also make certain that the sequence of the sections in the file remains unchanged.

You can optionally specify any pre-installation and/or post-installation commands or scripts that you wish anaconda to execute.

When you are done with editing the file, execute the *ksvalidator* command to check for any syntax errors and typos.

> # **ksvalidator /tmp/ks.cfg**

Using the Kickstart Configurator

Execute the *system-config-kickstart* command in an X terminal or click Applications → System Tools → Kickstart to start the Kickstart Configurator program. If this tool is unavailable, install it using one of the package installation methods outlined in Chapter 06 "Package Management".

There are eleven configuration areas listed in the left window pane, each of which is briefly explained below:

Basic Configuration (Figure 7-23): allows you to enter basic information such as language, keyboard type, timezone, root password, whether to encrypt root password, whether to specify an installation key, target system architecture, plus three additional checkboxes at the bottom to enable or disable options – Reboot system after installation, Perform installation in text mode (graphical is default), and Perform installation in interactive mode.

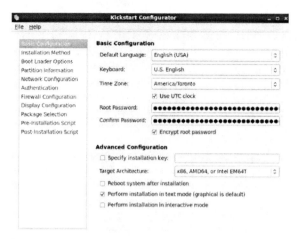

Figure 7-23 Kickstart Configurator – Basic Configuration

Installation Method (Figure 7-24): allows you to choose an installation method and an installation source.

Figure 7-24 Kickstart Configurator – Installation Method

BootLoader Options (Figure 7-25): allows you to choose bootloader-related settings. If you wish to specify a password for GRUB, you can do so here. Choose a location to place the bootloader, choices are MBR (preferred) and the */boot* partition. You can also specify parameters to pass to the default kernel.

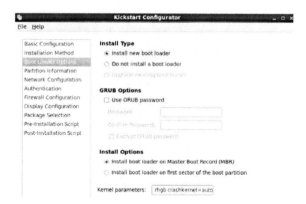

Figure 7-25 Kickstart Configurator – Bootloader Options

Partition Information (Figure 7-26): allows you to enter partition-related configuration information. "Clear Master Boot Record" removes all existing GRUB/bootloader information from the MBR (zerombr yes). The next set of three options enables you to remove all existing partitions including non-Linux partitions if there are any, remove all existing linux partitions only, and preserve existing partitions. "Initialize the disk label" removes existing disk partitioning label and updates it with the new partitioning scheme that you will choose here. Click Add at the bottom to define partition information such as partition type, file system type, size, mount point. Click RAID to create RAID partitions if required.

Figure 7-26 Kickstart Configurator – Partition Information

Network Configuration (Figure 7-27): allows you to configure one or more network interfaces. You can specify static IP information or let a DHCP server supply that information for you. If you are going to be performing a network boot / network install, you will need to enable BOOTP here.

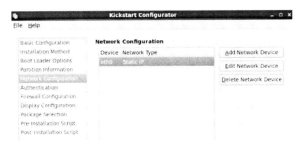

Figure 7-27 Kickstart Configurator – Network Configuration

Authentication (Figure 7-28): allows you to choose a user authentication type. Three choices – MD5, SHA512, and SHA256 – are available. You can also enable finger print reader support. In addition, you can enable NIS, LDAP, Kerberos 5, Hesiod, Samba, and Name Service Cache, and supply information as required for the authentication service that you wish to use.

Figure 7-28 Kickstart Configurator – Authentication

Firewall Configuration (Figure 7-29): allows you to activate or deactivate (disable or warn) SELinux, and enable or disable the firewall. If you enable the firewall, you will be able to select network services and specify ports to work through the firewall.

Figure 7-29 Kickstart Configurator – Firewall Configuration

Display Configuration (Figure 7-30): allows you to choose whether you want the graphical environment to be installed. Also, you can choose to have the Setup Agent started at the first system reboot after the installation has been completed. The other two options for the Setup Agent are to not run it and run it in reconfiguration mode. In the reconfiguration mode, the system will probe each hardware component during the boot process.

Figure 7-30 Kickstart Configurator – Display Configuration

Package Selection (Figure 7-31): allows you to select packages to be included in the installation.

Figure 7-31 Kickstart Configurator – Package Selection

Pre-Installation and Post-Installation Scripts: allow you to add pre- and/or post-installation scripts.

When you are done with the customization, go to File → Save File and specify a name such as *ks.cfg* and location such as */root* to save it.

Preparing a Bootable USB Flash Drive for Kickstart

Now that you have the kickstart configuration file */tmp/ks.cfg* ready, follow the procedure below to prepare a bootable USB flash drive for use with the kickstart installation.

I have purposely not described preparing a bootable CD media for kickstart as USB flash drives are more common and easy to handle than are the CDs/DVDs.

1. Download the *boot.iso* file from *rhn.redhat.com* from where you downloaded the RHEL DVD image, and place it in the */tmp* directory.
2. Plug in an empty USB flash drive and run the following command to copy the *boot.iso* file assuming that the partition for the USB drive is */dev/sdc1*. The following command will destroy the data on the USB drive.

   # dd if=/dev/zero of=/dev/sdc1 bs=1M count=100

3. Copy the *boot.iso* file to the USB drive:

   # dd if=/tmp/rhel-server-6.2-x86_64-boot.iso of=/dev/sdc

4. Copy the *ks.cfg* file to the USB drive:

   # dd oflag=append if=/tmp/ks.cfg of=/dev/sdc

If you do not wish to add the support for kickstart to a bootable USB flash drive, then skip step 4 from the above procedure. The first three steps are good to create a minimal bootable USB flash drive, which may be used to boot a system and then point it to a network server, a local hard drive, or the installation DVD for RHEL installation.

Beginning the Installation Process

Once a USB bootable media is ready and has a kickstart configuration file copied on it, go to the client system, plug in the bootable USB media in a port, start (or restart) the system, point the system to boot off the USB drive, press the ESC key when the "Welcome to Red Hat Enterprise Linux!" screen appears, and type the following at the boot: prompt:

ks=hd:sdc:/ks.cfg

The above procedure is also valid if you boot off the RHEL installation DVD.

If you have copied the *ks.cfg* file into the */var/www/html/rhel6* directory on the HTTP server at 192.168.2.200, type the following at the boot: prompt:

ks=http://192.168.2.200/rhel6/ks.cfg

If you have copied the *ks.cfg* file into the */var/ftp/pub/rhel6* directory on the FTP server at 192.168.2.200, type the following at the boot: prompt:

ks=ftp://192.168.2.200/rhel6/pub/ks.cfg

You can also use the Virtual Machine Manager to specify the kickstart file location without having to boot a virtual machine to the Linux boot: prompt. See URL Options in Figure 7-7.

Chapter Summary

In this chapter, you learned the concepts surrounding virtualization, KVM, hypervisor, virtual machines, and guests. You looked at the features and benefits associated with using virtualization.

You learned how to configure a RHEL system for providing network installation services using the FTP and HTTP protocols.

You reviewed the graphical virtual machine manager program and saw how to interact with it to perform various management functions pertaining to virtual machines. You created a virtual machine using the virtual machine manager program and installed RHEL in it with installation files located on a configured network installation server.

You studied the features and benefits associated with the kickstart method of hands-off installation. You customized a kickstart file for a test installation over the network using the configuration information provided in the file. You saw how to create a USB flash drive as a minimal boot media. You learned various ways of booting a client system and redirecting it to other sources for accessing the installation files.

Chapter Review Questions

1. What is the name of the package group that contains the qemu-kvm package?
2. What would the command "dd if=/dev/zero of=/dev/sdc1" do?
3. By default, the firstboot program is set to automatically run after an installation. True or False?
4. What three protocols are supported for configuring over the network installation servers?

---

5. Which directory is automatically shared when the default HTTP server is configured?
6. Which command can be used to start the virtual machine manager program?
7. What would the command "ks=http://192.168.2.200/rhel6/ks.cfg" do?
8. What is the default location where the virtual machine files are stored?
9. */boot* must be defined as an LVM logical volume. True or False?
10. Kickstart supports fully unattended installations. True or False?
11. What is the name of the file that is created during the RHEL installation and it captures all configuration information entered?
12. What is the name of the program that allows you to generate a kickstart file?
13. KVM is the default virtualization hypervisor software in RHEL6. True or False?
14. Which directory is automatically shared when the default FTP server is configured?
15. Which daemon must be running in order to connect to the hypervisor?
16. What argument would you supply with bootproto if you would like to obtain IP information from an available DHCP server?
17. Which package group includes the graphical virtual machine manager program?
18. By default, SELinux is set to enforcing during installation. True or False?
19. What is the name of the Kickstart Configurator program?
20. What would happen if you set the onboot option to no in the *ks.cfg* file?

Answers to Chapter Review Questions

1. The Virtualization package group contains the qemu-kvm package.
2. The *dd* command provided will wipe out information from the first sectors of the first partiton on the *sdc* drive.
3. False.
4. The three protocols supported for accessing over the network installation server are FTP, HTTP, and NFS.
5. The */var/www/html* directory.
6. The *virt-manager* command.
7. The *ks* directive provided will boot the system and use the kickstart configuration file located on the HTTP server in the */var/www/html/rhel6* directory.
8. The */var/lib/libvirt/images* directory.
9. False. */boot* cannot be part of LVM.
10. True.
11. The *anaconda-ks.cfg* file.
12. The name of the program is Kickstart Configurator.
13. True.
14. The */var/ftp/pub* directory.
15. The *libvirtd* daemon.
16. The dhcp argument.
17. The Virtualization Client package group contains the graphical virtual machine manager program.
18. True.
19. The *system-config-kickstart* command.
20. The network interface will not get activated automatically at system reboots.

Labs

Lab 7-1: Create a Virtual Network

Set up a virtual network called rhcsanet. Enable DHCP to provide IP addresses from 192.168.100.51 to 192.168.100.100 on subnet 255.255.255.0. Ensure that this virtual network use NATting on all physical interfaces.

Lab 7-2: Create a Virtual Storage Pool

Set up a new storage pool called rhcsastr. Use the directory filesystem as the storage type and it should be located in the */var/lib/libvirt/vstorage* directory.

Lab 7-3: Perform a Network Installation of RHEL6

Create a virtual machine and install RHEL6 in it. Use IP 192.168.3.200, netmask 255.255.255.0, gateway 192.168.3.1, nameserver 192.168.3.1, and hostname *outsider.example.net* (assign IP on the 192.168.122 subnet for installation and later change it). Use standard partitioning to create */boot* 200MB, lvol1 for / 4GB, lvol2 for swap 1GB, lvol3 for */usr* 4GB, lvol4 for */var* 2GB, lvol5 for */opt* 1GB, lvol6 for */tmp* 500MB, and lvol7 for */home* 500MB in that order. Select packages to support X Window, GNOME desktop, KDE desktop, and graphical administration tools. Create a local user account called *useroutsider* with all the defaults.

Lab 7-4: Perform a Kickstart Installation of RHEL6

Create a virtual machine. Install RHEL6 in it using kickstart as explained in this chapter. Create and use any boot media. Use IP 192.168.2.210 and hostname *testbox.example.com* (assign IP on the 192.168.122 subnet for installation and later change it). Use partitioning at will. Select necessary packages to support X Window and a graphical desktop.

Lab 7-5: Add Virtual Disks to Virtual Machines

Set up a new storage volume and allocate one 2GB disk to *insider.example.com* and one 2GB disk to testbox.example.com from this volume. Use storage settings as you wish.

Chapter 08

Linux Boot Process & Kernel Management

This chapter covers the following major topics:

- ✓ Run control levels and the system boot process
- ✓ Boot into specific run levels and interact with GRUB
- ✓ Kernel initialization and the Upstart process
- ✓ Understand sequencer, configuration, and initialization directories
- ✓ Overview of virtual console screens
- ✓ Understand and interpret system log files
- ✓ Manage services
- ✓ Linux rescue mode and how to use it
- ✓ Install a lost or corrupted bootloader and a file
- ✓ Linux kernel, versioning, and directory structure
- ✓ Manage kernel modules, and install and upgrade the kernel

This chapter includes the following RHCSA objectives:

12. Boot, reboot, and shut down a system normally
13. Boot systems into different runlevels manually
14. Use single-user mode to gain access to a system
16. Locate and interpret system log files
35. Configure systems to boot into a specific runlevel automatically
19. Start, stop, and check the status of network services
40. Configure network services to start automatically at boot
44. Update the kernel package appropriately to ensure a bootable system
45. Modify the system bootloader

$RHEL$ goes through multiple phases during startup. It starts selective services during its transition from one phase into another, and provides the administrator with an opportunity to interact with the bootloader. RHEL starts a number of services through configured scripts during a system boot.

A good understanding of the boot and shutdown processes makes it easy for the administrator to troubleshoot and fix boot and shutdown related issues such as unbootable systems, forgotten root password, lost critical system files, and a corrupted bootloader. You may have to boot the system in the rescue mode to fix some of these problems.

There are a number of log files on the RHEL system. Many of these files log information for specific services; however, the system log file captures general and specific log data for almost all services and applications that run on the system.

The kernel controls everything on the system that runs Linux. It controls the system hardware including memory, processors, disks, and I/O devices; it runs, schedules, and manages processes and service daemons; it enforces security and access controls; and so on. The kernel receives instructions from the shell, engages appropriate hardware resources, and acts as instructed. The kernel is comprised of several modules, and each module brings a unique functionality to the kernel. A new Linux kernel must be installed or an existing kernel must be upgraded when a need arises from an application or functionality standpoint.

Run Control Levels

RHEL changes run levels when a system shutdown occurs. Similarly, run levels are changed when a system boot occurs. The following sub-sections discuss system run control levels and how to manipulate them.

What are Run Control Levels?

System *run control* (rc) levels are pre-defined and determine the current state of the system. RHEL supports seven rc levels of which six are currently supported. Not all of them are commonly used though. The default rc level is 5 if X Window and desktop software are installed. Table 8-1 describes various run levels.

Run Level	Description
0	RHEL is down and the system is halted.
1	Single user state with all file systems mounted and SELinux activated. Scripts located in the */etc/rc.d/rc1.d* directory are executed. This run level is used for critical system administration tasks that cannot be performed in other run states.
2	Multi-user state. Most system and network services, except for NFS server and X window system, are running.
3	Multi-user state. All system and network services including the NFS server running.
4	Not implemented.
5	Fully operational multi-user state with X window and GUI desktop running. This is the default run level.
6	RHEL reboots.

Table 8-1 System Run Levels

Table 8-1 tells that run levels 0 and 6 are used for shutting down and rebooting the system. Run level 1 brings the system up into the single user mode where only the system administrator can log in and perform administrative tasks that require all or most system services in the stopped state. Network connectivity does not start at this run level. Run levels 2 and 3 provide multi-user accessibility into the system. The difference between the two lies in the number of network services started up. Run level 3 has all networking capabilities up, in contrast with run level 2 which has fewer. If X Window software is not installed, run level 3 becomes the default. Run level 4 is not implemented; it may be defined and used in a future release of the operating system. Run level 5 is the default run level at which RHEL is fully functional with all services including networking and X/GUI fully operational.

Checking Current and Previous Run Levels

To check the current and the previous run levels of the system, use the *who* command with the –r option:

```
# who –r
        run-level 5  2012-08-15 14:43              last=3
```

The output indicates that the system is currently running at run level 5 and its last run level was 3. The output also displays the time stamp of when this run level change occurred.

You may also use the *runlevel* command which displays the previous and current run levels:

```
# runlevel
3 5
```

Changing Run Levels

Run levels are also referred to as the *init* levels because the system uses the *init* command to modify the levels. The *shutdown* command may also be issued instead for changing run levels and shut down the system. Additionally, the *halt*, *reboot*, and the *poweroff* commands are available to stop the system.

The init or the telinit Command

The *init* or the *telinit* command are used to change the run levels. If the system is currently in run level 5 and you need to switch into run level 3, issue either of the following:

```
# init 3
# telinit 3
```

This command gracefully stops all services and daemons that should not be running at level 3. It does not affect any other running processes and services.

Similarly, by initiating *init 1* from run level 3, most system services and daemons are stopped for the system to transition into the single user state.

To stop all system services gracefully and bring the system to the halt state, pass 0 as an argument to the *init* command:

---

init 0

Likewise, to stop all system services gracefully, shut down the system, and reboot it to the default run level, supply 6 as an argument to the *init* command:

init 6

If the system is running at run level 1 and you wish to bring it up to 5, issue the following:

init 5

The shutdown Command

The *shutdown* command is preferred over the *init* command by many administrators. This command stops all services, processes, and daemons in a sequential and consistent fashion as does the *init* command. It broadcasts a message to all logged in users and waits for one minute, by default, for users to save their work and log off, after which time it begins stopping services, processes, and daemons. It will unmount file systems and proceeds as per the options specified at the command line.

The following examples show options and arguments that may be supplied with *shutdown*:

# **shutdown –r 300**	(broadcasts a message, waits for 5 minutes, then stops all services gracefully, shuts the system down, and reboots it to the default run level).
# **shutdown –r 0**	(broadcasts a message, begins stopping services immediately and gracefully, shuts the system down, and reboots it to the default run level).
# **shutdown –r now**	(same as "shutdown –r 0").
# **shutdown –h 20**	(broadcasts a message, waits for 20 minutes, and then brings the system down gracefully to the halt or power off state).
# **shutdown –H 20**	(broadcasts a message, waits for 20 minutes, and then brings the system down gracefully to the halt state).
# **shutdown –P 20**	(broadcasts a message, waits for 20 minutes, and then brings the system down gracefully to the power off state).
# **shutdown –k**	(broadcasts a message, disables user logins, but does not actually bring the system down).

The *shutdown* command actually calls the *init* command behind the scenes to perform run level changes. You may use the *init* command instead. The only two features not available with *init* compared to *shutdown* are that *init* does not broadcast a message and does not wait for a period of time. It starts the run level change process instantly.

The halt, reboot, and poweroff Commands

The *halt*, *reboot*, and the *poweroff* commands without any options perform the same action that the *shutdown* command would perform with "–H now", "–r now", and "–P now", respectively. You may specify the –f option with any of these commands to halt, reboot, or poweroff the system immediately. This option calls the *kill* command with signal 9 and terminates all running processes right away forcing the system to go down quickly; however, it introduces the risk of damaging application files and file system structures. It is not recommended to use this option from any multi-user run level. Two of the three commands – *halt* and *reboot* – may be executed with the –p option to power off the system as well. The –v option is also available with the three commands for verbose messages.

Linux Boot Process

RHEL goes through the *boot* process when the system is powered up or reset, with the boot process lasting until a login prompt appears. A step-by-step boot process is presented concisely as well as in detail in this section.

In A Nutshell

A summary of the boot process from powering up or resetting a system to when you see a login prompt is provided below:

- ✓ Power on external devices.
- ✓ Power on the system (or start the virtual machine).
- ✓ The UEFI (*Unified Extensible Firmware Interface*) or the BIOS (*Basic Input/Output System*) kicks in and runs the *Power On Self Test* (POST) on the system hardware components such as processor, memory, and I/O, and initializes them. See the next sub-section for details on BIOS and UEFI.
- ✓ The UEFI/BIOS locates the boot device and loads the initial bootloader program called GRUB (*Grand Unified Bootloader*) into memory. GRUB is the default and the only bootloader program in RHEL6. It typically resides on the *Master Boot Record* (MBR), which is located on the very first sector of the boot device, or in the */dev/sda1* (*/boot*) partition (MBR is a region on the disk where partition information is stored). GRUB code located in the MBR is referred to as *stage 1* on systems using BIOS. On the UEFI-based systems, this code is stored in the EFI system partition.
- ✓ GRUB stage 1 calls GRUB *stage 1.5* on BIOS-based systems, and consults the */boot/grub/grub.conf* file for boot directives such as available kernels, their locations, associated ramdisk images, the root file system locations, and the default kernel to boot. On the UEFI-based systems, the */boot/efi/EFI/redhat/grub.conf* file is consulted for the same directives.
- ✓ GRUB stage 1 or stage 1.5 calls and loads GRUB *stage 2*, which presents a list of available bootable operating systems in a graphical screen. Several commands are available at this point that may be executed from the grub> shell prompt for performing tasks such as choosing a non-default boot device and booting the system interactively.
- ✓ GRUB stage 2 calls the default kernel (kernel file name begins with *vmlinuz-* followed by the version) located in the */boot* file system, loads it into memory, and decompresses and initializes it. It also loads required drivers and modules by copying one or more appropriate initramfs images into memory, and the contents of the root directory into the ramdisk.
- ✓ The kernel initializes the memory and configures hardware components as it finds, locates and decompresses the initramfs images(s) into */sysroot*, and loads necessary drivers. It then initializes any configured software RAID and LVM virtual devices, creates a root device, and mounts the root file system in read-only mode on */*.
- ✓ The kernel calls the *init* command from the ramdisk and transfers the control over to it to initiate the system initialization process. *init* is the Upstart process management daemon and is also referred to as the *first process* as it always possesses PID 1. In other words, *init* is the parent of all the processes that are started on the system.

- ✓ *init* calls the */etc/rc.d/rc.sysinit* script to perform several configuration tasks.
- ✓ *init* then looks in the */etc/event.d* directory to determine how to establish the system in each run level.
- ✓ *init* reads the source function library located in the */etc/rc.d/init.d/functions* file to set up how to start, kill, and determine the PID of a program.
- ✓ *init* references the *inittab* file in */etc* to determine the default run level to boot into.
- ✓ *init* calls the */etc/rc.d/rc* script to execute all startup scripts needed to bring the system to the default run level.
- ✓ *init* finally runs the */etc/rc.d/rc.local* script and spawns getty messages that display the login prompt.

UEFI / BIOS Initialization Phase

The BIOS is a small memory chip in the computer that stores system date and time, list and sequence of boot devices, I/O configuration, etc. This information may be customized as desired.

Depending on the computer manufacturer, you need to press a key to enter the BIOS setup. Pressing the Esc key displays a boot menu where you can choose a source to boot the system off. The computer goes through the hardware initialization phase that involves detecting and diagnosing peripheral devices. It runs the POST on the devices as it finds them, installs drivers for the graphics card and the attached monitor, and begins displaying system messages on the video hardware. It discovers a usable boot device, loads the GRUB bootloader program into memory, and passes the control over to it. Boot devices on newer computers support booting from optical devices, USB flash and external disks, network, and other media.

The UEFI is a new architecture-independent standard that computer manufacturers have widely adopted in their new hardware offerings as a pre-boot environment to an operating system such as RHEL. This mechanism provides enhanced boot and runtime services, and superior features such as speed, over the legacy BIOS. It has its own device drivers, is able to mount and read extended type file systems, includes UEFI-compliant application tools, and may contain one or more bootloader programs. It comes with a boot manager that allows you to choose a different source for booting the system. Most computer manufacturers have customized the features for their own platform. You may find varying menu interfaces among other differences.

The GRUB Phase

The GRUB bootloader is stored in the MBR or the */boot* file system in BIOS-based systems, and in the EFI partition in UEFI-based systems. It only supports the ext2 type file system, as well as ext3 and ext4 but without journaling. You can interact with GRUB to perform any non-default boot-related tasks such as booting into one of the run levels described in Table 8-1 and booting with a non-default device such as an optical or a USB media. You need to press a key before the timeout expires to interrupt the autoboot process and interact with GRUB. If you wish to boot the system using the default boot device with all the configured default settings, do not press any key, as shown below, and let the system go through the autoboot process.

Booting Red Hat Enterprise Linux (2.6.32-220.el6.x86_64) in 5 seconds...

If you choose to interrupt the autoboot process, you will get to the GRUB menu where you can perform a number of tasks such as searching for alternate boot devices, performing a manual boot from a non-default boot device or kernel, viewing or altering boot configuration, and booting into a non-default run level. Figure 8-1 displays the main GRUB menu.

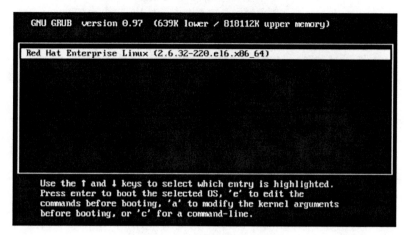

Figure 8-1 GRUB Menu

Three commands are available in the main menu along with the Enter key, which boots the highlighted operating system. The commands are listed and explained in Table 8-2.

Command	Description
e	Allows you to enter the edit menu.
a	Allows you to modify the kernel string.
c	Takes you to the grub> command prompt.

Table 8-2 GRUB Main Menu Commands

Figure 8-2 shows a picture of the boot directives displayed when *e* is typed in the GRUB main menu. If a password is set to enter GRUB menu, then press *p* and enter the password.

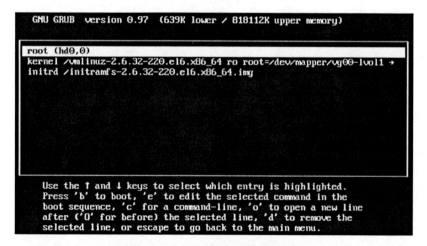

Figure 8-2 GRUB Edit Menu

Several commands are available in the GRUB edit menu and are listed and described in Table 8-3.

Command	Description
b	Boots the operating system.
e	Allows you to edit the highlighted directive.
c	Takes you to the grub> command prompt.
o	Opens a new line after the highlighted entry.
O	Opens a new line before the highlighted entry.
d	Deletes the highlighted line entry.
Esc	Takes you back to the main menu.

Table 8-3 GRUB Edit Menu Commands

Booting into Specific Run Levels

RHEL can be booted into one of several run levels, as identified in Table 8-1, from the GRUB menu. Choosing to boot into a non-default run level may be achieved either by modifying the initdefault directive in the */etc/inittab* file or entering an appropriate level at the bootloader prompt. Usually, booting into a non-default run level is done to perform administrative tasks that cannot be done properly in a multi-user system state. Such tasks may include backing up / and */usr* file systems, and troubleshooting a specific problem. For example, if the default run level is set to 5 but the system cannot start the X window system, you might want to boot it into run level 3 in order to fix the X window startup issue. You may want to boot the system into run level 1 (single user mode) to fix any issues such as changing a forgotten *root* password.

The following text explains how to boot the system into a non-default run level:

Booting into Single User Mode

Follow the instructions below to boot into this mode:

1. Press any key before the countdown ends to interact with GRUB.
2. Press *a* to go to the append mode.
3. Press Spacebar and type *1, s,* or *single* and then press Enter to boot.
4. Perform any required maintenance tasks.
5. Execute the *exit* command when done to boot into the default run level. Alternatively, issue the *reboot* or the *shutdown* command to restart the system into the default run level.

Booting into Run Level 2 or 3

Booting the system into run level 2 or 3 starts most system services. Follow the instructions below:

1. Press any key before the countdown ends to interact with GRUB.
2. Press *a* to go to the append mode.
3. Press Spacebar and enter the desired run level and then press Enter to boot.

Exercise 8-1: Reset Lost root User Password

In this exercise, you will boot the system into single user mode and reset the password for the *root* user (assuming that the *root* password has been lost and no *root* user login session is currently running). After the password reset, reboot the server to the default run level and validate the password.

1. Reboot the system if it is already running:

    # **reboot**

2. Press a key to interact with GRUB.
4. Press *a* to go to the append mode.
5. Press Spacebar and type *s* and then press Enter.
6. At the command prompt, run the following and change the *root* password:

    # **passwd root**

7. Execute the *exit* command to boot into the default run level.
8. When the system is up, try logging in with the new *root* password to validate the change.

Understanding GRUB Configuration File

It is important that you comprehend the contents of the GRUB configuration file *grub.conf* located in the */boot/grub* directory on BIOS-based systems and in the */boot/efi/EFI/redhat* directory on UEFI-based systems. This file has symlinks from */etc/grub.conf* and */boot/grub/menu.lst* files. Here is an excerpt from the default file:

nl /boot/grub/grub.conf
. . . .
```
    10  default=0
    11  timeout=5
    12  splashimage=(hd0,0)/grub/splash.xpm.gz
    13  hiddenmenu
    14  title Red Hat Enterprise Linux (2.6.32-220.el6.x86_64)
    15      root (hd0,0)
    16      kernel /vmlinuz-2.6.32-220.el6.x86_64 ro root=/dev/mapper/vg00-lvol1 rd_NO_LUKS rd_NO_MD
rd_NO_DM rd_LVM_LV=vg00/lvol1 rd_LVM_LV=vg00/lvol2 LANG=en_US.UTF-8 SYSFONT=latarcyrheb-sun16
KEYBOARDTYPE=pc KEYTABLE=us rhgb  quiet
    17      initrd /initramfs-2.6.32-220.el6.x86_64.img
```

The line contents from the file are explained below:

✓ Line #1 to #9: Comments.
✓ Line #10: Sets the default kernel, as defined with the corresponding "title" (found on line #14), to boot. If the value is set to 1, it would point to the next kernel associated with "title", and so on for subsequent values.
✓ Line #11: Gives you 5 seconds to interrupt the autoboot process to interact with GRUB.
✓ Line #12: Sets the location of the graphical GRUB screen. The value indicates the first partition (*/boot*) on the first boot disk located at */boot/grub/splash.xpm.gz*.
✓ Line #13: Instructs the system to hide the GRUB menu until a key is pressed to interrupt the autoboot process.
✓ Line #14: Specifies the name or short description of an available kernel.
✓ Line #15: Sub-entry that points to the location of the boot partition as defined in the */boot/grub/device.map* file.

```
# cat /boot/grub/device.map
(hd0)    /dev/sda
```

✓ Line #16: Sub-entry that provides the kernel directive and the arguments to be supplied with it during the boot process. It includes the kernel file name and its location with respect to *boot*. It indicates to load the kernel read-only from the root logical volume */dev/mapper/vg00-lvol1* (it may instead have a 128-bit UUID number associated with the root file system). It states not to detect any LUKS-encrypted file systems or software RAID partitions. It provides names for the logical volumes associated with the root file system and the swap. It specifies the language, font, keyboard type, and the keyboard mapping to be used. The last two entries instruct to boot RHEL in Red Hat Graphical Boot mode with all non-critical boot messages hidden.

✓ Line #17: Sub-entry that initializes the RAM disk so the kernel starts up and mounts an initial root file system from */boot/initramfs-2.6.32-220.el6.x86_64.img*. This file is created when the kernel is built. It may also be recreated if needed using the *mkinitrd* command.

Managing GRUB

There are a number of commands accessible at the grub> prompt for GRUB and other pre-boot management tasks. To access the grub> command prompt, reboot your RHEL system and stop the autoboot process. Enter *c* to go to the grub> prompt. Press the TAB key at the prompt to see the list of available commands (Figure 8-3). Usage of some of these commands is demonstrated in this sub-section.

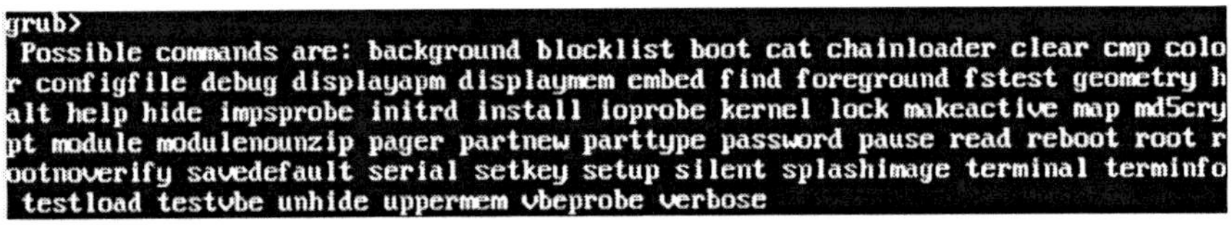

Figure 8-3 GRUB Management Commands

To find which disk and partition contains the *grub.conf* file in the */boot/grub* directory:

```
grub> find /grub/grub.conf
(hd0,0)
```

To view the contents of the *grub.conf* file:

```
grub> cat /grub/grub.conf
```

To obtain information about the root partition:

```
grub> root
(hd0,0): Filesystem type is ext2fs, partition type 0x83
```

To load the root partition that is resident in the first partition on the first hard drive:

```
grub> root (hd0,0)
Filesystem type is ext2fs, partition type 0x83
```

To list files in the */boot* file system, type the following and then press the TAB key:

grub> **kernel /**
  Possible files are: lost+found grub efi .vmlinuz-2.6.32-220.el6.x86_64.hmac Sys
  Tem.map-2.6.32-220.el6.x86_64 config-2.6.32-220.el6.x86_64 symvers-2.6.32-220.el
  6.x86_64.gz vmlinuz-2.6.32-220.el6.x86_64 initramfs-2.6.32-220.el6.x86_64.img

Kernel Initialization Phase

GRUB initializes the kernel and loads it into memory. Several messages appear on the console during the kernel initialization depending on the hardware and system configuration. These messages are logged to the */var/log/dmesg* file and may be viewed later with a *more* on the file or the *dmesg* (diagnostic messages) command. The messages include information about kernel version, memory, CPU, console, SELinux status, disks, root logical volume or partition, network interfaces, swap logical volume/partition, as well as any unusual events that occur in the system and may prove helpful in troubleshooting.

An excerpt from the *dmesg* file is shown below:

cat /var/log/dmesg
Initializing cgroup subsys cpuset
Initializing cgroup subsys cpu
Linux version 2.6.32-220.el6.x86_64 (mockbuild@x86-004.build.bos.redhat.com) (gc
c version 4.4.5 20110214 (Red Hat 4.4.5-6) (GCC)) #1 SMP Wed Nov 9 08:03:13 EST 2011
Command line: ro root=/dev/mapper/vg00-lvol1 rd_NO_LUKS LANG=en_US.UTF-8 rd_LVM_
LV=vg00/lvol2 rd_NO_MD rd_LVM_LV=vg00/lvol1 SYSFONT=latarcyrheb-sun16 rhgb crash
kernel=auto quiet KEYBOARDTYPE=pc KEYTABLE=us rd_NO_DM
KERNEL supported cpus:
  Intel GenuineIntel
  AMD AuthenticAMD
  Centaur CentaurHauls
BIOS-provided server RAM map:
  BIOS-e820: 0000000000000000 – 000000000009fc00 (usable)
. . . .

The Linux kernel is modular. During loading in memory, it loads the device drivers for the installed hardware and supporting software modules for components such as the ext4 filesystem type. The *lsmod* command can be used to list the loaded modules:

lsmod
Module	Size	Used by
autofs4	26888	3
sunrpc	243758	1
ipt_REJECT	2383	2
. . . .

The First Process Phase (Upstart)

The kernel calls the */sbin/init* command and transfers the control over to it to initiate the system initialization process, known as the *first process*. *init* calls the */etc/rc.d/rc.sysinit* script to perform configuration tasks such as initializing the hardware clock, loading kernel modules, configuring kernel parameters, setting the keyboard mapping, setting the hostname, remounting the root file system in read/write mode, checking and mounting local file systems, starting local swap spaces, and activating SELinux. It then looks in the */etc/event.d* directory to determine how to establish the system in each run level. It processes jobs stored in this directory whenever a particular event, such as start of the system, occurs. *init* reads the source function library located in the */etc/rc.d/init.d/functions* file to set up how to start, kill, and determine the PID of a program. It looks in the */etc/inittab* file to determine the default run level for the system to boot into. The only uncommented line in this file indicates that the default run level is set to 5:

> # **cat /etc/inittab**
>
> . . . .
>
> id:5:initdefault:

Sequencer, Configuration, and Initialization Directories

init then calls the */etc/rc.d/rc* script to execute all startup scripts needed to bring the system to the default run level. This script locates the services for startup in the sequencer directories */etc/rc.d/rc#.d*, gets configuration information from the startup configuration files located in the */etc/sysconfig* directory, and starts them up from the initialization directory */etc/rc.d/init.d*. It finally runs the */etc/rc.d/rc.local* script and spawns getty messages that display the login prompt. For run level 5, the */etc/X11/prefdm* script is called to bring up the default display manager as defined in the */etc/sysconfig/desktop* file.

Here is the list of files in the */etc/rc.d* directory:

> # **ll /etc/rc.d**
> total 60
> drwxr-xr-x. 2 root root 4096 Aug 18 09:19 init.d
> -rwxr-xr-x. 1 root root 2617 Oct 7 2011 rc
> drwxr-xr-x. 2 root root 4096 Aug 18 13:07 rc0.d
> drwxr-xr-x. 2 root root 4096 Aug 18 13:07 rc1.d
> drwxr-xr-x. 2 root root 4096 Aug 18 13:07 rc2.d
> drwxr-xr-x. 2 root root 4096 Aug 18 13:07 rc3.d
> drwxr-xr-x. 2 root root 4096 Aug 18 13:07 rc4.d
> drwxr-xr-x. 2 root root 4096 Aug 18 18:52 rc5.d
> drwxr-xr-x. 2 root root 4096 Aug 18 13:07 rc6.d
> -rwxr-xr-x. 1 root root   220 Oct 7 2011 rc.local
> -rwxr-xr-x. 1 root root 19029 Oct 7 2011 rc.sysinit

The following shows contents of the *rc5.d* sequencer directory. Notice that all scripts are mere symlinks to the actual start/stop scripts located in the initialization directory.

```
# ll /etc/rc.d/rc5.d
lrwxrwxrwx. 1 root root 20 Aug 18 09:14 K01certmonger -> ../init.d/certmonger
lrwxrwxrwx. 1 root root 23 Aug 18 09:14 K01matahari-host -> ../init.d/matahari-host
lrwxrwxrwx. 1 root root 26 Aug 18 09:14 K01matahari-network -> ../init.d/matahari-network
. . . . . . . .
```

There are six sequencer directories – *rc0.d, rc1.d, rc2.d, rc3.d, rc4.d, rc5.d*, and *rc6.d* –
corresponding to the six run levels, and contain two types of scripts: *start* and *kill*. The names of the
start scripts begin with an uppercase S and that for the kill scripts with an uppercase K. These
scripts are symbolically linked to the actual start/kill scripts located in the */etc/rc.d/init.d* directory.
Each start/kill script contains start, stop, restart, reload, force-reload, condrestart, try-restart, and
status functions corresponding to service start, stop, stop and start, configuration file re-read, force
configuration file re-read, restart only if it is already running, retry restart, and status check.

When the system comes up, the S scripts are executed one after the other in the ascending
numerical sequence from the default run level directory. Similarly, when the system goes down, the
K scripts are executed one after the other in the descending numerical sequence from the default run
level directory.

The following list from the */etc/rc.d/init.d* directory shows the actual initialization scripts:

```
# ll /etc/rc.d/init.d
. . . . . . . .
-rwxr-xr-x. 1 root root 1866 Nov 1 2011 wpa_supplicant
-rwxr-xr-x. 1 root root 3555 Sep 21 2011 xinetd
-rwxr-xr-x. 1 root root 4799 Jun 24 2010 ypbind
```

The following lists configuration files for the start/kill scripts placed in the */etc/sysconfig*
configuration directory:

```
# ll /etc/sysconfig
-rw-r--r--. 1 root root 183 Sep 23 2011 udev
-rw-r--r--. 1 root root 644 Nov 1 2011 wpa_supplicant
-rw-------. 1 root root 376 Sep 21 2011 xinetd
. . . . . . . .
```

Understanding Virtual Console Screens

As explained in Chapter 01 "Local Installation", there are six virtual console screens available
during the graphical mode RHEL installation and you can access them by pressing a combination of
keys. The same number of virtual screens are available once RHEL has been installed and is up and
running. Virtual consoles are set up and activated by the Upstart process by referencing the
/etc/sysconfig/init and */etc/init/start-ttys.conf* files.

The System Log Files

System logging is performed to keep track of messages generated by the kernel, daemons,
commands, user activities, and so on. The daemon *rsyslogd* is responsible for capturing kernel
messages and events, as well as for all other activities. This daemon is multi-threaded and supports

enhanced filtering, encryption protected relaying of messages, and various configuration options. It is started when the system enters run level 2 and runs the *etc/rc.d/rc2.d/S12rsyslog* script and terminated when the system changes run level to 1, 0, or 6, or shuts down, and calls the *etc/rc.d/rc#.d/K88rsyslog* script. These scripts are symbolically linked to the *etc/rc.d/init.d/rsyslog* file, which includes the start and stop functions corresponding with the service startup and termination. The *rsyslogd* daemon reads its configuration file *etc/rsyslog.conf* at startup. The default port used by the *rsyslogd* daemon is 514 and it may be configured to use either the UDP or the TCP protocol.

The *rsyslog* service is modular. This implies that the modules listed in its configuration file are dynamically loaded in the kernel as and when needed. Each module brings a new functionality to the system when loaded.

The *rsyslogd* daemon can be started or stopped manually. Here is how:

```
# service rsyslog start
Starting system logger:                  [ OK ]
# service rsyslog stop
Shutting down system logger:             [ OK ]
```

A PID is assigned to the daemon when it is started and is stored in the */var/run/syslogd.pid* file.

The System Log Configuration File

The system log configuration file is located in the */etc* directory and is called *rsyslog.conf*. The uncommented line entries from the file are shown below and explained subsequently. The section headings have been added to separate the directives in each section.

```
# grep –v ^# /etc/rsyslog.conf
#### MODULES ####
$ModLoad imuxsock.so  # provides support for local system logging (e.g. via logger command)
$ModLoad imklog.so    # provides support for kernel logging (previously done by rklogd)
#### GLOBAL DIRECTIVES ####
$ActionFileDefaultTemplate RSYSLOG_TraditionalFileFormat
#### RULES ####
*.info;mail.none;authpriv.none;cron.none    /var/log/messages
authpriv.*                                  /var/log/secure
mail.*                                      /var/log/maillog
cron.*                                      /var/log/cron
*.emerg                                     *
uucp,news.crit                              /var/log/spooler
local7.*                                    /var/log/boot.log
```

The default system log configuration file contains three sections: Modules, Global Directives, and Rules. The Modules section includes two modules: imuxsock.so and imklog.so. These modules are specified with the ModLoad directive and are loaded when required. The imuxsock.so module provides support for the local system logging via the *logger* command and the imklog.so module provides the kernel logging support.

The Global section contains only one active directive, which instructs the *rsyslogd* daemon to save the captured messages in the old, traditional way. Directives defined in this section apply to the *rsyslog* daemon.

Under the Rules section, each line entry consists of two fields. The left field is called *selector* and the right one is referred to as *action*. The selector field is further divided into two sub-fields, which are separated by a dot. The left sub-field, called *facility*, represents various system process categories that generate messages. Multiple facilities can be defined with each facility separated from the other by the semicolon character. The right sub-field, called *priority*, represents severity associated with the message. The action field determines the destination to send the message to.

Some of the facilities are kern, authpriv, mail, and cron. The asterisk character represents all of them.

Similarly, there are multiple priorities such as emergency (emerg), alert (alert), critical (crit), error (err), warning (warning), notice (notice), informational (info), debug (debug), and none. The sequence provided is in the descending criticality order. The asterisk character represents all of them. If a lower priority is selected, the daemon will log all messages of the service at that and higher levels.

The first line entry instructs the *rsyslog* daemon to capture and store informational messages from all services to the */var/log/messages* file but not to capture and store any messages for the mail, authentication, and cron services.

The second, third, and fourth line entries command the daemon to capture and log all messages generated by the authentication, mail, and cron facilities to the */var/log/secure*, */var/log/maillog*, and */var/log/cron* files, respectively.

The fifth line indicates that the *rsyslogd* daemon will display all emergency messages on the terminals of all logged in users.

The sixth line shows two facilities, separated by comma, that are defined with the same priority. These facilities tell the daemon to capture critical messages generated by uucp and news facilities and log them to the */var/log/spooler* file.

The last line entry causes the daemon to capture and store boot messages to the */var/log/boot.log* file.

Maintaining and Managing Log Files

In RHEL, all system log files are stored in the */var/log* directory. Run the *ll* command on the directory to list the contents. The following shows only the key log files:

```
# ll /var/log
-rw-------. 1 root root  78301 Sep 14 15:40 anaconda.log
-rw-r--r--. 1 root root   3379 Oct 22 12:06 boot.log
-rw-------. 1 root utmp   1152 Oct 20 20:51 btmp
-rw-------. 1 root root  52645 Oct 23 08:01 cron
-rw-------. 1 root root   1615 Oct 22 12:06 maillog
-rw-------. 1 root root 494794 Oct 23 08:07 messages
-rw-------. 1 root root  11586 Oct 23 08:07 secure
-rw-------. 1 root root  12203 Oct 22 08:18 yum.log
```

The output indicates that there are different log files for different services. Depending on the number of messages generated and captured, log files may fill up the file system quickly where the directory is located. Also, if a log file grows to a very large size, it becomes troublesome to load and read it.

In RHEL6, a script called *logrotate* in the */etc/cron.daily* directory runs the *logrotate* command once every day to rotate log files by sourcing the */etc/logrotate.conf* file and the configuration files located in the */etc/logrotate.d* directory. These configuration files may be modified to perform additional tasks such as removing, compressing, and emailing the identified log files.

Here is what the */etc/cron.daily/logrotate* file contains:

 # cat /etc/cron.daily/logrotate
```
/usr/sbin/logrotate /etc/logrotate.conf >/dev/null 2>&1
EXITVALUE=$?
if [ $EXITVALUE != 0 ]; then
    /usr/bin/logger -t logrotate "ALERT exited abnormally with [$EXITVALUE]"
fi
exit 0
```

The following shows an excerpt from the */etc/logrotate.conf* file:

 # cat /etc/logrotate.conf
```
# rotate log files weekly
weekly
# keep 4 weeks worth of backlogs
rotate 4
# create new (empty) log files after rotating old ones
create
# use date as a suffix of the rotated file
dateext
# uncomment this if you want your log files compressed
#compress
/var/log/wtmp {
    monthly
    create 0664 root utmp
        minsize 1M
    rotate 1
}
/var/log/btmp {
    missingok
    monthly
    create 0600 root utmp
    rotate 1
}
```

The */etc/logrotate.d* directory includes a number of configuration files for individual log files. Log files for services listed in this directory are rotated automatically. Some of the key configuration files in the */etc/logrotate.d* directory are listed below:

```
# ll /etc/logrotate.d
-rw-r--r--. 1 root  root  185 Feb  7  2012 httpd
-rw-r--r--. 1 root  root  165 Nov  8  2011 libvirtd
-rw-r--r--. 1 root  root  121 Jun 14 15:39 setroubleshoot
-rw-r--r--. 1 root  root   71 Oct 25  2011 subscription-manager
-rw-r--r--. 1 root  root  210 Aug  2  2011 syslog
-rw-r--r--. 1 root  root  188 Mar  2  2012 vsftpd
-rw-r--r--. 1 root  root  100 Sep 21  2011 yum
```

The Boot Log File

Log information for each service startup at system boot is captured in the */var/log/boot.log* file. You may review this file after the system has been booted up to see which services were started successfully and which were not. Here is a sample from the file:

```
# cat /var/log/boot.log
. . . . . . . .
Starting atd:                          [ OK ]
Starting Red Hat Network Daemon:       [ OK ]
Starting rhsmcertd 240 1440            [ OK ]
```

The System Log File

The default system log file is */var/log/messages*. This is a plain text file and may be viewed with any file display utility such as *cat*, *more*, *less*, *head*, or *tail*. This file may be viewed in real time using the *tail* command with the –f switch. This file captures the time stamp, the hostname, the daemon name, the PID of the process, and a short description of what is being logged.

The following shows some sample entries from the file:

```
# tail /var/log/messages
. . . . . . . .
Aug 19 03:28:02 server kernel: imklog 4.6.2, log source = /proc/kmsg started.
Aug 19 03:28:02 server rsyslogd: [origin software="rsyslogd" swVersion="4.6.2" x-pid="1208" x-
info="http://www.rsyslog.com"] (re)start
```

Managing Services

There are several system and network services that RHEL provides, which can be configured to start automatically at each system reboot. These services can also be configured to remain disabled at system boot. Moreover, the automatic start and stop can be defined at individual run levels. For example, you may want the NFS client service to start only if the server enters run level 5, but remain inactive at 2 and 3. Besides, you can view the operational status of a service and force it to reload its configuration if desired. A service can also be restarted manually.

RHEL provides the *chkconfig* and *service* commands to manage services at the command prompt. The *chkconfig* command gives the ability to display service startup settings, and set a service to start or stop at appropriate run levels. The *service* command allows you to start, stop, check the operational status of, and restart a service. This command also enables you to force the service to reload its configuration if there has been any change done to it.

Furthermore, a menu-driven program and a graphical tool are also available for service management, and are explained later in this section.

Exercise 8-2: List, Disable, and Enable a Service

In this exercise, you will list the start/stop settings for the *ntpd* service. You will then disable it so that it does not start when the system transitions into run level 4. You will then validate the new configuration for *ntpd* as well as check the settings for all the services available on the system. Finally, you will enable the service to autostart at all multiuser run levels, disable it to remain off at all run levels, and validate the new settings.

1. Issue the *chkconfig* command and list the current start/stop settings for the *ntpd* service:

   # **chkconfig --list ntpd**
   ntpd   0:off 1:off 2:on 3:on 4:on 5:on 6:off

   The output indicates that *ntpd* is set to start at run levels 2, 3, 4, and 5.

2. Run *chkconfig* and turn *ntpd* off for run level 4:

   # **chkconfig --level 4 ntpd off**

3. Execute *chkconfig* and list the new settings for *ntpd*:

   # **chkconfig --list ntpd**

4. Execute *chkconfig* again and list settings for all services available on the system:

   # **chkconfig --list**
   NetworkManager   0:off 1:off 2:on   3:on 4:on   5:on 6:off
   abrt–ccpp   0:off 1:off 2:off 3:on 4:off 5:on 6:off
   abrt–oops   0:off 1:off 2:off 3:on 4:off 5:on 6:off
   . . . .

5. Issue *chkconfig* to enable *ntpd* to start at all multiuser run levels:

   # **chkconfig ntpd on**

6. Run *chkconfig* to disable *ntpd* to remain stopped at all multiuser run levels:

   # **chkconfig ntpd off**

7. Confirm the new settings:

   # **chkconfig --list ntpd**
   ntpd   0:off 1:off 2:off 3:off 4:off 5:off 6:off

Exercise 8-3: Check Status of, Start, and Stop a Service

In this exercise, you will check whether the *ntpd* service is running. You will then stop it and recheck the running status. You will then start and then restart the service, and validate the new running status. Check the running status for all the services available on the system.

1. Issue the *service* command to check whether the *ntpd* service is currently running:

    # **service ntpd status**
    ntpd (pid 18068) is running...

2. Issue the *service* command and stop the *ntpd* service:

    # **service ntpd stop**
    Shutting down ntpd: [OK]

3. Issue the *service* command to check whether the *ntpd* service is down:

    # **service ntpd status**
    ntpd is stopped

4. Issue the *service* command and start the *ntpd* service:

    # **service ntpd start**
    Starting ntpd: [OK]

5. Issue the *service* command to stop and start the *ntpd* service in one go:

    # **service ntpd restart**
    Shutting down ntpd: [OK]
    Starting ntpd: [OK]

6. Issue the *service* command to check the operational status of all the services on the system:

    # **service --status-all**
    abrtd (pid 1608) is running...
    abrt-dump-oops (pid 1616) is running...
    acpid (pid 1398) is running...
    . . . .

Managing Services via a Menu-Driven Program

RHEL provides a menu-driven program called *ntsysv* for service management. Here is an example that demonstrates the use of this tool.

To enable several services at run levels 2 and 3, run the command as follows. It will bring up a window as shown in Figure 8-4 that will list all available services. Select the ones you want enabled and press OK.

ntsysv --level 23

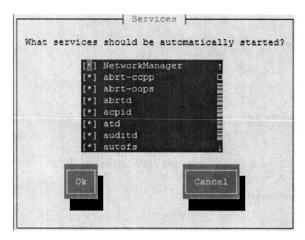

Figure 8-4 Service Management via Menu-Driven Program

By specifying run levels at the command prompt this way, you are directing *ntsysv* to modify start/stop only for those specific run levels for the services you are going to be selecting.

Managing Services via Service Configuration Utility

Figure 8-5 displays the Red Hat Service Configuration Utility interface, which is invoked via either the *serviceconf* or the *system-config-services* command in an X terminal window. Alternatively, you can invoke the program as System → Administration → Services.

Figure 8-5 Service Management via Service Configuration Utility

Click Customize and choose one or more run levels to configure service start/stop. In the left window pane, scroll up or down to look for the desired service and click Enable or Disable as needed. This ensures that the service will start or stop at the appropriate run levels. As well, you can highlight a service and start, stop, or restart it by clicking an appropriate button.

Linux Rescue Mode

Linux *rescue mode* is a special boot mode for fixing critical issues, such as reinstalling a corrupted stage 1 bootloader or accessing a problematic */boot* file system, that have rendered the system unbootable. In order to get into the rescue mode, the system needs to be booted with a RHEL installation DVD, ISO image, local Linux hard drive, or an image available over the network via NFS or HTTP. If you select NFS or HTTP, you will be required to configure a network interface and supply the location of the image.

The rescue mode presents you with four options – Continue, Read-Only, Skip, and Advanced – and allows you to choose one of them. Continue mounts the local file systems in read/write mode under the */mnt/sysimage* directory; Read-Only mounts them in read-only mode; Skip allows you to access the shell prompt with no file systems mounted, run diagnostics, or reboot the system; and Advanced is for adding any iSCSI or FCoE SAN devices, but you must have had a network interface configured in the previous step. Choosing any of the four options eventually presents you with the command prompt to run administrative commands for solving a boot issue.

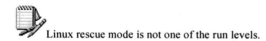 Linux rescue mode is not one of the run levels.

A minimal set of commands are available in the rescue mode to help troubleshoot and resolve problems. These commands include the *vi* and *nano* editors as well as tools to manage file and directory operations, software packages, disk partitioning and file systems, processes and kernel, archiving and compression, and network interfaces and connectivity. All rescue mode boot messages are logged to the */tmp/anaconda.log* file and system activities to the */tmp/syslog* file.

Exercise 8-4: Boot RHEL into the Rescue Mode

In this exercise, you will boot the system into the rescue mode using the RHEL installation DVD with networking disabled and file systems mounted in read-write mode in the chroot environment.

1. Boot the system with the installation DVD.
2. Choose *Rescue Installed System* from the boot menu, Figure 8-6, and press Enter.

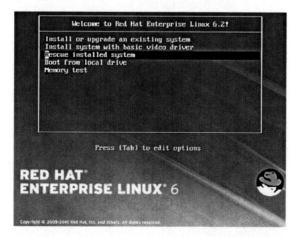

Figure 8-6 Booting into Linux Rescue Mode

3. Select a language and keyboard on subsequent screens.
4. Choose Local CD/DVD and press OK.
5. Choose no when asked whether you want network interfaces to start in the rescue mode.
6. Choose Continue and press Enter. Figure 8-7 shows the four options available and a short description of each of them.

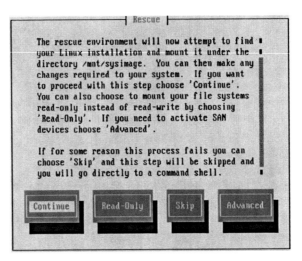

Figure 8-7 Linux Rescue Mode Options

7. Continue searches for a RHEL installation and, if found, mounts the / file system on */mnt/sysimage* in read/write mode. See Figure 8-8. Press OK to move on.

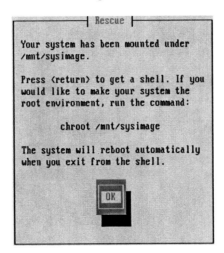

Figure 8-8 Linux Rescue Mode – Continue Option

8. Press OK again to go to the command prompt. Execute the *df* command to view the mounted file systems. See Figure 8-9.

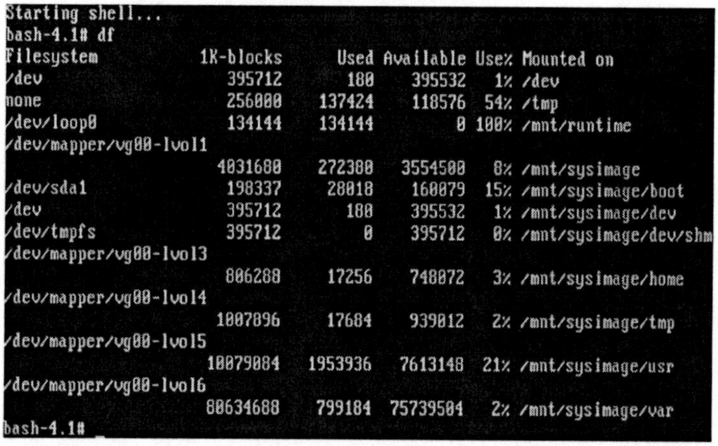

```
Starting shell...
bash-4.1# df
Filesystem              1K-blocks      Used Available Use% Mounted on
/dev                       395712       180    395532   1% /dev
none                       256000    137424    118576  54% /tmp
/dev/loop0                 134144    134144         0 100% /mnt/runtime
/dev/mapper/vg00-lvol1
                          4031680    272380   3554500   8% /mnt/sysimage
/dev/sda1                  198337     28018    160079  15% /mnt/sysimage/boot
/dev                       395712       180    395532   1% /mnt/sysimage/dev
/dev/tmpfs                 395712         0    395712   0% /mnt/sysimage/dev/shm
/dev/mapper/vg00-lvol3
                           806288     17256    748072   3% /mnt/sysimage/home
/dev/mapper/vg00-lvol4
                          1007896     17684    939012   2% /mnt/sysimage/tmp
/dev/mapper/vg00-lvol5
                         10079084   1953936   7613148  21% /mnt/sysimage/usr
/dev/mapper/vg00-lvol6
                         80634688    799184  75739504   2% /mnt/sysimage/var
bash-4.1#
```

Figure 8-9 Linux Rescue Mode – File Systems Mounted on /mnt/sysimage

9. Run the *chroot* command to make all file systems appear as if they are mounted directly under /, as shown in Figure 8-10.

> # **chroot /mnt/sysimage**

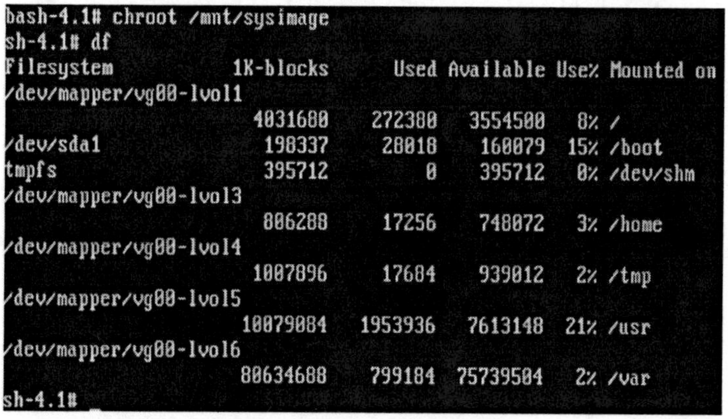

```
bash-4.1# chroot /mnt/sysimage
sh-4.1# df
Filesystem              1K-blocks      Used Available Use% Mounted on
/dev/mapper/vg00-lvol1
                          4031680    272380   3554500   8% /
/dev/sda1                  198337     28018    160079  15% /boot
tmpfs                      395712         0    395712   0% /dev/shm
/dev/mapper/vg00-lvol3
                           806288     17256    748072   3% /home
/dev/mapper/vg00-lvol4
                          1007896     17684    939012   2% /tmp
/dev/mapper/vg00-lvol5
                         10079084   1953936   7613148  21% /usr
/dev/mapper/vg00-lvol6
                         80634688    799184  75739504   2% /var
sh-4.1#
```

Figure 8-10 Linux Rescue Mode – File Systems Mounted Normally

Exercise 8-5: Reinstall a Corrupted Bootloader with grub-install

In this exercise, you will boot the system into the rescue mode and reinstall the corrupted stage 1 bootloader on the */dev/sda* disk using the *grub-install* command.

1. Boot the system into the rescue mode with all file systems mounted read/write in the chroot environment.
2. Issue the *grub-install* command to reinstall the bootloader on the boot disk's MBR:

> # **grub-install --root-directory=/ /dev/sda**

```
sh-4.1# grub-install --root-directory=/ /dev/sda
Installation finished. No error reported.
This is the contents of the device map //boot/grub/device.map.
Check if this is correct or not. If any of the lines is incorrect,
fix it and re-run the script `grub-install'.

# this device map was generated by anaconda
(hd0)     /dev/sda
```

Figure 8-11 Linux Rescue Mode – Reinstall a Corrupted Bootloader

3. Issue *exit* twice to go back. Select Reboot to restart the system.

Exercise 8-6: Reinstall a Corrupted Bootloader from the grub Shell

In this exercise, you will boot the system into the rescue mode and reinstall the corrupted stage 1 bootloader on the */dev/sda* disk using the *setup* command at the grub> prompt.

1. Boot the system into the rescue mode with all file systems mounted read/write in the chroot environment.
2. Issue the *grub* command to invoke the grub shell:

```
# grub
grub>
```

3. Run the *root* command to set the current root device:

```
grub> root (hd0,0)
root (hd0,0)
 Filesystem type is ext2fs, partition type 0x83
```

4. Execute the *setup* command to install the bootloader on the */dev/sda* disk:

```
grub> setup (hd0)
setup (hd0)
 Checking if "/boot/grub/stage1" exists... no
 ........
 Running "install /grub/stage1 (hd0) (hd0)1+27 p (hd0,0)/grub/stage2 /grub/grub.conf"... succeeded
Done.
```

5. Issue *quit* to exit out of the grub shell, and then *exit* twice to go back. Select Reboot to restart the system.

Exercise 8-7: Install a Lost File

A key system file or command can get corrupted or lost, and prevent a successful system boot up. Some of the necessary files and commands that must be available in order for the system to boot normally are */etc/passwd*, */etc/inittab*, */etc/fstab*, */sbin/init*, */bin/mount*, and */bin/bash*. If any of these files is deleted, it will need to be reinstalled, created, copied, or restored from backup.

In this exercise, you will boot the system into the rescue mode and restore the */bin/mount* command from the DVD.

1. Boot the system into the rescue mode with all file systems mounted read/write in the non-chroot environment.
2. Copy the *mount* command from */bin* to the */mnt/sysimage/bin* directory:

   # cp /bin/mount /mnt/sysimage/bin

3. Confirm the file copy:

   # ls –l /mnt/sysimage/bin/mount
   -rwsr-xr-x. 1 root root 76056 Sep 16 2011 /mnt/sysimage/bin/mount

4. Issue the *exit* command twice to go back. Select Reboot to restart the system.

Linux Kernel

The Linux kernel is a set of software components called *modules* that work together as a single unit to enable programs, services, and applications to run smoothly and efficiently on the system. Modules are basically *device drivers* used for controlling hardware devices such as controller cards and peripheral devices, as well as software components such as LVM, file systems, and RAID. Some of these modules are static to the kernel while others are loaded dynamically when needed.

When the system is booted, the kernel is loaded into memory with all static modules. The dynamic modules are loaded automatically when required. A Linux kernel that comprises of static modules only is referred to as the *monolithic* kernel; a Linux kernel that also includes dynamic modules is known as the *modular* kernel.

A monolithic kernel is typically larger in size than a modular kernel since it includes all components that may or may not be required at all times. This means the monolithic kernel needs to occupy more physical memory making it less efficient.

A modular kernel, on the other hand, is made up of mere critical and essential components, and loads dynamic modules automatically as and when needed. This means the modular kernel occupies less physical memory as compared to a monolithic kernel, making it faster and more efficient in terms of overall performance, and less vulnerable to crashes. Another benefit of having a modular kernel is that software driver updates only require the associated module to be recompliled without a system reboot. It does not need an entire kernel recompile.

RHEL distributions are available with kernels that are designed to support diverse processor architectures such as 32-bit and 64-bit Intel/AMD (i386/i686), 64-bit PowerPC, and IBM System z and 390 mainframes. The *uname* command with the –m option lists the architecture of the system. During the RHEL installation, anaconda automatically chooses the regular kernel that best suits the processor architecture. Files related to the kernel architecture are installed in the */boot* directory. The regular kernel supports no more than 4GB of physical memory on a system with one or more processors. Additional kernels are available for each supported architecture, which provide support for systems with multiple processors and physical memory beyond 4GB (up to 3TB), and allow for constructing several virtual machines on a single physical system. Optional kernel packages such as kernel-devel, kernel-debug, and kernel-headers may also be loaded to get additional device drivers, debug tools, and header information, respectively. RHEL distribution also has source codes available for supported kernels.

The default kernel installed during the installation is usually adequate for most system needs; however, it requires a rebuild when a new functionality is added or removed. The new functionality may be introduced by installing a new kernel, upgrading the existing kernel, installing a new hardware device, or changing a critical system component. Likewise, an existing functionality that is no longer required may be removed to make the kernel smaller resulting in improved performance and reduced memory utilization.

To control the behavior of the modules, and the kernel in general, several tunable parameters are set that define a baseline for the kernel functionality. Some of these parameters must be tuned to allow certain applications and database software to be installed smoothly and function properly.

RHEL allows you to generate and store several custom kernels with varied configuration and required modules, but only one of them is active at a time. Other kernels may be loaded by interacting with the GRUB.

Checking and Understanding Kernel Version

To check the version of the running kernel on the system, run the *uname* or the *rpm* command:

uname −r
2.6.32–220.el6.x86_64
rpm −q kernel
kernel–2.6.32–220.el6.x86_64

The output indicates that the kernel version currently in use is 2.6.32-220.el6.x86_64. An anatomy of the version information is displayed in Figure 8-12 and explained below.

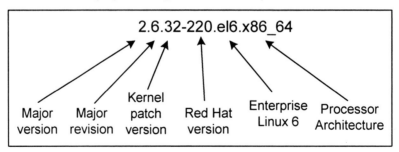

Figure 8-12 Anatomy of Kernel Version

From left to right:

✓ (2) indicates that this is the second major version of the Linux kernel. The major number changes when there are significant alterations, enhancements, and updates to the previous major version.
✓ (6) indicates that this is the sixth major revision of the second major version.

- ✓ (32) indicates that this is the thirty second patched version of this kernel with minor bug and security hole fixes, minor enhancements, and so on.
- ✓ (220) indicates that this is the two hundred and twentieth version of Red Hat customized kernel.
- ✓ (el6) indicates that this kernel is for Red Hat Enterprise Linux 6.
- ✓ (x86_64) indicates that this kernel is for the 64-bit architecture computers.

A further analysis designates that 2.6.32 holds the general Linux kernel version information and the subsequent numbers and letters represent the Red Hat specific information.

Understanding Kernel Directory Structure

Kernel files are stored at different locations in the directory hierarchy of which three locations – */boot*, */proc*, and */lib/modules* – are of significance, and are explained below.

The /boot File System

The */boot* file system is created at system installation time and its purpose is to store kernel-related information including current working kernel and associated files. This file system also stores any updated or modified kernel data. An *ll* on */boot* displays the following information:

```
# ll /boot
-rw-r--r--. 1 root  root     100943 Nov  9 2011 config-2.6.32-220.el6.x86_64
drwxr-xr-x. 3 root  root       1024 Aug 18 09:18 efi
drwxr-xr-x. 2 root  root       1024 Aug 22 12:24 grub
-rw-r--r--. 1 root  root   15860985 Aug 18 09:21 initramfs-2.6.32-220.el6.x86_64.img
. . . . . . . .
-rwxr-xr-x. 1 root  root    3938800 Nov  9 2011 vmlinuz-2.6.32-220.el6.x86_64
```

The output indicates that the current kernel is *vmlinuz-2.6.32-220.el6.x86_64*, its boot image is stored in the *initramfs-2.6.32-220.el6.x86_64.img* file and configuration in the *config-2.6.32-220.el6.x86_64* file.

A sub-directory */boot/grub* contains GRUB information as shown below:

```
# ll /boot/grub
-rw-r--r--. 1 root  root      63 Aug 18 09:22 device.map
-rw-r--r--. 1 root  root   13380 Aug 22 12:24 e2fs_stage1_5
-rw-r--r--. 1 root  root   12620 Aug 22 12:24 fat_stage1_5
-rw-r--r--. 1 root  root   11748 Aug 22 12:24 ffs_stage1_5
-rw-------. 1 root  root     813 Aug 18 09:22 grub.conf
. . . . . . . .
```

The key file in */boot/grub* is *grub.conf*, which maintains a list of available kernels and defines the default kernel to boot.

The /proc File System

/proc is a virtual file system and its contents are created in memory at system boot and destroyed when the system goes down. Underneath this file system lie current hardware configuration and status information. A directory listing of */proc* is provided below:

```
# ll /proc
dr-xr-xr-x. 7  root    root      0 Aug 22 14:14 1
dr-xr-xr-x. 7  root    root      0 Aug 22 14:14 10
dr-xr-xr-x. 7  root    root      0 Aug 22 14:14 1020
. . . . . . . .
```

This file system contains several files and sub-directories. Some sub-directory names are numerical and contain information about a specific process with process ID matching the sub-directory name. Within each sub-directory, there are files and further sub-directories, which include information such as memory segment specific to that particular process. Other files and sub-directories contain configuration data for system components. If you wish to view configuration information for a specific item such as the CPU or memory, *cat* the contents of *cpuinfo* and *meminfo* files as shown below:

```
# cat /proc/cpuinfo
processor     : 0
vendor_id     : AuthenticAMD
cpu family    : 16
. . . . . . . .
# cat /proc/meminfo
MemTotal:      791428 kB
MemFree:       398148 kB
. . . . . . . .
```

The data stored under */proc* is referenced by a number of system utilities including *top*, *ps*, *uname*, and *vmstat*, to display information.

The /lib/modules Directory

This directory holds information about kernel modules. Underneath it are located sub-directories specific to the kernels installed on the system. For example, the *ll* output on */lib/modules* below shows that there is only one kernel on this system:

```
# ll /lib/modules
drwxr-xr-x. 7 root root 4096 Aug 18 09:20 2.6.32-220.el6.x86_64
```

Now issue the *ll* command on the kernel sub-directory *2.6.32-220.el6.x86_64*:

```
# ll /lib/modules/2.6.32-220.el6.x86_64
lrwxrwxrwx. 1 root root      46 Aug 18 09:15 build -> ../../../usr/src/kernels/2.6.32-220.el6.x86_64
drwxr-xr-x. 2 root root    4096 Nov  9  2011 extra
drwxr-xr-x. 11 root root   4096 Aug 18 09:15 kernel
-rw-r--r--. 1 root root 529526 Aug 18 09:20 modules.alias
. . . . . . . .
```

There are several files and a few sub-directories at this directory location. These files and sub-directories hold module specific information.

One of the key sub-directories is */lib/modules/2.6.32-220.el6.x86_64/kernel/drivers* where modules categorized in groups are stored in various sub-directories as shown in the listing below:

```
# ll /lib/modules/2.6.32-220.el6.x86_64/kernel/drivers
drwxr-xr-x. 3 root  root  4096 Aug 18 09:15 acpi
drwxr-xr-x. 2 root  root  4096 Aug 18 09:15 ata
drwxr-xr-x. 2 root  root  4096 Aug 18 09:15 atm
. . . . . . . .
```

Several module categories exist, such as ata, bluetooth, cdrom, firewire, input, net, pci, scsi, serial, usb, and video. These categories contain driver modules to control hardware components associated with them.

Managing Kernel Modules

Managing modules involve tasks such as listing, building, installing, loading, and unloading modules. It also includes the task to add modules to the initial ram disk. These tasks are explained in the following sub-sections.

Listing Modules

RHEL provides the *lsmod* command to view currently loaded modules. Alternatively, you can view them by displaying the contents of the */proc/modules* file. Both show module names, their sizes, how they are being used, and any dependent modules.

```
# lsmod
Module          Size     Used by
autofs4         26888    3
sunrpc          243758   1
ipt_REJECT      2383     2
. . . . . . . .
# cat /proc/modules
autofs4         26888 3 - Live 0xffffffffa03bc000
sunrpc          243758 1 - Live 0xffffffffa0367000
ipt_REJECT      2383 2 - Live 0xffffffffa0341000
. . . . . . . .
```

Building and Installing New Modules

RHEL detects hardware devices and loads appropriate modules automatically; however, there may be instances when a device is left undetected or added online (as in the case of SAN disk allocation). In such a situation, execute the *depmod* command to force the system to scan the hardware, find appropriate modules for the new devices, create required module dependencies, and update the */lib/modules/2.6.32-220.el6.x86_64/modules.dep* file, in addition to creating and updating several corresponding map files in the */lib/modules/2.6.32-220.el6.x86_64* directory.

```
# depmod –v
/lib/modules/2.6.32-220.el6.x86_64/kernel/arch/x86/kernel/cpu/cpufreq/powernow-k8.ko needs
"cpufreq_frequency_table_target": /lib/modules/2.6.32-220.el6.x86_64/kernel/drivers/cpufreq/freq_table.ko
. . . . . . . .
```

Here is a listing of the module files in the *lib/modules/2.6.32-220.el6.x86_64* directory:

```
# ll /lib/modules/2.6.18-92.el5 | grep modules
-rw-r--r--. 1 root root    529526 Aug 18 09:20 modules.alias
-rw-r--r--. 1 root root    509216 Aug 18 09:20 modules.alias.bin
-rw-r--r--. 1 root roo t     1310 Nov  9 2011 modules.block
. . . . . . . .
```

Loading and Unloading Modules

After a new module has been built and installed, execute the *modprobe* command to load it in memory. This command ensures that any dependent modules are also loaded prior to loading the specified module. The following example loads the disk mirroring module called *dm_mirror*:

```
# modprobe dm_mirror
```

To unload the *dm_mirror* module along with all unused dependent modules, run either the *modprobe* or the *rmmod* command:

```
# modprobe –r dm_mirror
# rmmod dm_mirror
```

If you wish to load all modules specific to a particular hardware category, use the *modprobe* command. Here is an example on how to load modules for usb hardware. The –l option lists modules being loaded and the –t option forces the command to load only the modules specific to the usb category.

```
# modprobe –lt usb
kernel/drivers/net/can/usb/ems_usb.ko
kernel/drivers/net/usb/catc.ko
kernel/drivers/net/usb/kaweth.ko
. . . . . . . .
```

To display information about a specific module such as dm_mirror, use the *modinfo* command:

```
# modinfo dm_mirror
filename:    /lib/modules/2.6.32-220.el6.x86_64/kernel/drivers/md/dm-mirror.ko
license:     GPL
author:      Joe Thornber
description: device-mapper mirror target
srcversion:  84C4FF236EC7949FFDA87E1
depends:     dm-region-hash,dm-mod,dm-log
vermagic:    2.6.32-220.el6.x86_64 SMP mod_unload modversions
```

Exercise 8-8: Add a Module to the Initial RAM Disk

As you know GRUB initializes the RAM disk so that the kernel starts up, mounts an initial root file system, and loads modules. The RAM disk is located in the compressed file */boot/initramfs-2.6.32-220.el6.x86_64.img* and contains module and device information. The RAM disk file is created when the kernel is built, and may also be created if you wish to add or remove support for a specific module.

In this exercise, you will add a module called Bluetooth to the initial RAM disk so that the support for Bluetooth devices becomes available after the next system reboot.

1. Change the directory into */boot* and list the ramdisk file:

   # **cd /boot && ll initramfs***
   -rw-r--r--. 1 root root 15860985 Aug 18 09:21 initramfs-2.6.32-220.el6.x86_64.img

2. Run the following to list the contents of the file:

   # **zcat initramfs* | cpio –itv**
   drwxr-xr-x 24 root   root   0 Aug 18 09:20 .
   drwxr-xr-x 2 root   root   0 Aug 18 09:20 initqueue-finished
   drwxr-xr-x 3 root   root   0 Aug 18 09:20 dev
   . . . .

3. Generate a new file called *initramfs-Bluetooth-`uname –r`* using the *mkinitrd* command:

   # **mkinitrd --with=Bluetooth initramfs-Bluetooth-`uname –r` `uname –r`**

4. Update the */boot/grub/grub.conf* file and change the initrd entry to reflect the new name:

   Initrd /initrd-Bluetooth-2.6.32-220.el6.x86_64

5. Reboot the system and validate the new module with the *lsmod* command:

   # **lsmod | grep Bluetooth**
   Bluetooth   31922 3

Installing and Upgrading the Kernel

Unlike handling other package installs and upgrades, installing and upgrading kernel packages require extra care as you might end up with an unbootable system. It is recommended to have a bootable media handy prior to starting either process. The install process adds a new kernel to the system leaving the existing kernel(s) intact, whereas, the upgrade process simply overwrites the existing kernel.

An upgraded kernel is usually required when there are deficiencies or bugs identified in the existing kernel hampering the kernel's smooth operation or a new version of an application needs to be installed on the system that requires a new version of the kernel. In either case, the new kernel addresses existing issues as well as adds bug fixes, security updates, new features, and support for additional hardware devices.

The process for installing and upgrading a kernel are similar. The *rpm* command may be used if you wish to perform these tasks manually, otherwise the *yum* command or the PackageKit tools are available to carry out these tasks hassle-free.

When using the *rpm* command, always install (–i) the new kernel even though you want to upgrade (–U) an existing kernel. This will ensure that you can go back to the previous kernel if needed.

Exercise 8-9: Install a New Kernel with rpm

In this exercise, you will install a new kernel with the *rpm* command with the assumption that a higher version such as kernel-2.6.32-220.4.1, along with its dependent package kernel-firmware-2.6.32-220.4.1 and its header file kernel-headers-2.6.32-220.4.1, have been downloaded and saved in the */tmp* directory. You will ensure that the existing kernel and its configuration remain intact.

1.   Run the *uname* command and check the version of the running kernel:

   # **uname –r**
   2.6.32–220.el6.x86_64

2.   Run the *rpm* command on the kernel files to install the new kernel:

   # **rpm –ivh /tmp/kernel-*.rpm**
   Preparing...   ### [100%]
      1.kernel–firmware   ### [50%]
      2.kernel   ### [100%]
      3.kernel–headers   ### [100%]

3.   Query all installed packages to confirm that the packages for the new kernel have been installed:

   # **rpm –qa | grep ^kernel**
   kernel–firmware–2.6.32–220.el6.noarch
   kernel–firmware–2.6.32–220.4.1.el6.noarch
   kernel–headers–2.6.32–220.el6.x86_64
   kernel–headers–2.6.32–220.4.1.el6.x86_64
   kernel–2.6.32–220.el6.x86_64
   kernel–2.6.32–220.4.1.el6.x86_64

   The output indicates that a higher kernel version 2.6.32-220.4.1 has been installed.

4.   *cat* the */boot/grub/grub.conf* file for an additional confirmation of the kernel:

 # cat /boot/grub/grub.conf

. . . .

title Red Hat Enterprise Linux Server (2.6.32-220.4.1.el6.x86_64)
    root (hd0,0)
    kernel /vmlinuz-2.6.32-220.4.1.el6.x86_64 ro root=/dev/mapper/vg00-lvol1 rd_NO_LUKS
LANG=en_US.UTF-8 rd_NO_MD rd_LVM_LV=vg00/lvol1 SYSFONT=latarcyrheb-sun16 rhgb
crashkernel=128M quiet rd_LVM_LV=vg00/lvol6 KEYBOARDTYPE=pc KEYTABLE=us rd_NO_DM
    initrd /initramfs-2.6.32-220.4.1.el6.x86_64.img

. . . .

5. Reboot the system so that the new kernel is loaded.
6. Run the *uname* command again to confirm loading of the new kernel:

 # uname –r
2.6.32-220.4.1.el6.x86_64

Exercise 8-10: Install a New Kernel with yum

In this exercise, you will install a new kernel using the *yum* command with the assumption that your system has access to the RHN repository. You will need to ensure that the existing kernel and its configuration do not get modified.

1. Run the *uname* command and check the version of the running kernel:

 # uname –r
2.6.32-220.4.1.el6.x86_64

2. Run the *yum* command to install the new kernel:

 # yum –y update kernel

. . . .

Resolving Dependencies
--> Running transaction check
---> Package kernel.x86_64 0:2.6.32-279.5.2.el6 will be installed
--> Processing Dependency: kernel-firmware >= 2.6.32-279.5.2.el6 for package: kernel-2.6.32-279.5.2.el6.x86_64
--> Running transaction check
---> Package kernel-firmware.noarch 0:2.6.32-220.4.2.el6 will be updated
---> Package kernel-firmware.noarch 0:2.6.32-279.5.2.el6 will be an update
--> Finished Dependency Resolution
Dependencies Resolved

```
================================================================================
 Package          Arch      Version            Repository            Size
================================================================================
Installing:
 kernel           x86_64    2.6.32-279.5.2.el6   rhel-x86_64-server-6    25 M
Updating for dependencies:
 kernel-firmware  noarch    2.6.32-279.5.2.el6   rhel-x86_64-server-6   8.7 M
Transaction Summary
```

```
-----------------------------------------------------------------------------------
Install       1 Package(s)
Upgrade       1 Package(s)
Total download size: 34 M
Downloading Packages:
. . . . . . . .
Running Transaction
  Updating   : kernel-firmware-2.6.32-279.5.2.el6.noarch       1/3
  Installing : kernel-2.6.32-279.5.2.el6.x86_64                2/3
  Cleanup    : kernel-firmware-2.6.32-220.4.2.el6.noarch       3/3
Installed products updated.
Installed:
  kernel.x86_64 0:2.6.32-279.5.2.el6
Dependency Updated:
  kernel-firmware.noarch 0:2.6.32-279.5.2.el6
Complete!
```

3. Query all installed packages to confirm that the packages for the new kernel have been installed:

   # **rpm –qa | grep ^kernel**
   ```
   kernel-2.6.32-220.4.1.el6.x86_64
   kernel-firmware-2.6.32-279.5.2.el6.noarch
   kernel-headers-2.6.32-220.4.1.el6.x86_64
   kernel-2.6.32-279.5.2.el6.x86_64
   ```

   The output indicates that a higher kernel version 2.6.32-279.5.2 has been installed.

4. *cat* the */boot/grub/grub.conf* file for an additional confirmation of the kernel:

   # **cat /boot/grub/grub.conf**
   ```
   . . . . . . . .
   default=0
   . . . . . . . .
   title Red Hat Enterprise Linux Server (2.6.32-279.5.2.el6.x86_64)
       root (hd0,0)
       kernel /vmlinuz-2.6.32-279.5.2.el6.x86_64 ro root=/dev/mapper/vg00-lvol1 rd_NO_LUKS
   LANG=en_US.UTF-8 rd_NO_MD rd_LVM_LV=vg00/lvol1 SYSFONT=latarcyrheb-sun16 rhgb
   crashkernel=128M quiet rd_LVM_LV=vg00/lvol6  KEYBOARDTYPE=pc KEYTABLE=us rd_NO_DM
       initrd /initramfs-2.6.32-279.5.2.el6.x86_64.img

   title Red Hat Enterprise Linux Server (2.6.32-220.4.1.el6.x86_64)
       root (hd0,0)
       kernel /vmlinuz-2.6.32-220.4.1.el6.x86_64 ro root=/dev/mapper/vg00-lvol1 rd_NO_LUKS
   LANG=en_US.UTF-8 rd_NO_MD rd_LVM_LV=vg00/lvol1 SYSFONT=latarcyrheb-sun16 rhgb
   crashkernel=128M quiet rd_LVM_LV=vg00/lvol6  KEYBOARDTYPE=pc KEYTABLE=us rd_NO_DM
       initrd /initramfs-2.6.32-220.4.1.el6.x86_64.img
   . . . . . . . .
   ```

5. Reboot the system so the new kernel is loaded.
6. Run the *uname* command again to confirm loading of the new kernel:

    # **uname –r**
    2.6.32–279.5.2.el6.x86_64

Chapter Summary

This chapter started off with a discussion of run levels and how to view and switch between them. You learned how to shutdown a RHEL system using the available tools.

You learned the system boot process that covered the pre-boot administration tasks, and the kernel and system initialization. Pre-boot administration included interacting with UEFI/BIOS and GRUB, booting into specific run levels, and an analysis of the bootloader configuration file; kernel initialization covered viewing messages generated during the system startup; and system initialization included an explanation of the virtual console screens and an understanding of the sequencer, configuration, and initialization directories, that were referenced to bring the system to a fully functional state. You examined how to configure one or more services to autostart at specific run levels using commands, a menu-driven program, and a graphical interface.

You reviewed system logging and analyzed its configuration file. You saw how and what actions could be associated with the selector fields. You looked at the concept of logrotate and saw how it could be used to rotate large log files.

You looked at the Linux rescue mode, and performed exercises on how to enter this mode and fix issues that had rendered the system unbootable.

You learned about the Linux kernel and the concept of modules that form the kernel. You performed an analysis on the kernel version and looked at key directories that hold kernel-specific information. You used commands to list, build, load, and unload modules, and display their information.

Finally, you installed new kernels using the *rpm* and *yum* commands.

Chapter Review Questions

1. Both BIOS and UEFI are used in newer computers. True or False?
2. What is the purpose of the *modinfo* command?
3. What would the *setup (hd0)* command do in the rescue mode and at the grub> prompt?
4. The *init* command may be used to rebuild a new kernel. True or False?
5. Which script is responsible to bring up the default display manager at system boot?
6. Which command would you run at the grub> prompt to get the root partition information?
7. By default, GRUB is stored in the master boot record. True or False?
8. What is the location of the *grub.conf* file in the UEFI-based systems?
9. What is the name of the upstart program?
10. How many sequencer directories are there by default?
11. Which command would allow you to edit a directive in the grub menu?
12. You must boot a system into the rescue mode in order to reset the *root* user's password. True or False?
13. Where would you change the default boot timeout value?
14. Which file stores the location information of the boot partition?

15. Which script is responsible for spawning getty messages?
16. What would the command *kernel /* do at the grub> command prompt?
17. Which two commands may be used to manage system services?
18. By default, where does the system mount file systems in the rescue mode?
19. What is the purpose of the *grub-install* utility?
20. What is the key difference between a monolithic and a modular kernel?
21. Which file does the system check to determine the default run level?
22. What two type of scripts do the sequencer directories contain?
23. What is the path to the directory that stores system initialization scripts?
24. Where are system configuration files stored?
25. Which command is responsible for setting up and activating virtual console screens?
26. What is the name for the boot log file?
27. Which two files would you want to view to obtain processor and memory information?
28. What would the command *service rsyslog restart* do?
29. What are two boot issues that you can fix from the rescue mode?
30. Define run control levels.
31. In which run level does the X window and the graphical desktop interface become available?
32. What two commands can be used to determine current and previous run levels?
33. What are the two key differences between the *shutdown* and the *init* commands?
34. Which command can you use to determine the kernel release information?
35. The *lsmod* command is used to rebuild modules. True or False?
36. What two commands can you use to unload a module?
37. You cannot use the *yum* command to upgrade a Linux kernel. True or False?
38. Which command would allow you to go to the append mode in grub?
39. What is the difference between the –U and the –i options with the *rpm* command?
40. What is the other name for the first process?

Answers to Chapter Review Questions

1. True.
2. The purpose of the *modinfo* command is to display information about the specified module.
3. The purpose of the command provided is used to reinstall the bootloader.
4. False.
5. The */etc/X11/prefdm* script.
6. The *root* command will provide root partition information at the grub prompt.
7. True.
8. The *grub.conf* file location in the UEFI-based systems is in the */boot/grub* directory.
9. The name of the upstart program is init.
10. There are seven sequencer directories by default.
11. The e command will allow editing a directive in the grub menu.
12. False. It could be done from the single user mode.
13. In the *grub.conf* file.
14. The *grub.conf* file stores the location information of the boot partition.
15. The */etc/rc.d/rc.local* script is responsible for spawning getty messages.
16. The command provided will list files in the */boot* file system.
17. The *service* and the *chkconfig* commands.
18. On the */mnt/sysimage* mount point.
19. The purpose of the *grub-install* command is to reinstall the bootloader.

20. A monolithic kernel loads all modules that may or may not be needed while the modular kernel loads modules when they are required.
21. The */etc/inittab* file.
22. The start and the kill scripts.
23. The */etc/rc.d/init.d* directory.
24. The */etc/sysconfig* directory.
25. The *init* command.
26. The */var/log/boot.log* file.
27. The *cpuinfo* and *meminfo* files in the */proc* file system.
28. The command provided will restart the *rsyslogd* process.
29. Reinstall a corrupted bootloader and reinstall a lost system file.
30. Run control levels determine the current state of the system.
31. The X window and the graphical desktop interface become available in run level 5.
32. The *who* command with the –r option and the *runlevel* commands.
33. The *init* command does not send a broadcast message and begins the shutdown process right away, whereas, the *shutdown* command sends a message as well as waits for the specified period of time.
34. The *uname* command.
35. False.
36. The *modprobe* or the *rmmod* command.
37. False.
38. The a command will allow to go to the append mode in grub.
39. The –U option would force the *rpm* command to upgrade the specified package or install it if it does not already installed, whereas the –i option would force the *rpm* command to install the package and fail if the package does not already exist.
40. The *init* process.

Labs

Lab 8-1: Modify the Default Run Level

Modify the default run level to 2 in the appropriate file(s) and reboot the system. Run a command to validate. Restore the run level back to what it was and reboot to validate.

Lab 8-2: Install a New Kernel

Check the current version of the kernel on your system. Download a higher version and install it. Reboot the system and ensure it is booted with the new kernel. Configure the system to boot with the old kernel and reboot it to validate.

Lab 8-3: Recover an Unbootable System

Remove the */bin/bash* file and reboot the system. Check if it boots up successfully into the default run level. You should be able to see issues with the boot process. Reboot the system into the Linux rescue mode and take appropriate measures to recover the lost file. Reboot to validate.

Chapter 09

Disk Partitioning

This chapter covers the following major topics:

✓ Disk management techniques
✓ Use fdisk and parted to create, modify, and delete partitions
✓ Understand LVM concepts, components, and structure
✓ Use LVM to initialize and uninitialize a physical volume; create, display, extend, reduce, and remove a volume group; and create, display, extend, reduce, and remove a logical volume

This chapter includes the following RHCSA objectives:

20. List, create, delete, and set partition type for primary, extended, and logical partitions
21. Create and remove physical volumes, assign physical volumes to volume groups, and create and delete logical volumes
24. Add new partitions and logical volumes, and swap to a system non-destructively
29. Extend existing unencrypted ext4-formatted logical volumes

$Data$ is stored on disk drives that are logically divided into partitions. A partition can exist on a portion of a disk, on an entire disk, or it may span multiple disks. Each partition may contain a file system, a raw data space, a swap space, or a dump space.

A file system is used to hold files and directories, a raw data space may be used for databases and other applications for faster access, a swap space is defined to supplement the physical memory on the system, and a dump space is created to store memory and kernel images after a system crash has occurred.

RHEL offers three solutions for creating and managing partitions on disks. These solutions include disk management using the *fdisk* and *parted* utilities, the *software RAID*, and the *Logical Volume Manager* (LVM) technique. All three may be used and can co-exist on a single disk. This chapter provides detailed coverage on standard partitioning and LVM; software RAID is beyond the scope.

During RHEL installation, a disk partitioning program called *disk druid* is invoked to carve up one or more available disks. This program allows you to use the standard partitioning, the software RAID partitioning, or the LVM partitioning technique, or a combination. The Disk Druid utility is available only during the installation process and cannot be run later.

In this chapter, exercises will be performed on host *server.example.com* which was installed in Chapter 07 "Virtualization & Network Installation". This host was allocated one 20gb disk for the OS and three 4gb disks for practice, and was installed with the default file system layout.

Disk Partitioning Using fdisk

The *fdisk* tool is commonly used to carve up disks on RHEL systems. This text-based, menu-driven program allows you to display, add, modify, verify, and delete partitions. It supports up to three usable primary partitions and one extended partition. Within the extended partition, additional logical partitions may be created. The *fdisk* partitioning information is stored in the *partition table,* which is located on the first sector of the disk, and is called MBR.

The *fdisk* utility is invoked on a disk device, but to determine the number of disks available on the system and to view the basic information about each disk and its partitioning, run the *fdisk* command with the –l option:

```
# fdisk –l
Disk /dev/vda: 21.5 GB, 21474836480 bytes
........
   Device Boot    Start    End    Blocks    Id  System
/dev/vda1  •       1       64     512000    83  Linux
/dev/vda2          64      2611   20458496  8e  Linux LVM

Disk /dev/vdb: 4294 MB, 4294967296 bytes
........
Disk /dev/vdc: 4294 MB, 4294967296 bytes
........
Disk /dev/vdd: 4294 MB, 4294967296 bytes
........
Disk /dev/vde: 4294 MB, 4294967296 bytes
........
```

Disk /dev/mapper/vg_server-lv_root: 19.3 GB, 19302187008 bytes

. . . .

Disk /dev/mapper/vg_server-lv_swap: 1644 MB, 1644167168 bytes

. . . .

The output indicates that there are five disks in the system. The first disk is */dev/vda* (21.5gb) and the remaining four disks are */dev/vdb*, */dev/vdc*, */dev/vdd*, and */dev/vde* (each is 4gb). It further shows that the first disk has two partitions: */dev/vda1* and */dev/vda2*. The first partition contains */boot* and the second one is an LVM physical volume comprising the root and swap logical volumes. The rest of the disks are empty.

vd represents the virtual disk. a represents the first disk, b represents the second disk, and so on. 1 represents the first partition on the disk, 2 represents the second partition on the disk, and so on. Therefore, */dev/vda1* represents the first partition on the first disk and */dev/vda2* represents the second partition on the same disk.

The main menu of the *fdisk* command can be invoked by specifying a specific disk with the command such as */dev/vda*:

fdisk /dev/vda
Command (m for help):

There are several sub-commands shown in the output if you type *m* and press the Enter key. Table 9-1 lists and describes some key sub-commands.

Command	Description
a	Toggles the bootable flag.
c	Toggles the DOS compatibility flag.
d	Deletes a partition.
l	Lists supported partition types.
m	Prints this menu.
n	Adds a new partition. There are two sub-options – primary (p) and extended (e) – available with this command. You can define up to 4 primary partitions where one of them must be an extended partition. Within the extended partition, you may create up to 12 logical partitions.
p	Prints the partition table information.
q	Quits without saving changes.
t	Changes a partition's system id.
u	Changes display/entry units.
v	Verifies the partition table.
w	Writes the partition table information to disk and quits.
x	Extra functionality available to experts only. Run *x* and then *m* to produce the list of sub-commands.

Table 9-1 fdisk Sub-Commands

Let us perform some exercises to understand the procedure on how to create, modify, and delete partitions with *fdisk*. In Chapter 10 "File Systems & Swap", you will create file system and swap space structures in these partitions.

Exercise 9-1: Create a Partition Using fdisk

In this exercise, you will create a 4GB primary partition on the */dev/vdb* disk using the *fdisk* utility for use as a file system, and confirm the creation.

1. Execute the *fdisk* command on the */dev/vdb* disk:

   # **fdisk /dev/vdb**

2. Choose *n* to add a new partition, *p* for the primary partition, and then 1 for the first primary partition on this disk:

   Command (m for help): **n**
   Command action
     e extended
     p primary partition (1-4)
   p
   Partition number (1-4): **1**

3. Enter the cylinder number where you wish the partition to begin. Choose the default cylinder number, which is 1:

   First cylinder (1-522, default 1): **1**

4. Specify +4GB as the partition size. (You can also enter an end cylinder number based on which *fdisk* will automatically calculate the size of the partition. Or, you can enter the size in KBs or MBs. The default is to use all available disk space):

   Last cylinder, +cylinders or +size{K,M,G} (1-522, default 522): **+4GB**

5. Execute *p* to print the updated partition table:

   Command (m for help): **p**
   Disk /dev/vdb: 4294 MB, 4294967296 bytes
   . . . .
     Device Boot   Start   End   Blocks   Id System
     /dev/vdb1   1   487   3911796 83 Linux

6. Write this information to the disk using the *w* command and exit out of *fdisk*:

   Command (m for help): **w**
   The partition table has been altered!
   Calling ioctl() to re-read partition table.
   Syncing disks.

7. Execute the *partprobe* command at the command prompt to force the kernel to re-read the updated partition table information:

   # **partprobe /dev/vdb**

8. Confirm the partition information using any of the following:

 # fdisk –l /dev/vdb
Disk /dev/vdb: 4294 MB, 4294967296 bytes

. . . .

Device Boot	Start	End	Blocks	Id	System
/dev/vdb1	1	487	3911796	83	Linux

 # grep vdb /proc/partitions

8	16	4194304	vdb
8	17	3911796	vdb1

Exercise 9-2: Create a Swap Partition Using fdisk

In this exercise, you will create a 2GB primary partition on the */dev/vdc* disk using the *fdisk* utility for use as a swap partition, and confirm the creation.

1. Invoke the *fdisk* command on the */dev/vdc* disk:

   # fdisk /dev/vdc

2. Choose *n* to add a new partition, *p* for the primary partition, and then 1 for the first primary partition on the disk:

   Command (m for help): **n**
   Command action
     e   extended
     p   primary partition (1-4)
   p
   Partition number (1-4): **1**

3. Enter the cylinder number where you wish the partition to begin. Choose the default cylinder number, which is 1:

   First cylinder (1-522, default 1): **1**

4. Specify +2GB as the partition size. (You can also enter an end cylinder number based on which *fdisk* will automatically calculate the size of the partition. Or, you can enter the size in KBs or MBs. The default is to use all available disk space):

   Last cylinder, +cylinders or +size{K,M,G} (1-522, default 522): **+2GB**

5. Execute *l* to list the known partition types; Linux swap (82) will be among them:

   Command (m for help): **l**

0 Empty	24 NEC DOS	81 Minix / old Lin	bf Solaris
1 FAT12	39 Plan 9	82 Linux swap / So	c1 DRDOS/sec (FAT-
2 XENIX root	3c PartitionMagic	83 Linux	c4 DRDOS/sec (FAT-
3 XENIX usr	40 Venix 80286	84 OS/2 hidden C:	c6 DRDOS/sec (FAT-
4 FAT16 <32M	41 PPC PReP Boot	85 Linux extended	c7 Syrinx

5 Extended	42 SFS	86 NTFS volume set	da Non-FS data
6 FAT16	4d QNX4.x	87 NTFS volume set	db CP/M / CTOS / .
7 HPFS/NTFS	4e QNX4.x 2nd part	88 Linux plaintext	de Dell Utility
8 AIX	4f QNX4.x 3rd part	8e Linux LVM	df BootIt
9 AIX bootable	50 OnTrack DM	93 Amoeba	e1 DOS access
a OS/2 Boot Manag	51 OnTrack DM6 Aux	94 Amoeba BBT	e3 DOS R/O
b W95 FAT32	52 CP/M	9f BSD/OS	e4 SpeedStor
c W95 FAT32 (LBA)	53 OnTrack DM6 Aux	a0 IBM Thinkpad hi	eb BeOS fs
e W95 FAT16 (LBA)	54 OnTrackDM6	a5 FreeBSD	ee GPT
f W95 Ext'd (LBA)	55 EZ-Drive	a6 OpenBSD	ef EFI (FAT-12/16/
10 OPUS	56 Golden Bow	a7 NeXTSTEP	f0 Linux/PA-RISC b
11 Hidden FAT12	5c Priam Edisk	a8 Darwin UFS	f1 SpeedStor
12 Compaq diagnost	61 SpeedStor	a9 NetBSD	f4 SpeedStor
14 Hidden FAT16 <3	63 GNU HURD or Sys	ab Darwin boot	f2 DOS secondary
16 Hidden FAT16	64 Novell Netware	af HFS / HFS+	fb VMware VMFS
17 Hidden HPFS/NTF	65 Novell Netware	b7 BSDI fs	fc VMware VMKCORE
18 AST SmartSleep	70 DiskSecure Mult	b8 BSDI swap	fd Linux raid auto
1b Hidden W95 FAT3	75 PC/IX	bb Boot Wizard hid	fe LANstep
1c Hidden W95 FAT3	80 Old Minix	be Solaris boot	ff BBT
1e Hidden W95 FAT1			

6. Execute *p* to print the updated partition table:

    ```
    Command (m for help): p
    Disk /dev/vdc: 4294 MB, 4294967296 bytes
    . . . . . . . .
        Device Boot    Start    End    Blocks      Id System
    /dev/vdc1          1        244    1959898+    83 Linux
    ```

7. Execute *t* to change the partition type of */dev/sdc1* to Linux swap (82):

    ```
    Command (m for help): t
    Selected partition 1
    Hex code (type L to list codes): 82
    Changed system type of partition 1 to 82 (Linux swap / Solaris)
    ```

8. Write this information to the disk using the *w* command and exit out of *fdisk*:

    ```
    Command (m for help): w
    The partition table has been altered!
    Calling ioctl() to re-read partition table.
    Syncing disks.
    ```

9. Execute the *partprobe* command at the command prompt to force the kernel to re-read the updated partition table information:

    # partprobe /dev/vdc

10. Confirm the partition information using any of the following:

Red Hat Certified System Administrator & Engineer

 # fdisk –l /dev/vdc
Disk /dev/vdc: 4294 MB, 4294967296 bytes

. . . .

```
Device Boot    Start    End    Blocks   Id System
/dev/vdc1         1      244   1959898+ 82 Linux swap / Solaris
```
 # grep vdc /proc/partitions
```
8    32   4194304 vdc
8    33   1959898 vdc1
```

Exercise 9-3: Delete a Partition Using fdisk

In this exercise, you will delete both partitions */dev/vdb1* and */dev/vdc1* that you created in the previous exercises, and confirm the deletion.

1. Execute the *fdisk* command on the */dev/vdb* disk:

   # fdisk /dev/vdb

2. Execute *p* to print the partition table:

   Command (m for help): **p**
   Disk /dev/vdb: 4294 MB, 4294967296 bytes

   . . . .
   ```
   Device Boot    Start    End    Blocks   Id System
   /dev/vdb1         1      487   3911796  83 Linux
   ```

3. Execute *d* to delete partition 1:

   Command (m for help): **d1**

4. Write this information to the disk using the *w* command and exit out of *fdisk*:

   Command (m for help): **w**
   The partition table has been altered!
   Calling ioctl() to re-read partition table.
   Syncing disks.

5. Execute the *partprobe* command at the command prompt to force the kernel to re-read the updated partition table information:

   # partprobe /dev/vdb

6. Confirm the deletion using any of the following:

   # fdisk –l /dev/vdb
   Disk /dev/vdb: 4294 MB, 4294967296 bytes

   . . . .
   # grep vda /proc/partitions
   ```
   8    16   4194304 vdb
   ```

7. Repeat steps 1 to 6 to delete the */dev/vdc1* partition.

Disk Partitioning Using parted

parted is another tool for slicing disks. This text-based, menu-driven program allows you to display, add, check, modify, copy, resize, and delete partitions. The main interface of the program looks like the following. It produces a list of sub-commands when you run *help* at the parted prompt:

```
# parted
GNU Parted 2.1
Using /dev/vda
Welcome to GNU Parted! Type 'help' to view a list of commands.
(parted) help
  align-check TYPE N                        check partition N for TYPE(min|opt) alignment
  check NUMBER                              do a simple check on the file system
  cp [FROM-DEVICE] FROM-NUMBER TO-NUMBER    copy file system to another partition
  help [COMMAND]                            print general help, or help on COMMAND
  mklabel,mktable LABEL-TYPE                create a new disklabel (partition table)
  mkfs NUMBER FS-TYPE                       make a FS-TYPE file system on partition NUMBER
  mkpart PART-TYPE [FS-TYPE] START END make a partition
  mkpartfs PART-TYPE FS-TYPE START END make a partition with a file system
  move NUMBER START END                     move partition NUMBER
  name NUMBER NAME                          name partition NUMBER as NAME
  print [devices|free|list,all|NUMBER]    display the partition table, available devices, free space, all found
partitions, or a particular partition
  quit                                      exit program
  rescue START END                          rescue a lost partition near START and END
  resize NUMBER START END                   resize partition NUMBER and its file system
  rm NUMBER                                 delete partition NUMBER
  select DEVICE                             choose the device to edit
  set NUMBER FLAG STATE                     change the FLAG on partition NUMBER
  toggle [NUMBER [FLAG]]                    toggle the state of FLAG on partition NUMBER
  unit UNIT                                 set the default unit to UNIT
  version                       display the version number and copyright information of GNU Parted
```

There are several sub-commands in the main menu. Table 9-2 lists and describes some key sub-commands.

Command	Description
check	Checks the specified file system.
cp	Copies a file system to another partition.
help	Displays the help on the specified command.
mklabel	Makes a new disk label.
mkfs	Makes the specified file system type.
mkpart	Makes a partition without a file system in it.
mkpartfs	Makes a partition with a file system in it.
move	Moves the partition number.

Red Hat Certified System Administrator & Engineer

Command	Description
name	Assigns a name to a partition.
print	Displays the partition table, a specific partition, or all devices.
quit	Quits parted.
rescue	Recovers a lost partition.
resize	Resizes the specified partition number and file system within it.
rm	Removes the specified partition.
select	Selects a device to edit.
set	Sets FLAG on the specified partition number.
toggle	Toggles the state of FLAG on the specified partition number.
unit	Sets default unit.

Table 9-2 parted Sub-Commands

At the *parted* command prompt, you can invoke the help on a specific sub-command. For example, to obtain help on *mklabel*, issue the following:

(parted) **help mklabel**
    mklabel,mktable LABEL-TYPE   Create a new disklabel (partition table)
  LABEL-TYPE is one of: aix, amiga, bsd, dvh, gpt, mac, msdos, pc98, sun, loop
(parted)

Let us create, modify, and delete partitions with *parted* to understand its usage. As noted earlier in this chapter, there are four disk drives – */dev/vdb*, */dev/vdc*, */dev/vdd*, and */dev/vde* – available on *server.example.com*, and the following exercises will be performed on this host.

Exercise 9-4: Create a Partition Using parted

In this exercise, you will create a 2GB primary partition on the */dev/vdc* disk using the *parted* utility for use as a file system, and confirm the creation.

1.   Execute the *parted* command on */dev/vdc*:

    # **parted /dev/vdc**
    GNU Parted 2.1
    Using /dev/vdc
    Welcome to GNU Parted! Type 'help' to view a list of commands.
    (parted)

2.   Assign a label to the disk with *mklabel*. This must be done on a new disk.

    (parted) **mklabel**
    New disk label type? **msdos**
    Warning: The existing disk label on /dev/vdc will be destroyed and all data on
    this disk will be lost. Do you want to continue?
    Yes/No? **yes**

There are several supported label types available that can be assigned to the disk. Some of them are msdos, bsd, and sun.

3. Create a partition of size 2GB using *mkpart*:

```
(parted) mkpart
Partition type? Primary/extended? primary
File system type? [ext2]? ext4
Start? 1
End? 2g
```

4. Execute *print* to verify:

```
(parted) print
. . . . . . . .
Number  Start    End      Size     Type     File system   Flags
  1     1049kB  2000MB  1999MB  primary
```

5. Exit out of *parted* and execute the *partprobe* command at the command prompt to force the kernel to re-read the updated partition table information:

   # **partprobe /dev/vdc**

6. Confirm the partition information using any of the following:

```
# parted /dev/vdc print
. . . . . . . .
Number  Start    End      Size     Type     File system   Flags
  1     1049kB  2000MB  1999MB  primary
# grep vdc /proc/partitions
  8    32   4194304 vdc
  8    33   1951744 vdc1
```

Exercise 9-5: Create a Swap Partition Using parted

In this exercise, you will create a 1GB primary partition on the */dev/vdc* disk using the *parted* utility for use as a swap partition, and confirm the creation.

1. Execute the *parted* command on */dev/vdc*:

   # **parted /dev/vdc**
   Using /dev/vdc
   Welcome to GNU Parted! Type 'help' to view a list of commands.
   (parted)

2. Create a partition of size 1GB using *mkpart*:

```
(parted) mkpart
Partition type? Primary/extended? primary
File system type? [ext2]? linux-swap
Start? 2001m
End? 1g
```

3. Execute *print* to verify:

```
(parted) print
. . . . . . . .
Number Start      End     Size    Type     File system  Flags
   1      1049kB  2000MB  1999MB  primary
   2      2001MB  3000MB   999MB  primary
```

4. Exit out of *parted* and execute the *partprobe* command at the command prompt to force the kernel to re-read the updated partition table information:

   # **partprobe /dev/vdc**

5. Confirm the partition information using any of the following:

   # **parted /dev/vdc print**

```
. . . . . . . .
Number Start      End     Size    Type     File system  Flags
   1      1049kB  2000MB  1999MB  primary
   2      2001MB  3000MB   999MB  primary
```
   # **grep vdc /proc/partitions**
```
   8    32   4194304 vdc
   8    33   1951744 vdc1
   8    34    976896 vdc2
```

Exercise 9-6: Repurpose a Partition Using parted

In this exercise, you will alter the partition type of the */dev/vdc1* partition for use as an LVM partition.

1. Execute the *parted* command on */dev/vdc*:

   # **parted /dev/vdc**

2. Run the *set* sub-command:

```
(parted) set
Partition number? 1
Flag to Invert? lvm
New state [ON]/off? on
```

3. Execute *print* to verify:

```
(parted) print
. . . . . . . .
Number Start     End     Size    Type     File system  Flags
   1     1049kB  2000MB  1999MB  primary               lvm
   2     2000MB  3000MB  1000MB  primary
```

---

4. Exit out of *parted* and execute the *partprobe* command at the command prompt to force the kernel to re-read the updated partition table information:

   # **partprobe /dev/vdc**

5. Confirm the partition information using any of the following:

   # **parted /dev/vdc print**

   . . . .
   ```
   Number Start    End     Size    Type     File system Flags
      1       1049kB 2000MB  1999MB  primary              lvm
      2       2000MB 3000MB  1000MB  primary
   ```
   # **grep vdc /proc/partitions**

Exercise 9-7: Delete a Partition Using parted

In this exercise, you will delete both partitions that you created in the previous exercises, and confirm the deletion.

1. Execute the *parted* command on the */dev/vdc* disk:

   # **parted /dev/vdc**

2. Execute *rm* to delete both partitions:

   (parted) **rm 1**
   (parted) **rm 2**

3. Execute *print* to verify:

   (parted) **print**

4. Exit out of *parted* and execute the *partprobe* command at the command prompt to force the kernel to re-read the updated partition table information:

   # **partprobe /dev/vdc**

5. Confirm the partition information using any of the following:

   # **parted /dev/vdc print**
   # **grep vdc /proc/partitions**

Partitioning Using the Graphical Disk Utility

Disk Utility is a graphical partition management tool and is available in the GNOME desktop. It may be invoked by running the *palimpsest* command at the shell prompt or choosing System →
Administration → Disk Utility from the GNOME desktop. Some of the key disk management functions that you can perform with this tool are to format a drive, measure read and write performance, unmount a volume (using the *umount* command), format a volume (using the *mkfs*

command), run a check on a file system (using the *fsck* command), edit a file system label (using the *e2label* command), and edit, delete, or create a partition (using the *fdisk* command). Figure 9-1 shows the main menu of the Disk Utility with one of the disks selected.

Figure 9-1 Graphical Disk Utility

Use the *rpm* command to check whether this utility is already installed on the system:

 # rpm –qa | grep disk-utility
gnome-disk-utility-2.30.1-2.el6.x86_64
gnome-disk-utility-libs-2.30.1-2.el6.x86_64
gnome-disk-utility-ui-libs-2.30.1-2.el6.x86_64

If this utility is not already installed, you can use one of the methods outlined in Chapter 06 "Package Management" to install it.

Disk Partitioning Using Logical Volume Manager

The Logical Volume Manager (LVM) solution is widely used for managing disk storage. LVM allows you to accumulate spaces taken from one or several partitions or disks (called *physical volumes*) to form a large logical container (called *volume group*), which is then divided into logical partitions (called *logical volumes*). RHEL6 includes version 2 of LVM called *LVM2* which is the default when you use the LVM technique to carve up disks. Figure 9-2 dipicts the LVM components.

The LVM structure is made up of three key virtual objects called physical volume, volume group, and logical volume. These virtual objects are further carved up in *physical extents* (PEs) and *logical extents* (LEs). The LVM components are explained in the following sub-sections.

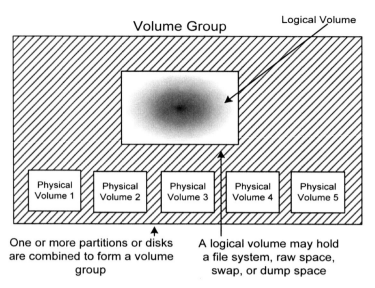

Figure 9-2 LVM Structure

Physical Volume

A physical volume (PV) is created when a standard partition, a software RAID partition, or an entire disk is brought under the LVM control after going through the initialization process which constructs LVM data structures on it. A partition or disk cannot be used under LVM if it is not initialized.

Currently, there is one physical volume on *server.example.com*, which was created during the RHEL installation. Run the *pvs* command to view it:

```
# pvs
  PV          VG            Fmt    Attr   PSize   PFree
  /dev/vda2   vg_serverlvm2 a--    19.51g  0
```

The output confirms that there is one physical volume */dev/vda2* of size 19.5GB in the *vg_server* volume group.

Volume Group

A volume group (VG) is created when at least one physical volume is added to it. The space from all physical volumes in a volume group is summed up to form a large pool of storage, which is then used to build one or more logical volumes. A volume group can have any name assigned to it at the time of its creation. For example, it may be called *vg01*, *vgora*, or *vgweb* so as to identify the type of information that it is created to store.

Currently, there is one volume group on *server.example.com*, which was created during the RHEL installation. Run the *vgs* command to view it:

```
# vgs
  VG          #PV   #LV   #SN   Attr    VSize   VFree
  vg_server   1     2     0     wz—n-   19.51g  0
```

The output confirms that there is one volume group *vg_server* on the system and it contains one physical volume.

Physical Extent

A physical volume is divided into several smaller logical pieces when it is added to a volume group. These logical pieces are known as physical extents (PEs). An extent is the smallest allocatable unit of space in LVM. At the time you create a volume group, you can either define the size of the PE or leave it to the default size of 4MB. This means an 8GB disk would contain approximately 2,000 PEs. Now all physical volumes added to this volume group will use the same PE size.

The following command displays the physical extent size used in the *vg_server* volume group:

```
# vgdisplay vg_server
. . . . . . . .
PE Size        4.00 MiB
. . . . . . . .
```

Logical Volume

A volume group consists of a pool of spaces taken from one or more physical volumes. This volume group space is divided into one or more logical partitions called logical volumes (LVs).

A logical volume can be expanded or shrinked, and can use space taken from one or several physical volumes inside the volume group.

The default naming convention for logical volumes is *lvol1, lvol2, lvol3*, and so on. However, you may use any naming scheme that you wish. For example, a logical volume may be called *system*, *undo*, or *table* so as to identify the type of information that it is created to store.

Currently, there are two logical volumes on *server.example.com* that were created during the RHEL installation. Run the *lvs* command to view them:

```
# lvs
LV         VG               Attr     LSize    Origin   Snap%   Move   Log     Copy%   Convert
lv_root    vg_server-wi-ao  17.98g
lv_swap    vg_server-wi-ao  1.53g
```

The output confirms that there is one volume group *vg_server* on the system and it contains two logical volumes.

Logical Extent

A logical volume is made up of extents called logical extents (LEs). Logical extents point to physical extents. The larger a logical volume is, the more LEs it will have.

The PE and LE sizes are usually kept the same within a volume group. However, a logical extent can be smaller or larger than a physical extent. The default LE size is also 4MB. The following command displays the information about the */dev/vg_server/lv_root* logical volume. The output does not indicate the LE size, however, you can convert the LV size in MBs and then divide the result by the Current LE count to get the LE size (which comes to 4MB in the following example).

```
# lvdisplay /dev/vg_server/lv_root
. . . . . . . .
LV Size        17.98 GiB
Current LE     4602
. . . . . . . .
```

Managing Disk Space Using LVM

Managing the disk space using LVM involves several tasks such as creating a physical volume, creating and displaying a volume group, creating and displaying a logical volume, extending a volume group, extending a logical volume, reducing a logical volume, renaming a logical volume, removing a logical volume, reducing a volume group, removing a volume group, and uninitializing a physical volume. You may use either the LVM commands or the graphical LVM Configuration tool to accomplish these tasks.

Table 9-3 lists and describes some key LVM commands.

Command	Description
pvck	Checks the integrity of a physical volume.
pvcreate	Initializes a disk or partition for LVM use.
pvdisplay	Displays the details of a physical volume.
pvresize	Resizes a physical volume.
pvmove	Moves data from one physical volume to another.
pvremove	Uninitializes a physical volume.
pvs	Lists physical volumes.
pvscan	Scans all disks containing physical volumes structures.
vgck	Checks the integrity of a volume group.
vgcreate	Creates a volume group.
vgdisplay	Displays the details of a volume group.
vgextend	Adds a physical volume to a volume group.
vgreduce	Removes a physical volume from a volume group.
vgrename	Renames a volume group.
vgremove	Removes a volume group.
vgs	Lists volume groups.
vgscan	Scans all disks containing volume groups and rebuilds cache.
lvcreate	Creates a logical volume.
lvdisplay	Displays the details of a logical volume.
lvextend	Extends the size of a logical volume.
lvm	Interactive logical volume management tool.
lvmdiskscan	Scans for devices with LVM structures.
lvreduce	Reduces a logical volume size.
lvremove	Removes a logical volume.
lvrename	Renames a logical volume.
lvresize	Resizes a logical volume.
lvs	Lists logical volumes.
lvscan	Scans all disks containing logical volume structures.

Table 9-3 LVM Commands

All the LVM commands listed in Table 9-3 support the –v option for verbosity. See the man pages of each command for additional options and usage.

As noted in the beginning of this chapter, there are four disks available on *server.example.com* for practice purposes. These are */dev/vdb*, */dev/vdc*, */dev/vdd*, and */dev/vde*. You will be using them in the following exercises.

Exercise 9-8: Create a Physical Volume and a Volume Group

In this exercise, you will initialize one disk */dev/vdb* and one partition */dev/vdc1* for use as physical volumes. You will then create a volume group called *vg01* and add both physical volumes to it. You will use the PE size 16MB for the volume group and display the details of the volume group and the physical volumes.

1. Initialize the */dev/vdb* disk and the */dev/vdc1* partition using the *pvcreate* command (create the partition if it does not already exist):

   # **pvcreate /dev/vdb /dev/vdc1**
   Writing physical volume data to disk "/dev/vdb"
   Physical volume "/dev/vdb" successfully created
   Writing physical volume data to disk "/dev/vdc1"
   Physical volume "/dev/vdc1" successfully created

2. Create *vg01* volume group using the *vgcreate* command and add */dev/vdb* and */dev/vdc1* physical volumes to it. Use the –s option to specify the PE size in MBs.

   # **vgcreate –s 16 vg01 /dev/vdb /dev/vdc1**
   Volume group "vg01" successfully created

3. Display the details of the *vg01* volume group using the *vgdisplay* command with and without the –v option:

   # **vgdisplay vg01**
   --- Volume group ---
   VG Name   vg01
   System ID
   Format   lvm2
   Metadata Areas   2
   Metadata Sequence No 1
   VG Access   read/write
   VG Status   resizable
   MAX LV   0
   Cur LV   0
   Open LV   0
   Max PV   0
   Cur PV   2
   Act PV   2
   VG Size   5.84 GiB
   PE Size   16.00 MiB

```
Total PE          374
Alloc PE / Size   0 / 0
Free  PE / Size   374 / 5.84 GiB
VG UUID           7IHLcv-jAc5-EIQR-lH4s-EskR-bfCx-jAaK15
```

vgdisplay –v vg01

```
. . . . . . . .
--- Physical volumes ---
PV Name           /dev/vdb
PV UUID           NISKj3-szbj-WcU9-H7Az-5ykR-fzSq-VlsToz
PV Status         allocatable
Total PE / Free PE   255 / 255


PV Name           /dev/vdc1
PV UUID           Vq0r4U-yS8M-srWe-zSBH-5XHg-8VPS-MOKhAe
PV Status         allocatable
Total PE / Free PE   119 / 119
```

The *vgdisplay* command shows that there are two physical volumes in the *vg01* volume group with 255 PEs in the first physical volume and 119 in the second, totaling 374 overall. The PE size is 16MB. Now if you multiply 374 by 16, you will get the total allocatable disk space in the volume group, which comes to 5.98GB. The last portion from the output under the "Physical volumes" heading displays the information about the physical volumes included in *vg01*.

4. Display the details of the physical volumes using the *pvdisplay* command:

pvdisplay /dev/vdb

```
   Using physical volume(s) on command line
   --- Physical volume ---
   PV Name           /dev/vdb
   VG Name           vg01
   PV Size           4.00 GiB / not usable 16.00 MiB
   Allocatable       yes
   PE Size           16.00 MiB
   Total PE          255
   Free PE           255
   Allocated PE      0
   PV UUID           NISKj3-szbj-WcU9-H7Az-5ykR-fzSq-VlsToz
```

pvdisplay /dev/vdc1

```
   Using physical volume(s) on command line
   --- Physical volume ---
   PV Name           /dev/vdc1
   VG Name           vg01
   PV Size           1.87 GiB / not usable 9.96 MiB
   Allocatable       yes
   PE Size           16.00 MiB
   Total PE          119
   Free PE           119
   Allocated PE      0
   PV UUID           Vq0r4U-yS8M-srWe-zSBH-5XHg-8VPS-MOKhAe
```

Once a disk or a standard partition is initialized and added to a volume group, both are treated identically. LVM does not differentiate between the two.

Exercise 9-9: Create a Logical Volume

In this exercise, you will create two logical volumes called *lvol0* and *oravol* in the *vg01* volume group. You will use 2GB for the *lvol0* logical volume and 3GB for the *oravol* logical volume from the available pool of space. You will display the details of the volume group and the logical volumes.

1. Create *lvol0* logical volume of size 2GB using the *lvcreate* command. Use the –L option to specify the logical volume size in MBs.

   # lvcreate –L 2000 vg01
   Logical volume "lvol0" created

The size may be specified in KBs (kilobytes), MBs (megabytes), GBs (gigabytes), TBs (terabytes), PBs (petabytes), EBs (exabytes), or LEs (logical extents), however, MB is the default.

2. Create *oravol* logical volume of size 3GB using the *lvcreate* command. Use the –l option to specify the logical volume size in number of LEs.

   # lvcreate –l 188 –n oravol vg01
   Logical volume "oravol" created

3. Execute the *vgdisplay* command on the *vg01* volume group with the –v option to get the details. You can also run the *lvdisplay* command on the */dev/vg01/lvol0* and */dev/vg01/oravol* to obtain the details of the two logical volumes.

   # vgdisplay –v vg01
   . . . .

Cur LV	2
Alloc PE / Size	313 / 4.89 GiB
Free PE / Size	61 / 976.00 MiB

   --- Logical volume ---

LV Name	/dev/vg01/lvol0
VG Name	vg01
LV UUID	7CkZwX–DphV–u2rj–rrjK–PK3M–pp9Q–pdZSp9
LV Write Access	read/write
LV Status	available
# open	0
LV Size	1.95 GiB
Current LE	125
Segments	1
Allocation	inherit
Read ahead sectors	auto
– currently set to	256
Block device	253:2

```
--- Logical volume ---
LV Name              /dev/vg01/oravol
VG Name              vg01
LV UUID              8MRiST-EbYb-i3QL-rDnJ-0qu1-3QBH-qE7xx1
LV Write Access      read/write
LV Status            available
# open               0
LV Size              2.94 GiB
Current LE           188
Segments             2
Allocation           inherit
Read ahead sectors   auto
- currently set to   256
Block device         253:3

--- Physical volumes ---
PV Name              /dev/vdb
. . . . . . . .
Total PE / Free PE   255 / 0

PV Name              /dev/vdc1
. . . . . . . .
Total PE / Free PE   119 / 61
```

If you want to understand where the *vgdisplay* command gets volume group information, go to the
/etc/lvm/backup directory and view the contents of the specific volume group file.

Exercise 9-10: Extend a Volume Group and a Logical Volume

In this exercise, you will add another disk */dev/vdd* to the *vg01* volume group to increase the pool of allocatable space. You will run the *pvcreate* command to initialize the */dev/vdd* disk prior to adding it to the volume group. You will increase the size of the *lvol0* logical volume from 2GB to 5GB using the *lvextend* command and the size of the *oravol* logical volume from 3GB to 4.2GB using the *lvresize* command. You will display the details of the volume group and the logical volumes.

1. Execute the *pvcreate* command on the */dev/vdd* disk to prepare it for use in the volume group:

    # **pvcreate /dev/vdd**
    Writing physical volume data to disk "/dev/vdd"
    Physical volume "/dev/vdd" successfully created

2. Issue the *vgextend* command on the */dev/vdd* disk to add it to the *vg01* volume group:

    # **vgextend vg01 /dev/vdd**
    Volume group "vg01" successfully extended

3. Run the *vgdisplay* command on the *vg01* volume group using the −v option to confirm the addition:

 # vgdisplay −v vg01

. . . .

Cur PV	3
Act PV	3
VG Size	9.83 GiB
PE Size	16.00 MiB
Total PE	629
Alloc PE / Size	313 / 4.89 GiB
Free PE / Size	316 / 4.94 GiB

. . . .

--- Physical volumes ---

. . . .

PV Name	/dev/vdd
PV UUID	XeBKX8-5D3Y-OBQF-RtNH-Nc6d-8y4P-hk2olI
PV Status	allocatable
Total PE / Free PE	255 / 255

The output indicates that there are now three physical volumes and the total size of the allocatable space has grown to 9.83GB. It also shows the increased number of physical extents. The added physical volume is listed at the bottom of the output under the "Physical volumes" heading.

4. Execute the *lvextend* command on the *lvol0* logical volume and specify either the absolute desired size for the logical volume or the additional amount you wish to add to it:

 # lvextend −L 5GB /dev/vg01/lvol0
 # lvextend −L +3GB /dev/vg01/lvol0
Extending logical volume lvol0 to 5.00 GiB
Logical volume lvol0 successfully resized

5. Execute the *lvresize* command on the *oravol* logical volume and specify either the absolute desired size for the logical volume or the additional amount you wish to add to it:

 # lvresize −L 4.2GB /dev/vg01/oravol
 # lvresize −L +1.2GB /dev/vg01/oravol
Rounding up size to full physical extent 1.20 GiB
Extending logical volume oravol to 4.70 GiB
Logical volume oravol successfully resized

6. Run the *lvdisplay* command to verify the new sizes. You can also use the *vgdisplay* command with the −v option to obtain this information.

 # lvdisplay /dev/vg01/lvol0

. . . .

LV Name	/dev/vg01/lvol0

. . . .

LV Size	5.00 GiB

Current LE   320

....
lvdisplay /dev/vg01/oravol

....
LV Name   /dev/vg01/oravol

....
LV Size   4.70 GiB
Current LE   301

....

Exercise 9-11: Reduce and Remove a Logical Volume

In this exercise, you will decrease the size of the *lvol0* logical volume from 5GB to 3.2GB using the *lvreduce* and the *lvresize* commands. You will then remove the *oravol* logical volume using the *lvremove* command. You will display the details of the volume group and the logical volumes.

1. Execute any of the following on the *lvol0* logical volume and specify either the absolute desired size for the logical volume or the amount that you wish to subtract from it. You will have to answer the "Do you really want to reduce lvol0?" question in the affirmative.

   # **lvreduce −L 3.2GB /dev/vg01/lvol0**
   # **lvreduce −L -1.8GB /dev/vg01/lvol0**
   # **lvresize −L 3.2GB /dev/vg01/lvol0**
   # **lvresize −L -1.8GB /dev/vg01/lvol0**
        WARNING: Reducing active logical volume to 3.20 GiB
        THIS MAY DESTROY YOUR DATA (filesystem etc.)
     Do you really want to reduce lvol0? [y/n]: **y**
        Reducing logical volume lvol0 to 3.20 GiB
        Logical volume lvol0 successfully resized

There is risk involved when reducing the size of a logical volume. There are chances that any data located on the logical extents being reduced might get lost. To be on the safe side, perform a backup of the data in the logical volume before proceeding.

2. Execute the *lvremove* command on the *oravol* logical volume to remove it. You will have to answer the "Do you really want to remove active logical volume oravol?" question in the affirmative.

   # **lvremove /dev/vg01/oravol**
     Do you really want to remove active logical volume oravol? [y/n]: **y**
        Logical volume "oravol" successfully removed

Removing a logical volume is a destructive task. Ensure that you perform a backup of any data in the target logical volume prior to deleting it. You will need to unmount the file system or disable swap in the logical volume. See Chapter 10 "File Systems & Swap" on how to unmount a file system and disable swap.

3. Run the *vgdisplay* command on the *vg01* volume group with the –v option to confirm space allocation and logical volume deletion:

```
# vgdisplay –v vg01
. . . . . . . .
LV Name              /dev/vg01/lvol0
. . . . . . . .
LV Size              3.20 GiB
Current LE           200
. . . . . . . .
```

Exercise 9-12: Reduce and Remove a Volume Group

In this exercise, you will reduce the *vg01* volume group by removing */dev/vdb*, */dev/vdc1*, and */dev/vdd* physical volumes from it using the *vgreduce* command and then remove the empty volume group.

1. Execute the *vgreduce* command on the *vg01* volume group to remove all three physical volumes from it:

```
# vgreduce vg01 /dev/vdb /dev/vdc1 /dev/vdd
Removed "/dev/vdb" from volume group "vg01"
Removed "/dev/vdc1" from volume group "vg01"
Removed "/dev/vdd" from volume group "vg01"
```

2. Execute the *vgremove* command on the *vg01* volume group to remove it:

```
# vgremove vg01
Volume group "vg01" successfully removed
```

 You can also use the –f option with the *vgremove* command to force the volume group removal even though it still contains any number of logical and physical volumes.

 Remember to proceed with caution whenever you perform reduce and remove operations.

Exercise 9-13: Uninitialize a Physical Volume

In this exercise, you will uninitialize the */dev/vdb*, */dev/vdc1*, and */dev/vdd* physical volumes by removing the LVM structural information from them using the *pvremove* command.

1. Issue the *pvremove* command on the */dev/vdb*, */dev/vdc1*, and */dev/vdd* physical volumes to uninitialize them:

```
# pvremove /dev/vdb /dev/vdc1 /dev/vdd
Labels on physical volume "/dev/vdb" successfully wiped
. . . . . . . .
```

Managing Disk Space Using the LVM Configuration Tool

The LVM Configuration tool is a graphical interface for performing most LVM management tasks that have been presented in the previous section. These tasks include:

- ✓ Displaying volume groups, logical volumes, and physical volumes, and their properties.
- ✓ Adding physical volumes to volume groups, or removing them.
- ✓ Creating or removing logical volumes.
- ✓ Displaying the physical to logical extent mapping.
- ✓ Migrating extents from one physical volume to another.
- ✓ Creating a snapshot of a logical volume.

Figure 9-3 displays the expanded view of the main window that pops up when the LVM Configuration tool *system-config-lvm* is invoked. This tool may also be invoked as System → Administration → Logical Volume Management. The main interface shows the physical and logical views of each configured volume group, and all physical and logical volumes within each volume group. It also shows any unallocated volumes (pvcreated disks and partitions but have not been assigned to a volume group) and uninitialized entities (disks and partitions that have not been pvcreated) available on the system. As you highlight an item in the left hand pane, its graphical representation is displayed in the middle window along with properties in the right pane. The middle window also shows actions available to be performed on a selected item when it is chosen.

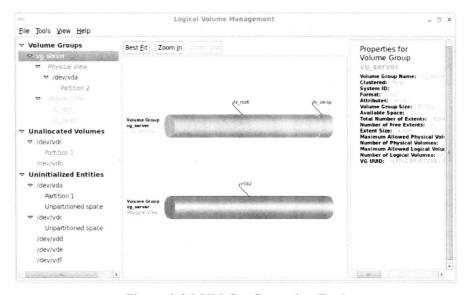

Figure 9-3 LVM Configuration Tool

Chapter Summary

This chapter started off with an overview of disk management tools available in RHEL6 and features and benefits associated with them. Later, it presented several exercises for slicing and managing disks using these tools.

You performed functions such as creating, displaying, modifying, and deleting partitions using the fdisk and parted utilities. You looked at the graphical Disk Utility that may be used to perform several disk management tasks.

You studied the concepts, components, and structure of LVM at length. You learned how to perform LVM management tasks using commands. You performed several tasks including converting disks and partitions into physical volumes; creating, displaying, extending, reducing, and removing volume groups; creating, displaying, extending, reducing, and removing logical volumes; and uninitializing physical volumes.

Finally, an overview of the LVM Configuration tool was presented which offers a graphical representation of LVM objects and allows you to administer them conveniently.

Chapter Review Questions

1. What are the two commands that you can use to reduce the number of logical extents from a logical volume?
2. A partition can be used as an LVM object. True or False?
3. Write the command to add physical volumes /dev/vdd1 and /dev/vdc to vg20 volume group.
4. Where is the partition table information stored by default?
5. What would vdd3 mean in a virtualized environment?
6. What is the name of the command line tool to invoke the LVM Configuration tool?
7. Why would you use the w command within fdisk?
8. What is the purpose of the partprobe command and why is it necessary to run this command after adding or modifying a partition in fdisk and parted?
9. What are the two commands that you can use to add logical extents to a logical volume?
10. Write the command to remove vg20 volume group along with any logical and physical volumes that it contains.
11. What is the default size of a physical extent in LVM?
12. Which of the three disk management solutions discussed in this chapter, is used most widely?
13. Which file in the /proc file system stores the in-memory partitioning information?
14. What is the default name of the first logical volume in a volume group?
15. Which command within fdisk would you use to change the type of a partition?
16. What is one difference between the pvs and pvdisplay commands?
17. What is the name of the command that you can issue to invoke the graphical Disk Utility?
18. When can a disk or partition be referred to as a physical volume?
19. A volume group can be created without any physical volume in it. True or False?
20. Where does the LVM store volume group details?
21. A single disk can be used by all three disk management solutions at a time. True or False?
22. Write the command to remove webvol logical volume from vg20 volume group.
23. It is necessary to create file system structures in a logical volume before it can be used to store files in it. True or False?
24. What is the name of the flag for swap in parted?
25. Physical and logical extents are typically of the same size. True or False?
26. What is the purpose of the pvremove command?
27. What would the command pvcreate /dev/vdd do?
28. A disk or partition can be added to a volume group without being initialized. True or False?
29. Write the command to create a logical volume called webvol of size equal to 100 logical extents in vg20 volume group.

30. What would the *fdisk* command do if it is executed with the –l option?
31. Write the command to remove */dev/vdd1* physical volume from *vg20* volume group.
32. Why would you use the –f option with the *lvremove* command?
33. Write the command to create a volume group called *vg20* with physical extent size set to 64MB and include a physical volume */dev/vdd* to the volume group.
34. Which command would you use to view the details of a volume group and logical and physical volumes within it?
35. What is the primary difference between the *mkpart* and *mkpartfs* sub-commands within *parted*?

Answers to Chapter Review Questions

1. The *lvreduce* and the *lvresize* commands.
2. True.
3. *vgextend vg20 /dev/vdd1 /dev/vdc*
4. The partition table information is stored on the Master Boot Record.
5. *vdd3* points to the third partition on the fourth virtual disk.
6. The *system-config-lvm* comand.
7. The *w* command saves any changes made within *fdisk*.
8. The purpose of this command is to instruct the kernel to update the in-memory copy of the updated disk partitions made with *fdisk* or *parted*.
9. The *lvextend* and the *lvresize* commands.
10. *vgremove –f vg20*
11. The default size of a PE in LVM is 4MB.
12. The Logical Volume Manager solution.
13. The *partitions* file.
14. *lvol0* is the default name for the first logical volume created in a volume group.
15. The *t* command.
16. The *pvs* command lists the physical volumes, whereas the *pvdisplay* command displays the details.
17. The *palimpsest* command.
18. After the *pvcreate* command has been executed on it successfully.
19. True, an empty volume group can be created.
20. In the */etc/lvm/backup* directory.
21. True, a single disk can be used by all three disk management solutions at a time.
22. *lvremove /dev/vg20/webvol*
23. True, it is mandatory.
24. The flag name for swap in *parted* is linux-swap.
25. True.
26. The *pvremove* command is used to uninitialize a physical volume.
27. This command will initialize the */dev/vdd* disk for use in a volume group.
28. False, a disk or partition must be initialized before it can be added to a volume group.
29. *lvcreate –l 100 –n webvol vg20*
30. The command provided will list all disks and partitions on the system.
31. *vgreduce vg20 /dev/vdd1*
32. With this option, the *lvremove* command does not prompt for a yes or no.
33. *vgcreate –s 64 vg20 /dev/vdd*
34. The *vgdisplay* command with the –v option.

35. The *mkpart* command creates a partition only, whereas the *mkpartfs* command not only creates a partition but also creates file system structures in it.

Labs

Lab 9-1: Manage Partitioning with fdisk

Create three 1GB primary partitions on the *vdc* disk using the *fdisk* utility and set appropriate flags to use them as a file system, a swap space, and as an LVM partition, respectively. Use defaults or your own ideas for missing information.

Lab 9-2: Manage Partitioning with parted

Create three 500MB primary partitions on the *vdd* disk using the *parted* utility and set appropriate flags to use them as a file system, a swap space, and as an LVM partition, respectively.

Lab 9-3: Manage Storage with LVM

Initialize *vdb* and *vde* disks for use in LVM. Create a volume group called *vg10* and add both physical volumes to it. Ensure that *vg10* uses 32MB as the physical extent size. Add three logical volumes called *lvol0*, *swapvol*, and *oravol* to the volume group with sizes 1.2GB, 2.4GB, and 2.1GB, respectively. Create *vdd4* and *vdd5* partitions of size 1GB each and initialize them for use in LVM. Add both new physical volumes to *vg10*. Create an additional logical volume called *lvol1* using 30 LEs. Add 200MB to *oravol* logical volume and remove 150MB from *lvol0*.

Chapter 10

File Systems & Swap

This chapter covers the following major topics:

✓ Overview of file systems

✓ Create and manage extended and LUKS-encrypted file systems

✓ Mount and unmount extended, LUKS-encrypted, removable, and network file systems manually and automatically

✓ Check and repair file system structures

✓ Set and view enhanced file permissions using Access Control List (ACL)

✓ Understand swapping and paging; how to create, activate, and delete swap

This chapter includes the following RHCSA objectives:

22. Create and configure LUKS-encrypted partitions and logical volumes to prompt for password and mount a decrypted file system at boot

23. Configure systems to mount file systems at boot by UUID or label

24. Add new partitions and logical volumes, and swap to a system non-destructively

25. Create, mount, unmount, and use ext2, ext3, and ext4 file systems

26. Mount, unmount, and use LUKS-encrypted file systems

27. Mount and unmount CIFS and NFS network file systems

28. Configure systems to mount ext4, LUKS-encrypted, and network file systems automatically

31. Create and manage Access Control Lists (ACLs)

A file system is a logical container that holds files and directories. A file system can be encrypted to safeguard the data that it holds. Each file system must be connected to the root of the directory hierarchy in order to be accessible. This is typically done automatically when the system boots up, however, it can be done manually as well. Each file system created has a unique UUID associated with it which can be used to mount it. Alternatively, a label can be assigned to a file system and then used for mounting and unmounting it. A file system space may fill up quickly depending on the file system's usage. There are tools available that help keep an eye on the space utilization. A file system may become corrupted if it is not properly unmounted due to a system crash or some other similar reasons. In such a situation, it is necessary to run a check utility and fix any issues found with the file system.

Removable and network file systems may be mounted (and unmounted) on a RHEL system and be accessed in a similar fashion as a local file system. This can be done manually using the same commands that are used for mounting and unmounting local file systems. Alternatively, RHEL provides the AutoFS service that may be configured to mount removable and network file systems automatically without having to enter any commands explicitly.

Access Control Lists allows the administrator to enforce extended security attributes on a file or directory. These attributes are in addition to the existing Linux access rights. A directory can have default ACL settings applied to it to allow multiple users to share its contents without having to change permissions on new files and sub-directories created within it.

Swapping and paging are used on every Linux system. In fact, the RHEL installation cannot proceed unless a swap space is defined. Swapping and paging provide a mechanism to move pages of data from the physical memory to the swap region and back as and when required.

Understanding and Managing File System

A *file system* is a logical container that is used to store files and directories. Each file system is created in a separate partition or logical volume. A typical RHEL system usually has numerous file systems. During installation, only two file systems are created by default: / and */boot*, though you can choose custom partitioning and construct separate containers to store dissimilar information. Typical additional file systems created during the installation are */home*, */opt*, */tmp*, */usr*, and */var*. / and */boot*; these are special file systems without which a system cannot boot.

Storing dissimilar data in separate file systems versus storing all data in a single file system offers the following advantages. You can:

✓ Make a specific file system accessible or inaccessible to users independent of other file systems. This hides or reveals information contained within that file system.
✓ Perform file system repair activities on individual file systems.
✓ Keep dissimilar data in separate file systems.
✓ Optimize or tune each file system independently.
✓ Grow or shrink a file system independent of other file systems.

Moreover, some backup tools such as *dump* works only at the file system level.

Types of File System

There are several different types of supported file systems in RHEL that may be categorized in three groups: disk-based, network-based, and memory-based. Disk-based file systems are typically created on hard drives, CD/DVD drives, floppy disks, USB flash drives, USB portable drives, and so on. Network-based file systems are basically disk-based file systems but they are shared over the network to become network accessible. Memory-based file systems such as */proc* are virtual. They are created at system startup and destroyed when the system goes down. Table 10-1 lists and explains various common disk- and network-based file system types supported in RHEL6.

File System	Type	Description
ext2	Disk	The 2nd generation of the extended file system. The 1st generation is no longer supported.
ext3	Disk	The 3rd generation of the extended file system. It supports journaling for faster recovery, offers superior reliability, supports file systems of sizes up to 16TB, and supports files of sizes up to 2TB. This type of file system uses a series of contiguous physical blocks on the hard disk called *extents*, which results in improved read and write performance and reduced fragmentation. From a recovery standpoint, each metadata update is written in its entirety to the *journal* after it has been completed. The system looks in the journal of each file system following a reboot after a system crash has occurred, and recovers the file system rapidly using the updated structural information stored in its journal.
ext4	Disk	The 4th generation of the extended file system. It supports all features of the ext3 file system type in addition to a larger file system size of up to 1EB and a bigger file size of up to 16TB. ext4 is the default file system type in RHEL6.
ISO 9660	Disk	This is used for CD and DVD file systems.
MS-DOS	Disk	This is used for pre-Windows 95 file system formats.
NTFS	Disk	This is used for post-Windows 95 file system formats on hard disks, USB drives, and floppy disks. It supports security at the user login level.
VFAT	Disk	Same as NTFS but does not support security at the user login level. It supports FAT16 and FAT32 types as well.
NFS	Network	Network File System. A directory or file system shared over the network for access by other Linux systems. See Chapter 19 "Network File System" for details.
AutoFS	Network	Auto File System. A directory or file system shared over the network and set to mount and unmount automatically on a remote system.
CIFS	Network	Common Internet File System (a.k.a. Samba). A directory or file system shared over the network for access by Windows and other Linux systems. See Chapter 20 "Samba & FTP" for details.

Table 10-1 File System Types

This chapter covers ext2, ext3, and ext4 file systems at length. It also covers mounting and unmounting ISO 9660, CIFS, and NFS file systems. Memory-based, VFAT, MS-DOS, and NTFS file systems are beyond the scope.

Extended File Systems

Extended file systems have been supported in RHEL for years. The first generation is obsolete and is no longer supported. The second and third generations have been available in RHEL for a long time. The third generation was the first in the series that supported what is called *journaling*. The fourth generation is the latest in the series and is superior to the previous generations.

The structure of an extended file system is built in a partition or logical volume when the file system is created. This structure is divided into two sets. The first set, which is very small, holds the file system's metadata information and the second set, which is almost the entire partition or the logical volume, stores the actual data.

The metadata information includes the *superblock* which keeps vital file system structural information such as the type, size, and status of the file system, and the number of data blocks in it. Since the superblock holds such critical information, a copy of it is automatically stored at various locations throughout the file system. The superblock at the beginning of the file system is referred to as the *primary superblock* and all others as *backup superblocks*. If for any reasons the primary superblock is corrupted or lost, the file system becomes inaccessible. The metadata also contains the *inode table* which maintains a list of the *index node* (inode) numbers. Each inode number is assigned to a file when the file is created, and holds the file related information such as its type, permissions, ownership, group membership, size, and last access/modification time. The inode also holds and keeps track of the pointers that point to the actual data blocks where the file contents are located.

The ext3 and ext4 file systems are journaling file systems that have several benefits over the conventional ext2 file system. One major advantage is their ability to recover quickly after a system crash. Both ext3 and ext4 file systems keep track of their structural (metadata) changes in a *journal*. Each structural update is written in its entirety to the journal after it has completed. The system looks in the journal of each file system after it has been rebooted following a crash, and recovers the file system rapidly using the updated structural information stored in its journal.

In contrast with ext3, the ext4 file system supports very large file systems of sizes up to 1EB (ExaByte) and files of sizes up to 16TB. Additionally, ext4 uses a series of contiguous physical blocks on the hard disk called *extents*. This technique improves the performance of very large files and reduces the fragmentation.

LUKS-Encrypted File System

LUKS (*Linux Unified Key Setup*) is an encryption method for securing the data stored in a block device such as a partition or a logical volume. After a partition or logical volume has been encrypted with LUKS, a passphrase (or a key) is required to decrypt the data. You will have one master key and one or more distinct user keys given to individual users to decrypt the master key and get access to the data. LUKS uses the device mapper kernel subsystem for encryption management. During RHEL installation, the anaconda installer program offers you an opportunity to set encryption on file systems. If you select that option, you will be prompted to enter the passphrase at each system reboot. Refer to Chapter 01 "Local Installation" for details.

File System Administration Commands

Managing file systems involves creating, manually mounting, labeling, viewing, extending, reducing, unmounting, automatically mounting, modifying attributes of, and removing a file system. There are several commands available to help perform these tasks. Table 10-2 lists and describes them.

Command	Description
blkid	Displays block device attributes.
cryptsetup	Sets up LUKS encryption on a file system.
df	Displays file system utilization in detail.
du	Calculates disk usage of directories and file systems.
dumpe2fs	Displays the metadata information of an extended file system.
e2fsck	Checks and repairs an extended file system.
e2label	Modifies the label on an extended file system.
findfs	Finds a file system by label or UUID.
fsck	Frontend to the *e2fsck*, *fsck.ext2*, *fsck.ext3*, and *fsck.ext4* commands.
fsck.ext2	Equivalent to *fsck –t ext2*.
fsck.ext3	Equivalent to *fsck –t ext3*.
fsck.ext4	Equivalent to *fsck –t ext4*.
fuser	Lists and terminates processes using a file system.
lsof	Lists open files.
mke2fs	Creates an extended file system.
mkfs	Frontend to the *mke2fs*, *mkfs.ext2*, *mkfs.ext3*, and *mkfs.ext4* commands.
mkfs.ext2	Equivalent to *mkfs –t ext2*.
mkfs.ext3	Equivalent to *mkfs –t ext3*.
mkfs.ext4	Equivalent to *mkfs –t ext4*.
mount	Mounts a file system for user access. Also displays currently mounted file systems.
resize2fs	Resizes an extended file system.
tune2fs	Tunes an extended file system attributes.
umount	Unmounts a file system.

Table 10-2 File System Management Commands

Mounting and Unmounting a File System

The *mount* command is used to attach a file system to the root of the directory hierarchy. This command is executed after a file system has been constructed and it needs to be made accessible to users. For this purpose, a directory is created, which is referred to as the *mount point*, and then the file system is mounted on top of it. The *mount* command is also used to mount other types of file systems such as optical and network file systems. This command adds an entry to the */etc/mtab* file and the kernel adds an entry to the */proc/mounts* file after a file system has been successfully mounted.

A mount point should be empty when an attempt is made to mount a file system on it, otherwise, the contents of the mount point will hide. As well, the mount point must not be in use or the mount attempt will fail.

Several options are available with the *mount* command. Some of them are described in Table 10-3.

Option	Description
async (sync)	All file system I/O to occur asynchronously (synchronously).
atime (noatime)	Updates (does not update) the inode access time for each access.
auto (noauto)	Mounts (does not mount) the file system when the –a option is specified.
defaults	Accepts all default values (rw, suid, dev, exec, auto, nouser, and async.
dev (nodev)	Interprets (does not interpret) the device files on the file system.
exec (noexec)	Permits (does not permit) the execution of a binary file.
owner	Allows the file system owner to mount the file system.
remount	Remounts an already mounted file system.
ro (rw)	Mounts a file system read-only (read/write).
suid (nosuid)	Enables (disables) running setuid and setgid programs.
user (nouser)	Allows (disallows) a normal user to mount a file system.
users	Allows all users to mount and unmount a file system.

Table 10-3 mount Command Options

The opposite of the *mount* command is the *umount* command, which is used to detach a file system from the root of the directory hierarchy and make it inaccessible to users. This command removes the corresponding entry from the */etc/mtab* file and the kernel removes its entry from the */proc/mounts* file.

Examples on the usage of the *mount* and *umount* commands are provided in Exercises later in this chapter.

Determining the UUID of a File System

An extended file system created in a standard partition or a logical volume has a UUID (*Universally Unique IDentifier*) assigned to it at the time of its creation (using the *mkfs* command or one of its variants). Assigning a UUID makes the file system unique since there will likely be several file systems. The biggest benefit with using UUID is that it always stays persistent across system reboots. The UUID is used by default in RHEL6 in the */etc/fstab* file for any file system that is created by the system in a standard partition.

The system attempts to mount all file systems listed in the */etc/fstab* file at every system reboot. Each file system has an associated device file and UUID, but may or may not have a corresponding label. The system checks for the presence of each file system's device file, UUID, or its label, and then attempts to mount it.

The */boot* file system, for instance, is located in a partition and the device file associated with this partition on *server.example.com* is */dev/vda1*. The *tune2fs* command with the –l option or the *blkid* *command* can be issued to determine the UUID of */dev/vda1*. Also you can *grep* for boot in the */etc/fstab* file.

```
# tune2fs –l /dev/vda1
Filesystem UUID: d77e99bd-720b-455a-ab16-892217141156
# blkid
/dev/vda1: UUID="d77e99bd-720b-455a-ab16-892217141156" TYPE="ext4"
# grep boot /etc/fstab
UUID=d77e99bd-720b-455a-ab16-892217141156 /boot    ext4    defaults  1 2
```

The above output indicates that the UUID of */boot* is used in the */etc/fstab* file instead of its partition name which is */dev/vda1*. A discussion on the */etc/fstab* file is provided in a later sub-section.

A UUID is also assigned to a file system created in a logical volume, however, it need not be used in the */etc/fstab* file as the device files for physical volumes and logical volumes do not change. They remain persistent across system reboots.

Labeling an Extended File System

A label may be used instead of a UUID to keep the file system association with its device file persistent across system reboots. The */boot* file system, for instance, is located in a partition and the device file associated with this partition on *server.example.com* is */dev/vda1*. The *tune2fs* command with the –l option can be issued to determine the label of */dev/vda1*. This command uses the term "volume name" instead of the term "label" in its output as indicated below:

```
# tune2fs –l /dev/vda1 | grep volume
Filesystem volume name:  <none>
```

The above output shows that there is currently no label assigned to the */boot* file system. A label is not needed for a file system if you are using its UUID, however, you can still apply one using the *e2label* command. The following example creates and applies the label "bootvol" to the */dev/vda1* device file which has the */boot* file system in it:

```
# e2label /dev/vda1 bootvol
```

Execute any of the following commands to confirm the label on the file system:

```
# e2label /dev/vda1
bootvol
# tune2fs –l /dev/vda1 | grep volume
Filesystem volume name:  bootvol
```

Now this label may be used for the */boot* entry instead of the UUID in the */etc/fstab* file. A discussion on the */etc/fstab* file is provided in the next sub-section.

A label may also be applied to a file system created in a logical volume, however, it need not be used in the */etc/fstab* file as the device files for physical volumes and logical volumes do not change. They remain persistent across system reboots.

Automatically Mounting a File System at System Reboots

File systems defined in the */etc/fstab* file are mounted automatically at each system reboot. This file must contain proper and complete information for each listed file system. An incomplete or inaccurate entry might leave the system in an undesirable state. Another benefit with adding entries to this file is that if a file system needs to be mounted manually, you can specify its mount point, device file, UUID, or label with the *mount* command, and the *mount* command will obtain rest of the information from this file.

The default *fstab* file contains entries for file systems that are created at the time of installation. On *server.example.com* for instance, this file contains the following information:

```
# cat /etc/fstab
/dev/mapper/vg_server-lv_root                          /         ext4    defaults         1 1
UUID=d77e99bd-720b-455a-ab16-892217141156              /boot     ext4    defaults         1 2
/dev/mapper/vg_server-lv_swap                          swap      swap    defaults         0 0
tmpfs                                                  /dev/shm  tmpfs   defaults         0 0
devpts                                                 /dev/pts  devpts  gid=5,mode=620   0 0
sysfs                                                  /sys      sysfs   defaults         0 0
proc                                                  /proc     proc    defaults         0 0
```

There are six columns per row and they are explained below:

- ✓ The first column defines the physical or virtual device where the file system is resident, or its associated UUID or label.
- ✓ The second column defines the mount point for the file system. In the case of a swap partition or logical volume, you will see an entry by the name "swap" in this column.
- ✓ The third column specifies the type of file system such as ext3, ext4, tmpfs, devpts, sysfs, proc, vfat, or iso9660. In the case of a swap partition or logical volume, you will see an entry by the name "swap" in this column.
- ✓ The fourth column specifies any options to use when mounting the file system. Some of these options are listed and described in Table 10-3 earlier.
- ✓ The fifth column specifies whether to write the data to the file system when the system shuts down. A value of 0 disables it and a value of 1 enables it.
- ✓ The last column indicates the sequence number in which to run the *fsck* (file system check and repair) utility on the file system. By default, the sequence number is 1 for /, 2 for */boot* and other physical file systems, and 0 for memory-based, remote, and removable file systems.

If you wish to mount all the file systems listed in the *fstab* file manually or after adding a new entry to it, run the *mount* command with the –a option:

```
# mount –a
```

The above command will only mount the file systems listed in the *fstab* file that are not already mounted.

Reporting File System Space Usage

File system space usage reporting involves checking the used and available file system space. The *df* (disk free) command is used for this purpose. It reports details of file system blocks and lists each file system with its device file; total, used, and available blocks; percentage of used blocks; and the mount point. By default, the *df* command displays the output in KBs if no option is specified. However, you can specify the –m option to view the output in MBs or the –h option to view the information in human readable format.

Run the *df* command with the –h option:

```
# df –h
Filesystem          Size   Used  Avail  Use%  Mounted on
/dev/mapper/vg_server-lv_root
                    18G    2.8G   15G   17%   /
tmpfs              387M  272K  387M    1%   /dev/shm
/dev/vda1          485M   33M  428M    7%   /boot
```

With the –h option, the command shows the count in KBs, MBs, or GBs, as appropriate. Try running the *df* command with the other options as well as with the "–t ext3" and the "–t ext4" options.

Estimating File Space Usage

A file, directory, or an entire file system may be estimated for space usage. The *du* (disk usage) command is used for this purpose. It reports a summary of space occupied by directories and file systems. By default, this command displays the output in KBs if no option is specified. However, you can specify the –m option to view the output in MBs or the –h option to view in human readable format.

Run the *du* command with the –h option:

```
# du –h /boot
276K   /boot/grub
246K   /boot/efi/EFI/redhat
248K   /boot/efi/EFI
250K   /boot/efi
13K    /boot/lost+found
22M    /boot
```

With the –h option, the command shows the count in KBs, MBs, or GBs, as appropriate. Try running the *du* command with the other options and also try it with the –s option.

Analyzing Space Usage with Disk Usage Analyzer

The Disk Usage Analyzer is a graphical tool that may be used to perform file system space analysis. This tool can be started by running the *baobab* command or choosing Applications → System Tools → Disk Usage Analyzer. Figure 10-1 shows the screen that pops up when this tool is invoked.

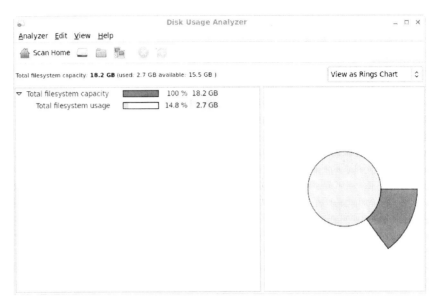

Figure 10-1 Disk Usage Analyzer

This tool scans local or remote directories and file systems, and reports on usage in terms of number of objects, used capacity, etc. in a tree or ring chart.

If the Disk Usage Analyzer tool is not already installed, use one of the package installation methods discussed in Chapter 06 "Package Management" and install the baobab package to get this tool.

Repairing a Damaged File System

The structure of a file system could be damaged when an abnormal system shutdown or crash occurs. To maintain file system integrity, a utility called *fsck* (or one of its variants *fsck.ext2*, *fsck.ext3*, *fsck.ext4*, or *e2fsck*) is used. This utility is called automatically when a reboot occurs following an abnormal system shutdown or crash. It performs multiple checks on file system structures, reports any inconsistencies as it finds, and attempts to fix them automatically. If an inconsistency cannot be resolved, it prompts for user intervention. The *fsck* command can also be executed manually on a file system at the command line, but you have to ensure that the file system is unmounted before you attempt to run this utility on it. This command expects a "yes" or "no" response as it proceeds with trying to correct inconsistencies.

During the check, if *fsck* encounters that the primary superblock is corrupted, it prompts for user intervention. You will need to run the *dumpe2fs* command to list backup superblock locations and then specify one of the locations with the *fsck* command to replace the primary superblock.

While checking a file system, *fsck* may encounter a file with a missing name. It moves the file to the *lost+found* directory located in that file system. This file is known as an *orphan* file and is renamed to correspond to its inode number. You need to figure out the actual name of the file and its original location. Use the *file* command to determine the file's type. If it is a text file, use *cat* or *more* to view its contents; otherwise, use the *strings* command to view the legible contents in it. You can move the file to its correct directory location if you determine its whereabouts.

Exercise 10-1: Create and Mount Extended File Systems

In Chapter 09 "Disk Partitioning", you learned about the management of disks using the standard and the LVM partitioning techniques. In order for a partition or logical volume to be able to store files and directories, it must be initialized as a file system; otherwise, it can only be used as a raw partition or swap. In RHEL, the command that is used to construct an ext2, ext3, or ext4 file system is *mkfs*. As noted earlier, the *mkfs* command is the frontend to the *mke2fs*, *mkfs.ext2*, *mkfs.ext3*, and *mkfs.ext4* commands.

In Lab 9-1, you created the *vdc1* partition of size 1GB; in Lab 9-2, you created the *vdd1* partition of size 500MB; and in Lab 9-3, you created *vg10* volume group with four physical volumes *vdb*, *vde*, *vdd4*, and *vdd5* and four logical volumes *lvol0* (1.05GB), *lvol1* (960MB), *swapvol* (2.4GB), and *oravol* (2.3GB).

In this exercise, you will initialize *vdc1* as an ext2 file system, *vdd1* as an ext3 file system, and */dev/vg10/lvol0* as an ext4 file system on *server.example.com*. You will create */ex101a*, */ex101b*, and */ex101c* mount points and manually mount the three file systems using their device names. You will run appropriate commands to confirm their sizes as 1GB, 500MB, and 1.05GB, respectively. You will append entries to the */etc/fstab* file for the *vdc1* file system using its UUID, the *vdd1* file system using its label, and the *lvol0* file system. You will unmount them manually using their mount points. You will then reboot the system and run appropriate commands to verify that all three file systems have been mounted automatically.

1. Run the *mkfs* command with the –t ext2 option on the */dev/vdc1* partition to create ext2 file system structures:

        # mkfs –t ext2 /dev/vdc1
        mke2fs 1.41.12 (17–May–2010)
        Filesystem label=
        OS type: Linux
        Block size=4096 (log=2)
        Fragment size=4096 (log=2)
        Stride=0 blocks, Stripe width=0 blocks
        61184 inodes, 244306 blocks
        12215 blocks (5.00%) reserved for the super user
        First data block=0
        Maximum filesystem blocks=251658240
        8 block groups
        32768 blocks per group, 32768 fragments per group
        7648 inodes per group
        Superblock backups stored on blocks:
            32768, 98304, 163840, 229376

        . . . .

2. Run the *mke2fs* command on the */dev/vdd1* partition to create ext3 file system structures:

        # mke2fs –t ext3 /dev/vdd1
        mke2fs 1.41.12 (17–May–2010)
        Filesystem label=
        OS type: Linux

Block size=4096 (log=2)
Fragment size=4096 (log=2)
Stride=0 blocks, Stripe width=0 blocks
122160 inodes, 488368 blocks
24418 blocks (5.00%) reserved for the super user
First data block=0
Maximum filesystem blocks=503316480
15 block groups
32768 blocks per group, 32768 fragments per group
8144 inodes per group
Superblock backups stored on blocks:
        32768, 98304, 163840, 229376, 294912

. . . .

3.   Run the *mkfs.ext4* command on the */dev/vg10/lvol0* logical volume to create ext4 file system structures:

 # mkfs.ext4 /dev/vg10/lvol0
mke2fs 1.41.12 (17-May-2010)
Filesystem label=
OS type: Linux
Block size=4096 (log=2)
Fragment size=4096 (log=2)
Stride=0 blocks, Stripe width=0 blocks
67680 inodes, 270336 blocks
13516 blocks (5.00%) reserved for the super user
First data block=0
Maximum filesystem blocks=276824064
9 block groups
32768 blocks per group, 32768 fragments per group
7520 inodes per group
Superblock backups stored on blocks:
        32768, 98304, 163840, 229376

. . . .

4.   Create */ex101a*, */ex101b*, and */ex101c* mount points:

 # mkdir /ex101a /ex101b /ex101c

5.   Mount the */dev/vdc1*, */dev/vdd1*, and */dev/vg10/lvol0* file systems using the *mount* command:

 # mount /dev/vdc1 /ex101a
 # mount /dev/vdd1 /ex101b
 # mount /dev/vg10/lvol0 /ex101c

6.   Use the *df* command with the –h option and confirm that the sizes for the three file systems are 1GB, 500MB, and 1.05GB:

```
# df –h
. . . . . . . .
/dev/vdc1                   940M  1.2M  891M  1% /ex101a
/dev/vdd1                   463M  2.3M  437M  1% /ex101b
/dev/mapper/vg10-lvol0      1.1G  34M   954M  4% /ex101c
```

7. Determine the UUID for */dev/vdc1* using the *tune2fs* command:

   # tune2fs –l /dev/vdc1 | grep UUID
   Filesystem UUID: ce407ada-0cb3-4486-9764-2a5c35b91b48

8. Apply the label "exvdd1" to */dev/vdd1* using the *e2label* command:

   # e2label /dev/vdd1 exvdd1

9. Confirm that the label has been successfully applied to */dev/vdd1* using the *tune2fs* command:

   # e2label /dev/vdd1
   exvdd1

10. Open the */etc/fstab* file and append entries for all three file systems:

    ```
    # vi /etc/fstab
    UUID=ce407ada-0cb3-4486-9764-2a5c35b91b48    /ex101a ext2 defaults 1 2
    LABEL=exvdd1                                 /ex101b ext3 defaults 1 2
    /dev/vg10/lvol0                              /ex101c ext4 defaults 1 2
    ```

11. Unmount all three file systems using their mount points:

    # umount /ex101a /ex101b /ex101c

12. Reboot the system using the *shutdown* command:

    # shutdown –ry now

13. Run the *mount* command and check if both file systems are mounted:

    ```
    # mount | grep ex101
    /dev/vdc1 on /ex101a type ext2 (rw)
    /dev/vdd1 on /ex101b type ext3 (rw)
    /dev/mapper/vg10-lvol0 on /ex101c type ext4 (rw)
    ```

Exercise 10-2: Create and Mount a LUKS File System

In Lab 9-3, you created *vg10* volume group with four physical volumes *vdb*, *vde*, *vdd4*, and *vdd5* and four logical volumes *lvol0* (1.05GB), *lvol1* (960MB), *swapvol* (2.4GB), and *oravol* (2.3GB).

In this exercise, you will initialize */dev/vg10/lvol1* as a LUKS-encrypted file system and then create an ext4 file system in it. You will create */lvol1_luks* mount point and manually mount the file

---

system using its device name. You will run appropriate commands to confirm its size as 960MB. You will create a passkey file and assign that file to the file system. You will add appropriate entries to the *etc/crypttab* and *etc/fstab* files for the file system using its LVM device file for automatic mounting at reboots. You will unmount the file system manually using its mount point. You will then reboot the system and run appropriate commands to verify that the LUKS-encrypted file system has been mounted automatically.

1. Check whether the LUKS encryption packages are installed:

   # **rpm –qa | grep luks**
   cryptsetup-luks-libs-1.2.0-6.el6.x86_64
   cryptsetup-luks-1.2.0-6.el6.x86_64

   Install the packages if they are not already installed using one of the procedures outlined in Chapter 06 "Package Management".

2. Run the *cryptsetup* command on the *lvol1* logical volume to initialize it. You must say YES to proceed. Supply a passphrase twice.

   # **cryptsetup luksFormat /dev/vg10/lvol1**
   WARNING!

   This will overwrite data on /dev/vg10/lvol1 irrevocably.
   Are you sure? (Type uppercase yes): **YES**
   Enter LUKS passphrase:
   Verify passphrase:
   Command successful.

3. Run the *cryptsetup* command again to open the logical volume and assign it the name "lvol1_luks". Enter the passphrase that you have created in step 2. This command will create a device file for lvol1_luks in the */dev/mapper* directory.

   # **cryptsetup –v luksOpen /dev/vg10/lvol1 lvol1_luks**
   Enter passphrase for /dev/vg10/lvol1:

4. Execute the *ll* command on the */dev/mapper* directory and *grep* for *lvol1-luks* to confirm that the device file has been created:

   # **ll /dev/mapper | grep lvol1_luks**
   lrwxrwxrwx. 1 root root   7 Sep 18 08:53 lvol1_luks -> ../dm-6

5. Execute the *mkfs* command to construct the ext4 file system structures in the logical volume:

   # **mkfs –t ext4 /dev/mapper/lvol1_luks**
   mke2fs 1.41.12 (17-May-2010)
   Filesystem label=
   OS type: Linux
   Block size=4096 (log=2)
   Fragment size=4096 (log=2)

Stride=0 blocks, Stripe width=0 blocks
61312 inodes, 245248 blocks
12262 blocks (5.00%) reserved for the super user
First data block=0
Maximum filesystem blocks=251658240
8 block groups
32768 blocks per group, 32768 fragments per group
7664 inodes per group
Superblock backups stored on blocks:
    32768, 98304, 163840, 229376
. . . .

6. Create a mount point called */lvol1_luks* for this file system:

    # **mkdir /lvol1_luks**

7. Run the *mount* command to mount the file system manually:

    # **mount /dev/mapper/lvol1_luks /lvol1_luks**

8. Issue the *df* command to verify that the LUKS file system has been mounted and verify its size:

    # **df**
    . . . .
    /dev/mapper/lvol1_luks 965560 17588 898924 2% /lvol1_luks

9. Create a passkey file to store the passphrase so this file system is mounted automatically at each system reboot. Execute the *dd* command and write some random data to it:

    # **dd if=/dev/random of=/root/lvol1_luks_pass.key bs=32 count=1**
    0+1 records in
    0+1 records out
    16 bytes (16 B) copied, 5.7453e-05 s, 278 kB/s

10. Issue the *cryptsetup* command and add the passkey file location to the file system:

    # **cryptsetup luksAddKey /dev/vg10/lvol1 /root/lvol1_luks_pass.key**

11. Open the */etc/crypttab* file and add the following entry:

    lvol1_luks /dev/vg10/lvol1 /root/lvol1_luks_pass.key

12. Open the */etc/fstab* file and add the following entry for an automatic mounting of this file system at system reboots:

    /dev/mapper/lvol1_luks /lvol1_luks ext4 defaults 1 2

---

13. Reboot the system and run the *mount* command to check the mount status of the *lvol1_luks* file system:

 # mount | grep lvol1_luks
/dev/mapper/lvol1_luks on /lvol1_luks type ext4 (rw)

Exercise 10-3: Unmount and Remove File Systems

In this exercise, you will *cd* into the */ex101a* mount point and try to unmount the *vdc1* and *lvol0* file systems. You will take appropriate measures to unmount the *vdc1* file system if you get a "device busy" message. You will destroy both file systems using appropriate commands. You will delete the */ex101a* and */ex101c* mount points, and remove the corresponding file system entries from the */etc/fstab* file. You will then reboot the system and confirm that the system has been booted up without any issues. You will run appropriate commands after the system is up to verify that both file systems have been removed.

Removing a file system is a destructive operation. It wipes out all data from the file system. Always proceed with caution.

1. Run the *umount* command on */ex101a* and */ex101c* to unmount the file systems:

   # umount /ex101a /ex101c
       (In some cases useful info about processes that use
       the device is found by lsof(8) or fuser(1))

2. Issue the *fuser* command with the –cu options (c indicates the Process ID and u indicates the user owning the Process ID) to determine which user(s) and process(es) are using the */ex101a* mount point:

   # fuser –cu /ex101a
   /ex101a: 2322c(root)

3. Run the *fuser* command again but with the –ck options to terminate process 2322 which is reported in the previous step as being using the mount point:

   # fuser –ck /ex101a

4. Now you should be able to umount the file system at */ex101a*:

   # umount /ex101a

5. Execute the *fdisk* command on the */dev/vdc* disk and remove the *vdc1* partition:

   # fdisk /dev/vdc
   Command (m for help): **p**
     Device Boot   Start   End   Blocks   Id System
   /dev/vdc1   1   1939   977224+ 83 Linux
   Command (m for help): **d1**

Command (m for help): **w**

....

 # partprobe /dev/vdc

6. Execute the *lvremove* command on the */dev/vg10/lvol0* logical volume to remove it:

    # lvremove /dev/vg10/lvol0
    Do you really want to remove active logical volume lvol0? [y/n]: **y**
    Logical volume "lvol0" successfully removed

7. Remove the */ex101a* and */ex101c* mount points:

    # rmdir /ex101a /ex101c

8. Open the */etc/fstab* file in *vi* and remove the following two entries from it:

    UUID=432fa2bf–60d2–4bbe–8d26–5cd79c896f63   /ex101a ext2 defaults 1 2
    /dev/vg10/lvol0   /ex101c ext4 defaults 1 2

9. Reboot the system with the *shutdown* command:

    # shutdown –ry now

10. Issue the *mount* command and verify that the two file systems are not there:

    # mount | grep ex101

Administering File Systems Using the LVM Configuration Tool

The LVM Configuration tool may also be used to create, mount, and remove file systems. These tasks may be performed at the time of logical volume creation and removal.

To create a logical volume *lvol2* of size 2GB in the *vg10* volume group along with extended file system structures in it, execute *system-config-lvm* in an X terminal window or click System → Administration → Logical Volume Management in the GNOME desktop and go to Volume Groups → vg10 → vg10 Logical View. Click Create New Logical Volume. Enter a logical volume name and size. Click to choose a file system type. Check the boxes beside "Mount" and "Mount when rebooted", and specify a mount point. Click OK. This will create the specified logical volume and initialize it with the file system type. It will also create a mount point and mount the logical volume on it. Moreover, it will add an entry to the */etc/fstab* file so the file system is automatically mounted at each system reboot. See Figure 10-2.

To remove the *lvol2* logical volume in the *vg10* volume group along with the extended file system in it, execute *system-config-lvm* in an X terminal window or click System → Administration → Logical Volume Management in the GNOME desktop and go to Volume Groups → vg10 → vg10 Logical View → lvol2. Click Remove Logical Volume and then click Yes to remove the logical volume as well as the file system in it.

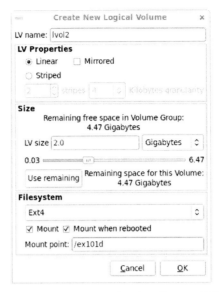

Figure 10-2 LVM Configuration Tool – Create LV & File System

Exercise 10-4: Repair a Damaged File System

In this exercise, you will unmount the */dev/vdd1* file system and run the file system check and repair utility on it. You will replace its primary superblock assuming that it is corrupted. You will then mount the file system back.

1. Unmount */dev/vdd1* using its mount point */ex101b*:

    # **umount /ex101b**

2. Execute the *fsck* command on */dev/vdd1*:

    # **fsck /dev/vdd1**
    fsck from util-linux-ng 2.17.2
    e2fsck 1.41.12 (17-May-2010)
    exvdd1: clean, 11/122400 files, 25991/488848 blocks

3. Run the *dumpe2fs* command to obtain a list of backup superblock locations:

    # **dumpe2fs /dev/vdd1 | grep superblock**
    dumpe2fs 1.41.12 (17-May-2010)
      Primary superblock at 1, Group descriptors at 2-3
      Backup superblock at 8193, Group descriptors at 8194-8195
      Backup superblock at 24577, Group descriptors at 24578-24579
    . . . .

4. Run the *dumpe2fs* command to obtain a list of backup superblock locations:

 # fsck –b 8193 /dev/vdd1

fsck from util-linux-ng 2.17.2

e2fsck 1.41.12 (17–May–2010)

exvdd1 was not cleanly unmounted, check forced.

Pass 1: Checking inodes, blocks, and sizes

Pass 2: Checking directory structure

Pass 3: Checking directory connectivity

Pass 4: Checking reference counts

Pass 5: Checking group summary information

exvdd1: ***** FILE SYSTEM WAS MODIFIED *****

exvdd1: 11/122400 files (0.0% non-contiguous), 25991/488848 blocks

5. Mount the file system back:

 # mount /ex101b

Mount and Unmount File Systems in Removable Media

You may need to mount, unmount, and check status of file systems in removable devices such as an optical or a USB device. The following examples demonstrate how to mount and unmount a CD/DVD and a USB flash drive media.

To mount a CD/DVD with device file *022/dev/cdrom* (or *022/dev/sr0*) formatted with ISO 9660 file system type on *022/cdrom0* in read-only mode, run the following:

 # mkdir /cdrom0
 # mount –t iso9660 –o ro /dev/cdrom /cdrom0

To mount a USB flash drive such as *022/dev/sdb1* formatted with the ext4 file system type on *022/usb0*:

 # mkdir /usb0
 # mount /dev/sdb1 /usb0

To check their status, use either the *df* or the *mount* command:

 # df

. . . .

/dev/sr0	3505504	3505504	0	100%	/cdrom0
/dev/sdb1	198337	56690	131407	31%	/usb0

 # mount

. . . .

/dev/sr0 on /cdrom0 type iso9660 (ro)

/dev/sdb1 on /usb0 type ext4 (rw)

To unmount the CD/DVD and the USB, issue the *umount* command and supply either their mount points or their associated device files as an argument. If the mount points are provided, the command gets the device file names from the *022/etc/mtab* file.

 # umount /cdrom0 (or umount /dev/cdrom or umount /dev/sr0)

 # umount /usb0 (or **umount /dev/sdb1**)

Exercise 10-5: Mount and Unmount a CIFS File System

For this exercise, it is assumed that a Samba server is already configured on *physical.example.com* and it is sharing the */project1* directory in read/write mode over the network with appropriate firewall rules and SELinux settings in place. It is also assumed that the share is configured for access to user *aghori* with password Welcome01 on *server.example.com*.

In this exercise, you will mount this Samba share (a.k.a. CIFS file system) on *server.example.com*, check the share's status, and then unmount it. You will add an entry for the share to the */etc/fstab* file and reboot the system to confirm that the share has been automatically mounted after the reboot. Refer to Chapter 20 "Samba & FTP" for details on Samba.

1. Check if the system has the smbclient package already installed to be able to use the Samba client functionality:

   # rpm –qa | grep smb
   gvfs-smb-1.4.3-12.el6.x86_64
   libsmbclient-3.5.10-114.el6.x86_64
   gnome-vfs2-smb-2.24.2-6.el6.x86_64

2. Create a mount point called */physical/cifs/remote*:

   # mkdir –p /physical/cifs/remote

3. Run the *smbclient* command with the –L option to list available resources from *physical.example.com* and specify the username and password with the –U option:

   # smbclient –L physical.example.com –U aghori%Welcome01

4. Edit */etc/fstab* and add the following entry for the remote file system. This is accomplished to ensure that the file system is automatically mounted when *server.example.com* is rebooted.

   //192.168.2.200/project1 /physical/cifs/remote cifs   credentials=/etc/samba/smbcred 0 0

5. Create the */etc/samba/smbcred* file and add the following user credentials to it:

   # vi /etc/samba/smbcred
   username=aghori
   password=Welcome01

6. Run any of the following to mount the Samba share:

   # mount –t cifs –o rw //physical.example.com/project1 /physical/cifs/remote
   # mount /physical/cifs/remote

7. Use the *df* command or the *mount* command to list the share and the *ll* command to list the contents.

8. Issue the *umount* command and supply the mount point as an argument to unmount the file system:

> # **umount /physical/cifs/remote**

9. Reboot the system:

> # **shutdown –ry now**

10. Check using the *df* or the *mount* command if the file system has been automatically remounted after the reboot.

Another way of accessing the Samba share is to go to Places → Home Folder → File → Open Location. Enter smb:// and an icon associated with connected workgroups and/or domains should appear. Click the icon to view a list of all systems with Samba shares available. Click *physical.example.com* to access the share.

Exercise 10-6: Mount and Unmount an NFS File System

In Exercise 6-3 of Chapter 06 "Package Management", you configured an NFS server on *physical.example.com* (192.168.2.200) and exported the */var/yum/repos.d/remote* directory.

In this exercise, you will mount, check status of, and unmount this NFS file system on *server.example.com*. You will add an entry for the file system to the */etc/fstab* file and reboot the system to confirm that the file system has been automatically mounted after the reboot. Refer to Chapter 19 "Network File System" for details on NFS.

1. Check if the system has the rpcbind and nfs-utils packages already installed in order to be able to use the NFS client functionality:

> # **rpm –qa | egrep 'rpcbind|nfs-utils'**
> nfs-utils-1.2.3–15.el6.x86_64
> rpcbind-0.2.0–9.el6.x86_64
> nfs-utils-lib-1.1.5–4.el6.x86_64

2. Create a mount point called */physical/nfs/remote*:

> # **mkdir –p /physical/nfs/remote**

3. Set the *rpcbind* and *nfslock* services to autostart at each system reboot, and validate:

> # **chkconfig rpcbind on**
> # **chkconfig nfslock on**
> # **chkconfig --list | egrep 'rpcbind|nfslock'**
> rpcbind   0:off 1:off 2:on 3:on 4:on 5:on 6:off
> nfslock   0:off 1:off 2:on 3:on 4:on 5:on 6:off

4. Start (or restart) the *rpcbind* and *nfslock* services, and check the running status:

```
# service rpcbind start
Starting rpcbind:                        [ OK ]
# service nfslock start
Starting NFS statd:                      [ OK ]
# service rpcbind status
rpcbind (pid  1463) is running...
# service nfslock status
rpc.statd (pid  1589) is running...
```

5. Edit */etc/fstab* and add the following entry for the remote file system. This is accomplished to ensure that the file system is automatically mounted when *server.example.com* is rebooted.

    192.168.2.200:/var/yum/repos.d/remote   /physical/nfs/remote nfs   ro,intr   0   0

6. Execute any of the following to mount the NFS file system:

    ```
    # mount –t nfs –o ro,intr 192.168.2.200:/var/yum/repos.d/remote \
    /physical/nfs/remote
    # mount /physical/nfs/remote
    # service netfs start
    Mounting other filesystems:         [ OK ]
    ```

7. Run the *df* command or the *mount* command and check the status of the remote file system.
8. Issue the *umount* command and supply the mount point as an argument to unmount the file system:

    ```
    # umount /physical/nfs/remote
    ```

9. Reboot the system:

    ```
    # shutdown –ry now
    ```

10. Using the *df* command or the *mount* command, check if the file system has been automatically remounted after the reboot.

Auto File System

In the previous two exercises and the sub-section, you learned how to mount removable and network file systems using the standard file system mount method (for brevity, I will use the term *resource* to represent CD/DVD, USB, NFS, and CIFS file systems wherever applicable). RHEL provides a way to mount these resources automatically when a removable medium is inserted in its drive (CD/DVD), plugged in to a port (USB), or an activity with a command such as *ls* or *cd* occurs on a mount point (NFS or CIFS). The removable file systems stay mounted but the network file systems get automatically unmounted if they are no longer accessed for a pre-defined period of time. This automatic resource mount facility is provided by the AutoFS (*Auto File System*) service.

Benefits Over the Standard Mount Method

There are several features and benefits associated with the AutoFS resource mount as compared to the standard resource mount, and are described below:

✓ AutoFS requires that the resources be defined in text configuration files called *maps*, which are located in the */etc* directory. In contrast, the standard mount information is defined in the */etc/fstab* file for each resource that needs to be mounted automatically at system reboots.

✓ AutoFS does not require the *root* privileges to mount resources which is required by the standard mount method.

✓ With AutoFS, the NFS client boot process never hangs if the NFS server is down or inaccessible. With the standard NFS mount, when a client system boots up and an NFS server listed in the */etc/fstab* file is unavailable, the client may hang until either the mount request times out or the NFS server becomes available.

✓ With AutoFS, an NFS file system is unmounted automatically if it is not accessed for five minutes by default. With the standard mount method, it stays mounted until it is either manually unmounted or the client shuts down.

✓ AutoFS supports wildcard characters and environment variables, which the standard mount method does not support.

✓ A special map is available with AutoFS that mounts all available NFS resources from a reachable NFS server when a user requests access to a resource on that server without explicitly defining each one of them. The standard mount method does not have any such feature available.

How AutoFS Works?

The AutoFS service consists of a daemon called *automount* that mounts configured resources automatically when inserted in their drives/ports or accessed. This daemon is invoked at system boot up. It reads the AutoFS master map and creates initial mount point entries in the */etc/mtab* file; however, the resources are not actually mounted at this time. When a CD/DVD medium is inserted in the drive, a USB key is plugged into a port, or a user activity occurs under one of the network file system mount points, the daemon actually mounts the requested resource at that time. If a network resource remains idle for a certain time period, *automount* unmounts it by itself.

AutoFS Configuration File

The configuration file for AutoFS is */etc/sysconfig/autofs*. This file is consulted when the AutoFS service is started or restarted. An excerpt from this file is shown below:

```
# cat /etc/sysconfig/autofs
MASTER_MAP_NAME="auto.master"
TIMEOUT=300
NEGATIVE_TIMEOUT=60
BROWSE_MODE="no"
MOUNT_NFS_DEFAULT_PROTOCOL=4
APPEND_OPTIONS="yes"
LOGGING="none"
OPTIONS=""
```

Several directives can be set in this file to modify the default behavior. Some of the key directives are shown above and described in Table 10-4.

Option	Description
MASTER_MAP_NAME	Defines the name of the master map.
TIMEOUT	Specifies, in seconds, the maximum idle time after which a resource is automatically unmounted.
NEGATIVE_TIMEOUT	Specifies, in seconds, a timeout value for failed mount attempts.
BROWSE_MODE	Defines whether maps are to be made browseable.
MOUNT_NFS_DEFAULT_PROTOCOL	Sets the default NFS version used.
APPEND_OPTIONS	Identifies additional options to the OPTIONS directive.
LOGGING	Specifies a logging level. Other options are verbose and debug.
OPTIONS	Defines global options.

Table 10-4 AutoFS Options

AutoFS Maps

The AutoFS service needs to know what resources to be mounted and from where. It also needs to know any specific options that you want it to use. This information is defined in AutoFS map files. There are four types of AutoFS maps: *master*, *special*, *direct*, and *indirect*.

The Master Map

The */etc/auto.master* file is the default master map as defined in the */etc/sysconfig/autofs* file with the MASTER_MAP_NAME directive. This map contains entries for special, direct, and indirect maps. Three sample entries are shown below:

```
# grep –v ^# /etc/auto.master
/net        –hosts
/-          /etc/auto.direct
/misc       /etc/auto.misc
```

The first entry is for a special map directing AutoFS to use the –hosts special map whenever a user attempts to access anything under */net*.

The second entry defines a direct map and points to the */etc/auto.direct* file for further information.

The last entry is for an indirect map notifying AutoFS to refer to the */etc/auto.misc* file for further information. The umbrella mount point */misc* will precede all mount point entries listed in the */etc/auto.misc* file.

The Special Map

The –hosts special map allows all NFS resources exported by all accessible NFS servers to get mounted under the */net* directory without explicitly mounting each one of them. The */etc/auto.net* file is executed to obtain a list of accessible servers and available exported resources. Accessing */net/<NFS_server>* will instruct AutoFS to mount all available resources from that NFS server. By default, the entry "/net –hosts" exists in the */etc/auto.master* file, and is enabled. This map is not recommended in an environment where there are many NFS servers exporting many resources as AutoFS will mount all available resources whether they are needed or not.

The Direct Map

The direct map is used to mount resources automatically on any number of unrelated mount points. Some key points to note when working with direct maps are:

- ✓ Direct mounted resources are always visible to users.
- ✓ Local and direct mounted resources can co-exist under one parent directory.
- ✓ Each direct map entry adds an entry to the /etc/mtab file.
- ✓ Accessing a directory containing many direct mount points mounts all resources.

Exercise 10-7: Access a Network Resource with AutoFS Direct Map

In Exercise 6-3 of Chapter 06 "Package Management", you configured an NFS server on *physical.example.com* (192.168.2.200) and exported the */var/yum/repos.d/remote* directory.

In this exercise, you will configure a direct map entry for this NFS resource on *server.example.com* to mount it automatically on */mnt* with AutoFS.

1. Check if the system has the autofs package already installed in order to be able to use the AutoFS functionality:

   # **rpm –qa | egrep autofs**
   autofs–5.0.5–39.el6.x86_64

2. Edit the */etc/auto.master* file and add the following entry if it does not already exist:

   /–   /etc/auto.direct

3. Create */etc/auto.direct* file and add the following entry to it:

   /mnt   physical.example.com:/var/yum/repos.d/remote

4. Execute the following to force the *automount* daemon to reload the maps:

   # **service autofs reload**

5. Run the *ll* command on the mount point */mnt* and then execute the *mount* command to verify that the resource is mounted and is accessible for use:

   # **ll /mnt**
   # **mount –v | grep nfs**
   physical.example.com:/var/yum/repos.d/remote on /mnt type nfs (addr=192.168.2.200)

The Indirect Map

The indirect map is used to automatically mount resources under one common parent directory. Some key points to note when working with indirect maps are:

- ✓ Indirect mounted resources only become visible after they have been accessed.
- ✓ Local and indirect mounted resources cannot co-exist under the same parent directory.
- ✓ Each indirect map puts only one entry in the /etc/mtab file.
- ✓ Accessing a directory containing many indirect mount points shows only the resources that are already mounted.

Exercise 10-8: Access a Network Resource with AutoFS Indirect Map

In Exercise 6-3 of Chapter 06 "Package Management", you configured an NFS server on *physical.example.com* (192.168.2.200) and exported the */var/yum/repos.d/remote* directory.

In this exercise, you will configure an indirect map entry for this NFS resource on *server.example.com* to mount it automatically under */misc* with AutoFS.

1. Run step 1 from Exercise 10-7.
2. Edit */etc/auto.master* and ensure that the following indirect map entry is defined:

   /misc   /etc/auto.misc

3. Edit */etc/auto.misc* file and add the following entry to it:

   yum   physical.example.com:/var/yum/repos.d/remote

4. Execute the following command to force the automount daemon to reload the maps:

   # **service autofs reload**

5. Run the *ll* command on the mount point */misc/yum* and then execute the *mount* command to verify that the resource is mounted and is accessible. AutoFS will create the mount point *yum* under */misc* and mount the remote file system on it.

   # **ll /misc/yum**
   # **mount –v | grep nfs**
   physical.example.com:/var/yum/repos.d/remote on /misc/yum type nfs (addr=192.168.2.200)

Accessing Removable Media with AutoFS Indirect Map

There are several entries defined in the */etc/auto.misc* file for automounting CD/DVD, floppy, and other removable media. Except for the CD/DVD, which is configured to be automounted on */misc/cd* mount point, automounting other media is disabled by default. You need to uncomment line entries for the media that you wish to be using and then force the *automount* service to reload the maps. An excerpt from the */etc/auto.misc* file is shown below:

```
# cat /etc/auto.misc
. . . . . . . .
cd              -fstype=iso9660,ro,nosuid,nodev  :/dev/cdrom
#linux          -ro,soft,intr       ftp.example.org:/pub/linux
#boot           -fstype=ext2        :/dev/hda1
#floppy         -fstype=auto        :/dev/fd0
#floppy         -fstype=ext2        :/dev/fd0
#e2floppy       -fstype=ext2        :/dev/fd0
#jaz            -fstype=ext2        :/dev/sdc1
```

Automounting User Home Directories

AutoFS allows you to use two special characters in your indirect maps. These special characters are * and &, and are used to replace the references to NFS servers and mount points.

For example, with user home directories located under */home* and exported by more than one NFS servers, the *automount* daemon will contact all available and reachable NFS servers concurrently when a user attempts to log on to the system. The daemon will only mount that specific user's home directory rather than the entire */home*. The indirect map entry for this type of substitution will look like:

```
*       -rw,soft,intr       &:/home/&
```

With this simple entry in place, there is no need to update any AutoFS configuration files if NFS servers with */home* exported are added or removed. Similarly, if user home directories are added or deleted, there will be no impact on AutoFS either.

The above entry can be placed in a separate map file such as */etc/auto.home*, in which case you will need to reflect the map name in the *auto.master* file.

```
/home           /etc/auto.home      --timeout=180
```

Reload the AutoFS maps after the above changes have been put in place.

Access Control List (ACL)

The *Access Control List* (ACL) provides an extended set of permissions that may be set on files and directories. These permissions are in addition to the standard Linux file and directory permissions discussed earlier in Chapter 04 "File Permissions, Text Editors, Text Processors & The Shell". The ACL allows you to define permissions for specific users and groups using either the octal or the symbolic notation of allocating permissions.

There are two commands – *getfacl* and *setfacl* – available to work with the ACL, the former is used to display the permission information and the latter sets, modifies, substitutes and deletes ACL entries. In addition, the *ll* command also displays the + sign right beside the permissions column if the ACL is set.

Before getting into any further details, look at Table 10-5 and see how ACL entries are used with the *setfacl* command.

ACL Entry	Description
u[ser]::perms	Standard Linux permissions for the owner.
g[roup]::perms	Standard Linux permissions for group members.
o[ther]:perms	Standard Linux permissions for public.
m[ask]:perms	Maximum permissions a specific user or a specific group can have on a file or directory. If this is set to rw- for example, then no specific user or group will have more permissions than read and write.
u[ser]:UID:perms (or u[ser]:username:perms)	Permissions assigned to a specific user. The user must exist in the /etc/passwd file.
g[roup]:GID:perms (or g[roup]:groupname:perms)	Permissions assigned to a specific group. The group must exist in the /etc/group file.

Table 10-5 ACL Entry Usage

Table 10-6 shows key options available with the *setfacl* command.

Option	Description
-b	Removes all ACL settings.
-d	Applies to the default ACL.
-k	Removes all default ACL settings.
-m	Modifies ACL settings.
-R	Applies recursively to all files and directories.
-x	Remove all ACL settings.

Table 10-6 setfacl Command Options

Exercise 10-9: Determine, Set, and Delete ACL Entries

In this exercise, you will create a file called *file1* as *user1* in *user1*'s home directory and check to see if there are any ACL settings on the file. You will check to ensure that the */home* file system, where *user1* has his home directory, is mounted with the acl support activated. You will apply ACL settings on the file for *user2* and allow him full access. You will review how the mask value on the file has changed. You will add *user4* to the ACL settings on the file. You will delete the settings for *user2* and then delete all other ACL settings from the file. You will apply new settings on the file for the owner, group members, public, and *user3*, and review the new mask value. Finally, you will remove all the settings from the file.

1. Log in as *user1* and create *file1*. Run the *ll* and the *getfacl* commands on the file and see if there are any ACL entries set on it.

   $ touch file1
   $ ll
   –rw-rw-r--. 1 user1 user1 0 Sep 21 11:35 file1

$ getfacl file1
file: file1
owner: user1
group: user1
user::rw-
group::rw-
other::r--

The output indicates that currently there are no ACL settings on the file. It also shows that the file owner is *user1* that belongs to group *user1*. The owner and the group members have read and write permissions and everyone else has read-only permission.

2. Log in as *root* on a different terminal (or switch into *root* with the *su* command) and check with the *tune2fs* command if the */home* file system is mounted with the acl support enabled (any file system created at the time of RHEL installation should have the acl support preset on it by default):

    # tune2fs –l /dev/vg00/lvol3 | grep acl
    Default mount options: user_xattr acl

The output indicates that the acl support is already activated. If not, run the following to remount */home* with acl and update the corresponding in the */etc/fstab* file:

    # mount –o remount,acl /home

3. Allocate read/write/execute permissions to *user2* with the *setfacl* command. Run this command as *user1*.

    $ setfacl –m u:user2:7 file1

4. Run the *ll* command to check if the + sign has appeared by the permission settings and the *getfacl* command to check the new ACL settings:

    $ ll
    -rwxrwxr--+ 1 user1 user1   0 Sep 21 11:35 file1
    $ getfacl file1
    user::rw-
    user:user2:rwx
    group::rw-
    mask::rwx
    other::r--

A row is added for *user2* showing rwx (7) permissions. Another row showing the mask is also added and is set to rwx (7) as well. The value of mask determines the maximum permissions assigned to a specific user or group. In this case, the maximum permissions allocated to *user2* is rwx and, therefore, mask is also set to rwx.

5. Run the *setfacl* command again and add *user4* with read/write permissions to *file1*. Confirm with *getfacl*.

```
$ setfacl –m u:user4:rw file1
$ getfacl file1
```
user::rw–
user:user2:rwx
user:user4:rw–
group::rw–
mask::rwx
other::r--

6. Now delete the ACL entries set for *user2* and confirm with *getfacl*:

```
$ setfacl –x u:user2 file1
$ getfacl file1
```
user::rw–
user:user4:rw–
group::rw–
mask::rw–
other::r--

Note that the maximum permissions are now reduced to read/write since that is what the maximum current permissions are for the specific user *user4*.

7. Delete all the ACL entries set on *file1* and confirm with *getfacl*:

```
$ setfacl –b file1
$ getfacl file1
```
user::rw–
group::rw–
other::r--

8. Now using the octal notation, apply ACL settings on *file1* with the owner to have rwx, group r, public none, and *user3* to have r-x:

```
$ setfacl –m u::7,g::4,o:0,u:user3:5 file1
$ getfacl file1
```
user::rwx
user:user3:r-x
group::r--
mask::r-x
other::---

9. Delete all the ACL entries set on *file1* and confirm with *getfacl*.

Default ACL

Sometimes it is imperative for several users that belong to different groups to be able to share the contents of a common directory. They want permissions set up on the directory in such a way that when files and sub-directories are created underneath, they inherit parent directory permissions.

This way the users do not have to modify permissions on each new file and sub-directory that is created there. Setting the default ACL on a directory fulfills this requirement.

The Default ACL can be described as the maximum discretionary permissions that can be allocated on a directory. Table 10-7 lists and describes how default entries are used with the *setfacl* command.

ACL Entry	Description
d[efault]:u:perms	Default standard Linux permissions for the owner set at the directory level.
d[efault]:u:UID:perms (or d[efault]:u:username:perms)	Default permissions for a specific user set at the directory level.
d[efault]:g:perms	Default standard Linux permissions for owner's group members set at the directory level.
d[efault]:g:GID:perms (or d[efault]:g:groupname:perms)	Default permissions for a specific group set at the directory level.
d[efault]:o:perms	Default permissions for public set at the directory level.
d[efault]:m:perms	Default maximum permissions a user or group can have when the user or a group member creates a file in a directory where the default ACL is set.

Table 10-7 Default ACL Entry Usage

Exercise 10-10: Set, Confirm, and Delete Default ACL Entries

In this exercise, you will create a directory */home/user3/project* as *user3* and set default ACL entries for *user1* and *user2* so that they both get read and write permissions on this directory. You will delete all the default entries at the end of the exercise. Make sure to execute step 2 from Exercise 10-9 if */home* does not already have acl support enabled.

1. Log in as *user3* and create a directory called *project* in his home directory. Run the *getfacl* command and see what the default permissions are on the directory.

    $ mkdir project
    $ getfacl project
    # file: project
    # owner: user3
    # group: user3
    user::rwx
    group::rwx
    other::r-x

    The output indicates that the default permissions on the directory are 775.

2. Allocate default read and write permissions to *user1* and *user2* with the *setfacl* command. Run this command as *user3* and use octal notation.

    $ setfacl –m d:u:user1:6,d:u:user2:6 project

```
$ getfacl project
user::rwx
group::rwx
other::r-x
default:user::rwx
default:user:user1:rw-
default:user:user2:rw-
default:group::rwx
default:mask::rwx
default:other::r-x
```

3. Delete all the default ACL settings from the directory and confirm:

```
$ setfacl –k project
$ getfacl project
user::rwx
group::rwx
other::r-x
```

Understanding and Managing Swap

Physical memory in the system is a finite temporary storage resource used for loading the kernel and data structures, and running user programs and applications. The system divides the memory into smaller pieces called *pages*. A page is typically 4KB in size.

The size of swap should not be less than the amount of physical memory; however, depending on application requirements, it may be twice or even larger. Run the *free* command to view how much physical memory is installed, used, and free in the system. You can use the –m or –g flag to list the values in MBs or GBs.

```
# free
                   total      used      free    shared   buffers    cached
Mem:             791432    408032    383400         0     19396    231420
-/+ buffers/cache:  157216    634216
Swap:           1605624         0   1605624
```

Alternatively, use the following to determine memory information:

```
# cat /proc/meminfo
MemTotal:      791432 kB
MemFree:       383408 kB
Buffers:        19416 kB
Cached:        238452 kB
SwapCached:         0 kB
Active:        195668 kB
. . . . . . . .
SwapTotal:    1605624 kB
SwapFree:     1605624 kB
. . . . . . . .
```

There is about 800MB (MemTotal) of total physical memory on this system. For performance enhancement reasons, the kernel uses as much memory as it can for caching data. As reads and writes occur constantly, the kernel struggles to keep the data in cache as pertinent as possible. The caching information is reported as the sum of the number of buffers and cached pages. The portion of the cache memory used by a certain process is released when the process is terminated, and is allocated to a new process as needed. The above output indicates that about 114MB (Inactive) of the total physical memory is available for use by new processes. The output also displays the total amount of configured swap (SwapTotal) and how much of it is currently in use and available (SwapFree).

Swap Space and Demand Paging

Swap space is a region on the physical disk used for demand paging purposes. When a program or process is spawned, it requires space in the memory to run and be processed. Although many programs can run concurrently, the physical memory cannot hold all of them at the same time. The kernel monitors the free physical memory. As long as it is below a high threshold, no paging occurs. When the amount of the free physical memory falls below that threshold, the system starts moving selected idle pages of data from physical memory to the swap space to make room to accommodate other programs. This is referred to as *page out*. Since the system CPU performs the process execution in a round-robin fashion, when the time comes for the paged out data to be executed, the CPU looks for that data in the physical memory and a *page fault* occurs resulting in moving the pages back to the physical memory from the swap space. The return of the paged out data to the physical memory is referred to as *page in,* and the entire process of paging data out and in is known as the *demand paging*.

RHEL systems with less physical memory but high memory requirements can become so busy with paging out and in that they do not have enough time to carry out other useful tasks causing the system performance to degrade. When this situation occurs, the system appears to be frozen. The excessive amount of paging that causes the system performance to go down is called *thrashing*.

When thrashing begins, or when the free physical memory falls below a low threshold, the system deactivates idle processes and prevents new processes from being initiated. The idle processes only get reactivated and new processes are only started when the system discovers that the available physical memory has climbed above the threshold level and thrashing has ceased.

Exercise 10-11: Create and Activate Swap Spaces

In Labs 9-1 and 9-2, you created two partitions of sizes 1GB (*vdc2*) and 500MB (*vdd2*) for use as swap and in Lab 9-3, you created a logical volume called *swapvol* of size 2.4GB in the *vg10* volume group for use as swap.

In this exercise, you will create swap structures in *vdc2* and *swapvol* using the *mkswap* command and enable swap in both of them using the *swapon* command. You will add entries to the */etc/fstab* file so that both swap regions are activated at each system reboot. Finally, you will reboot the system and confirm by using appropriate commands that the new swap regions have been successfully activated.

1. Create swap structures in the *vdc2* partition using the *mkswap* command:

```
# mkswap /dev/vdc2
Setting up swapspace version 1, size = 1024596 KiB
no label, UUID=91f551df-f15b-4fe8-97d7-62512ac25f2c
```

2. Create swap structures in the *swapvol* logical volume using the *mkswap* command:

```
# mkswap /dev/vg10/swapvol
Setting up swapspace version 1, size = 2457596 KiB
no label, UUID=7c065f9b-3637-47f5-9384-911d2271521a
```

3. Enable swapping in the partition and the logical volume using the *swapon* command:

```
# swapon /dev/vdc2
# swapon /dev/vg10/swapvol
```

4. Confirm activation of the new swap areas by running either the *swapon* command with the –s option or viewing the contents of the */proc/swaps* file:

```
# swapon –s
# cat /proc/swaps
/dev/dm-3            partition    2457592 0    -2
/dev/vde2           partition    1024592 0    -3
```

5. Run the *vmstat* command to display virtual memory statistics:

```
# vmstat
procs --------memory------- -swap- ---io--- -system- -----cpu-----
 r b  swpd free    buff cache  si so  bi  bo   in  cs  us sy id wa st
 0 0   0  400504  3428 231184 0  0  274  4   55  87  1 1 96 2 0
```

The command produced the output in six columns. The r and b under procs show the number of processes waiting for run time and in uninterruptible sleep; the swpd, free, buff, and cache under memory indicate amount of used virtual memory, idle memory, memory used as buffers, and cache; the si and so under swap determine amount of memory swapped in and out; the bi and bo under io display number of blocks in and out; the in and cs under system indicate numbers of interrupts and context switches per second; and us, sy, id, wa, and st identify percentages of total CPU time spent in running non-kernel code, kernel code, idle state, waiting for I/O, and stolen from a virtual machine.

The *vmstat* command can be run with the –s switch to display the output in a different format:

```
# vmstat –s
    231196  swap cache
   5087808  total swap
         0  used swap
   5087808  free swap
 . . . . . . . .
    319769 pages paged in
      5008 pages paged out
```

<pre>
        0   pages swapped in
        0   pages swapped out
  . . . .
</pre>

6. Finally, edit the */etc/fstab* file and add entries for both new swap areas so that they are activated at each system reboot:

<pre>
UUID=91f551df-f15b-4fe8-97d7-62512ac25f2c   swap   swap   defaults 0   0
/dev/vg10/swapvol   swap   swap   defaults 0   0
</pre>

7. Reboot the system and use the appropriate command after the system comes up to validate the activation of the new swap areas.

Exercise 10-12: Deactivate and Remove Swap Spaces

In this exercise, you will deactivate both swap regions that were added to the system in the previous exercise using the *swapoff* command. You will remove their entries from the */etc/fstab* file and reboot the system to confirm.

1. Deactivate swap in *vdc2* and *swapvol* using the *swapoff* command:

    # swapoff /dev/vdc2
    # swapoff /dev/vg10/swapvol

2. Edit the */etc/fstab* file and remove their entries.
3. Reboot the system and use the appropriate command after the system comes up to validate the activation of the new swap areas.

Chapter Summary

You learned about file systems, ACLs, and swap in this chapter. You reviewed file system concepts and types, and learned about extended and LUKS-encrypted file systems. You looked at various file system administration commands. You studied the concepts around mounting and unmounting file systems. You examined the UUID associated with file systems and saw how to apply labels to file systems. You analyzed the file system table and saw how to add entries to it to automatically activate file systems at system reboots. You looked at the tools available for reporting file system usage and estimating file space usage. You learned about the file system check and repair utility for finding and fixing issues related to unhealthy file systems. These topics on the file system topic were followed by several exercises on creating, mounting, unmounting, and removing extended and LUKS-encrypted file sytems using the commands and the LVM Configuration tool. Additional exercises were provided on repairing file systems and mounting and unmounting removable media and network file systems manually.

The next major topic in this chapter was about the AutoFS service. You learned about the concepts, features, benefits, and components associated with this service. You performed exercises to fortify your understanding of using this service to automatically mount removable file systems, and automatically mount and unmount network file systems.

Followed by a detailed coverage on AutoFS, the topic on ACLs was covered, which introduced you to the concepts and purpose of applying extended security attributes on files and directories. You performed exercises to strengthen the understanding.

Finally, you studied the concepts around swapping and paging, and how they worked. You performed exercises on creating, activating, viewing, deactivating, and removing swap spaces, as well as how to configure them to be automatically enabled at system reboots.

Chapter Review Questions

1. */etc* is a file system. True or False?
2. The other name for NFS is Samba. True or False?
3. ext4 is the default file system type in RHEL6. True or False?
4. AutoFS requires root privileges to automatically mount a network file system. True or False?
5. Write three differences between directories and file systems.
6. What type of information does the *blkid* command display?
7. Which command is used to configure LUKS-based encryption on a file system?
8. What is the difference between the *mkfs.ext4* and *mke2fs* commands?
9. What is the default timeout value for a file system before AutoFS unmounts it automatically?
10. What is the process of paging out and paging in known as?
11. */proc* is a memory based file system. True or False?
12. What would the command *mkswap /dev/vdc2* do?
13. Which two files contain entries for mounted file systems?
14. Write two commands that can be used to determine the UUID of a file system.
15. What would happen if you mount a file system on a directory that already contains files in it?
16. What would the luksOpen option do when used with the *cryptsetup* command?
17. A UUID is always assigned to a file system at its creation time. True or False?
18. What is the name of the AutoFS configuration file and where is it located?
19. What would the command *e2label /dev/vdd1 swap* do?
20. What would happen if you answer yes in lowercase letters when prompted by the *cryptsetup luksFormat /dev/vgtest/lvol1* command?
21. What is the command to start the LVM Configuration tool?
22. What is the purpose of using the *file* command?
23. What would the command *mount –t cifs –o ro //192.168.2.100/dir1 /dir1* do?
24. What would the entry " *   &:/home/& " in an AutoFS indirect map imply?
25. The difference between the primary and backup superblocks is that the primary superblock includes pointers to the data blocks where the actual file contents are stored whereas the backup superblocks don't. True or False?
26. What is the main difference between the ext2 and ext3 type file systems?
27. What would the command *setfacl –m d:u:user1:7,d:u:user4:6,d:o:4 dir* do?
28. What would the command *mkfs.ext4 /dev/vgtest/lvoltest* do?
29. What is the parent directory name where AutoFS mount a DVD medium?
30. Arrange the tasks in sequence: umount file system, mount file system, create file system, remove file system.
31. Which two files must contain appropriate entries for a LUKS-encrypted file system to automatically mount it at system reboots?
32. Which of the following two commands would you use to terminate all processes using the */ex101d* mount point: *fuser –cu* or *fuser –ck*?
33. The *parted* utility may be used to create LVM logical volumes. True or False?

34. What type of AutoFS map would have the "/- /etc/auto.media" entry in the *auto.master* file?
35. What would the *mount* command do with the –a switch?
36. Which command can you use to generate a list of superblock locations in a file system?
37. What two services must be running on the NFS client?
38. What would the command *df –t ext4* do?
39. What would happen if you try to apply ACL settings to a file that resides in a non-ACL activated file system?
40. What would the command *setfacl –m u::7,g::4,o:4,u:user2:7 file* do?
41. Name the four types of maps that AutoFS support.
42. What two commands can be used to determine the total and used physical memory and swap in the system?
43. Write the command to enable the swap space manually.
44. Which virtual file contains information about the swap?
45. The */etc/fstab* file can be used to activate swap spaces automatically at system reboots. True or False?
46. What are the commands to activate and deactivate swap spaces manually? Write two commands.
47. Which file does the *swapon –s* command refer to?
48. Name of the AutoFS daemon is *automountd*. True or False?

Answers to Chapter Review Questions

1. False.
2. False.
3. True.
4. False.
5. File systems can be mounted and unmounted, file systems can be extended or reduced, and file systems can be tuned independently.
6. The *blkid* command displays the block device file attributes.
7. The *cryptsetup* command.
8. No difference.
9. Five minutes.
10. The process is known as the demand paging.
11. True.
12. The command provided will create swap structures in the */dev/vdc2* partition.
13. The */etc/mtab* and the */proc/mounts* files.
14. The *tune2fs* and the *blkid* commands.
15. The files in the directory will hide.
16. The luksOpen option will instruct the *cryptsetup* command to open the specified LUKS-encrpted file system.
17. True.
18. The *autofs* file located in the */etc/sysconfig* directory.
19. The command provided will write the label "swap" to the */dev/vdd1* partition.
20. The command will not act as desired.
21. The command to start the LVM Configuration tool is *system-config-lvm*.
22. The *file* command displays the type of the specified file.
23. The command provided will mount the CIFS file system *dir1* in read-only mode on the */dir1* mount point from the system with IP 192.168.2.100.

24. This entry would instruct the AutoFS daemon to mount a user's home directory from whichever NFS server it is available from.
25. False.
26. ext3 uses journaling and ext2 does not.
27. The command provided will set default ACL of rwx for *user1*, rw for *user4* and r-only for everyone else on the *dir* directory.
28. The command provided will create ext4 file system structures in the */dev/vgtest/lvoltest* logical volume.
29. The */misc* directory.
30. Create file system, mount file system, unmount file system and remove file system.
31. The */etc/fstab* and */etc/crypttab* files.
32. The *fuser –ck* command.
33. False.
34. A direct map.
35. The command provided will mount all file systems listed in the */etc/fstab* file but are currently not mounted.
36. The *dumpe2fs* command.
37. The *nfslock* and *rpcbind* services.
38. The command provided will display all mounted file systems of type ext4.
39. You cannot apply ACL settings on a file or directory that resides in the file system with the acl option disabled.
40. The command provided will assign rwx permissions to the owner of *file* and to *user3*, and read-only permission to everyone else.
41. The master, special, direct, and indirect maps.
42. The *free* and *vmstat* commands.
43. The *swapon* command.
44. The */proc/swaps* file.
45. True.
46. The *swapon* and *swapoff* commands.
47. The */proc/swaps* file.
48. False.

Labs

Lab 10-1: Create and Mount Extended File Systems

Destroy all partitions and volume groups created on *vdb*, *vdc*, *vdd*, and *vde* drives in the previous labs and exercises.

Create a partition in *vdb* of size 1GB and initialize it with ext4 file system structures. Initialize the *vdc* disk, create a volume group called *vgtest* and include the *vdc* disk in it. Create *oravol* logical volume of size 1GB in the volume group and initialize it with ext4 file system structures. Create mount points of your choice, and mount both file systems manually. Apply any label to the file system created in *vdb1* and add both file systems to the *fstab* file using the label of *vdb1* and the LV name of *oravol*. Reboot the system and test if it boots up successfully and mounts both new file systems.

Lab 10-2: Create and Mount LUKS File Systems

Create *vdb2* and *dbvol* in *vgtest*. Initialize both for use as LUKS file systems. Assign them names of your choice. Create ext4 file system structures in them and mount them on mount points of your choice. Create separate passphrase files for both and add them to the respective file systems. Add entries in files so that both get automatically mounted at each system reboot without prompting for passphrases. Use defaults or your own ideas for missing information.

Lab 10-3: Apply ACL Settings

Create a file called *testfile* in *user1*'s home directory. Create a directory in *user2*'s home directory and call it *dir1*. Ensure that the */home* file system is mounted with acl option activated. Apply settings on *testfile* so that *user2* gets 7, *user3* gets 6, and *user4* gets 4 permissions. Apply default settings on *dir1* so that *user4* gets 7 and *user2* gets 5 permissions on it.

Lab 10-4: Create and Enable Swap

Create *vdb3* of size 1.5GB and *lvswap2* in *vgtest* of size 2.1GB. Initialize both for use as swap. Create swap structures in them and add entries in the file system table so that they get automatically activated at each system reboot. Use defaults or your own ideas for missing information.

Chapter 11

Users & Groups

This chapter covers the following major topics:

- ✓ Understand user authentication files
- ✓ Verify the consistency of user authentication files
- ✓ Lock user authentication files in order to edit them
- ✓ Create, update, and remove shadow password files
- ✓ Create, modify, and delete local user accounts
- ✓ Set password aging on local user accounts
- ✓ Use su and sudo commands
- ✓ Create, modify, and delete local group accounts
- ✓ Manage users, groups, and password aging with the Red Hat User Manager
- ✓ Understand user and system initialization files

This chapter includes the following RHCSA objectives:

  5. Log in and switch users in multiuser run levels
46. Create, delete, and modify local user accounts
47. Change passwords and adjust password aging for local user accounts
48. Create, delete, and modify local groups and group memberships

In order for an authorized person to gain access to the system, a unique *username* (aka *login name*) must be assigned and a user account must be created on the system. This user is assigned membership to one or more groups. Members of the same group have the same access rights on files and directories. Other users and members of other groups may or may not be given access to those files. User and group account information is recorded in several files. These files may be checked for inconsistencies, and edited manually, if necessary, by one administrator at a time. Password aging may be set on user accounts for increased access control. Users may switch into other users' accounts, including the *root* user account, provided they know their passwords. Regular users on the system may be allowed access to privileged commands by defining them appropriately in the configuration file related to the *sudo* command. Several user and system initialization files are involved when a user logs in.

Understanding User Authentication Files

RHEL supports three fundamental user account types: *root*, *normal*, and *service*. The *root* user possesses full powers on the system. It is the superuser or the administrator with full access to all services and administrative functions. This user is automatically created during RHEL installation. The normal users have user-level privileges. They cannot perform any administrative functions, but can run applications and programs that they are authorized to execute. The service accounts are responsible for taking care of the installed services. These accounts include apache, games, mail, and printing.

User account information is stored in four files: */etc/passwd, /etc/shadow, /etc/group*, and */etc/gshadow*. These files are updated when a user account is created, modified, or removed. The same files are referenced when a user attempts to log in to the system and, therefore, these files are referred to as user authentication files. These files are so critical that, by default, the system maintains their backup files called *passwd-*, *shadow-*, *group-*, and *gshadow-* in the */etc* directory. The *shadow* and the *gshadow* files as well as user administration commands that you are going to learn in this chapter are part of the shadow-utils package that is installed on the system at the time of RHEL installation.

Originally, user account information was stored only in the *passwd* and *group* files; however, it was realized over a period of time that user passwords located in the *passwd* file were not a safe storage location as it opened doors for security breaches. This realization led to the development of the shadow password (and related) files that are now part of the shadow-utils package in RHEL6.

The /etc/passwd File

The */etc/passwd* file contains vital user login data. Each line entry in the file contains information about one user account. There are seven fields per line entry, each separated by the colon (:) character. A sample entry from the file is displayed in Figure 11-1.

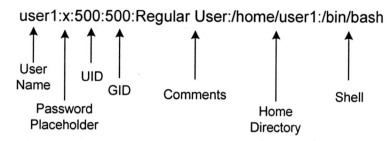

user1:x:500:500:Regular User:/home/user1:/bin/bash

User Name	UID	GID	Comments	Home Directory	Shell

Password Placeholder

Figure 11-1 The /etc/passwd File

Here is what is stored in each field:

- ✓ The first field contains the login name that a user uses to log in to the system. Usernames up to 255 characters including underscore and hyphen characters are supported, but the usernames should not use any special characters.
- ✓ The second field can contain an "x" (points to the */etc/shadow* file for the actual password), an asterisk "*" character (denotes that the account is disabled), or a combination of random letters, numbers, and special characters.
- ✓ The third field holds a unique number between 0 and approximately 2 billion. This number is known as the *User ID* (UID). User ID 0 is reserved for the *root* user, UIDs between 1 and 499 are typically reserved for system accounts and UIDs 500 and beyond are used for all other users. By default, RHEL begins assigning UIDs to new users starting at 500.
- ✓ The fourth field holds a number referred to as the *Group ID* (GID). This number corresponds with a group entry in the */etc/group* file. By default, RHEL creates a group for every new user by the same name as the username and the same GID as the user's UID. The GID defined in this field represents a user's primary group.
- ✓ The fifth field (called GECOS – General Electric Computer Operating System) optionally contains general comments about the user that may include the user's name, phone number, and location. This data may be viewed using the *finger* command and modified with the *chfn* command.
- ✓ The sixth field defines the absolute path to the user home directory. A *home* directory is the location where a user is placed after logging in to the system, and is typically used to store personal files for the user. The default location for user home directories is */home*.
- ✓ The last field contains the absolute path of the shell file that the user will be using as his primary shell after logging in. Common shells are bash (*/bin/bash*), dash (*/bin/dash*), and C (*/bin/csh*). The default shell assigned to users is the bash shell.

An excerpt from the *passwd* file is shown below:

 # cat /etc/passwd
root:x:0:0:root:/root:/bin/bash
bin:x:1:1:bin:/bin:/sbin/nologin
daemon:x:2:2:daemon:/sbin:/sbin/nologin
. . . .
user1:x:500:500::/home/user1:/bin/bash

Permissions on the /etc/passwd file should be 644 and the file must be owned by the *root* user.

The /etc/shadow File

The implementation of the shadow password mechanism in RHEL provides an added layer of user password security. With this mechanism in place, not only are the user passwords encrypted and stored at an alternative location in a more secure /etc/shadow file, but certain limits on user passwords in terms of expiration, warning period, etc. can also be implemented on a per-user basis. This is referred to as *password aging*. The *shadow* file is readable only by the *root* user, which makes the contents of the file concealed from everyone else.

With the shadow password mechanism active, a user is initially checked in the *passwd* file and then in the *shadow* file for authenticity.

The *shadow* file contains extended user authentication information. Each row in the file corresponds to one entry in the *passwd* file. There are nine fields per line entry in the *shadow* file and are separated by the colon (:) character. A sample entry from this file is exhibited in Figure 11-2.

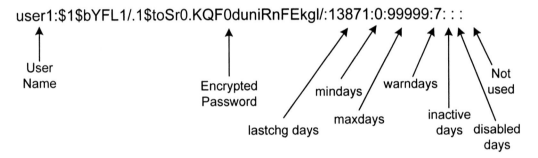

Figure 11-2 The /etc/shadow File

Here is what is stored in each field:

✓ The first field contains the login name as it appears in the /etc/passwd file.
✓ The second field contains a combination of random letters and numbers, which represents the user password in an encrypted form.
✓ The third field contains the number of days since the epoch time (January 01, 1970) when the password was last modified.
✓ The fourth field contains the minimum number of days that must elapse before the user password can be changed. This field can be altered with the *chage* command (the –m option) or the *passwd* command (the –n option).
✓ The fifth field contains the maximum number of days of password validity before the user starts getting warning messages to change it. This field may be altered with the *chage* command (the –M option) or the *passwd* command (the –x option).
✓ The sixth field contains the number of days the user gets warning messages to change password. This field may be altered with the *chage* command (the –W option) or the *passwd* command (the –w option).

✓ The seventh field contains the maximum allowable number of days of user inactivity. This field may be altered with the *chage* command (the –I option) or the *passwd* command (the –i option).

✓ The eighth field contains the number of days since the epoch time (January 01, 1970) after which the account expires. This field may be altered with the *chage* command (the –E option).

✓ The last field is reserved for future use.

An excerpt from the *shadow* file is shown below:

 # cat /etc/shadow
root:6vt/AbXKaEgvU9.us$RgcyMD7yIuhcZVPZdjh8WKTI96jicnFGnDqex8A5nwsDvGybpaCis48
DcC5a0C6wZO.wSoWI.na.ash8h3ylP.:15597:0:99999:7...
bin:*:15155:0:99999:7...
daemon:*:15155:0:99999:7...
. . . .
user1:1BqKmeXy3$6ziUK3CU.ImUpcvAhWxhj/:13871:0:99999:7...

 Permissions on the */etc/shadow* file should be 400 and the file must be owned by the *root* user.

The /etc/group File

The */etc/group* file contains the group information. Each row in the file stores one group entry. Each user is assigned at least one group, which is referred to as the user's primary group. In RHEL, a group name is the same as the user name that it is associated with by default. This group is known as the *User's Private Group* (UPG) and it safeguards the user's files from other users' access. There are four fields per line entry in the file and are separated by the colon (:) character. A sample entry from the file is exhibited in Figure 11-3.

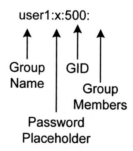

Figure 11-3 The /etc/group File

Here is what is stored in each field:

✓ The first field contains a unique group name, which must begin with a letter. By default, each user gets a unique group named the same as the user name. Other common groups may be created and users assigned to them. Group names up to 255 characters are supported.

✓ The second field is not typically used and is left blank. It may, however, contain an encrypted group-level password (copied and pasted from the */etc/shadow* file) or an "x" which indicates that the actual password is defined in the *etc/gshadow* file. You may set a password on a group if you want non-members to be able to change their group membership to this group using the *newgrp* command. The non-members must enter the correct password.

✓ The third field defines the GID, which is placed in the GID field of the *etc/passwd* file. By default, groups are created with GIDs starting at 500 and match the username that they are assigned to. Several users can be members of a single group simultaneously. Similarly, one user can be a member of several groups at the same time.

✓ The last field holds the usernames that belong to the group. Note that a user's primary group is defined in the *etc/passwd* file, and not in the *etc/group* file.

An excerpt from the *group* file is shown below:

cat /etc/group
root:x:0:root
bin:x:1:root,bin,daemon
daemon:x:2:root,bin,daemon
. . . .
user1:x:500:

 Permissions on the *etc/group* file should be 644 and the file must be owned by the *root* user.

The /etc/gshadow File

The shadow password implementation also provides an added layer of protection at the group level. With this mechanism activated, the group passwords are encrypted and stored at an alternative location in a more secure *etc/gshadow* file, which is only readable by the *root* user and, therefore, hides the contents from everyone else.

The *gshadow* file contains group encrypted password information. Each row in the file corresponds to one entry in the *group* file. There are four fields per line entry and are separated by the colon (:) character. A sample entry from this file is exhibited in Figure 11-4.

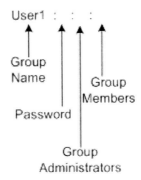

Figure 11-4 The /etc/gshadow File

Here is what is stored in each field:

✓ The first field contains the group name as appears in the *etc/group* file.

- ✓ The second field can contain a combination of random letters and numbers if a group password is set using the *gpasswd* command. These characters hold the password in an encrypted form. If the field contains the ! sign, it indicates that no users are allowed to access the group using the *newgrp* command; the !! signs indicate that a group password has never been set and no users are allowed to access the group using *newgrp*; and if the field is empty, it implies that only group member(s) can change into this group with *newgrp*.
- ✓ The third field lists usernames of group administrators that are authorized to add or remove members to and from this group with the *gpasswd* command.
- ✓ The last field holds usernames that belong to the group.

An excerpt from the *gshadow* file is shown below:

cat /etc/gshadow
root:::root
bin:::root,bin,daemon
daemon:::root,bin,daemon
sys:::root,bin,adm
. . . .
user1:!::

 Permissions on the */etc/gshadow* file should be 400 and the file must be owned by the *root* user.

Verifying Consistency of User Authentication Files

Sometimes inconsistencies occur in user authentication files. To check if the information in *passwd* and *shadow* files is valid and consistent, use the *pwck* command. This command checks and validates the number of fields in each line, login names, UIDs, GIDs, and existence of the login directory and shell, and reports any inconsistencies.

pwck
user 'adm': directory '/var/adm' does not exist
user 'uucp': directory '/var/spool/uucp' does not exist
. . . .
pwck: no changes

To verify if the information in the *group* and *gshadow* files is valid and consistent, use the *grpck* command. This command checks and validates the number of fields in each line and whether a user belonging to a group is missing from the *passwd* or the *shadow* file, and reports any inconsistencies.

Locking User Authentication Files

Although not recommended, periodically it is imperative to modify the *passwd* file by hand using an editor; however, if another user attempts to change his password while the *passwd* file is being edited, the end result is a successful password modification for the user and the *passwd* file is updated to reflect the change. Unfortunately, this change will be lost when the file is saved by the person who had opened it for modification with an editor.

To prevent such an unwanted situation from happening, use the *vipw* command to edit the *passwd* (and *shadow*) file. This command copies the *passwd* file temporarily as */etc/ptmp* and disables write access to it. When another user attempts to change his password while you are editing the file, the user will be denied permission. When you quit *vipw*, some automatic checks are performed on *ptmp* to validate the contents. If no errors are encountered, this file is saved as */etc/passwd*, otherwise, it will remain unchanged. At this point, you are then prompted with whether you want to modify the *shadow* file as well. If your response is yes, the *shadow* file will be opened as */etc/sptmp* and you will be able to edit it. After you quit that file, some automatic checks on *sptmp* will be performed to validate the contents. If no errors are encountered, the file is saved as */etc/shadow*, otherwise, it will remain unchanged. The files are now available again for any modifications and the other user should be able to change his password.

Alternatively, if you do not want to modify the *shadow* file by hand, you can simply run the *pwconv* command after manually modifying the *passwd* file to let the command reflect the changes in the *shadow* file.

To prevent a similar situation from happening to the *group* file, RHEL provides the *vigr* command to edit this file. This command copies the *group* file temporarily as */etc/gtmp* and disables write access to it. When another user attempts to modify the *group* file while you are editing it, the user will be denied permission. When you quit *vigr*, some automatic checks are performed on *gtmp* to validate the contents. If no errors are encountered, this file is saved as */etc/group*, otherwise, it will remain intact. At this point, you are prompted whether you want to modify the *gshadow* file as well. If your response is yes, the *gshadow* file will be opened as */etc/sgtmp* and you will be able to edit it. After you quit that file, some automatic checks on *sgtmp* will be performed to validate the contents. If no errors are encountered, the file is saved as */etc/gshadow*, otherwise, it will remain intact. The files are now available again for any modifications.

Alternatively, if you do not wish to modify the *gshadow* file by hand, you can simply run the *grpconv* command after manually modifying the *group* file to let the command reflect the changes in the *gshadow* file.

Creating, Updating, and Removing Shadow Password Files

There are typically four commands that can be used for creating, updating, or removing the *shadow* and *gshadow* files. These are listed and described in Table 11-1.

Command	Description
pwconv	Creates/updates the *shadow* file and moves the user passwords over from the *passwd* file.
pwunconv	Moves the user passwords back to the *passwd* file and removes the *shadow* file.
grpconv	Creates/updates the *gshadow* file and moves the group passwords over from the *group* file.
grpunconv	Moves the group passwords back to the *group* file and removes the *gshadow* file.

Table 11-1 Password Shadowing Commands

Here are a few examples that will show the usage of these commands.

To enable password shadowing if it is not already enabled, execute the *pwconv* command:

pwconv

This command works quietly and does not display any output unless there is a problem. It creates the *shadow* file with read-only permission for the *root* user:

ll /etc/shadow
-r-------- 1 root   root   1014   Sep   26   11:22   /etc/shadow

To enable password shadowing at the group level if it is not already enabled, execute the *grpconv* command:

grpconv

This command works quietly and does not display any output unless there is a problem. It creates the *gshadow* file with read-only permission for the *root* user:

ll /etc/gshadow
-r-------- 1 root   root   612   Sep   26   11:23   /etc/gshadow

To disable password shadowing on user and group passwords, and remove the shadow files, execute the *pwunconv* and *grpunconv* commands, respectively.

Managing User Accounts and Password Aging with Commands

Managing user accounts involves creating, assigning passwords to, modifying, and deleting user accounts. Managing password aging involves setting and modifying aging attributes on user accounts. You may use either the command line or the Red Hat User Manager to perform these tasks.

Table 11-2 lists and describes user account and password management commands.

Command	Description
useradd	Adds a user.
usermod	Modifies user attributes.
userdel	Deletes a user.
chage	Sets or modifies password aging attributes for a user.
passwd	Sets or modifies a user password, and some password aging attributes.

Table 11-2 User and Password Aging Commands

The useradd, usermod, and userdel Commands

Use the *useradd* command to create a user account. This command adds entries to the *passwd*, *group*, *shadow*, and *gshadow* files. The command creates a home directory for the user and copies the default user initialization files from the skeleton directory */etc/skel* into the user's home directory. There are several options available with the *useradd* command. Table 11-3 explains some of them.

Option	Description
–b (--base-dir)	Defines the absolute path to the base directory where user directories are to be created.
–c (-- comment)	Defines useful comments or remarks.
–d (--home-dir)	Defines the absolute path to the user home directory.
–e (--expiredate)	Specifies a date after which a user account is automatically disabled.
–f (--inactive)	Denotes maximum days of inactivity before a user account is declared invalid.
–g (--gid)	Specifies the primary group. If this option is not used, a group account matching the user name is created with GID matching the user's UID. If you wish to assign a different GID, specify it with this option. Make sure that the group already exists.
–G (--groups)	Specifies the membership for up to 20 supplementary groups. If this option is not specified, no supplementary groups will be added.
–k (--skel)	Specifies the location of the skeleton directory (default is */etc/skel*), which contains user initialization template files. These files are copied to the user's home directory when the user account is created. Three bash shell user initialization files – *.bash_profile*, *.bashrc*, and *.bash_logout* – are available in this directory by default. More information on these files is provided later in this chapter. You may customize these files or add more files to this directory so every new user gets all of them. Existing user home directories will not be affected by this change.
–M	Prevents the command from automatically creating a home directory for the user.
–m (--create-home)	Creates a home directory if it does not already exist.
–n	Prevents the command from automatically creating a private group for the user.
–o (--non-unique)	Means that the new user can share the UID with an existing user. When two users share a UID, both get identical rights on each other's files. This should only be done in specific situations.
–r	Creates a user account with a UID below 500 and with a never expiring password.
–s (--shell)	Defines the absolute path to the shell file.
–u (--uid)	Indicates a unique user ID. The first UID assigned is 500. If this option is not specified, the next available UID from */etc/passwd* file is used.
login	Specifies a login name to be assigned to the new user account.

Table 11-3 useradd Command Options

The command picks up the default values from the */etc/default/useradd* and */etc/login.defs* files. You can either view the */etc/default/useradd* file contents with the *cat* command or run the following to view them:

```
# useradd –D
GROUP=100
HOME=/home
INACTIVE=-1
EXPIRE=
SHELL=/bin/bash
SKEL=/etc/skel
CREATE_MAIL_SPOOL=yes
```

You may modify these defaults. For example, execute the following to change the default base directory to */usr/home* so that new user home directories are created there:

 # useradd –D –b /usr/home

Check the */etc/default/useradd* file again. The modification should be reflected in the file.

The other file */etc/login.defs* contains additional directives that set defaults for user mail directory, password aging, UID/GID limits, and so on. User and group management commands consult this file to obtain information that is not specified with the *useradd* command. A *grep* on the uncommented lines is shown below:

```
# grep –v ^# /etc/login.defs
MAIL_DIR            /var/spool/mail
PASS_MAX_DAYS       99999
PASS_MIN_DAYS       0
PASS_MIN_LEN        5
PASS_WARN_AGE       7
UID_MIN             500
UID_MAX             60000
GID_MIN             500
GID_MAX             60000
CREATE_HOME         yes
UMASK               077
USERGROUPS_ENAB     yes
ENCRYPT_METHOD      SHA512
```

These parameters include the mail directory location for the user (MAIL_DIR), password aging parameters for the user (PASS_MAX_DAYS, PASS_MIN_DAYS, PASS_MIN_LEN, and PASS_WARN_AGE), range of UIDs and GIDs to be allocated to new user and group accounts (UID_MIN, UID_MAX, GID_MIN and GID_MAX), and instructions for the *useradd* command to create a home directory for the user (CREATE_HOME), set the default umask for the user to 077 (UMASK), delete the user's group if it contains no more members (USERGROUPS_ENAB), and use the SHA512 algorithm for encrypting user passwords (ENCRYPT_METHOD).

You can make modifications to a user account with the *usermod* command. The syntax of this command is very similar to that of the *useradd* command. Majority of the options that this command accepts are identical to that of the *useradd* command's.

The *userdel* command is straightforward. It removes the specified user account from all the authentication files, and removes the user's home directory if the –r option is specified with the command.

The chage Command

The *chage* command is used to set and alter password aging parameters on a user account. Table 11-4 lists and describes key options available with this command.

Option	Description
–d (--lastday)	Specifies a date or number of days since the epoch time when the password was last modified. It corresponds to the third field in the *shadow* file.
–m (--mindays)	Specifies the minimum number of days that must elapse before the password can be changed. It corresponds to the fourth field in the *shadow* file.
–M (--maxdays)	Denotes the maximum days of validity of the password before a user starts getting warning messages to change password. It corresponds to the fifth field in the *shadow* file.
–W (--warndays)	Defines the number of days a user gets warning messages to change password. It corresponds to the sixth field in the *shadow* file.
–I (--inactive)	Defines the number of days of inactivity after a password has expired and before the account is locked. It corresponds to the seventh field in the *shadow* file.
–E (--expiredate)	Specifies a date or number of days since the epoch time on which the user account becomes deactivated. It corresponds to the eighth field in the *shadow* file.
–l	Lists password aging attributes set on a user account.

Table 11-4 chage Command Options

The passwd Command

The primary use of the *passwd* command is to set or modify a user's password; however, this command can be used to lock and unlock a user account and modify password aging parameters for a user account. Table 11-5 lists and describes some key options available with this command. You must log in as the *root* user to use these options.

Option	Description
–k	Activates an expired user account without changing its password.
–l	Locks a user account.
–u	Unlocks a user account.
–d	Deletes a user password without expiring the user account.
–n	Specifies the mumber of days that must elapse before the password can be changed. It corresponds to the fourth field in the *shadow* file.
–x	Denotes the maximum days of validity of the password before a user starts getting warning messages to change password. It corresponds to the fifth field in the *shadow* file.

Option	Description
–w	Defines the number of days a user gets warning messages to change password. It corresponds to the sixth field in the *shadow* file.
–i	Defines the number of days of inactivity after a password has expired and before the account is locked. It corresponds to the seventh field in the *shadow* file.
–S	Displays password-related information for the specified user account.

Table 11-5 passwd Command Options

Exercise 11-1: Create a User Account with Custom Values

In this exercise, you will create an account for user *aghori* with UID 504, home directory */home/aghori*, shell */bin/bash*, membership in group *aghori* with GID 504, and default initialization scripts copied to this user's home directory from */etc/skel*. You will assign the user a password and show the line entries from the *passwd*, *shadow*, *group*, and *gshadow* files for this user.

1.   Use the *useradd* command to create user account *aghori* with UID 504, home directory */home/aghori*, shell */bin/bash*, membership in group matching the username and UID with default initialization files copied over into this user's home directory from */etc/skel*:

   # useradd –u 504 –m –d /home/aghori –k /etc/skel –s /bin/bash aghori

2.   Create a password for *aghori* with the *passwd* command:

   # passwd aghori
   New password:
   Retype new password:
   passwd: all authentication tokens updated successfully.

3.   Run the *grep* command on the *passwd*, *shadow*, *group*, and *gshadow* files to check what the *useradd* command has added:

   # grep aghori /etc/passwd /etc/shadow /etc/group /etc/gshadow
   passwd:aghori:x:504:504::/home/aghori:/bin/bash
   shadow:aghori:6HkAggh3C$m4uHGBrupXpV1pn9nRlELq08lLxigUai/Ph0INO4Q9WYLYX/H5Ui5yHNj5
   OcJ/E7iDlwF4G3rT5JtFD4OXw0m0:15609:0:99999:7:::
   group:aghori:x:504:
   gshadow:aghori:!::

Exercise 11-2: Create a User Account with Default Values

In this exercise, you will create an account for user *bghori* with all the defaults defined in the */etc/default/useradd* file. You will assign the user a password and show the line entries from the *passwd*, *shadow*, *group*, and *gshadow* files for this user.

1.   Use the *useradd* command to create user account *bghori* with all the default values:

   # useradd bghori

2. Create a password for *bghori* with the *passwd* command:

    # **passwd bghori**
    New password:
    Retype new password:
    passwd: all authentication tokens updated successfully.

3. Run the *grep* command on the *passwd*, *shadow*, *group*, and *gshadow* files to check what the *useradd* command has added:

    # **grep bghori /etc/passwd /etc/shadow /etc/group /etc/gshadow**
    passwd:bghori:x:505:505::/home/bghori:/bin/bash
    shadow:bghori:6/Hh8bzoT$Vh2exK.ng.JeNyGPRXrOHL/y7wsmMQcJj2USVYKW7odVirCAn6XlFQQ5JEHB
    2sTQwuFcE1nfR9POKHUeDrAQe1:15609:0:99999:7:::
    group:bghori:x:505:
    gshadow:bghori:!::

Exercise 11-3: Set up Password Aging on User Accounts

In this exercise, you will configure password aging for user *aghori* using the *passwd* command. You will set mindays to 7, maxdays to 28, and warndays to 5. You will run the *chage* command to display the aging settings on this account. You will also configure password aging for user *bghori* using the *chage* command. You will set mindays to 10, maxdays to 30, warndays to 7, and expiry to December 31, 2013. You will run the *chage* command to display the aging settings on this account.

1. Use the *passwd* command to configure password aging for user *aghori* with mindays (–n option) set to 7, maxdays (–x option) set to 28, and warndays (–w option) set to 5:

    # **passwd –n 7 –x 28 –w 5 aghori**
    Adjusting aging data for user aghori.
    passwd: Success

2. Run the *chage* command with the –l option on user *aghori* to display the aging settings:

    # **chage –l aghori**
    . . . .
    Minimum number of days between password change   : 7
    Maximum number of days between password change   : 28
    Number of days of warning before password expires   : 5

3. Use the *chage* command to configure password aging for user *bghori* with mindays (–m option) set to 10, maxdays (–M option) set to 30, warndays (–W option) set to 7, and account expiry set to December 31, 2013:

    # **chage –m 10 –M 30 –W 7 –E 2013-12-31 bghori**

4. Run the *chage* command with the –l option on user *bghori* to display the aging settings:

 # chage –l bghori

. . . .

Account expires	₁ Dec 31, 2013
Minimum number of days between password change	₁ 10
Maximum number of days between password change	₁ 30
Number of days of warning before password expires	₁ 7

Exercise 11-4: Modify and Delete a User Account

In this exercise, you will modify certain parameters for user *aghori* and *bghori* using the *usermod* and *chage* commands, and then delete *aghori* using the *userdel* command. You will change this user's login name to *aghori1*, home directory to */home/aghori1*, and login shell to */bin/dash*. You will *grep* the *passwd* file for *aghori* to validate the new information. You will set a new expiry on user *aghori*'s account so that it expires on October 11, 2014. You will validate the change with the *chage* command. You will modify user *bghori* so that he is forced to change his password at next login, cannot modify his password within five days after he has changed his password, and his account never expires. You will validate the change with the *chage* command. You will lock the user *aghori1*'s account and then delete it.

1. Run the *usermod* command to modify user *aghori*'s login name to *aghori1*, home directory to */home/aghori1*, and login shell to */bin/dash*:

    # usermod –m –d /home/aghori1 –s /bin/dash –l aghori1 aghori

2. Execute the *grep* command to obtain the information for user *aghori1* from the *passwd* file:

    # grep aghori1 /etc/passwd
    aghori1₁x₁504₁504₁₁/home/aghori1₁/bin/dash

3. Issue the *usermod* command and set October 11, 2014 as the new expiry date on user *aghori1*'s account:

    # usermod –e 2014-10-11 aghori1

4. Use the *chage* command with the –l option to display the new aging information for user *aghori1*:

    # chage –l aghori1

    . . . .

Account expires	₁ Oct 11, 2014

    . . . .

5. Run the *chage* command to modify user *bghori* so that he is forced to change his password at next login (–d option), cannot modify his password within five days after he has changed his password (–m option), and his account never expires (–E –1 options):

    # chage –d 0 –m 5 –E -1 bghori

6. Use the *chage* command with the –l option to display the new aging information for user *bghori*:

chage –l bghori
Last password change   : password must be changed
Password expires   : password must be changed
Password inactive   : password must be changed
. . . .

7. Execute any of the following to lock user *aghori1*:

usermod –L aghori1
passwd –l aghori1
Locking password for user aghori1.
passwd: Success

8. Issue the *userdel* command to remove user *aghori1* along with his home and mail spool directories:

userdel –r aghori1

Switching Users

Even though you can log in to the system directly as *root*, it is recommended that you log in as yourself as a normal user and then use the *su* command to become *root* if necessary. This is a safer practice and ensures system security and protection. If you wish to disable your ability to log in to the system directly as *root*, comment all but the "console" entry in the */etc/securetty* file. This will force you to log in to the system with your user account and then be able to *su* into *root* if required.

For example, if user *bghori* needs to become *root*, he will have to know the *root* password:

$ **su**
Password:

If he specifies the dash (–) character with the *su* command, it will also execute the *root* user's initialization files to define the environment including the PATH variable:

$ **su –**
Password:

If he wants to switch into another user's account such as user *user1*, he will have to know that user's password:

$ **su – user1**
Password:

The rule of knowing another user's password to switch into does not apply to *root*. If *root* wishes to *su* into a user account, he is not prompted for the user password.

Doing as Superuser

RHEL offers a way for normal users to be able to run a set of privileged commands without having to know or enter the *root* user's password. This is done via a utility called *sudo* (superuser do). *sudo* provides protected access to administrative functions. Any normal user that requires access to one or more administrative commands is defined in the */etc/sudoers* file. This file may be edited using the *visudo* command, which creates */etc/sudoers.tmp* file and apply the changes there. When you save and quit, the contents are written to */etc/sudoers,* and */etc/sudoers.tmp* is removed. The syntax for user and group entries in the file is similar to the following example entries for user *user1* and group *dba*:

```
user1       ALL=(ALL)       ALL
%dba        ALL=(ALL)       ALL
```

These entries mean that *user1* and members of the *dba* group (group is represented by the % sign) have full privileges to run all administrative commands. Providing normal users or groups unlimited access to privileged commands is an undesirable configuration. The unlimited access should be restricted to the *root* user only.

Now, when *user1* or any member of the *dba* group executes a privileged command, he will be required to enter his own password:

$ sudo system-config-users
Password:

If you wish *user1* and members of the *dba* group not to be prompted for their own passwords, you can modify their entries in the *sudoers* file to look like:

```
user1       ALL=(ALL)       NOPASSWD:       ALL
%dba        ALL=(ALL)       NOPASSWD:       ALL
```

To restrict *user1* and the *dba* group members to be able to run only the *system-config-display* and *system-config-users* commands, modify the directives as follows:

```
user1       ALL=/usr/bin/system-config-display,/usr/bin/system-config-users
%dba        ALL=/usr/bin/system-config-display,/usr/bin/system-config-users
```

Alternatively, you can define a Cmnd_Alias directive and assign desired privileged commands to it. You can then assign the Cmnd_Alias directive to one or more users or groups as required. In the example below, one such directive called PACKAGE_CMD is defined that includes several package management commands. This directive is assigned to the *dba* group so that its members are able to perform package administration tasks.

```
Cmnd_Alias  PACKAGE_CMD = /usr/sbin/yum-config-software, /usr/sbin/rpm
%dba        ALL = PACKAGE_CMD
```

Similarly, if you require users that do not belong to a single group to be able to run package administration commands, use the User_Alias directive and assign the users to it. The following

---

example defines a directive called pkgadm and assigns several users to it. It is assumed that these users do not belong to a single group.

```
User_Alias    pkgadm = user1, user2, user3, user4, user5
%pkgadm       ALL = PACKAGE_CMD
```

The *sudo* command logs information of commands being executed to the */var/log/messages* file as the user who runs it (and not as *root*).

In this manner, you can define as many sets of commands and users/groups as you wish.

Managing Group Accounts

Managing group accounts involve creating and modifying groups, adding and deleting group administrators, and deleting groups. You can use either the commands or the Red Hat User Manager to perform these tasks.

Table 11-6 lists and describes group account management commands.

Command	Description
groupadd	Adds a group.
groupmod	Modifies group attributes.
groupdel	Deletes a group.
gpasswd	Adds group administrators, adds or deletes group members, assigns or revokes a group password, and disables access to a group via the *newgrp* command.

Table 11-6 Group Management Commands

The groupadd, groupmod, and groupdel Commands

Use the *groupadd* command to create a group account. This command adds entries to the *group* and *gshadow* files. There are certain options available with the *groupadd* command. Table 11-7 explains some of them.

Option	Description
–g (--gid)	Specifies the GID that you wish to assign to the group.
–o (--non-unique)	Means that the new group can share the GID with an existing group. When two groups share a GID, members of each group get identical rights on each other's files. This should only be done in specific situations.
–r	Creates a group account with a GID below 500.
groupname	Specifies a group name.

Table 11-7 groupadd Command Options

The command picks up the default values from the */etc/default/useradd* and */etc/login.defs* files.

You can make modifications to a group account with the *groupmod* command. The syntax of this command is very similar to that of the *groupadd* command. The majority of the options that this command accepts are identical to that of the *groupadd* command's. The *groupmod* command has the –n option which allows it to alter the name of an existing group.

The *groupdel* command is straightforward. It removes the specified group account from both *group* and *gshadow* files.

The gpasswd Command

The *gpasswd* command can be used to add group administrators, add or delete group members, assign or revoke a group password, and disable access to a group via the *newgrp* command. The *root* user can perform all of these tasks and the group administrator can perform only the last three. This command prompts to change the group password if invoked by *root* or the group administrator. The *gpasswd* command updates the */etc/group* and */etc/gshadow* files. There are several options available with this command that are listed and explained in Table 11-8.

Option	Description
–A	Adds one or more group administrators. Inserts an entry in the third field of the *gshadow* file.
–a	Adds a group member. Inserts an entry in the fourth field of both *group* and *gshadow* files.
–d	Deletes a group member.
–M	Substitutes all existing group members.
–R	Disables access to a group via the *newgrp* command.
–r	Revokes the password set on a group. Only group members can use the *newgrp* command to join the group.

Table 11-8 gpasswd Command Options

The command picks up the default values from the */etc/login.defs* file.

Exercise 11-5: Create, Modify, and Delete Group Accounts

In this exercise, you will create a group account called *linuxadm* with GID 5000. You will create another group account called *sales* with GID 5000. You will change the group name from *sales* to *mgmt*. You will change the GID for *linuxadm* from 5000 to 6000. You will add *user1* to *linuxadm* and verify with the *id* and *groups* commands.

1. Use the *groupadd* command to create group account *linuxadm* with GID 5000:

    # **groupadd –g 5000 linuxadm**

2. Issue the *groupadd* command to create group account *sales* with GID 5000:

    # **groupadd –o –g 5000 sales**

3. Execute the *groupmod* command to alter the name of group *sales* to group *mgmt*:

    # **groupmod –n mgmt sales**

4. Run the *groupmod* command to alter the GID of *linuxadm* to 6000:

    # **groupmod –g 6000 linuxadm**

5. Use the *usermod* command to add user *user1* to group *linuxadm* while retaining his existing memberships:

   # **usermod –a –G linuxadm user1**

6. Run the *id* command for *user1* for verification:

   # **id user1**
   uid=500(user1) gid=500(user1) groups=500(user1),6000(linuxadm)
   # **groups user1**
   user1 : user1 linuxadm

7. Execute the *groupdel* command and delete the *mgmt* group:

   # **groupdel mgmt**

Exercise 11-6: Manage Groups with gpasswd Command

In this exercise, you will add *user1* and *user2* as group administrators and *user2* and *user3* as members to the *linuxadm* group. You will substitute both existing group members with *user4*. You will assign a group password to the *linuxadm* group. You will log in as *user4* and list the current primary group membership for *user4*. You will switch the primary group to *linuxadm* and validate.

1. Use the *gpasswd* command to add *user1* and *user2* as administrators to the *linuxadm* group:

   # **gpasswd –A user1,user2 linuxadm**

2. Add *user2* and *user3* as members to the *linuxadm* group:

   # **gpasswd –a user2 linuxadm**
   # **gpasswd –a user3 linuxadm**

3. Substitute *user1* and *user2* with *user4* as a member to the *linuxadm* group:

   # **gpasswd –M user4 linuxadm**

4. Set a password on the *linuxadm* group:

   # **gpasswd linuxadm**

5. Log in as *user4* and run the *groups* command to list group membership for *user4*:

   # **groups**
   user4 user1 linuxadm

6. Run the *newgrp* command to change *user4*'s primary group to *linuxadm*:

   $ **newgrp linuxadm**

---

7. Run the *groups* command again to verify the new primary group for *user4*:

    $ groups
    linuxadm user4 user1

8. Execute either the *exit* command or press Ctrl+d to return to the original primary group.

Managing Users, Groups, and Password Aging with the Red Hat User Manager

The Red Hat User Manager is a graphical equivalent for performing most user and group administrative functions. This tool can be used to view, add, modify, and delete user and group accounts, set and modify password aging attributes, lock and unlock user accounts, and so on. It can be invoked with the *system-config-users* command or by clicking System → Administration → Users and Groups. The main menu (similar to what is shown in Figure 11-5) will appear.

Figure 11-5 Red Hat User Manager – Main Menu

The Add User and Add Group buttons across the top open up windows where you can fill in the information to add new user and group accounts. Figure 11-6 and 11-7 show their interfaces. The fields are self-explanatory.

Figure 11-6 Red Hat User Manager – Add New User

Figure 11-7 Red Hat User Manager – Add New Group

The Properties button at the top is activated when a user or group account is highlighted. It allows you to display or modify the highlighted user or group account. For User Properties, four tabs (User Data, Account Info, Password Info, and Groups) are available as shown in Figure 11-8. Select User Data to modify general user information; select Account Info to enable and set account expiration or lock the user; choose Password Info to enable and set password aging; and choose Groups to modify the user's group membership including its primary group.

Figure 11-8 Red Hat User Manager – Display / Modify User Properties

Highlight a group account under the Groups tab to display or modify its name or membership. For Group Properties, two tabs (Group Data and Group Users) are available as shown in Figure 11-9. Select Group Data to modify the group name and choose Group Users to modify group membership.

Figure 11-9 Red Hat User Manager – Display / Modify Group Properties

The Delete button at the top is highlighted when a user or group account is highlighted; the Delete button is used to remove the highlighted user or group from the system.

User Initialization Files

In Chapter 04 "File Permissions, Text Editors, Text Processors & The Shell" you used local and environment variables and modified the default command prompt and added useful information to it. In other words you modified the default shell environment to customize it according to your needs. The changes you made were lost when you logged off. What if you wanted to make those changes permanent so that each time you logged in they were there for you?

Modifications to the default shell environment can be stored in text files called the *initialization* files. These files are executed after a user is authenticated and before he gets the command prompt. There are two types of initialization files: *system-wide* and *per-user*.

System-wide Initialization Files

System-wide initialization files define general environment variables required by all or most users of the system. These files are maintained by the system administrator and can be modified to define any additional environment variables and customization needed by all system users.

Table 11-9 lists and describes system-wide initialization files for the bash shell users.

System-wide Files	Comments
/etc/bashrc	Defines functions and aliases, and sets the umask value, the command prompt, etc. It includes settings from the shell scripts located in the */etc/profile.d* directory.
/etc/profile	Sets environment variables such as PATH, USER, LOGNAME, MAIL, HOSTNAME, HISTSIZE, and HISTCONTROL for users and startup programs. It processes the shell scripts located in the */etc/profile.d* directory.
/etc/profile.d	Shell scripts located here are executed by the */etc/profile* file. This directory contains scripts for bash and C shell users.

Table 11-9 System-wide Initialization Files

Excerpts from the *bashrc* and *profile* files, and a list of files in the *profile.d* directory are displayed below:

```
# cat /etc/bashrc
# /etc/bashrc
# System wide functions and aliases
# Environment stuff goes in /etc/profile
if [ "$PS1" ]; then
  if [ -z "$PROMPT_COMMAND" ]; then
    case $TERM in
    xterm*)
. . . . . . . .
```

# cat /etc/profile
# /etc/profile
# System wide environment and startup programs, for login setup
# Functions and aliases go in /etc/bashrc
pathmunge () {
  case ":${PATH}:" in
    :"$1":)
      ;;
    *)
      if ["$2" = "after"] ; then
        PATH=$PATH:$1
      else
        PATH=$1:$PATH
      fi
  esac
}
. . . .

ll /etc/profile.d
-rw-r--r--. 1 root root 1133 Apr 28 2010 colorls.csh
-rw-r--r--. 1 root root 1143 Apr 28 2010 colorls.sh
-rw-r--r--. 1 root root   80 Oct 22 2010 cvs.csh
. . . .

Per-user Initialization Files

Per-user initialization files override or modify system defaults set by the system-wide initialization files. These files may be customized by individual users to suit their needs. By default, files located in the *etc/skel* directory are copied into the user's home directory when the user account is created.

You may create additional per-user initialization files in your home directory to define additional environment variables or set additional shell properties.

Table 11-10 lists and describes the per-user initialization files for the bash shell users.

Per-user Files	Comments
~/.bashrc	Defines functions and aliases. This file sources global definitions from the *etc/bashrc* file.
~/.bash_profile	Sets environment variables. This file sources the ~/.bashrc file to set functions and aliases.
~/.bash_logout	Defines any specific commands to be executed prior to logout.
~/.gnome2/	Directory that holds environment settings when GNOME desktop is started. Only available if GNOME is installed.
~/.kde/	Directory that holds environment settings when KDE desktop is started. Only available if KDE is installed.
~/.mozilla	Includes extensions and pluggings for the firefox browser. Only available if firefox is installed.

Table 11-10 Per-user Initialization Files

Excerpts from the *.bashrc*, *.bash_profile*, and *.bash_logout* files are displayed below:

```
# cat ~/.bashrc
# .bashrc
# Source global definitions
if [ -f /etc/bashrc ]; then
    . /etc/bashrc
fi
# User specific aliases and functions
# cat ~/.bash_profile
# .bash_profile
# Get the aliases and functions
if [ -f ~/.bashrc ]; then
    . ~/.bashrc
fi
# User specific environment and startup programs
PATH=$PATH:$HOME/bin
export PATH
# cat ~/.bash_logout
# ~/.bash_logout
```

Chapter Summary

In this chapter, you started off by building an understanding of *passwd, shadow, group,* and *gshadow* user authentication files. You looked at what these files contained, their syntax, and how to verify their consistency. You looked at ways to lock these files while being edited to avoid losing changes done by other users during that time.

You studied password shadowing and password aging. You learned about user management including creating, modifying, and deleting user accounts. You looked at how to set and modify password aging attributes on user accounts. You learned about tools to switch into another user's account and run privileged commands as a normal user. Likewise, you studied group management including creating, modifying, and deleting group accounts. You reviewed the Red Hat User Manager tool and saw how to manage user and group accounts in the graphical environment.

Finally, you learned about the system-wide and per-user initialization files, and what these files contained.

Chapter Review Questions

1. What is the name of the default backup file for *gshadow*?
2. What are the two commands for manually editing shadow password files exclusively?
3. What are the two commands for checking shadow password files consistency?
4. What would the command *useradd –D* do?
5. Name the four user authentication files.
6. Why would you use the –o option with the *groupadd* and *groupmod* commands?
7. The *passwd* file contains secondary user group information. True or False?
8. Which command would you use to add group administrators?
9. Name the two types of initialization files?

10. What would the command *passwd –l user10* do?
11. Name the three fundamental user account categories in RHEL.
12. What does the "x" in the password field in the *passwd* file imply?
13. Every user in RHEL gets a private group by default. True or False?
14. What would the *userdel* command do if it is run with the –r option?
15. UID 499 is reserved for normal users. True or False?
16. What is the first GID assigned to a group?
17. Write two command names for managing password aging.
18. What would the command *chage –E 2014-06-30 user10* do?
19. What is the name of the *sudo* configuration file?
20. What would the command *chage –l user5* do?
21. What is the first UID assigned to a user?
22. What is the difference between running the *su* command with and without the dash sign?
23. What would the command *passwd –n 7 –x 15 –w 3 user5* do?
24. What two commands are used to create and update the *shadow* and *gshadow* files?
25. What would the command *useradd user500* do?
26. What does the *gpasswd* command do?
27. Which command is used to change a user's primary group temporarily?
28. What would the command *chage –d 0 user60* do?
29. What four files are updated when a user account is created?
30. The */etc/bashrc* file contains shell scripts that are executed when a user logs in. True or False?

Answers to Chapter Review Questions

1. The name of the default backup file for *gshadow* is *gshadow-*.
2. The *vipw* and *vigr* commands.
3. The *pwck* and *grpck* commands.
4. The command provided will display the default values used when a user account is created or modified.
5. The *passwd, shadow, group,* and *gshadow* files.
6. The –o option lets the commands set a duplicate GID on a group.
7. False. The *passwd* file contains primary user group information.
8. The *gpasswd* command.
9. The two types of initialization files are system-wide and per-user.
10. The command provided will lock *user10*.
11. The three fundamental user account categories are root, normal, and system.
12. The "x" in the password field implies that the encrypted password is stored in the *shadow* file.
13. True.
14. The command provided will delete the specified user's home directory as well.
15. False.
16. The first GID assigned to a regular user is 500.
17. The *passwd* and *chage* commands.
18. The command provided will set June 30, 2014 as the expiry date on *user10*'s account.
19. The *sudo* configuration file is */etc/sudoers*.
20. The command provided will display password aging attributes for *user5*.
21. The first UID assigned to a regular user is 500.

22. With the dash sign the *su* command will process the specified user's initialization files, and it won't without this sign.
23. The command will set mindays to 7, maxdays to 15 and warndays to 3 on *user5*'s account.
24. The *pwconv* and *grpconv* commands.
25. The command provided will add *user500* with all pre-defined default values.
26. The *gpasswd* command is used to add group administrators, add or delete group members, assign and revoke a group password, and disable access to a group with the *newgrp* command.
27. The *newgrp* command.
28. The command provided will force *user60* to change his password at next login.
29. The *passwd*, *shadow*, *group*, and *gshadow* files are updated when a user account is created.
30. False.

Labs

Lab 11-1: Create and Modify a User Account

Create a user account called *user4000* with UID 4000, GID 5000, and home directory located in */usr*. GID 5000 should be assigned to group *sysadmin*. Assign this user a password and establish password aging attributes on this user account so that he cannot change his password within 4 days after setting it and his password remains valid for 15 days. This user should get warning messages for changing his password for 10 days before his account is locked. His account needs to expire on the 15th of February, 2014.

Lab 11-2: Modify a Group Account

Modify the GID from 5000 to 5010 for the *sysadmin* group. Add users *user1*, *user2*, and *user3* as members, and *user4* as the group administrator. Assign a password to this group and use the *newgrp* command as *user1* to validate that the password is working. Change this group's name to *dbadmin* and verify.

Chapter 12

Firewall & SELinux

This chapter covers the following major topics:

- ✓ Understand iptables firewall for host-based security control
- ✓ Modify firewall rules at the command prompt and via Firewall Configuration tool
- ✓ Describe Security Enhanced Linux
- ✓ Understand SELinux configuration file and security contexts
- ✓ Manage SELinux via commands and SELinux Configuration tool
- ✓ View SELinux-related alerts

This chapter includes the following RHCSA objectives:

50. Configure firewall settings using system-config-firewall or iptables
51. Set enforcing and permissive modes for SELinux
52. List and identify SELinux file and process context
53. Restore default file contexts
54. Use boolean settings to modify system SELinux settings
55. Diagnose and address routine SELinux policy violations

Running

a system in a networked or an Internet-facing environment requires that some measures be taken to tighten access to the system by identifying the type and level of security needed, and implementing it. Security features such as file and directory permissions, user and group level permissions, Access Control List, and shadow password and password aging mechanisms have been discussed in the previous chapters. This chapter covers firewall and SELinux in fair detail. In the RHCE section of this book, you will be using the knowledge gained in this chapter at an advanced level to modify firewall rules and SELinux settings to allow network services to work with these security services without any issues.

Firewall

A *firewall* is a protective layer that is configured between a private and a public network to segregate traffic. There are several types of firewalls one of which performs data packet filtering. A data packet is formed as a result of a process called *encapsulation* whereby the header information is attached to a message during the formation of a packet. The header includes information such as source and destination IP addresses, port, and type of data. Based on pre-defined *rules*, a firewall intercepts each inbound and outbound data packet, inspects its header information, and decides whether to allow the packet to pass through. A port is defined in the */etc/services* file for each network service available on the system, and are typically standardized across all network operating systems including RHEL. Some common services and the ports they listen on are: *ftp* on port 21, *ssh* on 22, *telnet* on 23, *postfix* on 25, *http* on 80, and *ntp* on port 123.

RHEL comes standard with a host-based packet-filtering firewall software called *iptables* that can be used to control the flow of data packets. Another major use of iptables in RHEL is to provide support for Network Address Translation (NAT), which is explained in detail in Chapter 14 "Advanced Firewall, TCP Wrappers, Usage Reporting & Remote Logging".

To check whether the iptables package is installed on the system, run the *rpm* command:

 # rpm –qa | grep iptables
iptables-1.4.7-4.el6.x86_64

The configuration file for the iptables firewall is the *iptables* file located in the */etc/sysconfig* directory and the */etc/rc.d/init.d/iptables* is the startup script, which reads the configuration file and loads the rules defined in it. The configuration file may be customized using a text editor, the *iptables* command, the *lokkit* command, or the Firewall Configuration tool (*system-config-firewall* or *system-config-firewall-tui*).

Understanding the iptables Configuration File

As mentioned, the */etc/sysconfig/iptables* file is the configuration file for the iptables firewall. This file contains all necessary rules needed to control the inbound and outbound traffic on the system. An excerpt from this file is shown below:

cat /etc/sysconfig/iptables
. . . .
*filter
:INPUT ACCEPT [0:0]
:FORWARD ACCEPT [0:0]
:OUTPUT   ACCEPT [0:0]
-A INPUT -m state --state ESTABLISHED,RELATED -j ACCEPT #Rule 1
-A INPUT -p icmp -j ACCEPT #Rule 2
-A INPUT -i lo -j ACCEPT #Rule 3
-A INPUT -m state --state NEW -m tcp -p tcp --dport 22 -j ACCEPT #Rule 4
-A INPUT -j REJECT --reject-with icmp-host-prohibited #Rule 5
-A FORWARD -j REJECT --reject-with icmp-host-prohibited
COMMIT

Iptables allows you to define tables containing groups of rules called *chains* with each table related to a different type of packet processing. In the output above, one default table called *filter* is defined, which includes three pre-defined chains (INPUT, FORWARD, and OUTPUT) with the INPUT chain containing several rules. Each inbound and outbound packet goes through at least one of the configured chains. Packets destined for the local system use the INPUT chain, packets originated from the local system use the OUTPUT chain, and packets that need to be routed to a different network use the FORWARD chain. Chains have a policy called *target*, which can have a value such as ACCEPT, REJECT, DROP, and LOG. The ACCEPT policy allows a packet to pass through, the REJECT policy throws a packet away and sends a notification back, the DROP policy simply throws a packet away without sending a notification, and the LOG policy sends packet information to the *rsyslogd* daemon for logging.

Also notice the string "state" in the file contents above. Some of the common states are NEW, ESTABLISHED, RELATED, and INVALID. The NEW state identifies packets that are not part of an existing communication, the ESTABLISHED state indicates that the packets are part of an existing communication, the RELATED state signifies packets generated in relation with some other existing communication, and the INVALID state identifies packets that do not match other states.

The first rule in the configuration file above will continue to accept all inbound connection requests, the second rule will accept incoming ICMP requests, the third rule will accept all inbound connection requests on the loopback interface, the fourth rule will accept all new matching TCP requests received on port 22, and the last two rules will reject any network packets from any source with the icmp-host-prohibited rejection reason sent back to the source system.

It is imperative that you modify the default iptables rules to match the requirements of packet filtering associated with application accessibility.

The iptables Command

The *iptables* command is used to modify the firewall rules on the system. This command supports several options, some of which are listed and described in Table 12-1.

Option	Description
−A	Appends one or more rules to a chain.
−D	Deletes one or more rules from a chain.
−F	Flushes a chain or a table.
−I	Inserts one or more rules in a chain.
−L	Displays currently loaded rules.
−N	Adds a new chain.
−R	Replaces a rule in a chain.
−X	Deletes a chain.
−d	Specifies a destination address.
--dport	Specifies the destination port number.
−i	Specifies a network interface to be used for inbound packets.
−j	Specifies where a packet will jump if it matches a rule.
−m	Specifies a matching name.
−n	Specifies not to resolve IP to hostname and port number to service name mappings.
−o	Specifies a network interface to be used for outbound packets.
−p	Defines a protocol as listed in the /etc/protocols file.
−s	Specifies the source address.
−t	Specifies the type of table. Default is filter, which contains rulesets used for packet filtering.
−v	Prints verbose output.

Table 12-1 iptables Command Options

Run the *iptables* command with the −L option to list the rules currently in place on the system:

```
# iptables −L
Chain INPUT (policy ACCEPT)
target    prot opt source          destination
. . . . . . . .
ACCEPT    all   -- anywhere        anywhere       state RELATED,ESTABLISHED
ACCEPT    icmp -- anywhere         anywhere
ACCEPT    all   -- anywhere        anywhere
ACCEPT    tcp  -- anywhere         anywhere       state NEW tcp dpt:ssh
REJECT    all   -- anywhere        anywhere       reject-with icmp-host-prohibited
Chain FORWARD (policy ACCEPT)
target    prot  opt source         destination
. . . . . . . .
REJECT    all   -- anywhere        anywhere       reject-with icmp-host-prohibited
Chain OUTPUT (policy ACCEPT)
target    prot  opt  source        destination
```

The above output indicates that the INPUT, FORWARD, and OUTPUT chains are set to ACCEPT all traffic as explained in the previous sub-section.

Managing iptables Service Start and Stop

The iptables service can be started, restarted, and stopped manually. It can also be configured to start automatically at specific run levels.

To start the iptables service:

> # **service iptables start**
> iptables: Applying firewall rules: [OK]

To restart the iptables service:

> # **service iptables restart**
> iptables: Flushing firewall rules: [OK]
> iptables: Setting chains to policy ACCEPT: filter [OK]
> iptables: Unloading modules: [OK]
> iptables: Applying firewall rules: [OK]

To stop the iptables service:

> # **service iptables stop**
> iptables: Flushing firewall rules: [OK]
> iptables: Setting chains to policy ACCEPT: filter [OK]
> iptables: Unloading modules: [OK]

To enable the iptables service to autostart at each system reboot:

> # **chkconfig iptables on**
> iptables   0:off 1:off 2:on 3:on 4:on 5:on 6:off

To check the status of the iptables service:

> # **service iptables status**
> Table: filter
> Chain INPUT (policy ACCEPT)
> . . . .

Exercise 12-1: Add and Manage iptables Rules

In this exercise, you will first delete all existing rules and then append rules to the filter table to allow inbound HTTP traffic on port 80, reject outbound ICMP traffic without sending a notification back, and forward all inbound traffic to 192.168.2.0/24 network. You will insert a rule to allow the first and subsequent incoming FTP connection requests on port 21 and append a rule to disallow all outgoing connection requests on port 25. You will save all these rules in the */etc/sysconfig/iptables* file and make necessary settings so that these rules are loaded each time you reboot the system in multiuser mode. You will restart the iptables service and run the command to ensure that the new rules have taken place.

1. Run the *iptables* command with the –F option to remove all existing rules:

iptables -F

2. Append a rule to the filter table to allow inbound HTTP traffic on port 80:

    # iptables –t filter –A INPUT –p tcp --dport 80 –j ACCEPT

3. Append a rule to reject outbound ICMP traffic without sending a notification back:

    # iptables –A OUTPUT –p icmp –j DROP

4. Append a rule to forward all inbound traffic to 192.168.2.0/24 network:

    # iptables –A FORWARD –d 192.168.2.0/24 –j ACCEPT

5. Insert a rule to allow the first and subsequent incoming FTP connection requests on port 21:

    # iptables –I INPUT –m state --state NEW –p tcp --dport 21 –j ACCEPT

6. Append a rule to disallow all existing and new outgoing connection requests on port 25:

    # iptables –A OUTPUT –m state --state NEW –p tcp --dport 25 –j DROP

7. Save the rules in the */etc/sysconfig/iptables* file and restart the firewall to activate the new rule:

    # service iptables save ; service iptables restart

8. Set the iptables firewall service to autostart at each system reboot, and validate:

    # chkconfig iptables on
    # chkconfig --list iptables
    iptables   0:off 1:off 2:on   3:on   4:on   5:on   6:off

9. Issue the *iptables* command and check whether the new rules have taken effect:

    # iptables –L

Exercise 12-2: Add and Manage More iptables Rules

In this exercise, you will first delete all existing rules and then insert/append rules to the filter table to allow inbound HTTP traffic on port 80 from 192.168.1.0/24, reject all inbound traffic from 192.168.3.0/24 on the *eth1* interface, reject outbound ICMP traffic to all systems on 192.168.3.0/24 except for the system with IP 192.168.3.3, forward all inbound traffic from 192.168.1.0/24 to 192.168.2.0/24, and delete the last rule added. You will save all these rules in the */etc/sysconfig/iptables* file and make necessary settings so that these rules are loaded each time you reboot the system in multiuser mode. You will restart the iptables service and run the command to ensure that the new rules have taken place.

1. Run the *iptables* command with the –F option to remove all existing rules:

iptables –F

2. Insert a rule to the filter table to allow inbound HTTP traffic on port 80 from 192.168.1.0/24 only:

iptables –I INPUT –s 192.168.1.0/24 –p tcp --dport 80 –j ACCEPT

3. Append a rule to reject all inbound traffic from 192.168.3.0/24 on the *eth1* interface:

iptables –A INPUT –s 192.168.3.0/24 –i eth1 –j DROP

4. Insert a rule to reject all outbound ICMP traffic to all systems on 192.168.3.0/24 except for the system with IP 192.168.3.3/24:

iptables –I INPUT –d ! 192.168.3.3/24 –p icmp –j DROP

5. Append a rule to forward all inbound traffic from 192.168.1.0/24 to 192.168.2.0/24:

iptables –A FORWARD –s 192.168.1.0/24 –d 192.168.2.0/24 –j ACCEPT

6. Delete the above rule:

iptables –D FORWARD –s 192.168.1.0/24 –d 192.168.2.0/24 –j ACCEPT

7. Save all the rules just added to the */etc/sysconfig/iptables* file:

service iptables save

Modifying iptables Rules Using the Firewall Configuration Tool

The Firewall Configuration tool in RHEL 6 allows you to manage the iptables firewall in a graphical environment. The main interface of this tool, as shown in Figure 12-1, is opened when you execute the *system-config-firewall* command in an X terminal window or with the System → Administration → Firewall in the GNOME desktop.

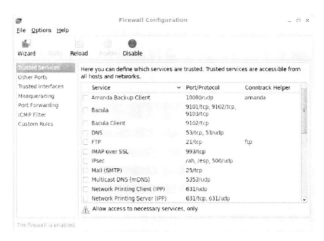

Figure 12-1 Firewall Configuration Tool

There are five buttons across the top: Wizard, Apply, Reload, Enable, and Disable. The Wizard button lets you configure firewall rules on a networked system by choosing to keep the existing configuration and modify it or load a default configuration and then customize it; the Apply button simply saves the changes to the */etc/sysconfig/iptables* file; the Reload button reloads the current iptables configuration; and the Enable and Disable buttons activate and deactivate the firewall.

In the left hand pane, there are seven categories: Trusted Services, Other Ports, Trusted Interfaces, Masquerading, Port Forwarding, ICMP Filter, and Custom Rules. Under Trusted Services, you can permit or block traffic for the specific application or service; under Other Ports, you can open a specific port associated with an existing service or specify a port or a range of ports to open; in Trusted Interfaces, you can view all visible network interfaces, add any missing interfaces, and choose one or more of these interfaces to have full access to the system; in Masquerading, you can configure to hide the IP information of your system on the Internet; under Port Forwarding, you can forward traffic for one port to another port on the same system or another system; under ICMP Filter, you can view a list of several types of ICMP messages and choose to disallow them from passing through the firewall; and under Custom Rules, you can add extra rules to the firewall.

The text version of this graphical tool can be started by running the *system-config-firewall-tui* command. It supports almost all the functions that are available in its graphical counterpart.

You can use one of these firewall configuration tools to customize the settings, but do not forget to make a backup of the */etc/sysconfig/iptables* file prior to running these tools to avoid losing existing data.

Security Enhanced Linux

Security Enhanced Linux (SELinux) is an implementation of the *Mandatory Access Control* (MAC) architecture developed by the *National Security Agency* (NSA), with the assistance of other organizations and Linux community, for enhanced, granular security controls in Linux. MAC provides an added layer of protection on top of the standard Linux *Discretionary Access Control* (DAC) security architecture which includes file and directory permissions, ACLs, and so on. With DAC in use, world-readable user home directories with potentially sensitive data could become the target of unauthorized access. SELinux limits this access by adding appropriate controls on the

---

home directories so that they can only be accessed by authorized processes and users. The added security controls are stored in *contexts* (or *labels*) and are applied on files, processes, and users.

SELinux controls are fine-grained meaning that if one service on the system is compromised, other services will remain intact.

SELinux Contexts for Files, Processes, and Users

SELinux contexts define security attributes set on individual files, processes, and users. There are generally four types of attributes associated with the contexts and are referred to as *subject*, *object*, *action*, and *level*. The subject attribute represents a process or a user, the object attribute identifies a role, the action attribute determines a type, and the level determines the SELinux security level in the form of sensitivity:category. RHEL 6 supports multiple sensitivity levels starting from s0 with 1024 different categories within each sensitivity level.

SELinux allows you to set security policies using *type enforcement* and *role-based access control* (RBAC). The type enforcement element identifies the domains for processes and the types for files. The RBAC, or simply a *role*, element controls which domains and types can be accessed. Each subject has an associated domain and each object within the domain has an associated type. Likewise, each subject has an associated role to ensure that the system and user processes are separated. By default, any process or service that is not explicitly permitted is denied.

Use the –Z option with the *ll*, *ps*, and *id* commands to view the current security contexts on files, processes, and users respectively. The following shows the four attributes currently set on the *install.log* file and the *Desktop* directory in the */root* directory:

```
# ll –Z /root
-rw-r--r--. root root system_u:object_r:admin_home_t:s0         install.log
drwxr-xr-x. root root unconfined_u:object_r:admin_home_t:s0      Desktop
```

The subject (user) field in the output provides an SELinux user name (system_u, unconfined_u, root, user_u, etc.) that is mapped to a Linux user (*root* in this case); the object (role) field defines an SELinux role (object_r, system_r, etc.); the action (type) field determines an SELinux type (admin_home_t, user_home_t, etc.); and the level (level) field determines the level of sensitivity & category used as defined in the */etc/selinux/targeted/setrans.conf* file. Contexts for all files in the system are stored in the */etc/selinux/targeted/contexts/files/file_contexts* file. Notice the dot next to the public execute permission in the first column of the output of the *ll* command. This dot tells you that the SELinux context is set on this file or directory.

By default, files created in a directory inherit the security contexts set on the directory. If a file is moved to another directory, it takes its security contexts with it, which may differ from the destination directory's contexts.

In the same manner, you can use the –Z option with the *ps* command to view the SELinux contexts for running processes:

```
# ps –eZ
LABEL                           PID     TTY     TIME      CMD
system_u:system_r:init_t:s0     1       ?       00:00:00 init
system_u:system_r:kernel_t:s0   2       ?       00:00:00 kthreadd
. . . . . . . .
```

The subject (user) field in the output provides an SELinux user name (system_u, unconfined_u, root, user_u, etc.) that is mapped to a Linux user; the object (role) field defines an SELinux role (object_r, system_r, etc.) associated with the system process such as daemons; the action (type) field determines what sort of protection is applied to the process (init_t, httpd_t, public_content_t, sshd_exec_t, kernel_t, etc.); and the level (level) field determines the level of sensitivity & category used. Any process that is unprotected will have the action set to unconfined_t.

Likewise, you can use the –Z option with the *id* command to view the security contexts associated with your Linux user account:

id –Z
unconfined_u:unconfined_r:unconfined_t:s0-s0:c0.c1023

If you run the above command as a normal user, you will get the same output. In RHEL, all Linux users, including *root*, run unconfined by default, which entails that they have full access to the system and are only restricted by the DAC restrictions. The above output indicates that the *root* user is mapped to the SELinux unconfined_u user (subject), running as unconfined_r role (role) in the unconfinred_t domain (type) with sensitivity level 0 (level).

SELinux user mappings with Linux users may be viewed using the *semanage* command. But before you are able to do that, ensure that packages policycoreutils and policycoreutils-python are installed on the system. These packages include several SELinux management commands such as *restorecon, chcon, secon, setfiles, semodule,* and *semanage.* You may also want to install the policycoreutils-gui package so that you can manage SELinux graphically using the SELinux Configuration tool *system-config-selinux.*

rpm –qa | grep policycoreutils
policycoreutils–python–2.0.83–19.24.el6.x86_64
policycoreutils–gui–2.0.83–19.24.el6.x86_64
policycoreutils–2.0.83–19.24.el6.x86_64

If any of these packages are missing on the system, use one of the package management commands explained in Chapter 06 "Package Management" to install it.

Execute the *semanage* command with the login –l arguments to view the default user mappings:

semanage login –l

Login Name	SELinux User	MLS/MCS Range
__default__	unconfined_u	s0-s0:c0.c1023
root	unconfined_u	s0-s0:c0.c1023
system_u	system_u	s0-s0:c0.c1023

The above output shows Linux users in the first column mapped to SELinux users in the second column, and the associated sensitivity & category level in the third column.

As stated above, all users including the *root* user run unconfined by default. If you want to switch a user such as *user1* into the staff_u role, run the *semanage* command. The following example first runs the *id* command with the –Z option to confirm that *user1* is running unconfined and then it switches the role:

```
$ id –Z
```
unconfined_u:unconfined_r:unconfined_t:s0-s0:c0.c1023
```
# semanage login –a –s staff_u user1
```

Now log off and log back in as *user1* and run the *id* command with the –Z option again to confirm the change:

```
$ id –Z
```
staff_u:staff_r:staff_t:s0-s0:c0.c1023

SELinux Configuration File

The key configuration file for SELinux is */etc/selinux/config* and the default contents of it are displayed below:

```
# cat /etc/selinux/config
# This file controls the state of SELinux on the system.
# SELINUX= can take one of these three values:
#    enforcing - SELinux security policy is enforced.
#    permissive - SELinux prints warnings instead of enforcing.
#    disabled - No SELinux policy is loaded.
SELINUX=enforcing
# SELINUXTYPE= can take one of these two values:
#    targeted - Targeted processes are protected.
#    mls - Multi Level Security protection.
SELINUXTYPE=targeted
```

The SELINUX directive in the file defines the activation mode for SELinux. Enforcing activates SELinux and allows or denies actions based on configured settings. Permissive activates SELinux, but permits all actions. It records all security violations; however, it does not hinder actions being taken. This mode is useful from a troubleshooting perspective, and developing or tuning SELinux security policy. The third option is to completely disable it. When activated in the enforcing mode, the SELINUXTYPE directive dictates the type of policy to be enforced. SELinux supports two policies: *targeted* and *mls* (*multi-level security*). With the targeted policy in place, you can modify SELinux restrictions set on files, processes, and users. The mls policy, on the other hand, allows you to tighten security at more granular levels. The default policy in RHEL6 is the targeted policy.

SELinux Administration Commands

Table 12-2 lists and explains some of the common SELinux administration commands.

Command	Description
chcon	Changes the SELinux security context on files. Changes do not survive system reboots.
getenforce	Displays the current SELinux mode of operation.
getsebool	Displays SELinux booleans and their values.
restorecon	Restores the default SELinux security contexts on files.
sealert	SELinux troubleshooting tool.

Command	Description
semanage	Manages SELinux policies. Changes survive system reboots.
sestatus	Displays the status of SELinux.
setenforce	Modifies the SELinux mode to enforcing or permissive.
setsebool	Modifies SELinux boolean values.
system-config-selinux	A graphical tool to administer SELinux.
togglesebool	Switches the value of a SELinux boolean.

Table 12-2 SELinux Administration Command

Checking and Modifying SELinux Activation Mode

The *getenforce* command is used to check the current SELinux activation mode:

 # getenforce
Enforcing

This command shows that the current activation mode is set to enforcing which implies that SELinux restrictions are in place.

The SELinux activation mode may be altered from enforcing to permissive using the *setenforce* command or toggling the switch in the */selinux/enforce* file. This change will take effect right away. Use one of the methods demonstrated below:

 # setenforce permissve
 # setenforce 0
 # echo 0 > /selinux/enforce

Alternatively, you can modify the SELINUX directive in the */etc/selinux/config* file and reboot the system:

 # vi /etc/selinux/config
SELINUX=permissive

SELinux cannot be disabled with the methods indicated above. You have to modify the SELINUX directive in the */etc/selinux/config* file or add the selinux directive to the */boot/grub/grub.conf* file as shown below. In either case, you will need to reboot the system.

 # vi /etc/selinux/config
SELINUX=disabled
 # vi /boot/grub/grub.conf
selinux=0

Now, if you wish to re-activate SELinux from disabled to either enforcing or permissive mode, you will need to change the SELinux directive in the */etc/selinux/config* file appropriately and need to remove the selinux directive from the */boot/grub/grub.conf* file. You must reboot the system in order for this change to take effect. The reboot will take longer than normal as SELinux will have to go through the process of adding the SELinux contexts on each system file.

Checking SELinux Status

You can check whether SELinux is enabled or disabled on the system using the *sestatus* command. This command also shows the policy being used if SELinux is enabled in the enforcing or permissive mode.

```
# sestatus
SELinux status:          enabled
SELinuxfs mount:         /selinux
Current mode:            enforcing
Mode from config file:   enforcing
Policy version:          24
Policy from config file: targeted
```

The above output indicates that SELinux is enabled in the enforcing mode and is using the targeted policy. With the –v option, the *sestatus* command also reports on the security contexts set on files and processes by consulting the */etc/sestatus.conf* file. The default contents of *sestatus.conf* is shown below:

```
# cat /etc/sestatus.conf
[files]
/etc/passwd
/etc/shadow
/bin/bash
/bin/login
/bin/sh
/sbin/agetty
/sbin/init
/sbin/mingetty
/usr/sbin/sshd
/lib/libc.so.6
/lib/ld–linux.so.2
/lib/ld.so.1
[process]
/sbin/mingetty
/sbin/agetty
/usr/sbin/sshd
```

Exercise 12-3: Modify SELinux File Contexts

In this exercise, you will change the user and type on the */root/anaconda-ks.cfg* file to user_u and public_content_t respectively, and verify. You will re-apply the contexts on the */root/anaconda-ks.cfg* file making sure that the contexts survive server reboots. You will copy the current contexts set on the */etc/passwd* file to the */etc/group* file. You will make a copy of the */etc/group* file and ensure that the copy preserves the contexts at the target location. You will modify the contexts on the */root* directory and restore the default contexts back on it.

---

1. Execute the *ll* command with the –Z option on the */root/anaconda-ks.cfg* file to determine the current SELinux contexts:

    # ll –Z /root/anaconda-ks.cfg
    –rw-------. root root system_u:object_r:admin_home_t:s0 /root/anaconda-ks.cfg

2. Run the *chcon* command and modify the contexts on the file to user_u and public_content_t:

    # chcon –vu user_u –t public_content_t /root/anaconda-ks.cfg
    changing security context of `/root/anaconda-ks.cfg'

3. Issue the *semanage* command and modify the contexts on the file to ensure that the new contexts survive a SELinux relabeling:

    # semanage fcontext –a –s user_u –t public_content_t /root/anaconda-ks.cfg

4. Run the *ll* command with the –Z option again on the file to validate the change:

    # ll –Z /root/anaconda-ks.cfg
    –rw-------. root root user_u:object_r:public_content_t:s0 /root/anaconda-ks.cfg

5. Issue the *chcon* command to copy the security contexts of */etc/passwd* to */etc/group*:

    # chcon –v --reference /etc/passwd /etc/group
    changing security context of '/etc/group'

6. Run the *cp* command to copy the */etc/group* file to the */tmp* directory making sure the contexts are preserved:

    # cp --preserve=context /etc/group /tmp

7. Run the *ll* command on both the source and the target to confirm that the contexts are the same:

    # ll –Z /etc/group
    –rw-r--r--. root root system_u:object_r:etc_t:s0   /etc/group
    # ll –Z /tmp/group
    –rw-r--r--. root root system_u:object_r:etc_t:s0   /tmp/group

8. Execute the *chcon* command on the */root* directory and modify its contexts to user_u and public_content_t:

    # chcon –vu user_u –t public_content_t /root
    changing security context of `/root'

9. Issue the *semanage* command and modify the contexts on the */root* directory to ensure that the new contexts survive a SELinux relabeling:

    # semanage –a –s user_u –t public_content_t /root

---

10. Issue the *ll* command with the –Z option again on the directory to validate the change:

> # **ll –dZ /root**
> dr-xr-x---. root root user_u:object_r:public_content_t:s0 /root

11. Run the *restorecon* command and revert the contexts to their default settings on the */root* directory, and confirm:

> # **restorecon –F /root**
> # **ll –dZ /root**
> dr-xr-x---. root root system_u:object_r:admin_home_t:s0 /root

Exercise 12-4: Modify SELinux User Contexts

In this exercise, you will map the Linux user *user2* with the SELinux user user_u, and verify. You will then remove the mapping.

1. Run the *semanage* command to map the Linux user *user2* with the SELinux user user_u:

> # **semanage login –a –s user_u user2**

2. Execute the *semanage* command again to verify the mapping:

> # **semanage login –l | grep user2**
> user2   user_u   s0

3. Log in as *user2* and run the *id* command with the –Z option to verify the context:

> $ **id –Z**
> user_u:user_r:user_t:s0

4. Issue the *semanage* command and remove the mapping between *user2* and user_u:

> # **semanage login –d user2**

SELinux Booleans

SELinux *booleans* allow you to modify some components of the SELinux policy without the need for restarting the server or recompiling the policy. Booleans can either be enabled or disabled, and their values are stored in the corresponding directive files located in the */selinux/booleans* directory. The directory listing for */selinux/booleans* is provided below:

> # **ll /selinux/booleans**
> -rw-r--r--. 1 root root 0 Oct 1 07:43 abrt_anon_write
> -rw-r--r--. 1 root root 0 Oct 1 07:43 abrt_handle_event
> -rw-r--r--. 1 root root 0 Oct 1 07:43 allow_console_login
> . . . .

Each file represents a boolean. A boolean value may be flipped using either the *setsebool* or the *togglesebool* command. For example, to toggle the value of the allow_console_login directive, use either of the following:

```
# setsebool allow_console_login 1
# togglesebool allow_console_login
allow_console_login: inactive
```

These commands make the change take effect right away without having to reboot the system, and store a 0 (disable) or a 1 (enable) in the */selinux/syslogd/allow_console_login* file. Alternatively, if you wish to make the change take effect right away as well as survive server reboots, use the –P option with the *setsebool* command, or use the *semanage* command:

```
# setsebool –P allow_console_login 1
# semanage boolean –m --on allow_console_login
```

Use the *getsebool* or the *semanage* command to verify:

```
# getsebool allow_console_login
allow_console_login --> on
# semanage boolean –l | grep allow_console_login
allow_console_login     (on, on) Allow direct login to the console device. Required for System 390
```

If you want to see a list of all available booleans on the system, use the –a option with the *getsebool* command or use the *semanage* command with the boolean –l arguments:

```
# getsebool –a
abrt_anon_write --> off
abrt_handle_event --> off
allow_console_login --> on

........
# semanage boolean –l
SELinux boolean        State    Default  Description
ftp_home_dir           (off, off) Allow   ftp to read and write files in the user home directories
smartmon_3ware         (off, off) Enable  additional permissions needed to support devices on 3ware controllers.
xdm_sysadm_login       (off, off) Allow   xdm logins as sysadm

........
```

Managing SELinux Using the SELinux Configuration Tool

The SELinux Configuration tool is a graphical tool that allows you to perform a number of configuration and management tasks including setting the SELinux activation mode and disabling it. The main menu, as shown in Figure 12-2, appears when you start this tool by running the *system-config-selinux* command in an X terminal window or choosing System → Administration → SELinux Management in the GNOME desktop.

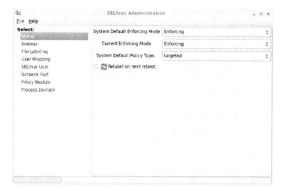

Figure 12-2 SELinux Configuration Tool – Main Menu

There are eight categories as listed in the left pane. You can click on a specific category to go there and view or modify settings. The first category Status shows the default and the current enforcing modes, as well as the default policy type. If you want the system to relabel the SELinux contexts on files at the next reboot, select the box next to "Relabel on next reboot". This should only be done when SELinux is enabled from the disabled state. The other seven categories are explained below.

The Boolean Category

Under this category you can turn a particular boolean on or off. You will be toggling many of these booleans as part of configuring services throughout this book.

Figure 12-3 SELinux Configuration Tool – Boolean

The File Labeling Category

This category allows you to modify SELinux security contexts on files. All modifications are stored in the */etc/selinux/targeted/contexts/files/file_contexts_local* file.

Figure 12-4 SELinux Configuration Tool – File Labeling

The User Mapping Category

This category allows you to modify extended attributes for regular and administrative users. The information displayed is the same that the *semanage* command with the login –l arguments showed earlier in this chapter.

Figure 12-5 SELinux Configuration Tool – User Mapping

The SELinux User Category

The SELinux User category allows you to set default roles for regular users (user_u), system users (system_u), unconfined users (unconfined_u), the *root* user, and a few other types of users.

Figure 12-6 SELinux Configuration Tool – User

The Network Port Category

This category lets you set SELinux port types with appropriate protocols and port numbers.

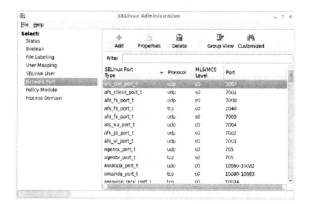

Figure 12-7 SELinux Configuration Tool – Network Port

The Policy Module Category

The Policy Module category displays versions associated with modules, allows you to add or remove modules, and enable module auditing.

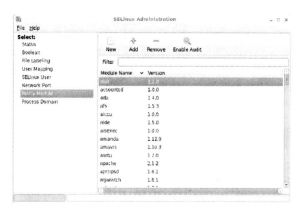

Figure 12-8 SELinux Configuration Tool – Policy Module

The Process Domain Category

This category displays the SELinux status, and allows you to switch between permissive and enforcing.

Figure 12-9 SELinux Configuration Tool – Process Domain

Viewing Alerts with the SELinux Troubleshooter

The SELinux troubleshooter is a tool that lets you view SELinux related issues conveniently in a graphical setting. It identifies issues, describes causes, and offers suggestions on fixing in addition to providing other associated assistance. This tool is invoked with the *sealert* command with the –b option or by choosing Applications → System Tools → SELinux Troubleshooter in the GNOME desktop. The setroubleshoot* packages need to be present on the system in order for you to be able to use this tool.

By default, all SELinux messages are logged to the system log file */var/log/messages* with an "avc" label and can be viewed with the SELinux Troubleshooter, and all SELinux denial messages are written to the */var/log/audit/audit.log* file. Messages in these files are logged by the *rsyslogd* and the *auditd* services, and both must be running to capture the messages. Run the *chkconfig* command and enable both services, if they are not already set, to start automatically at system reboots and the *service* command to start these services if they are not already running:

```
# chkconfig rsyslog on
# chkconfig auditd on
# service rsyslog restart
Shutting down system logger:     [ OK ]
Starting system logger:          [ OK ]
# service auditd restart
Stopping auditd:                 [ OK ]
Starting auditd:                 [ OK ]
```

Chapter Summary

In this chapter, you discussed two methods for enhanced system protection: firewall and SELinux. You reviewed the concepts of firewall. You looked at the firewall configuration file and understood entries placed in it. You studied the command and available options for listing and modifying firewall rules and starting and stopping the firewall. You also learned the graphical tool for firewall management.

You studied the concept and features of SELinux at length. You looked at how security contexts were associated with files, processes, and users, and you used commands to view and modify the contexts. You studied the SELinux configuration file which controls the state and defines the policy to be used. You learned several SELinux administrative commands and performed actions such as checking and modifying the activation mode and checking the status. You studied SELinux booleans and saw how to modify certain parts of the SELinux policy temporarily and persistently. You looked at the graphical SELinux configuration tool and reviewed each category briefly. Finally, you reviewed the SELinux Troubleshooter program and saw how SELinux related messages were logged.

Chapter Review Questions

1. What is the name of the command to start the graphical SELinux administration tool?
2. What would the command *iptables –A OUTPUT –p tcp --dport 25 –j DROP* do?
3. What are the two commands to display and modify the SELinux mode?
4. Iptables is a packet-filtering firewall. True or False?
5. SELinux is an implementation of discretionary access control. True or False?
6. Name the file where SELinux denial messages are logged?
7. What is the name of the graphical tool for firewall administration?
8. What would the command *iptables –F* do?
9. What is the name of the SELinux configuration file and where is it located in the system?
10. What is the difference between DROP and REJECT policies?
11. What does the ESTABLISHED state refer to in a firewall rule?
12. A firewall cannot be configured between two networks. True or False?
13. Which command can be used to change SELinux contexts permanently, *chcon* or *semanage*?
14. Which option is used with the *ps* command to view the security contexts on processes?
15. What would the command *restorecon –F /etc* do?
16. What is the name of the default SELinux policy used in RHEL6?
17. What is the name of the iptables configuration file and where is it located in the system?
18. What would the rule "-I INPUT -m state --state NEW -m tcp -p tcp --dport 22 -j ACCEPT" do?
19. What is the purpose of the command *sestatus*?
20. What is the name of the directory that stores SELinux boolean files?
21. Which two packages are required on the system to work with SELinux?
22. Place user, role, type, and level in the correct order.
23. What one task must be done to change the SELinux mode from enforcing to disabled?
24. What is the process of data packet formation called?
25. "service firewall start" will start the firewall service. True or False?
26. Which option with the *cp* command must be specified to preserve the SELinux contexts?
27. What would the command *semanage login –a –s user_u user10* do?
28. Name three commands that can be used to modify an SELinux boolean value?
29. If one service on the system running SELinux is compromised, all other services will be affected as well. True or False?
30. Place action, object, level, and subject in the correct order.
31. What would the command *iptables –L* do?
32. Name two methods to start the SELinux Troubleshooter program?

Answers to Chapter Review Questions

1. The name of the command to start the graphical SELinux administration tool is *system-config-selinux*.
2. This command will append a rule to the output chain to prevent TCP packets from leaving the system via port 25.
3. The *getenforce* and *setenforce* commands.
4. True.
5. False.
6. The SELinux denial messages are logged to the */var/log/audit/audit.log* file.
7. The graphical tool for firewall administration is called *system-config-firewall*.
8. The command provided will flush all the rules currently loaded on the system.
9. The configuration file name is selinux and it is located in the /etc/sysconfig directory.
10. The DROP policy sends a notification back while the REJECT policy does not.
11. The ESTABLISHED state refers to the connections that are already in place.
12. False.
13. The *semanage* command.
14. The –Z option.
15. This command will restore the default SELinux contexts on the */etc* directory.
16. The default policy used in RHEL6 for SELinux is the targetted policy.
17. The configuration file name is iptables and it is located in the */etc/sysconfig* directory.
18. The specified string will insert a rule in the iptables firewall to allow new traffic on port 22 to pass through without impacting any of the existing connections on this port.
19. The *sestatus* command displays the SELinux status.
20. The */selinux/booleans* directory.
21. The policycoreutils and policycoreutils-python packages.
22. The correct order is user, role, type, and level.
23. You will have to reboot the system.
24. The process of data packet formation is called encapsulation.
25. False.
26. The --preserve=context option.
27. This command will map Linux user user10 with SELinux user user_u.
28. The *semanage*, *setsebool*, and *togglesebool* commands.
29. False.
30. The correct order is subject, object, action, and level.
31. This command will display all loaded firewall rules on the system.
32. You can either start the SELinux Troubleshooter with the *sealert* command or from the GNOME desktop.

Labs

Lab 12-1: Add Firewall Rules

Insert a rule in the OUTPUT chain to allow UDP traffic to network 192.168.4.0/24 via port 100. Append a rule to the INPUT chain to disallow TCP traffic from network 192.168.3.0/24 via port 101 without sending a notification back to the requester, but allow from 192.168.3.33/24. Insert a rule in the INPUT chain to permit the established ssh communication channels to continue to function and allow for subsequent ssh requests to pass through the firewall as well.

Lab 12-2: Disable and Re-enable the SELinux Operating Mode

Check and make a note of the current SELinux operating mode. Modify the configuration file and set the mode to disabled. Reboot the system and validate the change. Restore the system back to the original SELinux mode and reboot to validate. Measure the time it takes for either reboot.

Chapter 13

Networking, Network Interfaces & DNS, DHCP and LDAP Clients

This chapter covers the following major topics:

✓ Basic networking including hostname, IP address, network classes, subnet, subnet mask, protocol, TCP/UDP, MAC addresses, and ARP
✓ Understand and use network interface configuration files and commands
✓ Configure a network interface and activate it
✓ Use ping and ip commands to test connectivity
✓ Manage network interfaces via Network Connections tool
✓ Understand and modify DNS client configuration files
✓ Automatically obtain IP assignments from a DHCP server
✓ LDAP concepts and terminology
✓ Understand and modify LDAP client configuration files
✓ Configure an LDAP client via the graphical Authentication Configuration tool

This chapter includes the following RHCSA objectives:

33. Configure networking and hostname resolution statically or dynamically
49. Configure a system to use an existing LDAP directory service for user and group information

A *network* is formed when two or more computers are interconnected to share resources and information. The computers may be interlinked via wired or wireless means, and a device such as a switch is required to interconnect several computers so that they can talk to one another on the network.

To comprehend how networks work and how network services are configured properly, some basic networking terminology first needs to be understood. For a system to be able to communicate with other systems on the network, one of its network interfaces must be configured with a unique IP address, hostname, and other essential network parameters. There are several files involved in the interface configuration and can be modified by hand or using a graphical tool. Once networking is configured, testing is done to ensure that the system is able to successfully communicate with other networked systems.

DNS client functionality needs to be established on the system so that it is able to communicate with systems on external networks. There are certain configuration files that need to be modified in order for this functionality to work. These files may be edited manually or graphically.

DHCP client functionality may need to be used on the system if configuring static IP assignments is not mandatory. This allows the system to pull required IP information from a configured DHCP server dynamically. There are certain ways to configure this ability that are discussed in this chapter.

LDAP client can be set up if a configured and available LDAP server is accessible for the client to be able to talk to it to obtain the needed directory information. There are certain configuration files that need to be modified in order for this functionality to work. These files may be edited manually or graphically.

Basic Networking

There are many basic concepts and terms that need to be grasped before you are able to configure network interfaces and perform advanced network client and server setups. Likewise, there are many configuration files and commands related to various network services that need to be thoroughly understood as well. Some of the concepts, terms, configuration files, and commands are explained in this chapter and many more will be introduced in the RHCE section of this book.

Hostname

A *hostname* is a unique alphanumeric name assigned to a system. It is normally allotted based on the purpose and primary use of the system. A hostname is defined in the startup configuration file */etc/sysconfig/network* via the HOSTNAME directive. A hostname can be changed temporarily using the *hostname* command or permanently by either modifying the HOSTNAME directive in the *network* file or running the Red Hat Network Configuration tool *system-config-network*.

The hostname of a system can be viewed using the *hostname* command:

hostname
server.example.com

If you need to change the hostname of the system to *server10*, for instance, on a temporary basis:

hostname server10
hostname
server10

And if you want to change the hostname of the system to *server10* on a permanent basis, edit the */etc/sysconfig/network* file and set the HOSTNAME directive to reflect the new hostname. You will need to reboot the system.

# **vi /etc/sysconfig/network**
HOSTNAME=server10
# **shutdown −ry now**

Whenever you modify the system hostname, make sure that you update the */etc/hosts* file to reflect the new hostname by changing the corresponding entry in that file.

Change the hostname back to *server.example.com* so that you are able to continue with exercises later in this chapter and the rest of the book.

IP Address

IP stands for *Internet Protocol* and represents a unique 32-bit software address that every single system on the network must have in order to communicate with other systems. IP addresses can be assigned on a temporary or permanent basis. Temporary addresses are referred to as the *dynamic* addresses and are typically leased from a DHCP server for a specific period of time. Permanent addresses, on the other hand, are called the *static* addresses and are set either manually or using a graphical tool. Permanent addresses are not changed unless there is a need.

To view IP addresses assigned to the network interfaces on your system, run the *ip* command with the addr argument:

ip addr
1: lo: <LOOPBACK,UP,LOWER_UP> mtu 16436 qdisc noqueue state UNKNOWN
   link/loopback 00:00:00:00:00:00 brd 00:00:00:00:00:00
   inet 127.0.0.1/8 scope host lo
   inet6 ::1/128 scope host
     valid_lft forever preferred_lft forever
2: eth0: <BROADCAST,MULTICAST,UP,LOWER_UP> mtu 1500 qdisc pfifo_fast state UP qlen 1000
   link/ether 52:54:00:28:e1:f6 brd ff:ff:ff:ff:ff:ff
   inet 192.168.2.201/24 brd 192.168.2.255 scope global eth0
   inet6 fe80::5054:ff:fe28:e1f6/64 scope link
     valid_lft forever preferred_lft forever

 You can also use the *ifconfig* command to obtain this information, but this command is now obsolete.

The above output indicates that there is currently one network interface called *eth0* configured on the system and it has 192.168.2.201 IP address assigned to it. The other interface called *lo* is the special purpose software interface reserved for use on every RHEL system. Its IP address is always 127.0.0.1 and it is referred to as the system's *loopback* (or *localhost*) address. This hostname is used by network applications to access the networking resources on the local system.

With the explosive growth of the Internet, the presence of an extremely large number of systems, and an ever increasing demand for additional addresses, the conventional IP address space that provides approximately 4.3 billion addresses, is almost used up. To meet the future demand, a new version of IP address space has been introduced and is now available for use. This new version is referred to as *IPv6* (IP version 6). By default, IPv6 is disabled via the NETWORKING_IPV6 directive as defined in the */etc/sysconfig/network* file.

Network Classes

An IP address is comprised of four octets that are divided into two portions: a *network* portion and a *node* portion. The network portion identifies the correct destination network and the node portion identifies the correct destination node on that network. Network addresses are classified into three classes referred to as class A, class B, and class C. Classes D and E are also defined and in use, however they are used in special environments such as scientific.

A node is any device on the network that has an IP address. Examples include computers, printers, routers, switches, hand-held devices, and so on.

Class A addresses are used for large networks with up to 16 million nodes. The network address range for class A networks is between 1 and 126. This class uses the first octet as the network portion and the rest of the octets as the node portion.

Class B addresses are used for mid-sized networks with up to 65 thousand nodes. The network address range for class B networks is between 128 and 191. This class uses the first two octets as the network portion and the other two octets as the node portion.

Class C addresses are used for small networks with up to 254 nodes. The network address range for class C networks is between 192 and 223. This class uses the first three octets as the network portion and the last octet as the node portion.

Subnetting and Subnet Mask

Subnetting is a method by which a large network address space can be divided into several smaller and more manageable logical sub-networks, commonly referred to as *subnets*. Subnetting usually results in reduced network traffic, improved network performance, and de-centralized and easier administration, among other benefits.

The following should be kept in mind when working with subnetting:

✓ Subnetting does not increase the number of IP addresses in a network. In fact, it reduces the number of useable IP addresses.
✓ All nodes in a given subnet have the same subnet mask.
✓ Each subnet acts as a separate network and requires a router to talk to other subnets.
✓ The first and the last IP address in a subnet are reserved. The first address points to the subnet itself and the last address is the broadcast address.

After a network address has been subnetted, you need to determine something called *subnet mask* or *netmask*. The subnet mask is the network portion plus the subnet bits. In other words, the subnet mask segregates the network bits from the node bits. It is used by routers to identify the start and end of the network/subnet portion and the start and end of the node portion of a given IP address.

The default subnet masks for class A, B, and C networks are 255.0.0.0, 255.255.0.0, and 255.255.255.0, respectively.

Protocol

A *protocol* is a set of rules governing the exchange of data between two networked nodes. These rules include how data is formatted, coded, and controlled. The rules also provide error handling, speed matching, and data packet sequencing. In other words, a protocol is a common language that all nodes on the network speak and understand. Some common protocols are TCP, IP, and ICMP. Protocols are defined in the */etc/protocols* file. An excerpt from the */etc/protocols* file is provided below:

```
# cat /etc/protocols
ip      0   IP      # internet protocol, pseudo protocol number
icmp    1   ICMP    # internet control message protocol
.........
```

TCP and UDP

TCP is reliable, connection-oriented, and point-to-point. When a stream of packets is sent to the destination node using the TCP protocol, the destination node checks for errors and packet sequencing upon its arrival. Each packet contains information such as the IP addresses for the source and the destination systems, port numbers, the data, a sequence number, and checksum fields. The TCP protocol at the source node establishes a point-to-point connection with the peer TCP protocol at the destination. When the packet is received by the receiving TCP, an acknowledgement is sent back. If the packet contains an error or is lost in transit, the destination node requests the source node to resend the packet. This ensures guaranteed data delivery and makes TCP reliable.

UDP, on the other hand, is unreliable, connectionless, and multi-point. If a packet is lost or contains errors upon arrival at the destination node, the source node is unaware of it. The destination node does not send an acknowledgement back to the source node. UDP is normally used for broadcast purposes.

Both TCP and UDP protocols use ports for data transmission between a client and its associated server program. Ports are defined in the */etc/services* file. Some common services and the ports they listen on are: *telnet* on port 23, *ftp* on 21, *postfix* on 25, *http* on 80, and *ntp* on 123. An excerpt from the */etc/services* file is shown below:

```
# cat /etc/services
# service-name   port/protocol  [aliases ...]   [# comment]
ftp              21/tcp
ftp              21/udp        fsp fspd
ssh              22/tcp                          # The Secure Shell (SSH) Protocol
.........
```

MAC Address and Address Resolution Protocol

MAC stands for *Medium Access Control* and represents a unique 48-bit address used to identify the correct destination node for data packets transmitted from the source node. The data packets include MAC addresses for both the source and the destination node. A network protocol called ARP, maps the MAC address to the destination node's IP address. The MAC address is also referred to as the *Ethernet, physical, hardware,* or *link layer* address.

You can use the *ip* command to list all network interfaces available on the system along with their Ethernet addresses:

> # **ip addr | grep –i ether**
> link/ether 52:54:00:28:e1:f6 brd ff:ff:ff:ff:ff:ff

As you are aware that IP and hardware addresses work hand in hand with each other and a combination of both is critical to identifying the correct destination node on the network. A protocol called *Address Resolution Protocol* (ARP) is used to enable IP and hardware addresses to work together. ARP determines the hardware address of the destination node when its IP address is already known.

ARP broadcasts messages over the network requesting each alive neighbor to reply with its link layer and IP addresses. The addresses received are cached locally by the node in a special memory area called the *ARP cache*, and can be viewed using the *ip* command with the neighbor argument:

> # **ip neighbor**
> 192.168.2.1   dev eth0 ll addr 00:23:51:87:f2:61 REACHABLE
> 192.168.2.102 dev em1 ll addr 00:11:95:ea:98:35 STALE
> 192.168.2.16 dev eth0 ll addr 38:59:f9:0d:43:60 NOARP

 You can also use the *arp* command to obtain this information, but this command is now obsolete.

The above output shows IP addresses of the neighbors in the first column followed by the interface that they are using, their link layer addresses, and their NUD (*Nieghbor Unreachability Detection*) state in the remaining columns. A neighbor can be in one of the four states: permanent, noarp, reachable, and stale. The permanent state indicates a valid entry that can only be deleted manually, the noarp state designates a valid entry that expires when it reaches the end of its lifetime, the reachable state shows a valid entry that expires when its reachability timeout expires, and the stale state specifies a valid bus suspicious entry.

Entries in the ARP cache are normally stored for 10 minutes and then removed. These entries can be modified or removed using the *ip* command. You can also add entries to the ARP cache with the *ip* command.

To remove an entry from the ARP cache for address 192.168.2.102:

> # **ip neighbor delete 192.168.2.102 nud stale dev eth0**

To add an entry to the ARP cache for address 192.168.2.102:

ip neighbor add 192.168.2.102 lladdr d4:be:d9:2e:2e:d9 dev eth0

Network Interfaces

Network interface cards are the hardware adapters that provide one or more Ethernet ports for network connectivity, and they are abbreviated as NICs. NICs may be built-in to the system board or are available as add-on adapters that are installed in system slots. NICs are available in one to four port configuration on a single adapter. Each NIC port may be configured with an appropriate IP address, subnet mask, and broadcast address using either commands or the Network Connection tool. Additional tasks such as activating and deactivating a network port manually and auto-activating at system reboots can also be performed using these tools.

RHEL includes several interface configuration files. The following sub-sections discuss the three essential files: */etc/sysconfig/network*, */etc/sysconfig/network-scripts/ifcfg-eth0*, and */etc/hosts*.

The /etc/sysconfig/network File

This is a crucial configuration file and it defines certain critical elements. This file contains at least the following two key directives:

NETWORKING=yes
HOSTNAME=server.example.com

A short description of these and other directives that may be defined in this file is provided in Table 13-1.

Directive	Description
GATEWAY	Defines the default gateway address. This directive can alternatively be defined in individual interface configuration files that are located in the */etc/sysconfig/network-scripts* directory.
GATEWAYDEV	Specifies the network interface device such as *eth0*, associated with the default gateway.
HOSTNAME	Sets the hostname of the system.
NETWORKING	Enables or disables IPv4 networking.
NETWORKING_IPV6	Enables or disables IPv6 networking.
NISDOMAIN	Defines the NIS domain name (if configured).

Table 13-1 Directives in /etc/sysconfig/network File

The /etc/sysconfig/network-scripts/ifcfg-eth0 File

This file holds the networking information for the first Ethernet interface on the system. For additional interfaces, the file names would be *ifcfg-eth1*, *ifcfg-eth2*, and so on. On many other systems, the Ethernet interface might be called *em1*, *br0*, or something else which would be reflected in the file name as well as in the file. Here are some of the more common directives from the *ifcfg-eth0* file:

```
# cat /etc/sysconfig/network-scripts/ifcfg-eth0
DEVICE=eth0
NM_CONTROLLED=yes
ONBOOT=yes
TYPE=Ethernet
BOOTPROTO=none
IPADDR=192.168.2.201
GATEWAY=192.168.2.1
DNS1=192.168.2.1
DEFROUTE=yes
IPV6INIT=no
NAME="System eth0"
UUID=5fb06bd0-0bb0-7ffb-45f1-d6edd65f3e03
NETMASK=255.255.255.0
USERCTL=no
HWADDR=52:54:00:28:E1:F6
```

These and other directives that may be defined in this file are described in Table 13-2.

Directive	Description
BOOTPROTO	Defines the boot protocol to be used. Values include "dhcp" to obtain the IP information from a DHCP server, "bootp" to boot off a network boot server and get the IP information from there, and "none" to use a static IP address defined in the file with the IPADDR directive.
BROADCAST	Specifies the broadcast IP address.
DEFROUTE	Defines the default route address.
DEVICE	Specifies the device name for the network interface.
DNS1	Places an entry for the first DNS server in the /etc/resolv.conf file if PEERDNS directive is set to "yes" in this file.
GATEWAY	See Table 13-1.
HWADDR	Describes the hardware address of the interface.
IPADDR	Specifies the IP address assigned to the interface.
IPV6INIT	Whether to enable IPv6 configuration for this interface.
NAME	Any description given to the interface.
NETMASK	Sets the netmask address.
NETWORK	Sets the network address.
NM_CONTROLLED	Whether the Network Manager utility is allowed to modify the configuration in this file. Default is yes.
ONBOOT	Whether to activate this interface at system boot.
PEERDNS	Whether to modify the DNS client resolver file /etc/resolv.conf. Default is "yes" if BOOTPROTO=dhcp is set.
USERCTL	Whether to allow non-root users to activate the interface.
UUID	The UUID associated with the interface.
TYPE	Specifies the type of the interface.

Table 13-2 Network Interface File Directives

There are other directives too that may be defined in this file depending on the type of interface connection used.

The /etc/hosts File

This file maintains the IP to hostname mapping for all the systems on the local network, and is typically used in the absence of DNS. The IP address configured on the interface needs to have a hostname assigned to it. In Chapters 01 and 07, you built four systems: *physical.example.com*, *server.example.com*, *insider.example.com*, and *outsider.example.net* with IP addresses 192.168.2.200, 192.168.2.201, 192.168.2.202, and 192.168.3.200, respectively. You can place these entries in the */etc/hosts* file as shown below:

192.168.2.200	physical.example.com	physical
192.168.2.201	server.example.com	server
192.168.2.202	insider.example.com	insider
192.168.3.200	outsider.example.net	outsider

Each line in the file contains an IP address in the first column followed by an official (or *canonical*) hostname in the second column. You may also define one or more aliases per entry such as *physical*, *server*, *insider*, and *outsider* in the above example. The official hostname and one or more aliases allow you to have multiple hostnames assigned to a single IP address. This allows users to access a system using any of its hostnames.

As indicated above, the */etc/hosts* file is typically used to access systems on the local network only. This file must be maintained on each individual system on the network and be edited whenever an update is required.

Network Interface Administration Commands

Network interfaces can be administered using commands or the graphical Network Connections tool. Table 13-3 lists and describes the commands.

Command	Description
ifconfig	Configures, activates, and deactivates an interface, and displays information about it. This command is now obsolete. Use the *ip* command instead.
ifdown	Deactivates an interface.
ifup	Activates an interface.
ip	Displays and administers interfaces, routing, etc. This command has replaced both *ifconfig* and *netstat* commands.
netstat	Displays network connections, routing tables, interface statistics, etc. This command is now obsolete. Use the *ip* command instead.
nm-connection-editor	The graphical Network Connections tool for network interface administration.
system-config-network-cmd	Displays details of configured interfaces.
system-config-network	The text equivalent of the *nm-connection-editor* tool.

Table 13-3 Network Interface Administration Commands

Exercise 13-1: Configure and Activate a Network Interface

In this exercise, you will configure a network interface called *eth1* on *server.example.com* with IP address 192.168.122.201, netmask 255.255.255.0, and broadcast 192.168.122.255. You will copy the existing *eth0* file in the */etc/sysconfig/network-scripts* directory as *eth1* and modify it with appropriate entries. You will ensure that this interface is activated at each system reboot. You will deactivate this interface at the command prompt and then reactivate it. You will reboot the system and verify the IP assignments for *eth1*. You will assign hostname *testserver.example.org* and alias *testserver* to this system and edit the */etc/hosts* file to add an entry for it. This will make *server.example.com* accessible via the *testserver.exmaple.org* hostname.

1. Run the *ip* command to configure *eth1*:

   # **ip addr add 192.168.122.201/24 broadcast 192.168.122.255 dev eth1**

2. Run the *ip* command again to verify the new IP assignments on *eth1*:

   # **ip addr**
   3: eth1: <BROADCAST,MULTICAST,UP,LOWER_UP> mtu 1500 qdisc pfifo_fast state UP qlen 1000
       link/ether 52:54:00:83:15:8d brd ff:ff:ff:ff:ff:ff
       inet 192.168.122.201/24 brd 192.168.122.255 scope global eth1
       inet6 fe80::5054:ff:fe83:158d/64 scope link
         valid_lft forever preferred_lft forever

3. Run the *cp* command to copy the *ifcfg-eth0* file as *ifcfg-eth1* in the */etc/sysconfig/network-scripts* directory:

   # **cd /etc/sysconfig/network-scripts; cp ifcfg-eth0 ifcfg-eth1**

4. Open the *ifcfg-eth1* file in the *vi* editor and modify the contents to look as per below:

   # **vi ifcfg-eth1**
   DEVICE=eth1
   NM_CONTROLLED=yes
   ONBOOT=yes
   TYPE=Ethernet
   BOOTPROTO=none
   IPADDR=192.168.122.201
   NAME="System eth1"
   NETMASK=255.255.255.0
   USERCTL=no
   HWADDR=52:54:00:83:15:8D

5. The interface is now up. Run the *ifdown* command to deactivate *eth1*:

   # **ifdown eth1**

6. Execute the *ifup* command to reactivate *eth1*:

Red Hat Certified System Administrator & Engineer

ifup eth1

7. Reboot the system:

   # shutdown –ry now

8. Run the *ip* command to verify the IP assignments:

   # ip addr

9. You can also run the *system-config-network-cmd* command to verify the IP assignments:

   ```
   # system-config-network-cmd | grep eth1
   DeviceList.Ethernet.eth1.AllowUser=False
   DeviceList.Ethernet.eth1.BootProto=none
   DeviceList.Ethernet.eth1.Device=eth1
   DeviceList.Ethernet.eth1.DeviceId=eth1
   DeviceList.Ethernet.eth1.Gateway=192.168.122.1
   DeviceList.Ethernet.eth1.HardwareAddress=52:54:00:83:15:8D
   DeviceList.Ethernet.eth1.IP=192.168.122.201
   DeviceList.Ethernet.eth1.IPv6Init=False
   DeviceList.Ethernet.eth1.NMControlled=True
   DeviceList.Ethernet.eth1.Netmask=255.255.255.0
   DeviceList.Ethernet.eth1.OnBoot=True
   DeviceList.Ethernet.eth1.PrimaryDNS=192.168.122.1
   DeviceList.Ethernet.eth1.Type=Ethernet
   HardwareList.Ethernet.eth1.Card.ModuleName=8139cp
   HardwareList.Ethernet.eth1.Description=Realtek Semiconductor Co., Ltd. RTL-8139/8139C/8139C+
   HardwareList.Ethernet.eth1.Name=eth1
   HardwareList.Ethernet.eth1.Status=system
   HardwareList.Ethernet.eth1.Type=Ethernet
   ProfileList.default.ActiveDevices.4=eth1
   ```

10. Open the */etc/hosts* file and append the following entry to it:

    ```
    # vi /etc/hosts
    192.168.122.201          testserver.example.org
    ```

Testing Network Connectivity with ping

After the new IP assignments have been applied to *eth1* on *server.example.com*, you need to test whether the IP address is pingable and accessible from other systems. Log on to *physical.example.com* or another RHEL system and run the *ping* (packet internet gropper) command for this purpose. The *ping* command is used to check the network connectivity at the IP level when the physical connectivity is ok and proper IP address and network assignments have been put in place. This command sends out a series of 64-byte *Internet Control Message Protocol* (ICMP) test packets to the destination IP address and waits for a response. With the –c option, you can specify the number of packets that you want to send.

The following will send three packets from *physical.example.com* to the 192.168.122.201 address that you have just configured on *server.example.com*:

```
# ping –c3 192.168.122.201
PING 192.168.122.201 (192.168.122.201) 56(84) bytes of data.
From 192.168.122.201 icmp_seq=1 ttl=255 time=2.40 ms
From 192.168.122.201 icmp_seq=2 ttl=255 time=1.21 ms
From 192.168.122.201 icmp_seq=3 ttl=255 time=2.06 ms


--- 192.168.122.201 ping statistics ---
3 packets transmitted, 3 received, 0% packet loss, time 2005ms
rtt min/avg/max/mdev = 1.216/1.893/2.403/0.501 ms
```

At the bottom of this output under the ping statistics, the number of packets transmitted, received, and lost has been shown. The packet loss should be 0% and the round trip time should not be too high for a healthy connection. In general, you can use this command to test connectivity with the system's own IP, the loopback IP, the default route, other systems on the local network, systems on a different network, and systems beyond the local network.

If *ping* fails in any of the situations, check if the network card is seated in the slot properly, its driver installed, network cable secured appropriately, IP and subnet mask values are set correctly, the default route configured right, and proper rules are defined in the firewall if it is activated. Verify the entries in the */etc/hosts* file and files in the */etc/sysconfig/network-scripts* directory.

Checking Network Packet Transfer with ip

The *ip* command reports the status, statistics, and routing information for active network interfaces on the system. When this command is executed with the –s link arguments, it displays the incoming and outgoing packet information. Examine the output of this command when you suspect an issue with an interface.

```
# ip –s link
. . . . . . . .
2: eth0: <BROADCAST,MULTICAST,UP,LOWER_UP> mtu 1500 qdisc pfifo_fast state UP qlen 1000
    link/ether 52:54:00:28:e1:f6 brd ff:ff:ff:ff:ff:ff
    RX: bytes    packets  errors   dropped overrun mcast
    109278895239782 0        0        0       0
    TX: bytes    packets  errors   dropped carrier  collsns
    3924952    44186    0        0       0        0
3: eth1: <BROADCAST,MULTICAST,UP,LOWER_UP> mtu 1500 qdisc pfifo_fast state UP qlen 1000
    link/ether 52:54:00:83:15:8d brd ff:ff:ff:ff:ff:ff
    RX: bytes    packets  errors   dropped overrun mcast
    143308     2128     0        0       0        0
    TX: bytes    packets  errors   dropped carrier  collsns
    142705     1446     0        0       0        0
```

Look at the rows that show the RX data above for the *eth0* and *eth1* interfaces. They depict the numbers of bytes and packets received via these interfaces and also show how many packets had errors or were dropped, among other information. Likewise, the TX data for both interfaces shows the same information for the transmitted packets.

Configuring Network Interfaces via Network Connections Tool

The Network Connections tool runs in both text and graphical modes and it is used for network interface configuration and administration. This tool can be invoked in the graphical mode by running the *nm-connection-editor* command in an X terminal window or by choosing System → Preferences → Network Connections in the GNOME desktop. Figure 13-1 shows what will appear when it is started on *server.example.com*.

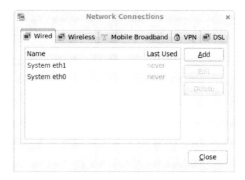

Figure 13-1 Network Connections Tool – Main Window

There are five tabs on the main window: Wired, Wireless, Mobile Broadband, VPN, and DSL with Add, Edit, and Delete buttons displayed on the right side. Each of these tabs shows the type of interfaces that are currently configured on the system. You can highlight an interface and click Edit to modify its properties or click Delete to remove it. You can add a new interface too. Clicking the Edit button after selecting the *eth1* interface will bring up the window shown in Figure 13-2. Go to the IPv4 Settings tab, Figure 13-3, to display or modify IP assignments as required.

Figure 13-2 Network Connections Tool – Edit a Wired Interface 1

Figure 13-3 Network Connections Tool – Edit a Wired Interface 2

The equivalent of the graphical Network Connections tool is the *system-config-network* command that runs in the text mode. The main menu of this text tool is displayed in Figure 13-4 below:

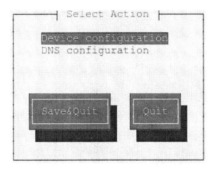

Figure 13-4 Network Connections Tool – Text Interface

You can press the Enter key on the Device Configuration to list all the visible interface. You can edit any of them and you can add a new interface.

The DNS Client

The DNS client functionality may be configured on the system provided a configured DNS server is available on the network to serve the client. In your case, the DNS services are offered by the device with IP address 192.168.2.1. To set up *server.example.com* as a DNS client, you need to configure three files */etc/nsswitch.conf, /etc/host.conf,* and */etc/resolv.conf* on it. There is no need to modify any SELinux booleans for DNS client functionality.

The /etc/nsswitch.conf File

In RHEL, more than one source may be employed for hostname lookups. For example, host information can be obtained from the */etc/hosts* file, DNS, NIS, or NIS+. Which sources to obtain the information from and in what order, is defined in the *name service switch* configuration file called *nsswitch.conf* located in the */etc* directory. This file is referenced if the */etc/host.conf* file does not exist.

The following example entry from this file consults the local */etc/hosts* file first and then DNS for hostname resolution:

    hosts: files   dns

There are four keywords available for use when more than one potential source is referenced in the *nsswitch.conf* file. These keywords are listed in Table 13-4 with their meanings.

Keyword	Meaning
SUCCESS	The information is found.
UNAVAIL	The source is down or not responding.
NOTFOUND	The information is not found.
TRYAGAIN	The source is busy, try again later.

Table 13-4 Name Service Source Status

Based on the status code for a particular source, one of two actions provided in Table 13-5, take place.

Action	Meaning
continue	Try the next source listed.
return	Do not try the next source.

Table 13-5 Name Service Source Actions

As an example, if the hosts entry looks like the following, the search will terminate when the required information is not found in DNS. The files source will be ignored.

    hosts: dns [NOTFOUND=return] files

Each keyword defined in Table 13-4 has a default action associated with it. If no keyword/action combination is specified, actions listed in Table 13-6 are assumed.

Keyword	Default Action
SUCCESS	return
UNAVAIL	continue
NOTFOUND	continue
TRYAGAIN	continue

Table 13-6 Name Service Source Default Actions

The /etc/host.conf File

This is a resolver configuration file and is referenced by the *dig* and *host* commands. A sample entry from the file is shown below:

 # cat /etc/host.conf
order hosts,bind

This line entry tells lookup commands to use the */etc/hosts* file for any name resolution first and then contact a DNS server.

The /etc/resolv.conf File

The *resolv.conf* file is the DNS resolver file where you can define three keywords, as described in Table 13-7.

Keyword	Description
domain	Specifies the default domain name to be searched for queries. This keyword is defined when there are multiple domains. It is not needed for a single domain environment.
nameserver	Specifies up to three DNS server IP addresses to be used one at a time in the order in which they are listed. If none specified, the local DNS server is assumed.
search	Specifies up to six domain names of which the first must be the local domain. The resolver appends these domain names one at a time in the order in which they are listed to the hostname specified when constructing queries.

Table 13-7 resolv.conf File

Configuring a DNS Client

Now that you have understood the DNS client configuration files, you can edit the *nsswitch.conf* and *resolv.conf* files and add the following entries to them:

 # vi /etc/nsswitch.conf
hosts: files   dns
 # vi /etc/resolv.conf
search   example.com   # local domain
nameserver   192.168.2.1   # IP address of a DNS server

If the client is configured to get the IP information from an available DHCP server, edit the associated network interface file in the */etc/sysconfig/network-scripts* directory and set the PEERDNS directive to no to prevent the DHCP server from supplying and overwriting DNS server information in the *resolv.conf* file.

You can use either the Network Connections tool or the *system-config-network* command to set or modify a client DNS entry. Refer to Figures 13-3 and 13-4.

The DHCP Client

Dynamic Host Configuration Protocol (DHCP) enables a system that acts as a *DHCP server* to provide IP address and other network parameters including subnet mask, default gateway, and DNS server IP to other systems on the network automatically. These other systems are referred to as *DHCP clients.*

When a system with DHCP client functionality enabled boots up or when you execute the *dhclient* command on the system, it broadcasts a DHCPDISCOVER message on the network requesting an available DHCP server to provide IP assignments. An available DHCP server receives this message and responds by sending a DHCPOFFER message back to the client, which includes all the required information that the client requires to configure itself. The client evaluates the information and returns a DHCPREQUEST message to the server requesting it to lease the IP address for a fixed amount of time. The server reserves the address for the client and confirms by sending a DHCPACK message back to the client. The client stores the lease information and configures itself with the IP assignments.

DHCP offloads you of IP configuration and maintenance tasks and it is an ideal solution for PC, workstation, and laptop users. The DHCP client functionality should not be used on systems that are used as servers or systems that require static IP settings.

Configuring a DHCP Client

Configuring the DHCP client functionality on a system is pretty straightforward. Perform the following steps on *insider.example.com* on the *eth1* interface with an assumption that *server.example.com* is offering DHCP services:

1. Edit the */etc/sysconfig/network-scripts/ifcfg-eth1* file and set the following directives:

    ```
    DEVICE=eth1
    BOOTPROTO=dhcp
    ONBOOT=yes
    PEERDNS=yes
    ```

2. Stop and restart the interface:

    ```
    # ifdown eth1
    # ifup eth1
    ```

3. Run the *ip* command and supply it with the addr argument and check if the network interface has picked up IP assignments from the DHCP server. With PEERDNS set to yes, the *resolv.conf* file will be populated with the nameserver information as well.

Alternatively, you can either use the Network Connections tool or the *dhclient* command to configure the DHCP client on the system.

Refer to Figures 13-3 and 13-4 to establish the DHCP client functionality via the Network Connections tool and its text equivalent *system-config-network.*

You can use the *dhclient* command and specify the interface name with it. This command broadcasts a message on the network and a configured DHCP server responds by returning the IP assignments, which the *dhclient* command uses to configure the interface.

 # dhclient eth1

All information is recorded in the system log file */var/log/messages*.

The LDAP Client

Lightweight Directory Access Protocol (LDAP) is a trivial, simplified networking protocol for obtaining directory information such as user authentication, email messaging, and calendar services over the network. This protocol is platform-independent, which makes it available on a variety of vendor hardware platforms running heterogeneous operating system software.

On large networks, maintaining user authentication information on every system becomes too cumbersome. For solving this problem, services such as LDAP are available. The LDAP server component is configured on a central server with all users defined in it. This server is accessible to all other networked systems. There is some client configuration that needs to be done on the systems so that they are able to authenticate users by consulting the LDAP server when the users attempt to log on to these systems.

On RHEL, an open source LDAP software product called OpenLDAP is used to provide both server and client functionalities. There are three packages that you require to configure an LDAP client on your system: openldap, openldap-clients, and nss-pam-ldapd. There are several configuration files that need to be modified to successfully configure an LDAP client on the system. These files include */etc/pam_ldap.conf*, */etc/openldap/ldap.conf*, */etc/nslcd.conf*, */etc/sysconfig/authconfig*, and */etc/nsswitch.conf*. The first file contains the PAM LDAP module information, the second file holds system-wide default LDAP client configuration directives, the third file sets directives for the LDAP nameservice daemon, the fourth file defines whether to use LDAP authentication, and the fifth file provides which source to consult for information and in which order.

LDAP Terminology

Before you look at LDAP client configuration files and how an LDAP client is configured, a grasp of the following key terms is essential.

Directory: A *directory* is a kind of a specialized database that stores information about objects such as people, profiles, printers, and servers. It organizes information in such a way that it becomes easier to find and retrieve needed information. It lists objects and gives details about them.

Entry: An *entry* is a building block for an LDAP directory and represents a specific record in it. In other words, an entry is a collection of information consisting of one or more attributes for an object. An LDAP directory, for instance, might include entries for employees, printers, and servers.

Attribute: An *attribute* contains two pieces of information: *attribute type* and *attribute values*, and is associated with one or more entries. The attribute type such as jobTitle represents the type of information the attribute contains. An attribute value is the specific information contained in that entry. For instance, a value for the jobTitle attribute type could be "director". Table 13-8 lists some common attribute types.

Attribute Type	Description
CommonName (cn)	A common name of an entry such as cn=John Doe
DomainComponent (dc)	The distinguished name (DN) of a component in DNS such as dc=example and dc=com.
Country (c)	A country abbreviation such as c=ca.
Mail (mail)	An email address.
Organization (o)	The name of an organization such as o=redhat.
OrganizationalUnit (ou)	The name of a unit within an organization such as ou=Printers.
Owner (owner)	The owner of an entry such as cn=John Doe, ou=Printers, dc=example, and c=ca.
Surname (sn)	A person's last name such as Doe.
TelephoneNumber (telephoneNumber)	A telephone number such as (123) 456-7890 or 1234567890.

Table 13-8 Common LDAP Attribute Types

Matching Rule: A *matching rule* matches the attribute value sought against the attribute value stored in the directory in a search and compare task. For example, matching rules associated with the telephoneNumber attribute could cause "(123) 456-7890" to match with either "(123) 456-7890" or "1234567890", or both.

Object Class: Each entry belongs to one or more *object classes*. An object class is a group of required and optional attributes that define the structure of an entry. For example, an organizationalUser object class may include commonName and Surname as required attributes and telephoneNumber, UID, streetAddress, and userPassword as optional attributes.

Schema: A *schema* is a collection of attributes and object classes along with matching rules and syntax, and other related information.

Distinguished Name: A *Distinguished Name* (DN) uniquely identifies an entry in the entire directory tree. It is similar in concept to the absolute pathname of a file in the Linux directory hierarchy.

Relative Distinguished Name: A *Relative Distinguished Name* (RDN), in contrast, represents individual components of a DN. It is similar in concept to the relative pathname of a file in the Linux directory hierarchy.

The /etc/pam_ldap.conf File

This file contains the LDAP PAM authentication module information. This module facilitates authentication, authorization, and password change on LDAP servers. This file has many directives that you may want to define per your requirements. Table 13-9 lists and describes some of the directives that are related to client configuration.

Directive	Description
host	Specifies the hostname or IP address of the LDAP server.
base	Defines the base DN to use for searches.
uri	Uniform Resource Identifier. Sets the URI scheme to be used to connect to the LDAP server.
ssl	Secure Sockets Layer. Specifies whether to use SSL/TLS (transport layer security).

Directive	Description
pam_password	Sets the password change protocol to use. Options are crypt, md5, and ad, among others.
tls_cacertdir	Defines the location where TLS certificates are stored.

Table 13-9 The /etc/pam_ldap File

The /etc/openldap/ldap.conf File

This file holds the default system-wide LDAP client values. Table 13-10 lists and describes some of them.

Directive	Description
HOST	Defines the hostname or IP address of the LDAP server.
BASE	Sets the base DN to use for searches.
URI	Specifies the URI scheme to be used to connect to the LDAP server.
TLS_CACERTDIR	Specifies the location where TLS certificates are stored.

Table 13-10 /etc/openldap/ldap.conf File

Exercise 13-2: Configure an LDAP Client

In this exercise, you will edit the relevant files to configure LDAP client services on *insider.example.com*. It is assumed for the purpose of this exercise that the LDAP service is configured on *server.example.com*. You will ensure that the client service is auto-started at each system reboot. You will run proper commands to test the client functionality.

1. Ensure that the following LDAP client software packages are installed on *insider.example.com*:

    # **rpm –qa | grep ldap**
    nss–pam–ldapd–0.7.5–14.el6_2.1.x86_64
    openldap–2.4.23–26.el6_3.2.x86_64
    openldap–clients–2.4.23–26.el6_3.2.x86_64

    If these packages are not already installed, use one of the methods provided in Chapter 06 "Package Management" to install them.

2. Edit the */etc/pam_ldap.conf* file and insert the following entries:

    # **vi /etc/pam_ldap.conf**
    host 192.168.2.201   # IP address of the ldap server (server.example.com)
    base dc=example,dc=com   # defines the default base DN
    uri ldap//192.168.2.201/   # sets the uri
    ssl start_tls   # encrypts passwords while in transit
    pam_password md5   # Use md5 PAM password management

3. Modify the */etc/openldap/ldap.conf* file and insert the following entries:

vi /etc/openldap/ldap.conf
URI ldap://192.168.2.201   # IP address of the ldap server (server.example.com)
BASE dc=example,dc=com   # default base DN
TLS_CACERTDIR /etc/openldap/cacerts   # TLS Certificate Authority (CA) certificate location

4. Open the */etc/nslcd.conf* file and append the following lines to define NSS lookups and PAM actions mapped to LDAP lookups:

    # vi /etc/nslcd.conf
    uri ldap//192.168.2.201/
    base d c=example,dc=com
    ssl yes
    tls_cacertdir /etc/openldap/cacerts

5. Edit the */etc/pam.d/system-auth-ac* file and append the following lines to direct all authentication requests to be served by the LDAP server:

    # vi /etc/pam.d/system-auth-ac
    auth   sufficient   pam_ldap.so use_first_pass
    account   [default=bad success=ok user_unknown=ignore] pam_ldap.so
    password   sufficient   pam_ldap.so use_authtok
    session   optional   pam_ldap.so

6. Alter the */etc/sysconfig/authconfig* file and set the USELDAP and USELDAPAUTH directives to yes:

    # vi /etc/sysconfig/authconfig
    USELDAP=yes
    USELDAPAUTH=yes

7. Open the */etc/nsswitch.conf* file and ensure that user authentication is first referenced in LDAP and then in the local files:

    # vi /etc/nsswitch.conf
    passwd: ldap files
    shadow: ldap files
    group: ldap files

8. Set the LDAP client facility to autostart at each system reboot, and validate:

    # chkconfig nslcd on
    # chkconfig --list nslcd
    nslcd   0:off 1:off 2:on 3:on 4:on 5:on 6:off

9. Start the LDAP client service on *insider.example.com*, and check the status:

    # service nslcd start
    Starting nslcd: [OK]

 # service nslcd status
nslcd (pid 3223) is running...

This completes the procedure to configure an LDAP client. Restart the *sshd* service and use the *ssh* command to log in to the system to test authentication. You can use the *ldapsearch* command to pull user information from the LDAP server and the *openssl* command to test TLS:

 # ldapsearch –x –Z
 # openssl s_client –connect server.example.com

Configuring an LDAP Client Using Authentication Configuration Tool

The LDAP client service can be set up using the graphical Authentication Configuration tool. To set up *insider.example.com* as an LDAP client of *server.example.com* LDAP server, execute *system-config-authentication* in an X terminal window (run *authconfig-tui* for its text equivalent) or choose System → Administration → Authentication in the GNOME desktop on *insider.example.com*. The Authentication Configuration tool screen will open up as shown in Figure 13-5.

Figure 13-5 Authentication Configuration Tool – LDAP Client Configuration

From the Identification & Authentication tab, select LDAP as the User Account Database under User Account Configuration and enter LDAP Search Base DN and LDAP Server information as shown in figure 13-5 above. Use TLS to encrypt connections and download a security certificate. Choose LDAP password as the Authentication Method.

Go to the Advanced Options tab, Figure 13-6, and choose MD5 as the Password Hashing Algorithm. Click Apply when done. This tool will perform the requested configuration updates in the proper configuration files.

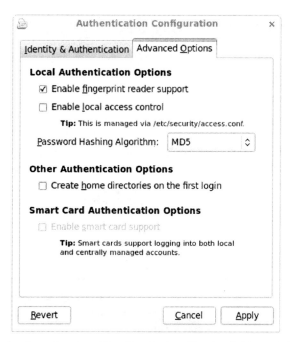

Figure 13-6 Authentication Configuration Tool – LDAP Client Options

This completes the setup for an LDAP client.

Chapter Summary

This chapter started off by providing an understanding of various essential network terms that built the foundation for networking topics in this chapter and future chapters in this book. Topics such as hostname, IP address, network classes, subnetting, subnet mask, protocol, and hardware address were covered.

The next section covered the basics of network interfaces and how to configure them to connect the system to the network. You looked at the configuration files involved and the directives defined in them. You learned how to assign network parameters to network interfaces using commands and graphical and textual tools. You reviewed how to edit files so that the interfaces could be activated whenever a system reboot occurred. You performed activation and deactivation of interfaces manually and defined IP addresses and hostnames in the */etc/hosts* database. You used tools to test network connectivity.

The next topic dealt with configuring the DNS client functionality. You studied the configuration files involved and configured the client service by editing files and using graphical and textual tools.

The next section covered the foundation of DHCP and how it worked. It then demonstrated how to configure the DHCP client service using commands and graphical and textual tools.

The last topic of this chapter provided an overview of LDAP and its terminology. It then demonstrated how to configure the DHCP client service using commands and graphical and textual tools.

Chapter Review Questions

1. What would the command *ip neighbor delete 192.168.1.210 nud stale dev eth1* do?
2. Which file is sourced during the system startup to set the system hostname?
3. What would the DNS resolver do if it finds the entry "dns [NOTFOUND=return] files" in *nsswitch.conf* file?
4. What is a relative distinguished name?
5. What is meant by the object Entry in LDAP terminology?
6. Which class of IP addresses has the least number of node addresses?
7. Which command can be used to obtain IP assignments from a DHCP server?
8. What is the use of *ifup* and *ifdown* commands?
9. Which command may you use to display the hardware address of a network interface?
10. Name the three key files for DNS client setup?
11. Which directory does the Network Manager store the interface configuration files?
12. Which file defines the protocols in the system?
13. List three benefits of using LDAP as a directory server.
14. What is the purpose of the ONBOOT directive in the interface configuration file?
15. The */etc/hosts* file maintains the hostname to hardware address mappings. True or False?
16. Which file contains service, port, and protocol mappings?
17. What would the *ip addr* command produce?
18. Which command can be run at the command prompt in an X terminal window to bring up the graphical Network Connections tool?
19. What is the equivalent of *netstat –i* in RHEL 6?
20. List any two benefits of subnetting.
21. What is the purpose of the *nsswitch.conf* file?
22. The *ip neighbor* command uses the RARP protocol to provide MAC to IP address mappings. True or False?
23. What would the command *system-config-authentication* do?
24. Name three key directives that are usually defined in the *resolv.conf* file.
25. What would happen to the system at boot up if the NETWORKING directive is set to no in the */etc/sysconfig/network* file?
26. What directive would you set in the interface configuration file if you want to obtain DNS information from a DHCP server?
27. Name the two commands that have been replaced by the *ip* command in RHEL 6?
28. Name any three configuration files that need to be modified for setting up an LDAP client.

Answers to Chapter Review Questions

1. The command provided will delete the entry for this IP from the ARP cache.
2. The */etc/sysconfig/network* file is sourced.
3. The DNS resolver would not continue to look for information in the *hosts* file if it is unable to find the requested information in DNS.
4. The relative distinguished name represents individual components of a DN.
5. The object Entry represents a specific record in the LDAP directory.
6. The C class has the least number of node addresses.
7. The *dhclient* command.
8. The purpose of these two commands is to activate and deactivate an interface.
9. The *ip* command.

10. The three files are *resolv.conf*, *host.conf*, and *nsswitch.conf*.
11. The */etc/sysconfig/network-scripts* directory.
12. The */etc/protocols* file.
13. Three benefits are access to uniform information, centralized storage for information, and user authentication.
14. The purpose of this directive is to tell the boot scripts whether to activate this interface.
15. False. This file maintains hostname to IP address mappings.
16. The */etc/services* file.
17. This command will display information about interfaces including IP and hardware address information.
18. The command to bring up the graphical Network Connections tool is *nm-connection-editor*.
19. The *ip neighbor* command is the equivalent for *netstat –i* in RHEL 6.
20. Better manageability and less traffic.
21. The *nsswitch.conf* file lists sources and their order for information lookup.
22. False. It uses ARP protocol.
23. This command will bring up the graphical Authentication Configuration tool.
24. The three directives are namserver, search, and domain.
25. No interfaces will be activated.
26. The PEERDNS directive.
27. The *ifconfig* and the *netstat* commands have been replaced by the *ip* command in RHEL 6.
28. The three files are */etc/nsswitch.conf*, */etc/pam_ldap.conf*, and */etc/openldap/ldap.conf*.

Labs

Lab 13-1: Change the Hostname of the System

Change the hostname of *insider.example.com* to *insider_new.example.com* by editing appropriate files. Reboot the system and verify the new name.

Lab 13-2: Assign a New IP to the System

Remove the current IP assignments for *eth0* on *insider_new.example.com* and replace them with something different but on the same subnet. Apply the new IP information using appropriate commands and test it using the *ip* and *ping* commands. Then edit appropriate files to make this modification persistent across reboots. Activate and deactivate the interface manually. Reboot the system and verify the new IP using appropriate commands.

Lab 13-3: Get IP Assignments from a DHCP Server

Remove the current IP assignments for *eth0* on *insider_new.example.com* and configure the interface so that it gets IP and DNS information from a DHCP server. Make the change permanent, reboot the system, and verify.

Section Two
RHCE

Chapter 14

Shell Scripting, Building an RPM Package, & iSCSI

This chapter covers the following major topics:

- ✓ What is shell scripting?
- ✓ Write scripts to display Linux kernel information, set local variables, use values of pre-defined environment variables, parse command outputs to variables, understand the usage of command line arguments, and the role of the shift command
- ✓ Execute and debug scripts
- ✓ Write interactive scripts
- ✓ Logical constructs, exit codes, test conditions, the if-then-fi construct, the if-then-else-fi construct, and the if-then-elif-fi construct
- ✓ Looping construct, test conditions, the for-do-done loop, and the while-do-done loop
- ✓ Build a custom RPM package
- ✓ Introduction to iSCSI and mount an iSCSI target persistently

This chapter includes the following RHCE objectives:

63. Use shell scripting to automate system maintenance tasks
60. Build a simple RPM that packages a single file
61. Configure a system as an iSCSI initiator that persistently mounts an iSCSI target

Shell scripts

Shell scripts are text files that contain Linux commands, control structures, and comments for the automation of lengthy tasks. There is nothing that you can place in the script that cannot be run at the command prompt. Scripts do not need to be compiled. In this chapter, many example scripts are provided and they are examined line-by-line. These example scripts begin with simple programs and advance to more complicated ones. As with any other programming language, the scripting skill develops over time as you read and write more and more scripts.

Software packages for RHEL are available in a special installable format known as RPM. Custom packages may be built from scratch according to specific requirements. Likewise, source codes for many packages may be obtained free of cost and modified. After the customization, the new package may be installed and distributed just like any other RPM package.

iSCSI is a storage networking protocol that is used to facilitate data transfer between computer systems and storage devices over an IP network using SCSI commands.

Shell Scripts

Shell scripts (a.k.a. *shell programs* or simply *scripts*) contain Linux commands and control structures for automating long, repetitive tasks such as managing packages and users; administering LVM and file systems; monitoring file system utilization; trimming log files; backing up and archiving files; removing core, temporary, and unnecessary files; starting and stopping databases and applications; and generating reports. Commands in the script are interpreted and run by the shell one at a time in the order in which they are listed in the script. Each line is executed as if it is typed and run at the command prompt. Control structures are utilized for creating and managing logical and looping constructs. Comments are also usually included to add general information about the script such as the author name, creation date, last modification date, purpose of the script, and its usage. If the script encounters an error during its execution, the error message is displayed on the screen.

Scripts covered in this chapter are written in the bash shell */bin/bash*, however they can be used in the Korn shell, but not all of them can run in the C shell without modifications.

You should use the *vi* or *nano* editor to create example shell scripts presented in this chapter. This will give you an opportunity to practice using these editors. The *nl* command is used to display the contents of the scripts along with associated line numbers. It is recommended to store these scripts in the */usr/local/bin* directory and add this directory to the PATH variable. The following example scripts are created in the */usr/local/bin* directory.

Displaying Linux Kernel Information

Let us create our first script called *sys_info.sh* and examine it line by line. Change the directory into */usr/local/bin* and use the vi editor to write the script. Type in what you see below excluding the line numbers:

nl sys_info.sh

```
1    #!/bin/bash
2    # The name of this script is sys_info.sh.
3    # The author of this script is Asghar Ghori.
4    # The script is created on October 28, 2008.
5    # This script is last modified by Asghar Ghori on October 14, 2012.
6    # This script should be located in the /usr/local/bin directory.
7    # The purpose of this script is to explain construct of a simple shell program.
8    echo "Display Linux Kernel Information"
9    echo "---------------------------------------------"
10   echo
11   echo "This machine is running the following kernel:"
12   /bin/uname –s
13   echo "This machine is running the following release of Linux kernel:"
14   /bin/uname –r
```

 Within vi, type :set nu to view line numbers associated with each line entry.

In this script, comments and commands are used, as follows:

The first line indicates in which shell the script will run. This line must start with the "#!" character combination followed by the full pathname to the shell file.

The next six lines contain comments: the script name, author name, creation time, last modification time, default location to store, and the purpose of this script. The # sign implies that anything written to the right of it is for informational purposes and will be ignored when the script is executed. Note that line 1 uses this character also but followed by the ! mark; this combination has a special meaning to the shell, which specifies the location of the shell file. Do not get confused between the two usages of the # sign.

Line number 8 has the first command of the script. The *echo* command prints on the screen whatever follows it. In this case, you will see "Display Linux Kernel Information" printed.
Line number 9 will underline the text "Display Linux Kernel Information".
Line number 10 has the *echo* command followed by nothing. This will insert an empty line in the output.
Line number 11 will print "The machine is running the following kernel:" on the screen.
Line number 12 will execute the *uname* command with the –s option and will return the name of the Linux kernel currently being used.
Line number 13 will print "This machine is running the following release of Linux kernel:".
Line number 14 will execute the *uname* command with the –r option and will return the release of the Linux kernel being run.

Executing a Script

The script created above does not have the execute permission since the default umask value for the *root* user is set to 0022, which allows read/write permissions to the owner, and read-only permission to group members and public. You can execute the script as follows while you are in */usr/local/bin*:

---

sh ./sys_info.sh

Alternatively, give the owner of the file the execute permission:

chmod +x sys_info.sh

Now you can run the script using either its relative path or the fully qualified path:

./sys_info.sh
/usr/local/bin/sys_info.sh

If the *usr/local/bin* directory is not set in the PATH variable, define it in the *etc/profile* file so that whoever logs on to the system has this path set. Alternatively, the path to this directory may be added to the *~/.bash_profile* file of the users who need access to this directory. The following shows how to set the path at the command prompt:

export PATH=$PATH:/usr/local/bin

Let us run *sys_info.sh* and see what the output will look like:

sys_info.sh
Display Linux Kernel Information

This machine is running the following kernel:
Linux
This machine is running the following release of Linux kernel:
2.6.32-220.el6.x86_64

Debugging a Shell Script

If you would like to use a basic debugging technique to understand why a script is not functioning the way it should, you can append –x to "#!/bin/bash" in the first line of the script so that it looks like "#!/bin/bash –x". Alternatively, you can execute the script as follows:

sh –x /usr/local/bin/sys_info.sh
+ echo 'Display Linux Kernel Information'
Display Linux Kernel Information
+ echo --

+ echo

+ echo 'This machine is running the following kernel:'
This machine is running the following kernel:
+ /bin/uname –s
Linux
+ echo 'This machine is running the following release of Linux kernel:'
This machine is running the following release of Linux kernel:
+ /bin/uname –r
2.6.32-220.el6.x86_64

With the + sign, the actual line from the script is echoed on the screen followed in the next line by what it would produce in the output.

Using Local Variables

You have dealt with variables previously and have seen how to use them. To recap, there are two types of variables: *local* (or *private*) and *global* (or *environment*). They are defined and used in scripts and at the command line in a similar manner.

Script *loc_var.sh* will define a local variable and display its value. Check the value of the variable again after the script execution is complete.

 # nl loc_var.sh
```
1   #!/bin/bash
2   # The name of this script is loc_var.sh.
3   # The author of this script is Asghar Ghori.
4   # The script is created on October 28, 2008.
5   # This script is last modified by Asghar Ghori on October 13, 2012.
6   # This script should be located in the /usr/local/bin directory.
7   # The purpose of this script is to explain how a local variable is defined and used in a shell
8   # program.
9   echo "Setting a Local Variable".
10  echo "-------------------- "
11  echo
12  SYS_NAME=server.example.com
13  echo "The hostname of this system is $SYS_NAME".
```

The following output will be generated when you execute this script:

 # loc_var.sh
Setting a Local Variable.
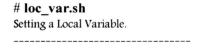

The hostname of this system is server.example.com.

Since it was a local variable and the script was run in a sub-shell, the variable had disappeared after the script execution was completed. The **echo $SYS_NAME** will display nothing.

Using Pre-Defined Environment Variables

The following script called *pre_env.sh* will display the values of two pre-defined environment variables SHELL and LOGNAME:

 # nl pre_env.sh
```
1   #!/bin/bash
2   # The name of this script is pre_env.sh.
3   # The author of this script is Asghar Ghori.
4   # The script is created on October 28, 2008.
5   # This script is last modified by Asghar Ghori on October 13, 2012.
```

6   # This script should be located in the /usr/local/bin directory.
7   # The purpose of this script is to explain how a pre-defined environment variable is used in a shell
8   # program.
9   echo "The location of my shell command is:"
10   echo $SHELL
11   echo "You are logged in as $LOGNAME".

The output will be:

pre_env.sh
The location of my shell command is:
/bin/bash
You are logged in as root.

Parsing Command Output

During the execution of a script, you can use the command substitution feature of the bash shell and store the output generated by the command into a variable. For example, the following script called *cmdout.sh* is the modified version of the *new_env.sh* script. You must enclose the *hostname* and *uname* commands in forward quotes.

nl cmdout.sh
1   #!/bin/bash
2   # The name of this script is cmdout.sh.
3   # The author of this script is Asghar Ghori.
4   # The script is created on October 28, 2008.
5   # This script is last modified by Asghar Ghori on October 14, 2012.
6   # This script should be located in the /usr/local/bin directory.
7   # The purpose of this script is to display how a command output is captured and stored in a
8   # variable.
9   echo "Parsing Command Output".
10   echo "---------------------- "
11   echo
12   SYS_NAME=`hostname`
13   OS_VER=`uname –r`
14   export SYS_NAME OS_VER
15   echo "The hostname of this system is $SYS_NAME".
16   echo "This system is running $OS_VER of RHEL Operating System kernel".

The output will be:

cmdout.sh
Parsing Command Output.

The hostname of this system is server.example.com.
This system is running 2.6.32-220.el6.x86_64 of RHEL Operating System kernel.

Using Command Line Arguments

Command line arguments (also called *positional parameters*) are the arguments specified at the command prompt with a command or script to be executed. The locations at the command prompt of the arguments as well as the location of the command, or the script itself, are stored in corresponding variables. These variables are special shell variables. Figure 14-1 and Table 14-1 help you understand them.

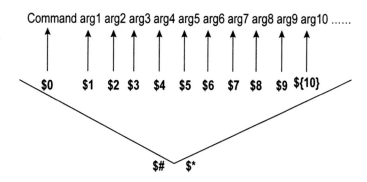

Figure 14-1 Command Line Arguments

Variable	Description
$0	Represents the command or script.
$1 to $9	Represents arguments 1 through 9.
${10} and so on	Represents arguments 10 and further.
$#	Represents the total number of arguments.
$*	Represents all arguments.
$$	Represents the PID of a running script.

Table 14-1 Command Line Arguments

The script *com_line_arg.sh* below will show the command line arguments that were supplied, a count of how many there were, the value of the first argument, and the process ID of the script itself:

```
# nl com_line_arg.sh
1    #!/bin/bash
2    # The name of this script is com_line_arg.sh.
3    # The author of this script is Asghar Ghori.
4    # The script is created on October 28, 2008.
5    # This script is last modified by Asghar Ghori on October 14, 2012.
6    # This script should be located in the /usr/local/bin directory.
7    # The purpose of this script is to explain the usage of command line arguments.
8    echo "There are $# arguments specified at the command line".
9    echo "The arguments supplied are: $*"
10   echo "The first argument is: $1"
11   echo "The Process ID of the script is: $$"
```

The result will be as follows when this script is executed with four arguments:

 # com_line_arg.sh toronto chicago london tokyo

There are 4 arguments specified at the command line.

The arguments supplied are: toronto chicago london tokyo

The first argument is: toronto

The Process ID of the script is: 27322

Shifting Command Line Arguments

The *shift* command is used to move command line arguments one position to the left. During this move, the first argument is lost. *com_line_arg_shift.sh* script below uses the *shift* command:

 # nl com_line_arg_shift.sh

```
1    #! /bin/bash
2    # The name of this script is com_line_arg_shift.sh.
3    # The author of this script is Asghar Ghori.
4    # The script is created on October 28, 2008.
5    # This script is last modified by Asghar Ghori on October 14, 2012.
6    # This script should be located in the /usr/local/bin directory.
7    # The purpose of this script is to show you impact of the shift command on command line
8    # arguments.
9    echo "There are $# arguments specified at the command line".
10   echo "The arguments supplied are: $*"
11   echo "The first argument is: $1"
12   echo "The Process ID of the script is: $$"
13   shift
14   echo "The new first argument after the first shift is: $1"
15   shift
16   echo "The new first argument after the second shift is: $1"
```

Let us execute the script with four arguments. Notice that a new value is assigned to $1 after each shift.

 # com_line_arg_shift.sh toronto chicago London tokyo

There are 4 arguments specified at the command line.

The arguments supplied are: toronto chicago london tokyo

The first argument is: toronto

The Process ID of the script is: 27336

The new first argument after the first shift is: chicago

The new first argument after the second shift is: london

Multiple shifts in a single attempt may be performed by furnishing the desired number of shifts to the *shift* command as an argument. For example, "*shift* 2" will carry out two shifts, "*shift* 3" will do three shifts, and so on.

Writing an Interactive Script

Interactive shell scripts are written when you want the script to prompt for an input and continue execution based on the input provided. The input is stored in a variable. The *read* command is used for reading the input and is usually preceded by the *echo* command to display a message telling what is expected as an input.

The *inter_read.sh* script below will list files and prompt for entry of a file name for removal. Notice the \c in the eleventh line of the script. This is an example of an *escape sequence*. It will tell the *echo* command not to send a carriage return and a line feed after displaying "Enter the name of file you want to remove:". When you enter a file name, you will be typing the file name on the same line right beside the colon character. Try running this script with and without the \c. Also, a pre-defined environment variable called PWD is used in the script to display your location in the directory tree.

 # nl inter_read.sh

```
1    #! /bin/bash
2    # The name of this script is inter_read.sh.
3    # The author of this script is Asghar Ghori.
4    # The script is created on October 28, 2008.
5    # This script is last modified by Asghar Ghori on October 14, 2012.
6    # This script should be located in the /usr/local/bin directory.
7    # The purpose of this script is to show you an example of how an interactive script works.
8    echo "Here is a list of all files in $PWD directory:"
9    /bin/ls –l
10   echo
11   echo –e "Enter the name of file you want to remove: \c".
12   read FILE
13   echo "Type 'y' to remove it, 'n' if you do not want to."
14   /bin/rm –i $FILE
15   echo "The file is removed".
```

Let us assume that you are logged in as *user1* when this script is run. Here is what the *inter_read.sh* script will do:

$ inter_read.sh

```
. . . . . . . .
–rw–rw–r––. 1 user1 user1    0 Sep 21 09:54 file1

Enter the name of file you want to remove: file1
Type y to remove it, n if you do not want to:
/bin/rm: remove regular file `testfile1'? y
The file is removed.
```

There are some other escape sequences such as \t for tab, \n for new line, \a for beep, \f for form feed, \r for carriage return, and \b for backspace that you can use in your shell scripts to increase readability. Try using them in the *inter_read.sh* script.

Additional escape sequences such as \h, \u, and \w may be used to display the hostname, username, and current working directory. For example, the following will change the primary command prompt to display the username, hostname, and current working directory location:

```
# PS1="\u@\h:\w :> $ "
root@server:~ :> $
```

Logical Constructs

In shell scripting there are times when a decision whether to perform an action needs to be made. In other words, a test condition (or logic) is defined in the script and based on the true or false status of the condition, the script decides what to do next.

The shell offers two logical constructs: the *if-then-fi* construct and the *case* construct. The if-then-fi construct has a few variations, which will also be covered in this chapter.

Before looking at the example scripts and seeing how logical constructs are used, let us discuss the *exit codes* and various *test* conditions. These will be employed in the example scripts later.

Exit Codes

Exit codes refer to the value returned when a program or script finishes execution. This value is based on the outcome of the program. If the program ran successfully, you will get a zero exit code, otherwise, a non-zero exit code will be received.

You may use the terms "zero status code", "zero exit code", "successful script completion", or "true value" interchangeably. Similarly, you may use the terms "non-zero status code", "failed script completion", or "false value" interchangeably. These values or codes are also referred to as the *return codes,* and are stored in a special shell variable called ?. Let us look at the following two examples to understand their usage:

```
# ls
file1 file2 file3 file4 file5 file6
# echo $?
0
# man
Usage: man [-M path] [-T macro-package] [ section ] name ...
or: man -k keyword ...
or: man -f file ...
# echo $?
1
```

In the first example, the *ls* command ran successfully and it produced the desired result, hence a zero return code. In the second example, the *man* command did not run successfully since it required an argument which was not supplied, therefore a non-zero return code. The *echo* command was then used in the above examples to display the value stored in the special variable ?.

Within a script you can define exit codes at different locations in order to help debug the script or the output by letting you know where exactly the script quit.

Test Conditions

Test conditions can be set on numeric values, string values, or files using the *test* command. You may instead enclose a test condition within the square brackets [] without using the *test* command explicitly. This is exhibited in the logical construct examples later.

Table 14-2 shows various testing condition operations that you can perform.

Operation on Numeric Value	Meaning
integer1 –eq integer2	Integer1 is equal to integer2.
integer1 –ne integer2	Integer1 is not equal to integer2.
integer1 –lt integer2	Integer1 is less than integer2.
integer1 –gt integer2	Integer1 is greater than integer2.
integer1 –le integer2	Integer1 is less than or equal to integer2.
integer1 –ge integer2	Integer1 is greater than or equal to integer2.

Operation on String Value	Meaning
string1–string2	Checks if the two strings are identical.
string1! –string2	Checks if the two strings are not identical.
–l string or –z string	Checks if string length is zero.
–n string	Checks if string length is non-zero.
string	Same as "–n string".

Operation on File	Meaning
–b file	Checks if file exists and is a block device file.
–c file	Checks if file exists and is a character device file.
–d file	Checks if file is a directory.
–e file	Checks if file exists.
–f file	Checks if file exists and is a normal file.
–g file	Checks if file exists and has setgid bit on it.
–L file	Checks if file exists and is a symbolic link.
–r file	Checks if file exists and is readable.
–s file	Checks if file exists and is non-zero in length.
–u file	Checks if file exists and has setuid bit set on it.
–w file	Checks if file exists and is writable.
–x file	Checks if file exists and is executable.
file1 –nt file2	Checks if file1 is newer than file2.
file1 –ot file2	Checks if file1 is older than file2.

Logic Operator	Meaning
!	Opposite result of expression.
–a	AND operator.
–o	OR operator.

Table 14-2 Test Conditions

Having described the exit codes and various test condition operations, let us take a look at a few example scripts and see how some of these may be utilized.

The if-then-fi Construct

The if-then-fi construct checks a condition for true or false. If the condition is true, an action is performed, otherwise, it gets you out of the if statement. The if statement ends with a fi. The syntax of this construct is:

```
if
    condition
then
    action
fi
```

It has been demonstrated earlier how to check the number of arguments specified at the command line. A shell script can be written to determine that number and print an error message if there are no command line arguments specified. The *if_then_fi.sh* will do that.

 # nl if_then_fi.sh

```
1    #!/bin/bash
2    # The name of this script is if_then_fi.sh.
3    # The author of this script is Asghar Ghori.
4    # The script is created on February 24, 2006.
5    # This script is last modified by Asghar Ghori on October 14, 2012.
6    # This script should be located in the /usr/local/bin directory.
7    # The purpose of this script is to explain the usage of if-then-fi logical construct.
8    if
9            [ $# –ne 2 ]        # You can also use it as "test $# –ne 2" in explicit mode.
10   then
11            echo "Error: Invalid number of arguments supplied."
12            echo "Usage: $0 source_file destination_file."
13   exit 2
14   fi
15   echo "Script terminated."
```

This script will display the following messages on the screen if executed without specifying exactly two arguments at the command line:

```
Error: Invalid number of arguments supplied.
Usage: if_then_fi.sh source_file destination_file
```

If you check the return code, a value of 2 will appear. This value reflects the exit code defined in the script on line number 13.

 # echo $?
```
2
```

Conversely, if you supply exactly two arguments, the return code will be 0 and the message will be:

```
Script terminated.
```
 # echo $?
```
0
```

The if-then-else-fi Construct

This construct is used when you want to execute one or the other action, depending on the result of the test condition. The general syntax of this structure is:

```
if
    condition
then
    action1
else
    action2
fi
```

If the result of the test condition is true, action1 is performed, otherwise action2 will be performed.

The following script called *if_then_else_fi.sh* will accept an integer value as an argument and tell if the value is positive or negative. If no argument is provided, it will display the usage of the script.

 # nl if_then_else_fi.sh
```
1    #!/bin/bash
2    # The name of this script is if_then_else_fi.sh.
3    # The author of this script is Asghar Ghori.
4    # The script is created on February 24, 2006.
5    # This script is last modified by Asghar Ghori on October 14, 2012.
6    # This script should be located in the /usr/local/bin directory.
7    # The purpose of this script is to explain the usage of if-then-else-fi logical construct.
8    if
9            [ $1 –gt 0 ]                    # You can also use it as "test  $1  –ge  0" in explicit mode.
10   then
11           echo  "$1 is a positive integer value".
12   else
13           echo  "$1 is a negative integer value".
14   fi
```

Run this script one time with a positive integer value and the next time with a negative value:

 # if_then_else_fi.sh 10
10 is a positive integer value.
 # if_then_else_fi.sh –10
–10 is a negative integer value.

The if-then-elif-fi Construct

This construct defines multiple conditions and whichever condition is met, the action associated with it is performed. The general syntax of the structure is:

```
if
    condition1
then
    action1
elif
    condition2
then
    action2
```

```
. . . . . . . . . . . .
. . . . . . . . . . . .
else
        action(n)
fi
```

The following script called *if_then_elif_fi.sh* is an enhanced version of the *if_then_else_fi.sh* script. It will accept an integer value as an argument and tells if it is positive, negative, or zero. If a non-integer value or no command line argument is supplied, the script will complain about it.

nl if_then_elif_fi.sh

```
1    #!/bin/bash
2    # The name of this script is if_then_elif_fi.sh.
3    # The author of this script is Asghar Ghori.
4    # The script is created on February 24, 2006.
5    # This script is last modified by Asghar Ghori on October 14, 2014.
6    # This script should be located in the /usr/local/bin directory.
7    # The purpose of this script is to explain you the usage of if-then-elif-fi logical construct.
8    if
9            [ $1 –gt 0 ]
10   then
11           echo "$1 is a positive integer value".
12           exit  1
13   elif
14           [ $1 –eq 0 ]
15   then
16           echo "$1 is zero integer value".
17           exit  2
18   elif
19           [ $1 –lt 0 ]
20   then
21           echo "$1 is a negative integer value".
22           exit  3
23   else
24           echo "$1 is not an integer value. Please supply an integer".
25           exit  4
26   fi
```

Run this script one time with a positive value, the second time with 0, the third time with a negative value, and the fourth time with a non-integer value. Each time you run it, check the value of the exit code to see where the script actually exited.

if_then_elif_fi.sh 10
10 is a positive integer value.
echo $?
1

if_then_elif_fi.sh 0
0 is zero integer value.

```
# echo $?
2

# if_then_elif_fi.sh -10
-10 is a negative integer value.

# echo $?
3

# if_then_elif_fi.sh abcd
abcd is not an integer value. Please supply an integer.
# echo $?
4
```

Looping Constructs

In shell scripting there are times when you need to perform a task on a number of given values, or until the specified condition becomes true or false. For example, if you wish to *pvcreate* a number of physical disks, you can either run the command on each disk one at a time or employ a loop to do it in one go. Likewise, based on a defined condition, you may want a program to continue to run until the condition becomes either true or false.

There are three constructs that implement looping: the for-do-done construct, the while-do-done construct and the until-do-done construct.

The *for-do-done* construct performs an operation on a list of given elements until the list is exhausted. The *while-do-done* construct performs an operation repeatedly based on the specified condition until the condition becomes false. The *until-do-done* construct does the opposite of what the while-do-done construct does. It performs an operation repeatedly based on the specified condition until the condition becomes true.

Test Conditions

The *let* command is used in looping constructs to check the condition every time a repetition is made. It compares the value stored in a variable against a pre-defined value. Each time the loop does an iteration, the value in the variable is altered. You can enclose the test condition within (( )) or double quotes " " instead of using the *let* command explicitly. This is exhibited in looping construct examples below.

Table 14-3 lists operators that can be used with the *let* command.

Having described various test condition operators, let us take a look at a few example scripts and see how some of these may be utilized.

Operator	Description
–	Unary minus.
!	Unary negation (same value but with a negative sign).
+	Addition.
–	Subtraction.
*	Multiplication.
/	Integer division.
%	Remainder.
<	Less than.
<=	Less than or equal to.
>	Greater than.
>=	Greater than or equal to.
=	Assignment.
==	Comparison for equality.
!=	Comparison for non-equality.

Table 14-3 *let* Operators

The for-do-done Loop

The for-do-done loop is executed on a list of elements until all the elements in the list are exhausted. Each element is assigned to a variable one after the other to be processed in the loop. The syntax of the for-do-done loop is:

```
for  VAR  in  list
do
      command block
done
```

The *for_do_done.sh* script below initializes the variable COUNT to 0. A list of elements is supplied to another variable ALPHABET. The for-do-done loop will read the ALPHABET value one after the other, assign it to the variable LETTER, and display the value on the screen. The *expr* command is an arithmetic processor and it is used simply to increment the COUNT by 1 at each iteration of the loop.

 # nl for_do_done.sh

```
1    #!/bin/bash
2    # The name of this script is for_do_done.sh.
3    # The author of this script is Asghar Ghori.
4    # The script is created on February 24, 2006.
5    # This script is last modified by Asghar Ghori on October 14, 2012.
6    # This script should be located in the /usr/local/bin directory.
7    # The purpose of this script is to explain you the usage of the for-do-done looping construct.
8    COUNT=0
9    ALPHABET="A B C D E F G H I J K L M N O P Q R S T U V W X Y Z"
10   for  LETTER in  $ALPHABET
```

```
11  do
12              COUNT=`/usr/bin/expr $COUNT + 1`
13              echo "Letter $COUNT is [$LETTER]"
14  done
```

The output of the script will be:

```
# for_do_done.sh
Letter 1 is [A]
Letter 2 is [B]
Letter 3 is [C]
Letter 4 is [D]
. . . . . . . .
Letter 24 is [X]
Letter 25 is [Y]
Letter 26 is [Z]
```

Another example is provided below. This script called *create_user.sh* can be used to create several user accounts. As each account is created, the value of $? is checked. If the value is 0, a message saying the account is created successfully will be displayed, otherwise the script will terminate. In case of a successful user creation, the *passwd* command will be invoked to assign a password to the user.

nl create_user.sh

```
1    #!/bin/bash
2    # The name of this script is create_user.sh.
3    # The author of this script is Asghar Ghori.
4    # The script is created on February 24, 2006.
5    # This script is last modified by Asghar Ghori on October 14, 2012.
6    # This script should be located in the /usr/local/bin directory.
7    # The purpose of this script is to create multiple user accounts in one go.
8    for USER in user10 user11 user12 user13 user14
9    do
10            echo "Creating account for user $USER".
11            /usr/sbin/useradd –m –d /home/$USER –s /bin/bash $USER
12            if [ $? = 0 ]
13                      then
14                                echo "$USER is created successfully."
15                      else
16                                echo "Failed to create user account $USER".
17                                exit
18             fi
19    /usr/bin/passwd $USER
20    done
```

The while-do-done Loop

The while-do-done loop checks for a condition and goes on executing a block of commands until the specified condition becomes false. The general syntax of the loop is:

```
while
    condition
do
    command block
done
```

The condition is usually an expression containing the *test* or the *let* command in either an implicit or an explicit mode, but they are normally used in implicit mode.

Let us look at the *while_do_done.sh* script below. The entire case statement is defined as a block of commands within the while-do-done loop here. When you choose one of the listed options, the command associated with that option will be executed. Once the command is finished executing, you will be prompted to press a key to go back to the menu. The loop will continue until you choose option 5 to terminate it.

 # nl while_do_done.sh

```
1    #!/bin/bash
2    # The name of this script is while_do_done.sh.
3    # The author of this script is Asghar Ghori.
4    # The script is created on February 24, 2006.
5    # This script is last modified by Asghar Ghori on October 14, 2012.
6    # This script should be located in the /usr/local/bin directory.
7    # The purpose of this script is to show you an example of how a menu driven program can
8    # be included in a while loop so it continues to run until a specific option from the menu is
9    # chosen to exit out.
10   while true
11   do
12           /usr/bin/clear
13           echo "                    Menu "
14           echo "    -----------------------------------------------------------"
15           echo "    [1] Display System's Current Date and Time".
16           echo "    [2] Display hostname of the system".
17           echo "    [3] Run the lspci command".
18           echo "    [4] Run the fdisk command with the –l option".
19           echo "    [5] Exit".
20           echo "    -----------------------------------"
21           echo
22           echo "Enter Your Choice [1-5] \c: "
23           read VAR
24           case $VAR in
25                   1) echo "Current System Date and Time is `/bin/date`"
26                   echo
27                   echo "Press any key to go back to the Menu......"
28                   read
```

```
29                      ;;
30                      2) /bin/hostname
31                      echo
32                      echo "Press any key to go back to the Menu......"
33                      read
34                      ;;
35                      3) /sbin/lspci
36                      echo "Press any key to go back to the Menu......"
37                      read
38                      ;;
39                      4) /sbin/fdisk –l
40                      echo "Press any key to go back to the Menu......"
41                      read
42                      ;;
43                      5) echo "Exiting ........."
44                      exit  0
45                      ;;
46                      *) echo "You have selected an invalid option".
47                      echo "Please choose a valid option".
48                      echo "Press any key to go back to the Menu......"
49                      read
50                      ;;
51          esac
52   done
```

Building a Custom RPM Package

The RPM software packages are in binary, installable format and they may be downloaded from sources such as the web, the installation DVD, and so on. Sometimes there is a need to build your own custom package based on a specific requirement in the environment. Source codes for many software packages are available from *ftp.redhat.com* and many other sites at no charge under the *GNU Public License* (GPL) license. The source code for many packages may be downloaded from the Red Hat website or elsewhere and customized for your requirements. There is a special directory hierarchy created for this purpose when you install a package containing the source code. Within that directory structure, you modify the code as per your requirements, update the specification file to include necessary guidelines for building the package, repackage the code to build an installable package, and distribute the package for installation.

In this section, you are going to build a simple RPM package containing a single file. You will install the required packages to support building a package, create the required directory structure, and go through the necessary steps to have a custom package ready for installation.

The Specification File

The *specification* file (or simply the *spec* file) defines necessary instructions for the *rpmbuild* command to successfully build a new package. This file also includes necessary information for the *rpm* command to install and remove this package.

The first few rows in the spec file, called the *preamble*, contain general information about the package. This includes the name, version, release, summary, group, license, source, and BuildRoot, and are described in Table 14-4. This information is read and displayed by the *rpm* command when queried with the –i option. Followed by the preamble, there are several sections that begin with the % sign and contain precise instructions for the *rpmbuild* command. Table 14-4 describes some of them as well.

Directive / Section	Description
Name	Specifies the name of the package to be called.
Version	Identifies the version of the package.
Release	Sets the number of times the package has been built at this version.
Summary	Describes a short description of the package.
Group	Specifies how this package is to be grouped with other packages.
License	Indicates the license type.
Source	Points to the source file name located in the SOURCES directory to be used for building the package.
BuildRoot	Indicates the directory where the package will be created during the build process.
%description	Provides a short description of the package.
%prep	Identifies any preparatory scripts to be run prior to building the source code.
%setup	This is a command macro that is used to move the source file from the SOURCES directory to the BUILD directory and decompress it. The –q option instructs the macro to run quietly with minimal output.
%build	Lists the commands for compiling the spec file and build the source code.
%install	Specifies the commands to be executed to copy the source file to the appropriate build directory tree locations.
%clean	Indicates the command to be executed after the completion of the package build to remove the temporary directory structure created during the build process.
%files	Specifies the files or directory structure to be included in the package.
%defattr	Sets the ownership and group membership for the package.
%changelog	Provides information as to who changed what and when in this package.

Table 14-4 SPEC File Description

Exercise 14-1: Build an RPM Package Containing One File

In this exercise, you will log in as *user1* and build an RPM package. You will create the necessary directory structure under the */home/user1* directory, copy the */etc/inittab* file for inclusion in the package, create a specification file with necessary directives, and build the package. You will install and query this package to verify that the package has been built correctly.

1. Run the *yum* command and install the rpm-build and rpmdevtools packages to get the necessary tools installed for building the custom package. The *yum* command will also pull and install several dependent packages.

    # **yum –y install rpm-build rpmdevtools**

2. Execute the *rpmdev-setuptree* command to create a sub-directory called *rpmbuild* and sub-directories BUILD, RPMS, SOURCES, SPECS, and SRPMS beneath it:

    $ rpmdev-setuptree

3. Verify the directory structure under *~/user1/rpmbuild*:

    $ ll rpmbuild
    drwxrwxr-x. 2 user1 user1 4096 Oct 20 20:19 BUILD
    drwxrwxr-x. 2 user1 user1 4096 Oct 20 20:19 RPMS
    drwxrwxr-x. 2 user1 user1 4096 Oct 20 20:19 SOURCES
    drwxrwxr-x. 2 user1 user1 4096 Oct 20 20:19 SPECS
    drwxrwxr-x. 2 user1 user1 4096 Oct 20 20:19 SRPMS

The *rpmbuild* command builds packages in the BUILD directory and stores binary RPMs in the RPMS directory. The SOURCES directory is where you store the source archive file for use in building the package and the SPECS directory is where you create a specification file for use during the package build process.

4. Create an archive of the */etc/inittab* file as *inittabpkg.tar.gz* in the SOURCES directory:

    $ tar cvzf rpmbuild/SOURCES/inittabpkg.tar.gz /etc/inittab

5. Verify the archive file in the SOURCES directory:

    $ ll rpmbuild/SOURCES
    -rw-r--r--. 1 root root 586 Oct 15 22:03 inittabpkg.tar.gz

6. Change into the SPECS directory and execute the *rpmdev-newspec* command to create a specification template file called *inittabpkg.spec* and verify it with the *ll* command:

    $ cd rpmbuild/SPECS && rpmdev-newspec inittabpkg
    Skeleton specfile (minimal) has been created to "inittabpkg.spec".
    $ ll
    -rw-r--r--. 1 root root 498 Oct 15 22:04 inittabpkg.spec

7. Open the *inittabpkg.spec* in the *vi* editor and modify the file to look like the following:

    $ vi inittabpkg.spec
    Name: inittabpkg
    Version: 1.0.0
    Release: 0%{?dist}
    Summary: Includes the /etc/inittab file with default contents
    Group: Default contents
    License: GPL
    Source: %{name}.tar.gz
    BuildRoot: /home/user1/rpmbuild/tmp/%{name}-%{version}
    %Description
    This package contains the /etc/inittab file with the default contents.

```
%prep
%setup -q -n etc
%files
/pkgs
%install
mkdir -p $RPM_BUILD_ROOT/pkgs
cp -R * $RPM_BUILD_ROOT/pkgs
%changelog
* Sun Oct 20 2012 Asghar Ghori version 1.0.0
- Initial RPM release containing /etc/inittab file
```

8. Build a binary package while you are in the SPECS directory by specifying the –bb options with the *rpmbuild* command:

$ rpmbuild –bb –v inittabpkg.spec
```
Executing(%prep): /bin/sh -e /var/tmp/rpm-tmp.0LQB32
+ umask 022
+ cd /home/user1/rpmbuild/BUILD
+ cd /home/user1/rpmbuild/BUILD
+ rm -rf etc
+ /bin/tar -xf -
+ /usr/bin/gzip -dc /home/user1/rpmbuild/SOURCES/inittabpkg.tar.gz
+ STATUS=0
+ '[' 0 -ne 0 ']'
+ cd etc
+ /bin/chmod -Rf a+rX,u+w,g-w,o-w .
+ exit 0
Executing(%install): /bin/sh -e /var/tmp/rpm-tmp.CRFMuR
+ umask 022
+ cd /home/user1/rpmbuild/BUILD
+ cd etc
+ mkdir -p /home/user1/rpmbuild/BUILDROOT/inittabpkg-1.0.0-0.el6.x86_64/pkgs
+ cp -R inittab /home/user1/rpmbuild/BUILDROOT/inittabpkg-1.0.0-0.el6.x86_64/pkgs
+ /usr/lib/rpm/check-rpaths /usr/lib/rpm/check-buildroot
+ /usr/lib/rpm/brp-compress
+ /usr/lib/rpm/brp-strip
+ /usr/lib/rpm/brp-strip-static-archive
+ /usr/lib/rpm/brp-strip-comment-note
Processing files: inittabpkg-1.0.0-0.el6.x86_64
Requires(rpmlib): rpmlib(CompressedFileNames) <= 3.0.4-1 rpmlib(PayloadFilesHavePrefix) <= 4.0-1
Checking for unpackaged file(s): /usr/lib/rpm/check-files /home/user1/rpmbuild/BUILDROOT/inittabpkg-
1.0.0-0.el6.x86_64
Wrote: /home/user1/rpmbuild/RPMS/x86_64/inittabpkg-1.0.0-0.el6.x86_64.rpm
Executing(%clean): /bin/sh -e /var/tmp/rpm-tmp.MaqDWu
+ umask 022
+ cd /home/user1/rpmbuild/BUILD
+ cd etc
+ rm -rf /home/user1/rpmbuild/BUILDROOT/inittabpkg-1.0.0-0.el6.x86_64
+ exit 0
```

This command will use the BUILD directory to create the binary package and store it in the RPMS directory.

9. Run the *ll* command to list the newly built rpm file:

    $ ll –R /home/user1/rpmbuild/RPMS
    /home/user1/rpmbuild/RPMS:
    total 4
    drwxr-xr-x. 2 user1 user1 4096 Dec 10 12:40 x86_64
    /home/user1/rpmbuild/RPMS/x86_64:
    total 4
    -rw-rw-r--. 1 user1 user1 2432 Dec 10 12:40 inittabpkg-1.0.0-0.el6.x86_64.rpm

10. Test install the package using the *rpm* command after the package has been created to make certain that it works. Run this command as the *root* user.

    # rpm –ivh /home/user1/rpmbuild/RPMS/x86_64/inittabpkg*.rpm
    Preparing...   ### [100%]
      1:inittabpkg   ### [100%]

11. Execute the *rpm* command as *root* on the package for verification:

    # rpm –qi inittabpkg

Name	: inittabpkg	Relocations: (not relocatable)
Version	: 1.0.0	Vendor: (none)
Release	: 0.el6	Build Date: Mon 10 Dec 2012 12:40:49 PM EST
Install Date : Mon 10 Dec 2012 12:44:18 PM EST		Build Host: insider.example.com
Group	: Default contents	Source RPM: inittabpkg-1.0.0-0.el6.src.rpm
Size	: 884	License: GPL
Signature	: (none)	
Summary	: Includes the /etc/inittab file with default contents	
Description :		

    This package contains the /etc/inittab file with the default contents.

Internet Small Computer System Interface

The *Internet Small Computer System Interface* (iSCSI) is a storage networking transport protocol that carries SCSI commands over IP-based networks and the Internet. This protocol enables data transfer between computer systems (called *iSCSI initiators*) and networked storage devices (called *iSCSI targets*). As this protocol works on IP networks, the distance between initiators and targets is irrelevant as long as the link between them is fast enough. Once the connection is established, any storage presented to the initiator from the target will appear as a local disk. This protocol typically communicates over port 3260.

The iSCSI initiator functionality becomes available when the iscsi-initiator-utils package is installed on the system. This package installs the iSCSI management command *iscsiadm*, configuration file */etc/iscsi/iscsid.conf*, startup script */etc/rc.d/init.d/iscsi* that logs into iSCSI targets at system startup, and the *iscsi* service management script */etc/rc.d/init.d/iscsi*.

Exercise 14-2: Configure an iSCSI Disk Device

In this exercise, you will install the iscsi-initiator-utils package on *server.example.com* (the initiator), set the *iscsi* service to autostart at each system reboot, start the *iscsi* service, discover available targets on the network, log into the available target devices, and create a file system of size 1GB using LVM in one of the disk device. You will add an entry for the file system to the */etc/fstab* file using the file system's UUID and mount the file system manually. You will reboot the system to ensure that the file system is automatically mounted. Assume that the name of the target is iscsi.example.com with IP 192.168.2.215.

1. Run the *yum* command to install the iscsi-initiator-utils package if it is not already installed:

   # **yum –y install iscsi-initiator-utils**

2. Set the *iscsi* service to autostart at each system reboot, and validate:

   # **chkconfig iscsi on**
   # **chkconfig --list | grep iscsi**
   iscsi   0:off 1:off 2:off 3:on 4:on 5:on 6:off

3. Use the *service* command and start the iSCSI service, and check the status:

   # **service iscsi start**
   # **service iscsi status**
   iscsi (pid 2143) is running...

4. Execute the *iscsiadm* command in the discovery mode (–m) to get a list of available targets (-t) from the specified host IP (–p):

   # **iscsiadm –m discovery –t st –p 192.168.2.215**

5. Issue the *iscsiadm* command again to log into all the disk devices exported on the target and enable all the targets to be accessed upon reboots:

   # **iscsiadm –m discovery –t st –l**

6. Execute the *fdisk* command with the –l option and you should be able to see the new iSCSI disks.

7. Run the *pvcreate* command on one of the disks and initialize it for use in LVM. Create a volume group and add the physical volume to it. Create a logical volume of size 1GB, construct file system structures in the logical volume, create a mount point, add an entry for the file system to the */etc/fstab* file using the file system's UUID, and manually mount the file system. Reboot the system and make sure that the file system is mounted automatically. Refer to Chapter 09 "Disk Partitioning" for the LVM tasks and Chapter 10 "File Systems & Swap" for the rest of the tasks in this step.

Chapter Summary

In this chapter, you learned the basics of scripting and how to create a custom RPM package. You were presented with several example shell scripts along with an analysis of each one of them. You looked at how to write and debug simple to advanced scripts. You analyzed each script line by line for better understanding.

The next major topic in this chapter provided you with step by step instructions on how to create a custom RPM package for installation in your environment.

Finally, you reviewed the basics of the iSCSI protocol and looked at how to discover iSCSI targets and create LVM and file system structures in them.

Chapter Review Questions

1. What would the condition file1 −nt file2 imply?
2. Which looping construct can be used to perform an action on listed items?
3. What is the function of the *shift* command?
4. You cannot control the startup of a database software via scripts. True or False?
5. What is the name of the directory where you would store the source file for repackaging?
6. The until-do-done loop continues to run until the specified condition becomes true. True or False?
7. What would the command *echo $?* do?
8. Which logical construct lets you specify multiple conditions?
9. When would you want to use an exit code in your script?
10. What would you modify in a shell script to run it in the debug mode?
11. What would be the impact of the *−d file1* condition in the logical construct?
12. What is the use of the *iscsiadm* command?
13. What does the *rpmbuild* command do?
14. What are the two fundamental types of logical constructs outlined in this chapter?
15. Which command would you use in a shell script to enable it to prompt for user input during execution?
16. What is the purpose of using the escape sequences in shell scripts?
17. What two types of constructs use test conditions?
18. What would != imply in a looping condition?
19. What comments may you want to include in a shell script? Write any six.
20. Under the GNU public license, any person can modify a package source code. True or False?
21. What are the three major components in a shell scripts?
22. The while-do-done loop continues to run until the specified condition becomes true. True or False?
23. What is one benefit of writing shell scripts?
24. What does the *echo* command do without any arguments?
25. What is the role of the spec file in the package rebuilding?
26. What would the command *sh −x /usr/local/bin/script1.sh* do?

Answers to Chapter Review Questions

1. This condition will check if *file1* is newer than *file2*.
2. The for-do-done loop.

---

3. The *shift* command moves an argument to the left.
4. False.
5. The SOURCES directory.
6. True.
7. This command will display the exit code of the last command executed.
8. The if-then-elif-fi construct.
9. The purpose of using the exit code in scripts is to determine where exactly the script quits.
10. You would add –x to the shell path.
11. This condition will be true if the specified filename is a directory.
12. The *iscsiadm* command is used for the administration of the iSCSI devices.
13. The *rpmbuild* command is used to build new packages from the source files.
14. The if-then-fi and case constructs.
15. The *read* command is used to prompt for user input.
16. Escape sequences are used in shell scripts for better readability.
17. The logical and the looping constructs.
18. != would check the value for non-equality.
19. The author name, creation date, last modification date, path, purpose, and usage.
20. True.
21. The three major components in a shell script are commands, control structures, and comments.
22. False.
23. One major benefit of writing shell scripts is to automate lengthy tasks.
24. The *echo* command will insert an empty line in the output when used without any arguments.
25. The *spec* file contains the instructions on how to build the specified package.
26. The command provided will execute the *script1.sh* script in the debug mode.

Labs

Lab 14-1: Write a Script to Find Specific Files

Write a script in the bash shell to find and display all files in the entire system with setgid bit enabled. The output should also display a count of such files. Include a statement at the beginning that says "Here are the files that have setgid enabled" and a statement at the end that says "There are <count of files> with setgid" (<count of files> should reflect a count of the files found).

Lab 14-2: Write a Script to List Packages

Write a script in the bash shell to list what packages are installed on the system using either the *rpm* or the *yum* command. Include a statement at the beginning that says "Packages currently installed on the system are:". The script should then look for the vsFTP package(s) and list them on the screen with a heading before the list that says "The vsFTP packages installed on the system are:".

Lab 14-3: Write a Script to Display SELinux Directives

Write a script in the bash shell to extract and list all uncommented directives in the SELinux configuration file */etc/sysconfig/selinux*. For each directive, the script should print "The current value of the <name of the directive> directive is <value>" on the screen. Include appropriate statements at the beginning and end of the output.

Lab 14-4: Create a Custom RPM Package

Create a custom RPM package called archtools-v1.0 and include the *cpio*, *tar*, and *star* commands in it.

## Chapter 15

Advanced Firewall, TCP Wrappers, Usage Reporting & Remote Logging

This chapter covers the following major topics:

- ✓ Basics of Network Address Translation and IP masquerading
- ✓ Overview of host-based security using TCP Wrappers
- ✓ Configure firewall rules to protect services and implement Network Address Translation
- ✓ Produce and deliver reports on processor, memory, disk, and network utilization
- ✓ Configure a system to log to a remote host
- ✓ Configure a system to accept logging form a remote system

This chapter includes the following RHCE objectives:

57. Use iptables to implement packet filtering and configure network address translation (NAT)
62. Produce and deliver reports on system utilization (processor, memory, disk, and network)
64. Configure a system to log to a remote system
65. Configure a system to accept logging from a remote system

Network address translation

Network address translation is a feature that enables a system on the internal network to access the Internet via an intermediary device. IP masquerading, in contrast, enables more than one systems on the internal network to access the Internet via an intermediary device. In either case, IP addresses of the systems on the internal network are concealed from the outside world, which only sees one IP address and that one IP address is of the intermediary device.

TCP Wrappers is a host-based service that allows you to control access into the system by network-based services. This access may be regulated at the user and host levels.

Monitoring and reporting utilization of processors, memory, disks, and network interfaces is the key in determining bottlenecks in the system. Monitoring helps identify any potential performance issues and the monitored data may be stored for future reference.

Log information generated on one system may be forwarded to and stored on a remote system. This makes the remote system a central repository for all messages generated on the system. The remote system may be configured to receive forwarded messages from several systems.

Advanced Firewall

RHEL comes standard with the iptables firewall software that lets you control the transfer of data packets at the host level. In addition, the iptables software also provides support for advanced features such as the *Network Address Translation* (NAT) and *IP masquerading*.

Network Address Translation and IP Masquerading

NAT is a function whereby a system on the internal network such as a home or corporate network, can access external networks such as other networks or the Internet, using a single, registered IP address configured on an intermediary device such as a router, a firewall, a combination of both, or a computer running RHEL or other operating system. IP masquerading is a variant of NAT and allows several systems on the internal network to access external networks using an intermediary device. With masquerading, requests originated from any of the internal systems appear to the outside world as being originated from the intermediary device. The intermediary device stores the IP addresses of the source systems in its cache, along with randomly-generated port numbers assigned to them, to keep traffic segregated for each system. These techniques save you from purchasing official IP addresses for each system on your internal network.

For NAT (and masquerading), you specify a destination NAT (DNAT) in the PREROUTING chain or a source NAT (SNAT) in the POSTROUTING chain. DNAT translates the destination IP address of an outbound packet and SNAT translates the source IP address of an outbound packet to the IP address of an intermediary device.

Exercise 15-1: Configure IP Masquerading

In this exercise, you will enable masquerading on *physical.example.com* with an internal address 192.168.2.200 configured on the *eth0* interface and an official address 198.202.11.15 on *eth1* for the Internet connectivity. You will use default network parameters for both interfaces. Refer to Chapter 13 "Networking, Network Interfaces & DNS, DHCP and LDAP Clients" on how to configure network interfaces.

1. Ensure *eth0* is configured correctly with 192.168.2.200 address.
2. Ensure *eth1* is configured correctly with 198.202.11.15 address.
3. Configure masquerading using the *iptables* command:

   # **iptables –t nat –A POSTROUTING –o eth1 –s 192.168.2.0/24 –j MASQUERADE**

4. Save the rule in the */etc/sysconfig/iptables* file and restart the firewall to activate the new rule:

   # **service iptables save ; service iptables restart**

5. Configure IP forwarding on *physical.example.com* so the inbound traffic from internal systems is forwarded to the registered address for Internet access. Edit the */etc/sysctl.conf* file and set the following directive to 1 so the value gets set at every system reboot:

   net.ipv4.ip_forward = 1

6. Execute the *sysctl* command with the –p option or the following for this change to take effect immediately:

   # **echo 1 > /proc/sys/net/ipv4/ip_forward**

7. Configure 198.202.11.15 as the default gateway address in the */etc/resolv.conf* file on systems on the internal network.

Now you should be able to access the Internet from any system on the internal network.

TCP Wrappers

TCP Wrappers is a host-based service that allows you to control access into a system from remote hosts sending service requests via the *xinetd* daemon. The access can be controlled at the user or host level by specifying which users from which hosts, networks, or domains are allowed or denied. The software package associated with the TCP Wrappers service is installed by default as part of the RHEL6 installation:

rpm –qa | grep wrappers
tcp_wrappers-7.6-57.el6.x86_64

Two files – *hosts.allow* and *hosts.deny* – located in the */etc* directory are critical to TCP Wrappers functionality. The default files contain no restrictions. The *.allow* file is referenced before the *.deny* file. If these files do not contain an entry for the requesting user or the host, network, or domain, TCP Wrappers grants the incoming request access into the system.

Here is how it works: when an *xinetd*-controlled client request comes in, *xinetd* consults the TCP Wrappers library */lib64/libwrap.so*, which scans the *.allow* file and instructs *xinetd* to start the server daemon for the client request if it finds a match in the file. If *.allow* does not contain a match, TCP Wrappers consults the *.deny* file. If a match is found, *xinetd* denies access to the client, otherwise, it starts the associated server daemon and allows the request in. TCP Wrappers entertains only the first match as it scans the files. Any changes made to either file take effect right away.

The format of both files is identical and is based on "what:who" as shown below:

<name of service daemon> : <list of clients>

The first column lists the name of a service daemon such as *telnetd*, *sshd*, and *ftpd*, and the second column specifies a user name, hostname, network address, or a domain name. Keywords ALL and EXCEPT may be used in either field to represent "open to all" and an "exception". The keyword LOCAL matches any hostnames or IP addresses that do not contain a leading dot, the keyword KNOWN matches any hostnames that can be resolved, and the keyword UNKNOWN is the opposite of KNOWN.

Several combinations with an explanation are provided in Table 15-1 for a better understanding.

Entry	/etc/hosts.allow	/etc/hosts.deny
ALL:ALL	All users from all hosts are allowed.	All users from all hosts are denied.
ALL:aghori	User *aghori* is allowed from all hosts.	User *aghori* is denied from all hosts.
ALL:aghori@physical.example.com	User *aghori* is allowed only from *physical.example.com*.	User *aghori* is denied only from *physical.example.com*.
ALL:.example.com	All users on *example.com* domain are allowed.	All users on *example.com* domain are denied.
ALL:192.168.2.	All users on the 192.168.2 network are allowed.	All users on the 192.168.2 network are denied.
sshd:ALL	All users from all systems are allowed *ssh* access.	All users from all systems are denied *ssh* access.
sshd:LOCAL	All users on the local network are allowed *ssh* access.	All users on the local network are denied *ssh* access.
telnetd:192.168.2.	All users on 192.168.2 network are allowed *telnet* access.	All users on 192.168.2 network are denied *telnet* access.
telnetd:192.168.2.0/16	All users on 192.168.2 network with netmask 255.255.0.0 are allowed *telnet* access.	All users on 192.168.2 network with netmask 255.255.0.0 are denied *telnet* access.
telnetd:192.168.2. EXCEPT 192.168.2.25	All users on the 192.168.2 network except for host 192.168.2.25, are allowed *telnet* access.	All users on the 192.168.2 network except for host 192.168.2.25, are denied *telnet* access.
telnetd:192.168.2. EXCEPT 192.168.2.25, 192.168.2.26	All users on the 192.168.2 network except for hosts 192.168.2.25 and .26, are allowed *telnet* access.	All users on the 192.168.2 network except from hosts 192.168.2.25 and .26, are denied *telnet* access.
telnetd,sshd:192.168.2.	All users on the 192.168.2 network are allowed *telnet* and *ssh* access.	All users on the 192.168.2 network are denied *telnet* and *ssh* access.

Entry	/etc/hosts.allow	/etc/hosts.deny
telnetd,sshd@192.168.0.201: 192.168.2.	All users on the 192.168.2 network are allowed *telnet* and *ssh* access via the interface configured with IP 192.168.0.201.	All users on the 192.168.2 network are denied *telnet* and *ssh* access via the interface configured with IP 192.168.0.201.
ALL EXCEPT sshd:192.168.2.	All users on the 192.168.2 network are allowed access to all but the *ssh* service.	All users on the 192.168.2 network are denied access to all but the *ssh* service.
ALL:.example.com EXCEPT 192.168.2.0/24	All users on the *example.com* domain except for those on the 192.168.2.0/24 network are allowed access to all the services.	All users on the 192.168.2.0/24 network except for those on the *example.com* domain are allowed access to all the services.

Table 15-1 TCP Wrappers hosts.allow and hosts.deny Files

TCP Wrappers logs information to the */var/log/secure* file.

System Usage Reporting

Monitoring system resources and reporting their utilization are important tasks of system administration. System resources include CPU, memory, disk, and network. There are several tools available and are used depending on what report you wish to generate for review. RHEL comprises of two software packages called sysstat and dstat that include tools to monitor the performance and usage of these resources, and generate reports. Moreover, numerous native tools in RHEL are available that administrators use to monitor and report resource utilization to determine where, if any, bottlenecks exist. These tools include *fdisk* and *df* for viewing current disk and file system utilization, *vmstat* for viewing virtual memory statistics, *top* for realtime viewing of CPU, memory, swap, and processes, and so on. See chapters in the RHCSA section for details on these tools.

The sysstat package includes commands such as *cifsiostat*, *iostat*, *mpstat*, *pidstat*, *sadf*, and *sar*, and the dstat package includes the *dstat* command. Table 15-2 summarizes these tools and their purposes.

Command	Description
cifsiostat	Reports read and write operations on CIFS file systems.
iostat	Reports CPU, device, partition, and NFS file system statistics.
mpstat	Reports activities for each available CPU.
pidstat	Reports statistics for running processes.
sadf	Displays the data in various formats gathered by the *sar* command.
sar	*system activity reporter*. Gathers and reports various system activities and stores the data in files in the raw form.
dstat	Reports CPU, disk, network, paging, and system statistics. More powerful and versatile than the *vmstat* and *iostat* commands combined.

Table 15-2 Commands in Sysstat Package

In addition to the commands listed above, the sysstat package also includes a startup script */etc/rc.d/init.d/sysstat*, two startup configuration files in the */etc/sysconfig* directory called *sysstat* and *sysstat.ioconf*, and a cron job file */etc/cron.d/sysstat*. Sysstat is a system status service executed automatically by the *cron* daemon. This service references the two startup configuration files for directions. The *sysstat* file contains only two directives: HISTORY and COMPRESSAFTER. The former sets a limit on the number of days (default is 7) to keep the log files and the latter determines the age (default is 10) of the log files after which to compress them. The *sysstat.ioconf* file contains the directives to gather disk I/O related information, and is typically left intact.

The sysstat service's cron job runs at regular intervals to gather resource utilization information. It stores the information in the raw format in log files located in the */var/log/sa* directory. By default, there are two cron entries in the file, and they are:

```
*/10 * * * * root /usr/lib64/sa/sa1 –S DISK 1 1
53 23 * * * root /usr/lib64/sa/sa2 –A
```

The first entry executes the *sa1* command with the –S DISK 1 1 options at every 10 minute interval. The –S option instructs the *sa1* command to write once and begin gathering information on the swap space one second after the command has been executed. Data logged by this command is stored in the */var/log/sa* directory in the sadd format (where sa represents the name of the command and dd the day of the month).

The second cron entry executes the *sa2* command every day at 23:53PM. With the –A option, the command captures the statistical information in the */var/log/sa/sardd* file (where sar represents the name of the command and dd the day of the month).

The *sadf* command is used to extract the data as desired from the */var/log/sa* directory and display on the screen for review. The data in this directory is the output of commands *sar*, *sa1*, and *sa2*. This data is stored in the raw format and cannot be read without the *sadf* tool. Let us first look at some of the key options available with this command. Table 15-3 summarizes them.

Option	Description
-d	Generates a report in a database-friendly format.
-D	Generates a report in a database-friendly format with the time expressed in seconds since the epoch time.
-e	Expresses the end time in the 24-hour format.
-p	Generates a report in the awk-friendly format.
-s	Expresses the start time in the 24-hour format.
-x	Generates a report in the XML-friendly format.

Table 15-3 sadf Command Options

Now let us use these options in some examples below.

The following example generates a report with data extracted from 9AM to 5PM on the fifteenth of the month:

```
# sadf –s 09:00:00 –e 17:00:00 /var/log/sa/sa15
server.example.com    600    1350306601    all    %user    0.09
server.example.com    600    1350306601    all    %nice    0.00
server.example.com    600    1350306601    all    %system 0.18
server.example.com    600    1350306601    all    %iowait  0.05
server.example.com    600    1350306601    all    %steal   0.00
. . . . . . . .
```

You can redirect the above output to a file for future review.

The following example generates a database-friendly report on cpu, memory, disk, and network by running the *sar* command with data from the twentieth of the month:

```
# sadf –d /var/log/sa/sa20 -- –urd –n DEV
# hostname;interval;timestamp;CPU;%user;%nice;%system;%iowait;%steal;%idle
server.example.com;600;2012-10-20 04:10.01 UTC;-1;0.04;0.00;0.02;0.08;0.00;99.86
server.example.com;600;2012-10-20 04:20.01 UTC;-1;0.03;0.00;0.02;0.05;0.00;99.91
server.example.com;600;2012-10-20 04:30.02 UTC;-1;0.02;0.00;0.01;0.06;0.00;99.91
. . . . . . . .
```

In the above example, the options specified after the two hyphen characters are the *sar* command options. Table 15-4 lists and describes some of the key *sar* command switches.

Option	Description
-d	Reports block devices.
-n	Reports network devices.
-P	Reports the specified number of CPU.
-r	Reports the memory usage.
-S	Reports the swap space utilization statistics.
-u	Generates a report on CPU usage.

Table 15-4 sar Command Options

The *dstat* command is a versatile replacement for the *vmstat*, *iostat*, and *ifstat* commands. This command displays real-time system resource utilization. Run this command with the –cdmn options to view the utilization of cpu (–c), disk (–d), memory (–m), and network (–n).

```
# dstat –cdmn
----total-cpu-usage---- -dsk/total- ------memory-usage----- -net/total-
usr sys idl wai hiq siq| read  writ| used  buff  cach  free| recv  send
  0   0 100   0   0   0|9855B 2145B| 283M 50.5M  334M  106M|   0     0
  0   0 100   0   0   0|   0     0 | 283M 50.5M  334M  106M| 138B  810B
  1   1  98   0   0   0|   0     0 | 283M 50.5M  334M  106M| 138B  346B
  0   0 100   0   0   0|   0     0 | 283M 50.5M  334M  106M| 230B  388B
```

Remote System Logging

System logging was covered in detail in Chapter 08 "Linux Boot Process & Kernel Management". Here, we will cover how to configure a local system so that it forwards all its log information to a remote loghost system. We will also see how to configure the remote system to be able to receive logs from the local system. This client / server logging setup proves useful in environments where there are many virtual machines running with RHEL as their guest operating system. For this setup to function properly, either the UDP or the TCP protocol may be used. The UDP protocol is connectionless and unreliable. In contrast, the TCP protocol is connection-oriented and reliable.

Exercise 15-2: Confiugre a System to Receive Log Messages

In this exercise, you will configure *physical.example.com* as a loghost to receive forwarded log messages from *server.example.com* and store them locally. You will use the UDP protocol and add a firewall rule to allow traffic to pass through port 514.

1. Open the */etc/rsyslog.conf* file in a text editor and uncomment the following two directives:

    $ModLoad imudp.so
    $UDPServerRun 514

2. Configure host-based access by allowing syslog traffic on port 514 to pass through the firewall:

    # **iptables –I INPUT –p udp –m udp --dport 514 –j ACCEPT**

3. Save the rule in the */etc/sysconfig/iptables* file and restart the firewall to activate the new rule:

    # **service iptables save ; service iptables restart**

4. Set the *rsyslog* service to autostart at each system reboot, and validate:

    # **chkconfig rsyslog on**
    # **chkconfig --list rsyslog**
    rsyslog   0:off 1:off 2:on 3:on 4:on 5:on 6:off

5. Restart the *rsyslog* service and check the running status:

    # **service rsyslog restart**
    # **service rsyslog status**
    rsyslog (pid 1235) is running...

Exercise 15-3: Configure a System to Forward Log Messages

In this exercise, you will configure *server.example.com* to forward all log messages to *physical.example.com*. You will use the UDP protocol and add a firewall rule to allow traffic to pass through port 514. You will generate a custom log message on *server.example.com* and check in the */var/log/messages* file on *physical.example.com* if the message is received to validate the log forwarding and receiving configuration.

1. Open the */etc/rsyslog.conf* file in a text editor and add the following to the end of the file:

    . @192.168.2.200:514

---

2. Set the *rsyslog* service to autostart at each system reboot, and validate:

   # **chkconfig rsyslog on**
   # **chkconfig --list rsyslog**
   rsyslog   0:off 1:off 2:on 3:on 4:on 5:on 6:off

3. Configure host-based access by allowing syslog traffic on port 514 to pass through the firewall:

   # **iptables –I INPUT –p udp –m udp --dport 514 –j ACCEPT**

4. Save the rule in the */etc/sysconfig/iptables* file and restart the firewall to activate the new rule:

   # **service iptables save ; service iptables restart**

5. Restart the *rsyslog* service:

   # **service rsyslog restart**
   Shutting down system logger: [OK]
   Starting system logger: [OK]

6. Execute the following command to submit a custom log message:

   # **logger –i "This is a test message from root on server.example.com"**

7. Log on to *physical.example.com* and tail the last line of the */var/log/messages* file for the above alert:

   # **tail –1 /var/log/messages**
   Oct 20 6:02:51 server root[10460]: This is a test message from root on server.example.com

Chapter Summary

This chapter covered a few important system administration topics. You learned the basics of network address translation and IP masquerading. You performed an exercise that demonstrated how to configure IP masquerading for connecting many systems on an internal network to the Internet without acquiring official IP addresses for each system.

You looked at the host-based access control service called TCP Wrappers. You saw the control files associated with this service. You studied a number of possible combinations that could be defined in the control files for finer access control into the system.

The third major topic was on monitoring and reporting system resource utilization. You reviewed several tools that could be used for these purposes. You used the *sar* command to capture performance data and used a different tool to display that data nicely.

The final topic in this chapter shed light on how to set up a loghost server and how to set up a system to send its log data to that loghost. You performed exercises to demonstrate how this setup worked.

Chapter Review Questions

1. Which command is used to display the performance data captured with the *sar* command?
2. What is the name of the syslog daemon in RHEL6?
3. Which command is used to configure IP masquerading in RHEL6?
4. What would TCP Wrappers do if the *hosts.deny* file contains ALL:.example.com entry?
5. Define Network Address Translation.
6. The *sar* command may be used to display data for a specified time range. True or False?
7. Where does TCP Wrappers stores its log information?
8. Which command in the dstat package displays real-time performance data for cpu, memory, disk, and network?
9. How many registered IP addresses would you need to obtain if all of your systems on the network are behind an intermediary device?
10. TCP Wrappers is a network-based service. True or False?
11. Name the two files that control the behaviour or TCP Wrappers?
12. What is the main configuration file for the *rsyslog* service?
13. In which file would you add the net.ipv4.ip_forward=1 entry so that it takes effect on every system reboot?
14. What would TCP Wrappers do if the *hosts.allow* file contains sshd:LOCAL entry?
15. What would *.info @192.168.122.32:9999 imply?
16. What would TCP Wrappers do if the *hosts.deny* file contains ALL EXCEPT telnetd:192.168.3 entry?

Answers to Chapter Review Questions

1. The *sadf* command is used to display the performance data collected with the *sar* command.
2. The syslog daemon in RHEL6 is called *rsyslogd*.
3. The *iptables* command.
4. TCP Wrappers will deny access to all systems except for the systems in the *example.com* domain.
5. NAT hides private IP addresses used on the internal network from the outside world.
6. False.
7. TCP Wrappers store its log information in the */var/log/secure* file.
8. The *dstat* command.
9. Only one.
10. False.
11. The */etc/hosts.deny* and */etc/hosts.allow* files.
12. The */etc/rsyslog.conf* file.
13. In the */etc/sysctl.conf* file.
14. The entry provided will allow *sshd* access from the systems on the local network.
15. The entry provided will forward all informational messages generated on the system to the loghost at 192.168.122.32 via port 9999.
16. The entry provided will deny access to all services except for the *telnetd* service from 192.168.3 network.

Labs

Lab 15-1: Configure TCP Wrappers to Secure Services

Edit one or both TCP Wrappers access control files as appropriate and allow *user1* and *user2* from 192.168.2 and *example.net* to be able to access the *sshd* and *vsftpd* services on the system. Prevent access to these services for *user3* and *user4* from 192.168.2.100 and *example.org*. Log in as *user1* and *user3* from RHEL systems (one with IP 192.168.2.90 and the other one with IP 192.168.2.100) and test the access. Observe the messages as they are logged to the */var/log/messages* and */var/log/secure* files.

Lab 15-2: Set Up a Central Loghost

Set up *server.example.com* as a central loghost to receive and store messages from *insider.example.com*. Configure the service to use the TCP protocol on port 1000. Test the functionality by generating an alert on *insider.example.com* and confirming its reception in the */var/log/message* file on *server.example.com*.

Chapter 16

Routing, Kerberos & DNS Server

This chapter covers the following major topics:

- ✓ Introduction to routing
- ✓ Manage routes using commands and a graphical tool
- ✓ Understand Kerberos authentication and configure a client
- ✓ Understand name resolution
- ✓ DNS concepts and components
- ✓ Configure a caching and a forwarding DNS server
- ✓ Verify DNS functionality with dig and host utilities

This chapter includes the following RHCE objectives:

56. Route IP traffic and create static routes
59. Configure a system to authenticate using Kerberos
DNS: Configure a caching-only name server
DNS: Configure a caching-only name server to forward DNS queries

Routing

Routing is the process of choosing paths on a network along which to send network traffic and routers are the devices that are employed for this purpose. In addition to the specialized hardware routing devices, a RHEL system may be configured to route traffic to other networks but to a lesser degree.

Kerberos is a client/server authentication protocol which works on the basis of tickets to allow systems communicating over a non-secure network to prove their identity to one another in a secure manner. With Kerberos setup in place, both the user and the server verify each other's identity.

Domain Name System is an OS- and hardware-independent service that is used for determining the IP address of a system by providing its hostname, and vice versa. DNS is used on the Internet and in corporate networks as the only choice for resolving hostnames. It is the de facto standard for hostname resolution.

Routing

Routing refers to the process of choosing a path over which to send a data packet. A routing device is required to perform the routing function, and this routing device can be a specialized and sophisticated hardware device called a *router* or it can be a RHEL system with more than one network interface ports available. The RHEL system can perform the routing function, but it will not be as sophisticated.

When systems on two separate networks communicate with each other, proper routes must be set up in order for the systems to be able to talk. For instance, if a system on network A sends a data packet to a system on network B, one or more routing devices is involved to route the packet to the correct destination network. The two networks can be located in the same data center or thousands of miles apart. Once the data packet reaches a router, the router selects the next router along the path toward the destination node. The packet passes from one router to another until it reaches the router that is able to deliver the packet directly to the destination system. Each router along the path is referred to as a *hop*.

One of three rules is applied in this mechanism to determine the correct route:

- ✓ If the source and destination systems are on the same network, the packet is sent directly to the destination system.
- ✓ If the source and destination systems are on two different networks, all defined routes are tried one after the other. If a proper route is found, the packet is forwarded to it, which then forwards the packet to the right destination system.
- ✓ If the source and destination systems are on two different networks but no routes are defined, the packet is forwarded to the *default router* (or the *default gateway*), which attempts to search for an appropriate route to the destination. If found, the packet is delivered to the destination system.

Routing Table

The *routing table* maintains information about available routes and their status. It is maintained by the kernel in the memory. The routing table can be viewed with the *ip* command:

```
# ip route
192.168.2.0/24    dev eth0   proto  kernel  scope  link  src  192.168.2.200
192.168.122.0/24 dev virbr0 proto  kernel  scope  link  src  192.168.122.1
169.254.0.0/16    dev eth0   scope link      metric 1002
default  via  192.168.2.1 dev eth0   proto static
```

The output is organized in multiple columns, which are explained in Table 16-1:

Column	Description
Network address	Determines the address of the destination network. The keyword default identifies the IP address for sending out traffic to other networks if a proper route is not found on the system.
dev	Name of the physical or virtual network interface to be used to send out traffic.
proto	Identifies the routing protocol as defined in the */etc/iproute2/rt_protos* file. The proto kernel implies that the route was installed by the kernel during autoconfiguration. The proto static means that the route was installed by you to override dynamic routing.
scope	Determines the scope of the destination as defined in the */etc/iproute2/rt_scopes* file. Values may be global, nowhere, host, or link.
src	Shows the source address for sending data to the destination.

Table 16-1 ip route Command Output Description

Managing Routes

Managing routes involves adding, modifying, or deleting a route, and setting the default route. The *ip* command or the Network Connections tool can be used to perform these tasks. Alternatively, you can create/edit interface specific *route-eth** files in the */etc/sysconfig/network-scripts* directory to add, modify, or delete route entries. Entries added with the *ip* command are lost when the system is rebooted; however, those added with the Network Connections tool or by editing the *route-eth** files are persistent across system reboots. The Network Connections tool automatically updates the corresponding *route-eth** files.

Exercise 16-1: Add Default and Static Routes

In this exercise, you will add a static route on *physical.example.com* to network 192.168.3.0 on *eth0* with the default class C netmask and gateway 192.168.2.1. You will also add a default route via 192.168.2.1. You will reboot the system and check whether the routes are still available. You will re-add the same routes by creating a file in the */etc/sysconfig/network-scripts* directory and adding entries to it. You will reboot the system and check whether the routes are still there. Finally, you will delete both routes.

1. Use the *ip* command and add a static route to network 192.168.3.0 on *eth0* with the default class C netmask and gateway 192.168.2.1:

    # ip route add 192.168.3.0/24 via 192.168.2.1 dev eth0

2. Run the *ip* command and add a default route on *eth0* via gateway 192.168.2.1:

> # **ip route add default via 192.168.2.1 dev eth0**

3. Reboot the system and run the *ip* command to check whether the new routes are there. You will not find them.

4. Create the file */etc/sysconfig/network-scripts/route-eth0* and insert the following entries:

> # **vi /etc/sysconfig/network-scripts/route-eth0**
> GATEWAY0=192.168.2.1
> NETMASK0=255.255.255.0
> ADDRESS0=192.168.3.0
> GATEWAY1=192.168.2.1
> NETMASK1=255.255.255.0
> ADDRESS1=default

5. Reboot the system or issue the *ifup-routes* command to activate the new routes. Execute the *ip* command to verify the new routes:

> # **/etc/sysconfig/network-scripts/ifup-routes eth0**
> # **ip route**

6. Run the *ip* command and delete both routes:

> # **ip route del default**
> # **ip route del 192.168.3.0/24**

7. Run the *ip* command again to confirm the deletion:

> # **ip route**

8. Remove the */etc/sysconfig/network-scripts/route-eth0* file:

> # **rm /etc/sysconfig/network-scripts/route-eth0**

Managing Routes via the Network Connections Tool

With the Network Connections tool, you can add a new route or modify and delete an existing one. Start the tool by running the *nm-connection-editor* command in an X terminal window or choosing System → Preferences → Network Connections in the GNOME desktop. Highlight an interface and click Edit to bring up the properties of the interface. Go to IPv4 Settings and click Routes to add or modify one or more routes. See Figure 16-1 below and also refer to Chapter 13 "Networking, Network Interfaces & DNS, DHCP and LDAP Clients" for more details on this tool.

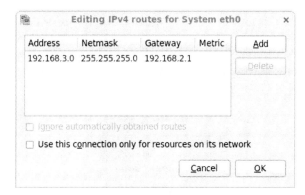

Figure 16-1 Network Connections Tool – Add/Delete Routes

Kerberos

Kerberos is a network authentication protocol that allows systems communicating over unsecure networks to identify themselves to one another securely. The default user authentication method used in RHEL requires a user to enter his password separately for each use of a network service. With this method, unencrypted user passwords are transmitted across unsecure networks. Kerberos, on the other hand, provides a secure authentication mechanism where a user registers only once and is trusted for all network services for a defined period of time, eliminating the frequent need of user password transmission over the unsecure network. In order for Kerberos to function as desired, a functional LDAP server is usually configured and made available on the network to provide the user account services.

With Kerberos, when a client connection is successfully established, a symmetric key (called *ticket*) is generated on a server that has *Authentication Services* (AS) and *Ticket Granting Services* (TGS) running. This server is referred to as the *Key Distribution Centre* (KDC) server or the Admin server and it has the Kerberos *kadmind* daemon running. This server provides Kerberos services to the clients that are in its *realm* (a realm is like a DNS domain). The ticket is valid only for a limited time period and it is granted to the client. The ticket includes credentials such as the client system's name, its IP address, a time stamp, a lifetime, and a random session key. After the ticket has been issued, the user does not have to re-enter his password until the ticket expires or the user logs out and logs back in. This ticket is stored on the client system and all Kerberos-aware service requests use this ticket rather than requiring the user to authenticate using a password. The Kerberos ticketing system relies heavily on resolving hostnames and on accurate timestamps to issue and expire tickets. Therefore, Kerberos requires adequate clock synchronization and a working DNS server to function correctly.

Kerberos uses port 88 by default so you need to ensure that this port is open in the firewall for the authentication traffic to pass through.

Configuring a Kerberos Client Using Authentication Configuration Tool

The Red Hat Authentication Configuration tool is a graphical tool that lets you configure the Kerberos client services on your system. You used this tool previously in Chapter 13 "Networking, Network Interfaces & DNS, DHCP and LDAP Clients" to configure an LDAP client. To configure a Kerberos client, invoke this tool by running the *authconfig-gtk* or the *system-config-authentication* command in an X terminal window (run *authconfig-tui* for its text equivalent) or clicking System → Administration → Authentication in the GNOME desktop. Enter information as shown in the Authentication Configuration section in Figure 16-2. The LDAP data in the User Account Configuration section is the same that you entered when setting up an LDAP client in Chapter 13.

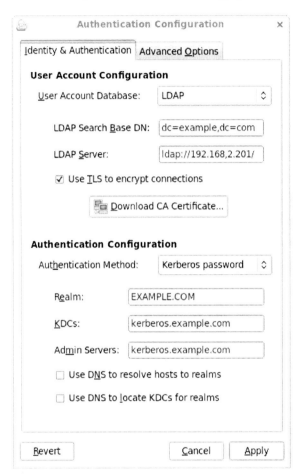

Figure 16-2 Authentication Configuration Tool – Kerberos Authentication Config

Choose Kerberos password as the authentication method for systems in the example.com realm. The KDC and the Admin services are assumed to be running on the same server called kerberos.example.com. Leave the other two options unchecked unless you have a working DNS available and want to resolve hostnames to realms and locate KDCs for realms using the DNS. Click Apply when done. This tool will perform the requested configuration updates in the */etc/nsswitch.conf* and */etc/sssd/sssd.conf* files, set the Kerberos client functionality to start at each system reboot, and will start the *sssd* daemon.

The */etc/nsswitch.conf* and */etc/sssd/sssd.conf* files will look like the following after the above change has occurred:

```
# grep sss /etc/nsswitch.conf
passwd:      files sss
shadow:      files sss
group:       files sss
netgroup:    files sss
```

The *grep* command above extracted the lines from the */etc/nsswitch.conf* file containing the sss entry, which implies that the user authentication information will now be obtained from the LDAP server just configured. This service is provided by the System Security Services Daemon (SSSD).

And the */etc/sssd/sssd.conf* file will have the following uncommented entries related to Kerberos:

```
ldap_id_use_start_tls = True
krb5_realm = EXAMPLE.COM
ldap_search_base = dc=example,dc=com
id_provider = ldap
krb5_kdcip = kerberos.example.com
auth_provider = krb5
chpass_provider = krb5
ldap_uri = ldap://192.168.2.201/
krb5_kpasswd = kerberos.example.com
cache_credentials = True
ldap_tls_cacertdir = /etc/openldap/cacerts
```

Table 16-2 explains these directives.

Directive	Description
krb5_realm	Specifies the Kerberos realm.
ldap_id_use_start_tls	Specifies to use TLS security.
ldap_search_base	Identifies the base DN to search for information.
id_provider	Specifies the authentication service to use.
ldap_uri	Identifies the LDAP server's IP address.
krb5_kdcip	Specifies the KDC hostname.
auth_provider	The name of the authentication provider for the realm.
chpass_provider	Specifies the name of the service to use for password change operations.
krb5_kpasswd	Specifies the hostname of the server that will handle password change operations.
cache_credentials	Determines if user credentials are to be cached locally.
ldap_tls_cacertdir	Identifies the directory location to store security certificates.

Table 16-2 LDAP / Kerberos Directives in the sssd.conf File

Domain Name System

Name resolution is a technique for determining the IP address of a system by providing its hostname. In other words, name resolution is a way of mapping a hostname to its IP address. Name resolution is used on the Internet and on corporate networks. When you enter the address of a website in a browser window, you are actually specifying the hostname of a remote machine that exists somewhere in the world. You may not know its exact location, but you do know its hostname. There is a complex web of hundreds of thousands of routers configured on the Internet. These routers maintain information about other routers closer to them. When you press the Enter key after typing in a website name, the hostname (the website name) is passed to a DNS server, which tries to get the IP address associated with the website's hostname. Once it gets that information, the request to access the website is forwarded to the web server hosting the website from one router to another until the request reaches the destination system. Determining an IP address by providing a hostname is referred to as *name resolution* (a.k.a. *name lookup* or *DNS lookup*), determining a hostname by providing an IP address is referred to as *reverse name resolution* (a.k.a. *reverse name lookup* or *reverse DNS lookup*), and the service employed to perform name resolution is called *Domain Name System* (DNS). DNS is platform-independent and is supported on a wide array of operating systems including RHEL. DNS is commonly recognized as *Berkeley Internet Name Domain* (BIND) in the Linux and UNIX worlds. In fact, BIND is an implementation of DNS on Linux and UNIX platforms, and was developed at the University of California, Berkeley. The terms DNS and BIND are used interchangeably throughout this chapter.

Name Resolution Approaches

There are two common methods employed for hostname resolution, and are explained below.

The /etc/hosts File

The */etc/hosts* file is typically used when there are not too many systems on the network. This file is maintained locally on individual systems, which implies that it must be updated on each system to maintain consistency whenever a system is added or removed, or its IP or hostname is modified.

Domain Name System

DNS is the de facto standard for name resolution used on the Internet and in corporate networks. Systems using DNS send name resolution requests to a DNS server instead of the *hosts* file depending on how the hosts directive is configured in the */etc/nsswitch.conf* file.

DNS Name Space and Domains

The DNS *name space* is a hierarchical organization of all the domains on the Internet. The root of the name space is represented by the dot character. The hierarchy right below the root is divided into top-level (first-level) domains such as com, gov, edu, mil, net, org, biz, tv, info, and two-character country-specific domains such as ca, uk, and fr. A DNS *domain* is a collection of one or more systems. Sub-domains fall under domains. For example, the com domain consists of second-level sub-domains such as redhat, hp, and ibm. Sub-domains can then be further divided into multiple, smaller third-level sub-domains, each of which may contain one or several systems. For example, *redhat.com* may contain a *ca.redhat.com* sub-domain. In short, any number of sub-domains can be defined within a domain.

Figure 16-3 exhibits the hierarchical structure of the DNS name space. It also shows various levels of domains.

At the deepest level of the hierarchy are the *leaves* (systems) of the name space. For example, a system *rhel01* in *ca.redhat.com* will be represented as *rhel01.ca.redhat.com*. If a dot is appended to the name to look like *rhel01.ca.redhat.com.*, it will be referred to as the *Fully Qualified Domain Name* (FQDN) for *rhel01*.

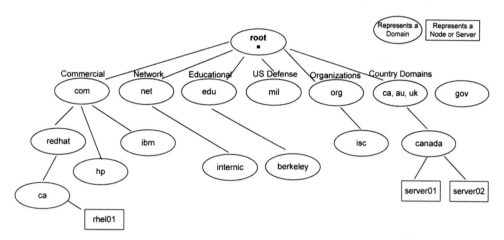 A system in the DNS name space may be a computer, a router, a switch, a network printer, or any other device with an IP address.

The hierarchical structure of the DNS enables the division of a single domain into multiple sub-domains with the management responsibility of each sub-domain delegated to different groups of administrators. This type of configuration allows each sub-domain to have its own DNS server with full authority on the information that the sub-domain contains. This distributed management approach simplifies the overall DNS administration in large and complex environments.

Figure 16-3 DNS Hierarchy

To run a server directly facing the Internet, you must get a domain name registered for it. Contact one of the accredited domain registrars licensed by the *Internet Corporation for Assigned Names and Numbers* (ICANN). Visit *www.icann.org* to obtain a list of licensed registrars or simply contact an ISP to get a domain registered.

DNS Zones and Zone Files

Every DNS server maintains complete information about the portion of the DNS name space it is responsible for. This information includes a complete set of authoritative data files. The portion of the name space for which a DNS server has a complete set of authoritative data files is known as the server's *zone* and the set of authoritative data files is known as the *zone files* or the *zone databases*.

DNS Roles

A role is a function that a system performs from a DNS standpoint. A system is typically configured to function as one of the three types of DNS servers, or as a client.

Master DNS Server

A *master DNS server* has the authority for its domain (or sub-domain) and contains that domain's data. Each domain must have one master server, which may delegate the responsibility of one or more sub-domains to other DNS servers, referred to as slave and caching DNS servers.

Slave DNS Server

A *slave DNS server* also has the authority for its domain and stores that domain's zone files; however, the zone files are copied over from the master server. When updates are made to the zone files on the master server, the slave server gets a copy of the updated files automatically. This type of DNS server is normally set up for redundancy purposes in case the master server fails, and for sharing master server's load. It is recommended to have at least one slave server per domain to supplement the master.

Caching DNS Server

A *caching DNS server* has no authority for any domains. It gets data from a master or slave server and caches it locally in its memory. Like a slave server, a caching server is normally used for redundancy and load-sharing purposes. The replies to queries from a caching server are normally quicker than the replies from either one of the other two types of servers. This is because a caching server keeps data in the memory and not on the disk. This type of DNS server is typically used by ISPs where hundreds of thousands of queries for name resolution hit the DNS servers every minute.

Forwarding DNS Server

A *forwarding DNS server* has no authority for any domains. It simply forwards an incoming query to a specified DNS server.

DNS Client

A *DNS client* has a few files configured that are used for resolving hostname queries by referencing information defined in them.

Overview of BIND Software Packages

There are several software packages that need to be installed on the system in order to configure it as a DNS server or as a DNS client, or both. Some of the packages are optional. Table 16-3 describes the packages.

Package	Description
bind	Contains the BIND software for configuring a master, slave, caching, or forwarding server.
bind-chroot	Includes the directory structure to cordon off BIND in case DNS is compromised. This package is not required for a normal BIND functionality, but recommended.
bind-devel	Consists of BIND development libraries.
bind-dyndb-ldap	Provides the support for dynamic updates to LDAP.
bind-libs	Contains supported library files for bind and bind-utils packages.
bind-utils	Comprises of tools such as *dig* and *host*.

Table 16-3 BIND Software Packages

Understanding DNS Server Configuration and Zone Files

When working with DNS servers, several configuration and zone files are involved. The syntax of most of them is similar. Table 16-4 describes the files.

File	Description
/usr/share/doc/bind-9.7*/sample/	Contains default configuration and zone file templates.
/etc/sysconfig/named	The BIND startup configuration file.
/etc/rc.d/init.d/named	The BIND startup script.
/etc/named.conf	Specifies the location of zones to be served.
/etc/named.iscdlv.key	Defines the standard DNS encryption key.
/etc/named.rfc1912.zones	Defines default zones for a caching server.
/etc/rndc.key	Contains the authentication key.

Table 16-4 DNS Configuration and Zone Files

Let us look at some of these files and see what kind of information they store.

The /etc/sysconfig/named File

This file defines the location of the root directory where the DNS zone files are stored, as well as any options to be passed to the *named* daemon when it is started, and so on. With chroot package installed, the default ROOTDIR is set to */var/named/chroot*. If you wish to store information elsewhere, you will need to specify that directory location here, create that directory, and move files in there. The default contents of this file in a chrooted setting are shown below:

```
# cat /etc/sysconfig/named
. . . . . . . .
ROOTDIR=/var/named/chroot
```

The /etc/named.conf File

This is the primary DNS configuration file and is read each time the DNS server daemon *named* is started or restarted at boot time, or manually after the system is up. This file provides the DNS server with the names and location of zone databases with respect to the ROOTDIR directive for all domains that this server is configured to serve. The file typically contains options, zone, and include statements. Each statement begins with a { and ends in };. Both forward and reverse zone statements can be defined. The default version of this file is for caching-only name server. A sample *named.conf* file is shown below followed by an explanation:

```
# cat /etc/named.conf
options {
listen-on port 53 { 127.0.0.1; 192.168.2.201; };
allow-query { 127.0.0.1; 192.168.2.0/24; };
allow-recursion { 127.0.0.1; 192.168.2.0/24; };
directory "/var/named";
dump-file "/var/named/data/cache_dump.db";
statistics-file "/var/named/data/named_stats.txt";
memstatistics-file "/var/named/data/named_mem_stats.txt";
pid-file "/var/run/named/named.pid";
```

```
        dnssec-enable yes;
        dnssec-validation yes;
        dnssec-lookaside auto;
        bindkeys-file "/etc/named.iscdlv.key";;
        };
        logging {
                channel default_debug {
                        file "data/named.run";
                        severity dynamic;
                };
        };
        zone "." IN {
                type hint;
                file "named.ca";
        };
        include "/etc/named.rfc1912.zones";
```

In this file, comments begin with the // characters or can be enclosed within /* and */. Options can be defined at both global (within the options statement) and individual levels (within individual zone statements) as required. Individual options override the global options. The listen-on directive defines port 53, which the DNS server uses to listen to queries on the localhost interface and the interface configured with the 192.168.2.201 address. The systems, or for that matter, networks or domains defined within { } and separated by the semicolon character are called *match-list*. The allow-query directive defines a match-list, which enables the specified systems to query the DNS server. You can restrict queries to one or more networks by specifying IP addresses of the networks. The recursion directive instructs the DNS server to act as a recursive server. The directory directive identifies the location of the zone files with respect to the ROOTDIR directive. The dump-file and statistics-file directives specify the location for the DNS dump and statistics files. The memstatistics-file specifies the location for the memory usage statistics file. The pid-file directive provides the file name and location to store the *named* daemon's PID. The next four directives dnssec-enable, dnssec-validation, dnssec-lookaside, and bindkeys-file are related to the DNS security. The logging function instructs to log debug messages of dynamic severity to the */var/named/data/named.run* file.

There is one zone statement defined in this file. The . zone defines the root servers for query as defined in the *named.ca* file whose location is with respect to the directory directive setting.

The include directive at the very bottom of the file specifies any additional files to be referenced by *named*. The default entry is for the *named.rfc1912.zones* file.

Exercise 16-2: Configure a Caching DNS Server

In this exercise, you will configure *server.example.com* as a caching DNS server. You will use the default *named.conf* file without any modifications, set the BIND service to autostart at system reboots, open a port in the firewall, and start the BIND service.

1. Check whether the required BIND software packages are installed:

```
# rpm –qa | grep ^bind-*
bind-9.7.3-8.P3.el6.x86_64
bind-libs-9.7.3-8.P3.el6.x86_64
bind-utils-9.7.3-8.P3.el6.x86_64
bind-dyndb-ldap-0.2.0-7.el6.x86_64
```

2. Run the *named-checkconf* command to ensure that there are no syntax errors in *named.conf*:

```
# named-checkconf –v
9.7.3-P3-RedHat-9.7.3-8.P3.el6
```

3. Allow only the BIND daemon to be able to read *named.conf* by altering the SELinux file context to named_conf_t using the *chcon* command:

```
# chcon –t named_conf_t named.conf
```

4. Issue the *semanage* command and modify the contexts on the file to ensure that the new contexts survive a SELinux relabeling:

```
# semanage fcontext –a –t named_conf_t /etc/named.conf
```

5. Set the BIND service to autostart at each system reboot, and validate:

```
# chkconfig named on
# chkconfig list named
named        0:off  1:off  2:on  3:on  4:on  5:on  6:off
```

6. Configure host-based access by allowing BIND traffic on port 53 to pass through the firewall:

```
# iptables –I INPUT –p tcp --dport 53 –j ACCEPT
```

7. Save the rule in the */etc/sysconfig/iptables* file and restart the firewall to activate the new rule:

```
# service iptables save ; service iptables restart
```

8. Start the *named* service and check the running status:

```
# service named start
Starting named:                    [ OK ]
# service named status
named (pid  21967) is running...
```

Exercise 16-3: Configure a Forwarding DNS Server

In this exercise, you will configure *server.example.com* as a forwarding DNS server. You will make a copy of the existing *named.conf* file and modify the original to include forwarding directives.

1. Run step 1 from Exercise 16-2.
2. Make a copy of the */etc/named.conf* file and change */etc/named.conf* to look like:

```
options {
listen–on port 53 { 127.0.0.1; 192.168.2.200 };
        directory "/var/named";
        forwarders { 192.168.2.1;
        forward only; }; }
```

The forwarders directive instructs the server to forward all hostname queries to systems in the order in which they are listed on the match-list. The forward directive instructs the server to contact only the match-list servers for query.

3. Repeat steps 2 to 8 from Exercise 16-2 to complete the setup for a forwarding server.

Exercise 16-4: Configure a Caching DNS Server with Forwarding Capability

In this exercise, you will add the forwarding DNS server capability to the caching DNS server configured in Exercise 16-2. You will edit the *named.conf* file and add the forwarding directive.

1. Run step 1 from Exercise 16-2.
2. Open the original */etc/named.conf* file and modify the options statement to look as follows:

```
options {
listen–on port 53 { 127.0.0.1; 192.168.2.200 };
        directory "/var/named";
        forwarders { 192.168.2.1;
```

Leave all other directives in the file intact.

3. Repeat steps 2 to 7 from Exercise 16-2 to complete the setup for a forwarding server.

Verifying DNS Functionality Using the dig Command

dig (domain information groper) is a hostname lookup utility. It queries the nameservers listed in the *resolv.conf* file for lookups. If the nameservers are unable to fulfill the query, it contacts one of the root DNS servers listed in the *named.ca* file and tries to obtain the information from there. This command references the *host.conf* file for source order determination.

To perform a forward lookup to obtain the IP address of *www.redhat.com*:

 # dig www.redhat.com

```
. . . . . . . .
;www.redhat.com.                IN      A
www.redhat.com.        60    IN      CNAME  wildcard.redhat.com.edgekey.net.
wildcard.redhat.com.edgekey.net. 4704 IN CNAME  wildcard.redhat.com.edgekey.net.globalredir.akadns.net.
wildcard.redhat.com.edgekey.net.globalredir.akadns.net. 42 IN CNAME e1890.b.akamaiedge.net.
e1890.b.akamaiedge.net.  18    IN      A       184.29.119.214
;; Query time: 21 msec
;; SERVER: 192.168.2.1#53(192.168.2.1)
;; WHEN:   Thu Nov 1 14:51:54 2012
;; MSG SIZE rcvd: 191
```

To perform a reverse lookup on *www.example.com*, use the –x option with the command:

dig –x www.example.com
....
;com.redhat.www.in-addr.arpa.   IN   PTR
;; Query time: 4 msec
;; SERVER: 192.168.2.1#53(192.168.2.1)
;; WHEN: Thu Nov 1 14:52:27 2012
;; MSG SIZE rcvd: 45

To query a name server directly, specify either the nameserver's hostname or IP address:

dig @127.0.0.1 www.redhat.com
dig @192.168.2.1 www.redhat.com
....
;www.redhat.com.   IN   A
www.redhat.com.   60   IN   CNAME wildcard.redhat.com.edgekey.net.
wildcard.redhat.com.edgekey.net. 4640 IN CNAME wildcard.redhat.com.edgekey.net.globalredir.akadns.net.
wildcard.redhat.com.edgekey.net.globalredir.akadns.net. 879 IN CNAME e1890.b.akamaiedge.net.
e1890.b.akamaiedge.net. 20   IN   A   184.29.119.214

Verifying DNS Functionality Using the host Command

host is another hostname lookup utility. It queries the nameservers listed in the *resolv.conf* file for lookups. If the nameservers are unable to fullfil the query, it contacts one of the root DNS servers listed in the *named.ca* file and tries to obtain the information from there. This command references the *host.conf* file for source order determination.

To perform a forward lookup to obtain the IP address of *www.redhat.com*:

host www.redhat.com
www.redhat.com is an alias for wildcard.redhat.com.edgekey.net.
wildcard.redhat.com.edgekey.net is an alias for wildcard.redhat.com.edgekey.net.globalredir.akadns.net.
wildcard.redhat.com.edgekey.net.globalredir.akadns.net is an alias for e1890.b.akamaiedge.net.
e1890.b.akamaiedge.net has address 184.29.119.214

To perform a reverse lookup to obtain hostname of the IP address 184.29.119.214:

host 184.29.119.214
214.119.29.184.in-addr.arpa domain name pointer a184-29-119-214.deploy.akamaitechnologies.com.

To obtain detailed information, use the –v option with the command:

host –v www.redhat.com
....
;www.redhat.com.   IN   A
www.redhat.com.   60   IN   CNAME wildcard.redhat.com.edgekey.net.
wildcard.redhat.com.edgekey.net. 10186 IN CNAME wildcard.redhat.com.edgekey.net.globalredir.akadns.net.
wildcard.redhat.com.edgekey.net.globalredir.akadns.net. 131 IN CNAME e1890.b.akamaiedge.net.

```
e1890.b.akamaiedge.net. 20     IN    A        184.29.119.214
Received 191 bytes from 192.168.2.1#53 in 112 ms
Trying "e1890.b.akamaiedge.net"
;; ->>HEADER<<- opcode: QUERY, status: NOERROR, id: 21776
;; flags: qr rd ra; QUERY: 1, ANSWER: 0, AUTHORITY: 1, ADDITIONAL: 0
;; QUESTION SECTION:
;e1890.b.akamaiedge.net.          IN    AAAA
........
Received 101 bytes from 192.168.2.1#53 in 28 ms
```

Chapter Summary

This chapter introduced you to the concepts of routing, secure authentication mechanism, and DNS. You learned how data packets were routed and how default and static routes were added to the system using commands and a graphical tool. You configured a Kerberos-based authentication client. You studied how name resolution worked. You learned about DNS name space, domains, zones, zone files, roles, and software packages. You ran exercises to set up a caching and a forwarding DNS server. You used commands to verify their functionality.

Chapter Review Questions

1. Name two commands for hostname lookup?
2. Which command can be used to view the routing table?
3. What would *passwd: files sss* imply?
4. DNS is also referred to as BIND. True or False?
5. What is the function of the KDC?
6. What would the command *ip route delete 192.168.4.0/24* do?
7. What is the name of the file where you would add routes for the *eth0* interface?
8. Which two types of DNS servers do not have authority for any domains?
9. Can you use the Network Connections tool to add and delete routes. Yes or No?
10. What would the command *host 10.22.112.22* do?
11. Kerberos is a local authentication protocol. True or False?
12. Name one of the commands that you can issue to run the Authentication Configuration tool?
13. Name the four DNS server roles.
14. What is the purpose of the forward directive?
15. Name the hardware device that is used to move data packets between networks.
16. Which is the main configuration file for DNS?
17. What would the command *ip route* do?
18. Define name resolution in one sentence.
19. What is the name of the file where you list the DNS servers for query?
20. What is a default route?
21. Is the DNS name space hierarchical or flat?
22. What is a realm in the Kerberos terminology?
23. What is a fully qualified domain name?
24. What is a good method for hostname resolution on a local network?
25. What is the function of the forwarders directive in the *named.conf* file?

Answers to Chapter Review Questions

1. The *host* and *dig* commands.
2. The *ip route* command.
3. It will look for user credentials in the local passwd file and then in Kerberos.
4. True.
5. KDC is reponsible for generating tickets.
6. This command will delete the specified route.
7. The name of the file to add routes for *eth0* would be called *route-eth0*.
8. The caching and forwarding DNS servers.
9. Yes.
10. This command will perform a reverse lookup on the IP address.
11. False.
12. The *system-config-authentication* command.
13. The four DNS server roles are master, slave, caching, and forwarding.
14. The purpose of the forward directive is to command the server to contact the specified nameservers for query and fall back to a root nameserver if an answer cannot be found.
15. The hardware device that is used to move data packets between networks is called a router.
16. The */etc/named.conf* file.
17. This command will display the routing table.
18. Name resolution resolves hostname to IP address, and vice versa.
19. The */etc/resolv.conf* file.
20. The default route is used to route packets when no other routes are present.
21. Hierarchical.
22. A realm is like a domain in Kerberos.
23. A fully qualified domain name is a hostname with its domain name added to it.
24. The use of the */etc/hosts* file.
25. The purpose of the forwarders directive is to instruct the server to forward all hostname queries to systems in the order in which they are listed on the match-list.

Labs

Lab 16-1: Populate the hosts File

Populate the */etc/hosts* file so that it has FQDN and alias entries for *insider*, *server*, and *physical* systems for IP addresses defined in Chapter 01.

Lab 16-2: Configure DNS Client Service

Update DNS client configuration files on *insider.example.com* to include appropriate nameserver, domain, and search directives.

Lab 16-3: Set Up a Caching-Only DNS Server with Forwarding Ability

Set up the caching-only DNS service on *physical.example.com* and configure it to relay incoming queries to *insider.example.com*.

## Chapter 17

Internet Services, Kernel Parameters & NTP

This chapter covers the following major topics:

✓ Introduction to the Internet services
✓ The role of the Internet services daemon
✓ Internet services configuration files
✓ Enable and disable xinetd-controlled services
✓ Overview of kernel parameters
✓ Modify kernel parameters
✓ NTP concepts and components
✓ Configure an NTP server, peer, and client
✓ Use ntpdate to update system clock
✓ Query NTP servers

This chapter includes the following RHCE objectives:

58. Use /proc/sys and sysctl to modify and set kernel runtime parameters
NTP: Synchronize time using other NTP peers

RHEL Internet services enable a system to function as a provider of one or more services

to remote client systems over the network. These services include allowing users on the remote systems to log in, transfer files, send and receive mail, execute commands without logging in, get a list of logged in users, synchronize time, and so on.

The behaviour of the default kernel installed during the installation may be changed by modifying the values of one or more of its parameters. There are several kernel parameters that may be tuned for a smooth installation and proper operation of certain applications and database software.

The Network Time Protocol service maintains the clock on the system and keeps it synchronized with a more accurate and reliable source of time. Providing accurate and uniform time for systems on the network allows time-sensitive applications such as backup software, job scheduling tools, and billing systems to perform correctly and precisely. It also aids the logging daemon to capture information accurately in the log files.

The Internet Services

When services are accessed and used over the network, two software pieces called *client* and *server* are involved. The client program requests for a service running on a remote system. The server program on the remote system responds to and serves the client request, thus establishing a client/server communication channel. A single system can act as both a server and a client. It can provide services to other systems and may use their services as a client.

Server programs are started in one of two ways: via startup scripts located in sequencer directories or via the master server program daemon called *xinetd* (the *extended internet daemon*). *xinetd* is also referred to as the *super-server*. A detailed discussion on the startup scripts and sequencer directories was covered in Chapter 08 "Linux Boot Process & Kernel Management".

The *xinetd* daemon itself is started by one of the startup scripts when the system boots up into run level 3. This daemon reads its configuration file */etc/xinetd.conf* and sits in the memory listening on ports listed in the */etc/services* and */etc/rpc* files for services defined and enabled in the */etc/xinetd.d* directory. It waits for a client request for one of the *xinetd*-controlled services. When one such request arrives on a particular port, this daemon starts the server program corresponding to that port. It hooks the client and server pieces up, gets itself out of that communication, and starts listening on that port again on behalf of that server program. Every service uses a unique port number as defined in the *services* or the *rpc* file.

Common Internet Services

There are two major categories of the Internet services: one developed at the University of California at Berkeley, referred to as the *Berkeley* services, and the other developed for the US Department of Defense's *Advanced Research Projects Agency*, referred to as the *ARPA* services.

The Berkeley services include BIND, sendmail, *finger, rexec, rcp, rlogin, rsh, rup, ruptime, rwho,* etc. These services were primarily developed to run on the UNIX platform; however, most of them today are also available on non-UNIX platforms such as Linux and Microsoft Windows. The names of most Berkeley services begin with an "r" and, therefore, they are also referred to as r commands.

The ARPA services, on the other hand, include *ftp* and *telnet*, which can be used on both Linux and non-Linux platforms.

Table 17-1 lists and describes some common Internet services.

Service	Description
fingerd	Enables the *finger* command to display information about users on local and remote systems. The *fingerd* daemon is started by *xinetd* and it uses port 79.
ftpd	Enables file transfer with other Linux and non-Linux systems. The client program is *ftp* or one of its variants. The *ftpd* daemon is started by *xinetd* and it uses port 21.
named	Enables hostname-to-IP and IP-to-hostname resolution. The *named* daemon responds to lookup commands such as *dig* and *host*.
ntpd	Keeps the system clock in sync with a more reliable and accurate time source. This service uses port 123.
rcp	Enables you to perform a file copy. This protocol uses port 469.
rexecd	Enables executing a command on a remote system. The daemon is started by *xinetd*.
rsync	Enables you to perform a file copy using the differential method. This daemon is started by *xinetd* and it uses port 873.
sendmail	Sendmail is a widely used mail transport agent on large networks and the Internet. It uses port 25.
telnetd	Enables a user to log in to a remote system. The client program is *telnet*. *telnetd* is started by *xinetd* and it uses port 23.
tftpd	Works with *bootps* to enable diskless systems and Kickstart clients to transfer files containing boot and other configuration data. The client program is *tftp*. *tftpd* is started by *xinetd* and it uses port 69.

Table 17-1 Internet Services

The xinetd Daemon, /etc/xinetd.conf File and /etc/xinetd.d Directory

The *xinetd* daemon is the master daemon for many Internet services. This daemon is started automatically at system boot via the */etc/rc.d/init.d/xinetd* script. It listens for client connection requests on behalf of the services listed in the */etc/xinetd.d* directory and in the */etc/xinetd.conf* file over the ports defined in the *services* and *rpc* files, and starts up the relevant server process when a request arrives. Any time a change is made to *xinetd.conf* or a file in the *xinetd.d* directory, or a file is added to or removed from the *xinetd.d* directory, you will need to reload the configuration using the following command:

service xinetd reload

And, if the daemon needs to be started, stopped, or restarted manually, execute the above and pass the start, stop, or restart argument as appropriate. You also need to ensure that the daemon is configured to autostart at each system reboot by running the following command:

chkconfig xinetd on

The default contents of the *xinetd.conf* file look similar to the following:

cat /etc/xinetd.conf

. . . .

defaults
{
\# The next two items are intended to be a quick access place to
\# temporarily enable or disable services.
\#
\#   enabled   -
\#   disabled   -
\# Define general logging characteristics.
        log_type   = SYSLOG daemon info
        log_on_failure = HOST
        log_on_success = PID HOST DURATION EXIT
\# Define access restriction defaults
\#
\#   no_access   -
\#   only_from   -
\#   max_load   = 0
        cps   = 50 10
        instances   = 50
        per_source = 10
\# Address and networking defaults
\#
\#   bind   -
\#   mdns   = yes
        v6only   = no
\# setup environmental attributes
\#
\#   passenv   -
        groups   = yes
        umask   = 002
\# Generally, banners are not used. This sets up their global defaults
\#
\#   banner   -
\#   banner_fail -
\#   banner_success -
}
includedir /etc/xinetd.d

Directives defined in the default section of the file are applicable to all services listed in the *xinetd.d* directory; however, any directive that is also defined in an *xinetd.d* directory configuration file, overrides it. Here is an explanation of key directives defined in the *xinetd.conf* file:

✓   The defaults directive encloses all the default settings that are available to *xinetd*-controlled services at startup. The begin and end for this directive are marked with the curly braces { }.
✓   The enabled and disabled directives should not be uncommented unless you wish to perform some sort of testing, in which case you will need to specify names of services with these directives and uncomment them as appropriate. This will enable or disable services based on the setting of the disable directive in individual service scripts defined in the *xinetd.d*

directory. When the testing is finished, comment them out so that the enable and disable decision is made based on the disable directive setting defined within each individual script in the *xinetd.d* directory.

✓ The logging related information is controlled via three directives: log_type, log_on_failure, and log_on_success. The log_type directive directs the daemon to use *rsyslogd* for capturing connection information messages generated by the daemon facility in the */var/log/messages* file. The log_on_failure directive specifies to capture hostname or IP address of the system from where a failed login attempt is made. The log_on_success directive defines to capture the process ID of the process, the hostname or IP address, the duration of the service connection, and the service termination time if the connection was established successfully. An example successful entry from the *messages* file is shown below for reference:

```
Nov 10 12:19:39 physical xinetd[2815]: EXIT: ssh status=1 pid=2889 duration=192 (sec)
```

✓ The next set of directives defines the default access restrictions. With the no_access and only_from directives, you can specify one or more hostnames or IP addresses associated with hosts, networks, or domains from where you want incoming service requests allowed or denied. This level of access may also be defined at the individual service level in the *xinetd.d* directory scripts, and is evaluated after access is granted by TCP Wrappers. The max_load directive defines a load average number. After this number has been reached, the *xinetd* daemon no longer accepts new connections. The cps directive limits incoming connections to 50 per second. If this limit is surpassed, the daemon waits for 10 seconds before restarting to entertain new requests. The instances and per_source directives control the number of active connections per service; the default is set to 50 from not more than 10 sources.

✓ The bind, mdns, and v6only directives set address and networking defaults. With the bind directive, you can specify the IP address of one of the system's interfaces to be used for listening for incoming requests; the mdns directive is unavailable for now, and the third directive can be set to either yes to restrict to the use of IPv6 addresses only or no to allow both IPv4 and IPv6 addresses.

✓ The next set of directives – passenv, groups, and umask – defines environmental attributes. With passenv, global variables are specified and their values are used by the service when an incoming request attempts to establish a connection into that service. The groups directive can have one or more groups set as defined in the */etc/group* file, which provides the owner of a starting service access to those groups. The umask directive sets the default user mask.

✓ The includedir directive indicates the directory where service configuration files are located. The default is */etc/xinetd.d*.

The following output lists files located in the *xinetd.d* directory for each service currently installed and managed by *xinetd*:

 # ll /etc/xinetd.d

```
. . . . . . . .
-rw-r--r--. 1 root root    332 May 20  2009 rsync
-rw-------. 1 root root 1212 Jul   25 06:40 tcpmux-server
-rw-------. 1 root root 1149 Jul   25 06:40 time-dgram
-rw-------. 1 root root 1150 Jul   25 06:40 time-stream
```

All services are disabled by default. Here are the contents of the *rsync* service file with some added directives:

cat /etc/xinetd.d/rsync
```
# default: off
# description: The rsync server is a good addition to an ftp server, as it \
#       allows crc checksumming etc.
service rsync
{
    disable           = yes
    flags             = IPv6
    socket_type       = stream
    wait              = no
    user              = root
    server            = /usr/bin/rsync
    server_args       = --daemon
    log_on_failure  +  = USERID
}
```

There are eight directives defined as shown in the output above. The disable directive disables (yes) or enables (no) the service. The flags directive specifies how the service is to be used. The default is REUSE, which forces the service to be used continuously. The socket_type directive defines the type of socket to be used for communication; the wait directive with "no" value allows several application connections to occur concurrently; the user directive specifies the owner of the service; the server and server_args directives specify the absolute path of the service and any associated arguments; and the log_on_failure directive was explained earlier in the *xinetd.conf* file explanation.

Enabling an xinetd-controlled Service

Use the *chkconfig* command to enable an xinetd-controlled service to autostart at each system reboot. This command changes the value for the disable directive from yes to no in the */etc/xinetd.d/rsync* file.

The /etc/services and /etc/rpc Files

The two key files that define service names and associated ports and protocols that they use are the */etc/services* and */etc/rpc* files. Line entries for some of the Internet services are extracted from the *services* file and displayed below. The output shows service names in the first column along with port numbers and protocols in the second column, aliases in the third column, and comments in the last column.

cat /etc/services
```
. . . . . . . .
ftp         21/tcp
ftp         21/udp      fsp fspd
ssh         22/tcp                  # The Secure Shell (SSH) Protocol
ssh         22/udp                  # The Secure Shell (SSH) Protocol
telnet      23/tcp
telnet      23/udp

. . . . . . . .
```

Line entries for some RPC-based services are extracted from the *rpc* file and displayed below. It shows service names in the first column, followed by port numbers and associated aliases.

 # cat /etc/rpc

```
. . . . . . . .
rstatd       100001  rstat rup perfmeter rstat_svc
rusersd      100002  rusers
. . . . . . . .
```

Kernel Parameters

There are hundreds of kernel parameters whose values affect the overall behaviour of the kernel, and the running system in general. Runtime values of these parameters are stored in files in the */proc/sys* directory and can be altered on the fly by changing associated files. This change will remain in effect until either the value is readjusted or the system is rebooted. This temporary change can be made using either the *sysctl* or the *echo* command. To make the change persistent across reboots, the values will have to be defined in the */etc/sysctl.conf* file.

The default uncommented lines from the *sysctl.conf* file are displayed below:

 # cat /etc/sysctl.conf
```
net.ipv4.ip_forward = 0
net.ipv4.conf.default.rp_filter = 1
net.ipv4.conf.default.accept_source_route = 0
kernel.sysrq = 0
kernel.core_uses_pid = 1
net.ipv4.tcp_syncookies = 1
net.bridge.bridge-nf-call-ip6tables = 0
net.bridge.bridge-nf-call-iptables = 0
net.bridge.bridge-nf-call-arptables = 0
kernel.msgmnb = 65536
kernel.msgmax = 65536
kernel.shmmax = 68719476736
kernel.shmall = 4294967296
```

In addition to these kernel parameters, there are many others that are automatically set when the kernel is loaded at system boot. Run the *sysctl* command with the –a option to list the currently loaded parameters:

 # sysctl –a
```
kernel.sched_child_runs_first = 0
kernel.sched_min_granularity_ns = 1000000
kernel.sched_latency_ns = 5000000
kernel.sched_wakeup_granularity_ns = 1000000
. . . . . . . .
```

Exercise 17-1: Modify Kernel Parameters

In this exercise, you will set the value of the net.ipv4.tcp_abc kernel parameter to 5. You will activate the new value to take effect right away and also ensure that the value is persistent across system reboots.

1. Run the *sysctl* command to view the current value of the net.ipv4.tcp_abc kernel parameter. You can also *cat* the contents of the */proc/sys/net/ipv4/tcp_abc* file where the value of this parameter is stored.

   # **sysctl net.ipv4.tcp_abc**
   net.ipv4.tcp_abc = 0

2. Use the *sysctl* command to change the value from 0 to 5. You can also use the *echo* command and directly change the value in the parameter file.

   # **sysctl –w net.ipv4.tcp_abc=10**
   net.ipv4.tcp_abc = 5
   # **echo 5 > /proc/sys/net/ipv4/tcp_abc**

3. Edit the */etc/sysctl.conf* file and append the following line entry for this value to remain persistent across system reboots:

   net.ipv4.tcp_abc = 5

4. Run the *sysctl* command again to load the new value if it has not been loaded previously:

   # **sysctl –p**
   net.ipv4.tcp_abc = 10

Network Time Protocol

Network Time Protocol (NTP) is a networking protocol for syncing the system clocks with a more accurate and reliable time source. Having precise time on the networked systems allows time-sensitive applications such as billing applications and network logging, to function accurately. NTP uses the UDP protocol over port 123. In order to understand NTP, a discussion of NTP components and roles is imperative.

Time Source

A *time source* is a server that synchronizes its time with *Universal Coordinated Time* (UTC). Care should be taken when choosing a time source for a network and preference be given to a time server that is physically close and takes the least amount of time to send and receive NTP packets.

The most common time sources used are the local system clock, an Internet-based public time server, and a radio clock.

Local System Clock

You can arrange for one of the RHEL systems to function as a provider of time for other systems. This requires the maintenance of correct time on this local server either manually or automatically via the *cron* daemon. Keep in mind, however, that since this server is using its own system clock, it has no way of synchronizing itself with a more reliable and accurate external time source. Therefore, using a local system that relies on its own clock as a time server is the least recommended option.

Internet-Based Public Time Server

Several public time servers that can be employed for the provision of correct time are available via the Internet. One of the systems on the network must be connected to one or more such time servers. To make use of such a time source, a port in the firewall may need to be opened to allow for the flow of NTP traffic. Internet-based public time servers are typically operated by government organizations and universities. This option is more popular than using a local time server.

Several public time servers are available for access on the Internet. Visit *www.ntp.org* for a list.

Radio Clock

A radio clock is considered the most accurate provider of time. Some popular radio clock methods include *Global Positioning System* (GPS), *National Institute of Science and Technology* (NIST) radio broadcasts in the Americas, and DCF77 radio broadcasts in Europe. Of these, GPS-based sources are the most accurate. In order to use them, some hardware must be added to one of the local systems on the network.

Stratum Levels

As you are aware, there are numerous time source selection choices available to synchronize the system time with. These time sources are categorized into *stratum levels* based on their reliability and accuracy. There are 15 stratum levels ranging from 1 to 15 with 1 being the most accurate. The radio clocks are at stratum 1 as they are the most accurate. Stratum 1 time sources, however, cannot be used on a network directly. Therefore, one of the systems on the network at stratum 2, for instance, needs to be configured to get time updates from a stratum 1 server. The stratum 2 server then acts as the primary source of time for secondary servers and/or clients on the network. It can also be configured to provide time to stratum 3 or lower-reliability time servers.

If a secondary server is also configured to get time from a stratum 1 server directly, it will act as a peer to the primary server.

NTP Roles

A role is a function that a system performs from an NTP standpoint. A system can be configured to assume one or more of the following roles:

Primary NTP Server

A *primary* NTP server gets time from one of the time sources mentioned above and provides time to one or more secondary servers or clients, or both. It can also be configured to broadcast time to secondary servers and clients.

Secondary NTP Server

A *secondary* NTP server receives time from a primary server or directly from one of the time sources mentioned above. It can be used to provide time to a set of clients to offload the primary server or as a redundant time server when the primary server becomes inaccessible. Having a secondary server is optional, but highly recommended. It can be configured to broadcast time to clients and peers.

NTP Peer

An NTP *peer* provides time to an NTP server and receives time from that server. They usually work at the same stratum level. Both primary and secondary servers can be peers of each other.

NTP Client

An NTP *client* receives time from either a primary or a secondary time server. A client can be configured in one of the following ways:

- ✓ As a *polling* client that contacts a primary or secondary NTP server directly to get time updates to synchronize its system clock with.
- ✓ As a *broadcast* client that listens to time broadcasts by a primary or secondary NTP server. A broadcast client binds itself with the NTP server that responds to its requests and synchronizes its clock with it. The NTP server must be configured in broadcast mode in order for the broadcast client to be able to bind to it.

NTP Configuration File

The key configuration file for the NTP service is */etc/ntp.conf*. Some key directives from this file are displayed below:

```
driftfile /var/lib/ntp/drift
restrict default kod nomodify          notrap nopeer noquery
restrict 192.168.1.0 mask 255.255.255.0 nomodify notrap
server 0.rhel.pool.ntp.org
server 1.rhel.pool.ntp.org
server 2.rhel.pool.ntp.org
broadcast 192.168.1.255 autokey        # broadcast server
broadcastclient                        # broadcast client
server 127.127.1.0                     # local clock
fudge  127.127.1.0   stratum  10
peer
```

Table 17-2 describes these directives.

Directive	Description
driftfile	Specifies the location of the driftfile (default is */var/lib/ntp/drift*). This file helps the *ntpd* daemon keep track of local system clock accuracy.
restrict	Sets restrictions on incoming NTP queries. Several default options are defined with the restrict directive including kod, nomodify, notrap, nopeer, and noquery. The kod option prevents the server from being attacked with the *kiss of death* packets, which might cause the server to crash; the nomodify option disallows any modification attempts by other NTP servers; the notrap option disables control messages from being captured; the nopeer option prevents other NTP servers from establishing a peer relationship with this server; and the noquery option disallows all queries coming in from other NTP servers for time synchronization.
	The second restrict directive sets the default restrictions on time requests that are coming in from systems on the 192.168.2 network with the default class C netmask. It disallows modification attempts and disables kiss of death packets. The limited option is necessary with kod in order to activate it.
server	Specifies the hostname or IP address of the time server.
broadcast	Specifies the hostname or IP address of the time server that broadcasts time on the network.
broadcastclient	Enables (yes) or disables (no) a client to receive time updates from a broadcast server.
fudge	Defines the stratum level.
peer	Specifies hostname or IP address of the peer NTP server.

Table 17-2 Description of /etc/ntp.conf Directives

Exercise 17-2: Configure an NTP Server

In this exercise, you will set up *server.example.com* as an NTP server syncing time to its own local clock and providing time to client systems on the network. You will use the IP 127.127.1.0 which is reserved for this purpose. You will define this time source at stratum 10. You will open port 123 in the firewall.

1. Check whether the NTP software package is installed:

    # rpm –qa | grep ^ntp
    ntp-4.2.4p8-2.el6.x86_64
    ntpdate-4.2.4p8-2.el6.x86_64

    Install the packages if they are not already installed using one of the procedures outlined in Chapter 06 "Package Management".

2. Replace all the uncommented lines in the */etc/ntp.conf* file with the following:

    driftfile /var/lib/ntp/drift
    restrict default kod nomodify notrap nopeer noquery
    server 127.127.1.0
    fudge 127.127.1.0 stratum 10

3. Ensure that the */etc/sysconfig/clock* file has a proper timezone value defined in it:

   # cat /etc/sysconfig/clock
   ZONE="America/Toronto"

4. Set the NTP service to autostart at each system reboot, and validate:

   # chkconfig ntpd on
   # chkconfig --list ntpd
   ntpd   0:off   1:off   2:on   3:on   4:on   5:on   6:off

5. Configure host-based access by allowing NTP traffic on UDP port 123 to pass through the firewall:

   # iptables –I INPUT –p udp --dport 123 –j ACCEPT

6. Save the rule in the */etc/sysconfig/iptables* file and restart the firewall to activate the new rule:

   # service iptables save ; service iptables restart

7. Start the *ntpd* daemon and check the running status:

   # service ntpd start
   # service ntpd status
   ntpd (pid 3241) is running...

Exercise 17-3: Configure an NTP Peer

In this exercise, you will configure *physical.example.com* as a peer of NTP server *server.example.com*.

1. Perform step 1 from Exercise 17-2.
2. Replace all the uncommented lines in the */etc/ntp.conf* file with the following:

   driftfile /var/lib/ntp/drift
   restrict default kod nomodify notrap nopeer noquery
   peer 192.168.2.200

3. Run steps 3 to 7 from Exercise 17-2 to complete the peer setup.

Exercise 17-4: Configure an NTP Client

In this exercise, you will configure *insider.example.com* as an NTP client of *server.example.com*.

1. Perform step 1 from Exercise 17-2.
2. Replace all the uncommented lines in the */etc/ntp.conf* file with the following:

```
driftfile /var/lib/ntp/drift
restrict default kod nomodify notrap nopeer noquery
server 192.168.2.201
```

3.   Run steps 3 to 7 from Exercise 17-2 to complete the NTP client setup.

Configuring an NTP Client Using the Date/Time Properties Tool

The NTP client service can be set up using the graphical Date/Time Properties tool. To set up *insider.example.com* as an NTP client of *server.example.com* NTP server, execute *system-config-date* in an X terminal window or choose System → Administration → Date & Time in the GNOME desktop on *insider.example.com*. The Date/Time Properties tool screen will open up as shown in Figure 17-1.

Figure 17-1 Date/Time Properties Tool – NTP Client Configuration

Select Synchronize date and time over the network and add the IP address of *server.example.com* under NTP Servers. Remove the three public NTP servers listed there by default. Leave other options to their default settings. Click OK when done.

This completes the setup for an NTP client.

Updating System Clock Manually

You can run the *ntpdate* command anytime to bring the system clock close to the time on an NTP server. The *ntpd* daemon must not be running when this command is executed to obtain the desired results. Run *ntpdate* manually and specify either the hostname or the IP address of the time server. For example, to bring the clock on *insider.example.com* at par with the clock on *server.example.com* instantly, run the command on *insider.example.com* as follows:

 # ntpdate server.example.com
2 Nov 15:55:55 ntpdate[7313]: adjust time server server.example.com offset –0.020116 sec

Querying NTP Servers

NTP servers may be queried for time synchronization and server association status using the *ntpq* command. This command sends out requests to and receives responses from NTP servers. The command may be run in an interactive mode.

Run *ntpq* with the –p option to print a list of NTP servers known to the system along with a summary of their status:

```
# ntpq –p
     remote          refid            st  t  when  poll  reach  delay  offset  jitter
===========================================================================
*192.168.2.201  server.example.com   2   u   110   512   377   24.22  –0.201  0.000
```

Output resulted above is explained in Table 17-3.

Column	Description
remote	Shows IP addresses or hostnames of NTP servers and peers. Each IP/hostname may be preceded by one of the following characters: * Indicates the current source of synchronization. # Indicates the server selected for synchronization, but distance exceeds the maximum. o Displays the server selected for synchronization. + Indicates the system considered for synchronization. x Designated false ticker by the intersection algorithm. . Indicates the systems picked up from the end of the candidate list. - Indicates the system not considered for synchronization. Blank Indicates the server rejected because of high stratum level or failed sanity checks.
refid	Shows the sources of synchronization.
st	Shows the stratum level of the servers.
t	Shows available types: l = local (such as a GPS clock), u = unicast, m = multicast, b = broadcast, and - = netaddr (usually 0).
when	Shows time, in seconds, since a response was received from the server.
poll	Shows a polling interval. Default is 64 seconds.
reach	Shows the number of successful attempts to reach the server. The value 001 indicates that the most recent probe was answered, 357 indicates that one probe was not answered, and the value 377 indicates that all recent probes had been answered.
delay	Shows how long, in milliseconds, it took for the reply packet to return in response to a query sent to the server.
offset	Shows the difference in time, in milliseconds, between the server and the client clocks.
jitter	Shows how much the offset measurement varies between samples. This is an error-bound estimate.

Table 17-3 ntpq Command Output Description

Chapter Summary

This chapter introduced you to the Internet services, kernel tuning, and the Network Time Protocol service.

You touched upon the common Internet services and the super-server daemon. You analyzed various configuration files and reviewed the directives defined in them. You looked at how to activate and deactivate the Internet services.

You reviewed kernel parameters and saw how to modify their values using different methods.

You learned the concept of NTP and benefits associated with using it. You examined the components that are critical to the overall working of the NTP setup. You performed exercises to configure an NTP server, a peer, and a client. Finally, you saw how to update the system clock manually and query NTP servers.

Chapter Review Questions

1. What would the command *sysctl –p* do?
2. What is the name of the super-server daemon?
3. What are the three common sources for obtaining time?
4. Kernel parameters cannot be modified in multiuser mode. True or False?
5. Which directory contains the Internet service configuration files that are xinetd-controlled?
6. Which protocol and port does the NTP service use?
7. Can an NTP server be configured to broadcast time on the network? True or False?
8. What is the purpose of setting the include directive in the *xinetd* configuration file?
9. Write the difference between an NTP peer and an NTP client?
10. Write the command that would list the current source of time synchronization.
11. What is the name of the file where you can define kernel parameters?
12. What is the purpose of a driftfile?
13. What would the command *chkconfig xinetd on* do?
14. Define stratum levels.
15. What two commands can you use to alter a kernel parameter's value?
16. What would the command *ntpdate server.ntp.org* do?
17. What two files contain information about services and the ports that they listen on?
18. What would the command *sysctl –w net.ipv4.tcp_abc=3* do?
19. What is the de facto service used on the Internet and corporate networks for time synchronization?
20. Which graphical tool may be used to configure an NTP client?

Answers to Chapter Review Questions

1. This command will load the kernel parameters defined in the */etc/sysctl.conf* file.
2. The name of the super-server daemon is *xinetd*.
3. The local clock, a public time server, and the radio clock.
4. False.
5. The */etc/xinetd.d* directory.
6. The NTP service uses the UDP protocol and port 123.
7. True.

---

8. With the include directive you can specify the location of the *xinetd*-controlled service files.
9. An NTP peer provides time while an NTP client obtains time from it.
10. The *ntpq* command with the –p switch.
11. The */etc/sysctl.conf* file.
12. The driftfile is used for keeping track of local system clock accuracy.
13. This command will set the autostart flag to on for the *xinetd* service.
14. Stratum levels determine the reliability and accuracy of a time source.
15. You can use either the *sysctl* or the *echo* command.
16. This command will instantly synchronize the system's time with the specified server's.
17. The two files are */etc/services* and */etc/rpc*.
18. This command will write the specified kernel parameter and its value in the */etc/sysctl.conf* file.
19. The NTP service.
20. The Date/Time Properties tool.

Labs

Lab 17-1: Modify a Kernel Parameter

Modify the value of the kernel parameter called kernel.msgmax to 131072 and ensure that the value survive system reboots.

Lab 17-2: Configure the Network Time Protocol Service

Obtain a list of public time servers from *www.ntp.org* and configure *insider.example.com* as a peer of one of them. Configure NTP client service on *physical.example.com* to sync its clock with the NTP time server configured on *insider.example.com*. Run appropriate commands to validate the bonding between the NFT client and the NTP server.

Chapter 18

Electronic Mail

This chapter covers the following major topics:

- ✓ Introduction to electronic mail
- ✓ How the mail system works
- ✓ Understand Postfix configuration files and key directives
- ✓ Configure Postfix as a mail and a relay server
- ✓ Verify Postfix functionality
- ✓ Configure Dovecot as a non-secure and a secure mail server
- ✓ Verify Dovecot functionality

This chapter includes the following RHCE objectives:

SMTP: Configure a mail transfer agent (MTA) to accept inbound email from other systems

SMTP: Configure an MTA to forward (relay) email through a smart host

Email

$Email$ is an essential part of an operating system. Like other operating systems, RHEL also supports this functionality. There are several mail software programs supported on RHEL that allow you to configure a system as a server or client for sending and receiving email. These software programs include Postfix and sendmail as outgoing mail servers, Dovecot as an incoming mail server, and several client programs.

In RHEL6, *Postfix* is the default mail server for sending mail and *Dovecot* is the default mail server for receiving mail.

Mail System

A fully functional mail system requires four key elements – *Mail Transport Agent* (MTA), *Mail Submission Agent* (MSA), *Mail Delivery Agent* (MDA), and *Mail User Agent* (MUA). An MTA is responsible for transporting a message from a sending mail server and another MTA is responsible for accepting the message at a receiving mail server. The primary protocol used between the mail servers is called *Simple Mail Transport Protocol* (SMTP). The most widely used MTA at the sending end is called *sendmail*, with Postfix and Exim being alternatives providing increased security and improved performance over sendmail. An example of an MTA responsible for receiving mail messages is Dovecot, which invokes an MDA to forward the message to a mail spool location and store it there until an MUA at the receiving end fetches the message into the inbox of the user via either the *Post Office Protocol* (POP) or the *Internet Message Access Protocol* (IMAP). Examples of MDA are *Local Delivery Agent* (LDA), *procmail*, and *maildrop*, with LDA being part of the Dovecot software. An MUA is a mail client program that is used to write and read mail messages, and it submits the mail messages to an MTA via an MSA. Examples of an MSA are Postfix and sendmail. An MUA uses either POP or IMAP to access mail. Examples of MUA are *mail, mailx, elm, mutt, fetchmail, evolution, thunderbird, squirrelmail, Microsoft Outlook*, and *eudora*.

How Does the Mail System Work?

Here is how the mail system works. A user composes a mail message using an MUA on his computer. The user presses the Send button and the message is submitted through an MSA to a configured outgoing MTA via SMTP port 25. The MTA checks the destination address in DNS and locates an associated MX record for the destination system's domain. The message is eventually transported to the receiving MTA, which invokes an MDA to forward the message to a mail spool location and store it there until an MUA at the receiving end pulls it into the inbox of the user. If the message is destined to a local user, the MTA simply places it into the mailbox of that user without engaging DNS and the message from there is retrieved by the user via an MUA.

Depending on the location of the sending and receiving MUAs, the message may be transported on a large internal corporate network or over the Internet using tens or hundreds of routing devices. In either case, a DNS server is employed for name resolution. For small networks, the *hosts* file is sufficient.

In the remainder of this chapter, Postfix and Dovecot are covered at length. Other tools mentioned above are beyond the scope of this book.

Postfix

Postfix is sendmail compatible; however, it is easier to configure and administer, more modular, and provides better performance. It includes features such as pattern matching for filtering out unwanted mail and support for setting up several virtual mail domains on a single system. Postfix uses the SMTP protocol via port 25 for mail transfer.

Postfix Daemons

The primary daemon for Postfix is known as *master*, which is located in the */usr/libexec/postfix* directory. This daemon starts several helper daemons such as *nqmgr*, *pickup*, and *smtpd* to work as agents to carry out tasks on its behalf as necessary. Each daemon has a small, well-defined function and operates with only the minimal amount of privilege necesaary to carry out that function. The *nqmgr* daemon is a queue manager program and is responsible for mail transmission, relay, and local delivery; the *pickup* daemon transfers mail messages from the */var/spool/mqueue* directory to the */var/spool/mail* directory; and the *smtpd* daemon directs the incoming mail to the */var/spool/mail* directory. These daemons work together to manage the entire Postfix email system.

Postfix Configuration Files

Postfix configuration files are located in the */etc/postfix* directory with the exception of the *aliases* file, which is located in */etc*. There are several of them and are explained in the following sub-sections.

The /etc/aliases File

This file provides a mechanism to forward mail for local recipients to other local or remote addresses. Postfix references this file to determine alias settings. An excerpt of the default contents from this file is displayed below:

```
# cat /etc/aliases
. . . . . . . .
# General redirections for pseudo accounts.
bin:         root
daemon:      root
adm:         root
lp:          root
sync:        root
shutdown:    root
. . . . . . . .
# Person who should get root's mail
#root:       marc
```

The above output indicates that messages generated by system users will be forwarded to the *root* user. You can set up something similar for other users too. For example, if there was a user account named *user10* on this system and was receiving mail, but has now been removed, you can forward that user's mail to one or more other local or remote users. Modify the *aliases* file and add the following for *user10* so that his mail is forwarded to a local user and a remote user:

---

user10:   aghori, aghori@server.example.com

You can create a mailing group of several users so that when mail is sent to that group, it will go to all group members. Here is how you would create a mailing group called *linuxadmteam* with members *aghori*, *user1*, *user2*, and *user3*. Add the following line to the *aliases* file:

linuxadmteam: aghori, user1, user2, user3

After any modifications have been done to the *aliases* file, the Postfix service needs to be informed of those updates. For this purpose, the contents of this file are converted into the */etc/aliases.db* database file by either restarting the Postfix service or executing the *newalises* command.

The /etc/postfix/access File

This file may be used to establish user-based and host-based security for Postfix by setting up limits in the form of "pattern action" on users, hosts, domains, and networks. Each pattern has an associated action such as REJECT and OK. The following shows some examples.

To allow access to a single system:

192.168.122.20 OK

To allow access to systems in a domain:

example.com OK

To reject access to systems on a network:

192.168.3 REJECT

The contents of this file are converted into the *access.db* database for Postfix use using the *postmap* command.

 An alternative to this access control is through the use of the iptables firewall.

The /etc/postfix/canonical File

This file is used to establish limits on incoming mail from other users, domains, and networks, and it is configured with limits in the form of "pattern result". The following shows some examples.

To forward email sent to a local user *aghori* to aghori@example.com:

aghori aghori@example.com

To forward all mail destined for example.org domain to another domain example.com:

@example.org @example.com

The contents of this file are converted into the *canonical.db* database for Postfix use using the *postmap* command.

The /etc/postfix/generic File

This file is used to establish limits on outcoming mail to other users, domains, and networks. The syntax is identical to that of the *canonical* file.

The /etc/postfix/main.cf File

The *main.cf* file is the primary Postfix configuration file, which defines global settings for the Postfix operation. These settings include host, domain, and network to be served; mail owner; network interfaces to listen on; etc. Some key directives from this file are shown below. Note that the order in which the directives appear in the file is important.

```
queue_directory = /var/spool/postfix
myhostname = server.example.com
mydomain = example.com
myorigin = $myhostname
inet_interfaces = localhost
mydestination = $myhostname, localhost.$mydomain, localhost
mynetworks = 192.168.2.0/24, 127.0.0.0/8
relayhost = $mydomain
alias_maps = hash:/etc/aliases
alias_database = hash:/etc/aliases
queue_directory = /var/spool/postfix
command_directory = /usr/sbin
daemon_directory = /usr/libexec/postfix
mail_owner = postfix
mailq_path = /usr/bin/mailq.postfix
setgid_group = postdrop
```

These directives are explained in Table 18-1.

Directive	Description
queue_directory	Specifies the location of the Postfix queue.
myhostname	Sets the FQDN of the Postfix server.
mydomain	Specifies the domain name of the Postfix server.
myorigin	Specifies the domain name that the outgoing mail appears to have originated from.
inet_interfaces	Specifies network interfaces to be used for incoming mail.
mydestination	Specifies the domains that the Postfix server accepts mail from.
mynetworks	Specifies IP addresses of trusted networks.
relayhost	Identifies the hostname of another mail server to forward mail to. This sets up the server as a smart host.
alias_maps	Specifies the aliases database used by the local delivery agent.
aliases_database	Specifies the aliases database that is generated with the *newaliases* command.
queue_directory	Specifies the location of the spooled or received email.

Directive	Description
command_directory	Specifies the location of the Postfix commands.
daemon_directory	Specifies the location of the Postfix daemons.
mail_owner	Defines the owner of the Postfix queue and the owner of Postfix daemons.
mailq_path	Identifies the path of the Postfix *mailq* command.
setgid_group	Sets the group for mail submission and queue management.

Table 18-1 Key main.cf Directives

The *postconf* command is a Postfix configuration utility that can be used to display and modify the *main.cf* file contents. For example, to display the default and non-default settings, run this command with the –d and the –n option, respectively:

# **postconf –d**
2bounce_notice_recipient = postmaster
access_map_defer_code = 450
access_map_reject_code = 554

. . . .

# **postconf –n**
alias_database = hash:/etc/aliases
alias_maps = hash:/etc/aliases
command_directory = /usr/sbin
config_directory = /etc/postfix
daemon_directory = /usr/libexec/postfix
data_directory = /var/lib/postfix
debug_peer_level = 2
html_directory = no
inet_interfaces = localhost
inet_protocols = all
mail_owner = postfix
mailq_path = /usr/bin/mailq.postfix
manpage_directory = /usr/share/man
mydestination = $myhostname, localhost.$mydomain, localhost
newaliases_path = /usr/bin/newaliases.postfix
queue_directory = /var/spool/postfix
readme_directory = /usr/share/doc/postfix-2.6.6/README_FILES
sample_directory = /usr/share/doc/postfix-2.6.6/samples
sendmail_path = /usr/sbin/sendmail.postfix
setgid_group = postdrop
unknown_local_recipient_reject_code = 550

Rather than editing the *main.cf* file for modifying directives, you can alter a directive value directly from the command prompt using the *postconf* command. The following shows an example which sets the queue_directory directive to */var/spool/defaultmail*:

postconf –e queue_directory=/var/spool/defaultmail

The /etc/postfix/master.cf File

This file defines configuration rules for incoming email. It contains settings to control the *master* daemon. It lists service daemons needed by Postfix and starts them as required. An excerpt from this file is shown below:

```
# cat /etc/postfix/master.cf
. . . . . . . .
# service      type     private   unpriv    chroot    wakeup   maxproc   command + args
#              (yes)    (yes)     (yes)     (never)   (100)
# -----------------------------------------------------------------------------------------
smtp           inet     n         -         n         -        -         smtpd
pickup         fifo     n         -         n         60       1         pickup
cleanup        unix     n         -         n         -        0         cleanup
. . . . . . . .
```

For each service listed in the first column, there are seven associated columns, which define how the service will be used. These columns are listed and explained in Table 18-2.

Column	Description
service	Name of the service.
type	Transport mechanism to be used.
private	Whether the service is to be used by Postfix only.
unpriv	Whether the service is to be run by non-*root* users.
chroot	Whether the service is to be chrooted for the mail queue.
wakeup	Wake up interval for the service.
maxproc	Maximum number of threads the service can start.
command	Command associated with the service plus any arguments.

Table 18-2 master.cf File Contents

The /etc/postfix/relocated File

This file contains user information who are no longer on a domain. This file is syntactically identical to the *canonical* and *generic* files. The following shows an example.

To forward an email from one domain to another for user *aghori*:

aghori@example.com asghar_ghori2002@example.net

The contents of this file are converted into the *relocated.db* database for Postfix use using the *postmap* command.

The /etc/postfix/transport File

This file establishes mappings between email addresses and message deliver transports and next-hop destinations in the transport:nexthop format. An example of transport is local or smtp and that for nexthop a hostname or domain name. For example, the following entry sets up a mapping between a user email address and a host:

aghori@example.com smtp:insider.example.com

The contents of this file are converted into the *transport.db* database for Postfix use using the *postmap* command.

The /etc/postfix/virtual File

This file may be used to redirect mail intended for one user to one or more other email addresses. The following two examples redirect mail for user *aghori* to user *root*:

aghori root
aghori@example.com root

The contents of this file are converted into the *virtual.db* database for Postfix use using the *postmap* command.

Exercise 18-1: Configure Postfix to Accept Inbound Mail from Other Systems

In this exercise, you will configure *server.example.com* to function as a Postfix mail server to serve systems on the local network (192.168.2.0). You will activate SELinux booleans and configure appropriate host-based access rules. You will set and verify the default mail server program.

1. Check whether the Postfix software package is installed:

    # **rpm –qa | grep postfix**
    postfix-2.6.6-2.2.el6_1.x86_64

    Install the package if it is not already installed using one of the procedures outlined in Chapter 06 "Package Management".

2. Set the Postfix service to autostart at each system reboot, and validate:

    # **chkconfig postfix on**
    # **chkconfig --list postfix**
    postfix   0:off 1:off 2:on 3:on 4:on 5:on 6:off

3. Ensure that DNS or the */etc/hosts* file is configured properly.
4. Activate the SELinux boolean for Postfix and verify the activation:

    # **setsebool –P allow_postfix_local_write_mail_spool=1**
    # **getsebool allow_postfix_local_write_mail_spool**

5. Configure host-based access by allowing SMTP traffic on TCP port 25 to pass through the firewall for systems on the 192.168.2.0 network:

    # **iptables –I INPUT –m tcp –p tcp –s 192.168.2.0/24 --dport 25 –j ACCEPT**

6. Save the rule in the */etc/sysconfig/iptables* file and restart the firewall to activate the new rule:

Red Hat Certified System Administrator & Engineer

service iptables save ; service iptables restart

7. Open the *main.cf* configuration file in a text editor and set or modify the following directives:

```
myhostname = server.example.com
mydomain = example.com
myorigin = $mydomain
inet_interfaces = all
mynetworks = 192.168.2.0/24, 127.0.0.0/8
queue_directory = /var/spool/postfix
```

8. Run the *postconf* command to check for any syntax errors in the *main.cf* file:

postconf check

9. Run the *postconf* command to review the changes made to the postfix configuration in the *main.cf* file:

```
# postconf -n
alias_database = hash:/etc/aliases
alias_maps = hash:/etc/aliases
command_directory = /usr/sbin
config_directory = /etc/postfix
daemon_directory = /usr/libexec/postfix
data_directory = /var/lib/postfix
debug_peer_level = 2
html_directory = no
inet_interfaces = localhost
inet_protocols = all
mail_owner = postfix
mailq_path = /usr/bin/mailq.postfix
manpage_directory = /usr/share/man
mydestination = $myhostname, localhost.$mydomain, localhost
mydomain = example.com
myhostname = server.example.com
mynetworks = 192.168.2.0/24, 127.0.0.0/8
myorigin = $mydomain
newaliases_path = /usr/bin/newaliases.postfix
queue_directory = /var/spool/postfix
readme_directory = /usr/share/doc/postfix-2.6.6/README_FILES
sample_directory = /usr/share/doc/postfix-2.6.6/samples
sendmail_path = /usr/sbin/sendmail.postfix
setgid_group = postdrop
unknown_local_recipient_reject_code = 550
```

10. Start the Postfix service and check the running status:

```
# service postfix start
Starting postfix:          [ OK ]
```

 # service postfix status
master (pid 18196) is running...

11. Configure Postfix as the default mail server by running the *alternatives* command:

> **# alternatives --config mta /usr/sbin/sendmail.postfix**

12. Verify that Postfix has been configured as the default mail server:

> **# alternatives --display mta | grep current**
> link currently points to /usr/sbin/sendmail.postfix

Exercise 18-2: Configure Postfix to Relay Mail Through a Smart Host

In this exercise, you will configure *server.example.com* to relay mail coming in from 192.168.2.0 network and *example.com* domain to a smart host *insider.example.com*. You will disallow mail from systems on the 192.168.4.0 network and a system with IP address 192.168.2.210.

1. Run steps 1 to 7 from Exercise 18-1.
2. Open the *main.cf* configuration file and set or modify the following directives:

> relayhost = insider.example.com

3. Edit the */etc/postfix/access* file and add the hostnames or IP addresses of the remote systems, domains, or networks to be allowed or denied access to relay their mail using this Postfix server:

> 192.168.2   RELAY
> 192.168.2.210   REJECT
> example.com   RELAY
> 192.168.4   REJECT

4. Repeat steps 9 to 12 from Exercise 18-1.

Logging Mail Messages

All mail messages are logged to the */var/log/maillog* file. These messages are related to starting and stopping of the Postfix service, successful and unsuccessful user connection attempts, and mail sent and disallowed. The *maillog* is a plain text file and may be viewed with any file viewing utility such as *head, tail, cat, more,* or *vi.* The following shows an excerpt from the file:

 # tail /var/log/maillog
Nov 6 12:45:51 physical postfix/postfix-script[18384]: starting the Postfix mail system
Nov 6 12:45:51 physical postfix/master[18385]: daemon started -- version 2.6.6, configuration /etc/postfix

Exercise 18-3: Verify Mail Server Functionality

In this exercise, you will mail test messages to verify the functionality of the mail server configured in Exercise 18-1. You will send test messages locally to user *aghori* and remotely to user *aghori* on *server.example.com*. You will log in as user *aghori* on the local and the remote system and verify if the messages have been received.

1. Mail a message to a local user *aghori* with the subject "Local Test Mail" to test local mail functionality:

   # **date | mail –s "Local Test Mail" aghori**

2. Mail a message to a remote user aghori@physical.example.com with the subject "Remote Test Mail" to test network mail functionality:

   # **date | mail –s "Remote Test Mail" aghori@server.example.com**

3. Execute the *mailq* or the *postqueue* command to check whether the messages have been sent:

   # **mailq**
   # **postqueue –p**

4. Log in as user *aghori* on the destination systems and run the *mail* command to check the messages (messages are received in the */var/spool/mail/aghori* file):

   $ **mail**

Dovecot

Dovecot is an open source mail server software which is used to handle incoming mail. It uses the POP and IMAP protocols, and supports their secure cousins as well. The current version of POP is 3 and that of IMAP is 4, and their secure versions are referred to as POP3s and IMAPs. The secure versions use SSL encryption over TCP. The primary difference between the two protocols is that POP3 deletes the mail after it has been retrieved, whereas, IMAP4 keeps it on the mail server unless it is specifically deleted. IMAP4 also supports improved login mechanism in addition to a few other advantages that it has over POP3.

Dovecot uses TCP Wrappers for access control.

Dovecot Configuration File

The key configuration file for Dovecot is located in the */etc/dovecot* directory and it is called *dovecot.conf*. This file contains scores of directives that you can set based on requirements. Some of the key directives are listed and explained in Table 18-3.

Directive	Description
base_dir	Directory location to store runtime information.
protocols	Specifies one or more protocols to be served. Options are pop3 (port 110), pop3s (port 995), imap (port 143), and imaps (port 993).
listen	Identifies the IP address and port number to listen for non-secure incoming requests in case there are more than one network interfaces. No need to specify a port number if using the default port.
ssl_disable	Specifies the IP address and port number to listen for secure incoming requests in case there are more than one network interfaces. No need to specify a port number if using the default port.
syslogd_facility	Syslog facility to be used for logging via the *rsyslogd* daemon.
mail_location	Location of user inboxes.

Table 18-3 dovecot.conf Directives

Exercise 18-4: Configure a Non-Secure POP3 Dovecot Server

In this exercise, you will configure *insider.example.com* as a non-secure Dovecot server for receiving mail. You will use the pop3 protocol on the 192.168.2.202 address on the default port number with user mailboxes located in the */var/spool/mail* directory.

1. Check whether the Dovecot software package is installed:

    # **rpm –qa | grep dovecot**
    dovecot-2.0.9-2.el6_1.1.x86_64

    Install the package if it is not already installed using one of the procedures outlined in Chapter 06 "Package Management".

2. Set the Dovecot service to autostart at each system reboot, and validate:

    # **chkconfig dovecot on**
    # **chkconfig --list dovecot**
    dovecot   0:off 1:off 2:on 3:on 4:on 5:on 6:off

3. Ensure that DNS or the */etc/hosts* file is configured properly.
4. Configure host-based access by allowing traffic on TCP port 110 to pass through the firewall for systems on the 192.168.2.0 network:

    # **iptables –I INPUT –m tcp –p tcp –s 192.168.2.0/24 --dport 110 –j ACCEPT**

5. Save the rule in the */etc/sysconfig/iptables* file and restart the firewall to activate the new rule:

    # **service iptables save ; service iptables restart**

6. Open the *dovecot.conf* configuration file in a text editor and set or modify the following directives:

> protocols = pop3
> listen = 192.168.2.202
> ssl_disable = yes
> mail_location = /var/spool/mail

7. Open the TCP Wrappers' */etc/hosts.allow* file and add the following rule to it:

> dovecot : 192.168.2.202

8. Start the Dovecot service and check the running status:

> # **service dovecot start**
> Starting Dovecot Imap: [OK]
> # **service dovecot status**
> dovecot (pid 9117) is running...

Exercise 18-5: Configure a Secure IMAP Dovecot Server

In this exercise, you will configure *insider.example.com* as a secure Dovecot server for receiving mail. You will use the imaps protocol on the 192.168.2.202 address on the default port number with user mailboxes located in the */var/spool/mail* directory and custom SSL certificate information located in the */etc/pki/dovecot* directory.

1. Perform steps 1 to 5 from Exercise 18-4. Open port 993 for the imaps protocol in step 5.
2. Open the *dovecot.conf* configuration file in a text editor and set or modify the following directives:

> protocols = imaps
> listen = 192.168.2.202
> ssl_disable = no
> ssl_cert_file = /etc/pki/dovecot/certs/dovecot.pem
> ssl_key_file = /etc/pki/dovecot/private/dovecot.pem
> mail_location = /var/spool/mail

3. Open the *dovecot-openssl.cnf* configuration file in a text editor and set or modify the following directives:

> # **cat /etc/pki/dovecot/dovecot-openssl.cnf**

C=CA	# A two-letter country code
ST=Ontario	# A state or province name
L=Toronto	# A location or city name
O=example	# An organization or a company name
OU=IMAP server	# An organizational unit within the organization or company
CN=imap.example.com	# A common name for the mail server
emailAddress=postmaster@example.com	# The email address of an administrator

4. Rename the default certificate and key files:

cd /etc/pki/dovecot/certs && mv dovecot.pem dovecot.pem.old
cd /etc/pki/dovecot/private && mv dovecot.pem dovecot.pem.old

5. Generate a new private key based on the above configuration:

/usr/share/doc/dovecot-1.0.7/examples/mkcert.sh
Generating a 1024 bit RSA private key
....................++++++
...............................++++++
writing new private key to '/etc/pki/dovecot/private/dovecot.pem'
-----
subject= /C=CA/ST=Ontario/L=Toronto/O=example/OU=IMAP
server/CN=imap.example.com/emailAddress=postmaster@example.com
SHA1 Fingerprint=D1:48:22:36:D5:21:7B:DF:65:B2:03:F2:81:21:5D:8F:91:53:2A:8C

6. Execute steps 7 and 8 from Exercise 18-4.

Exercise 18-6: Verify Dovecot Functionality

In this exercise, you will verify the Dovecot functionality configured in Exercises 18-4 and 18-5 using simple tools such as *telnet* and *mutt*.

1. Run the *telnet* command and enter the username and associated password. Install the package for telnet if it is not already installed.

telnet insider pop3
Trying 192.168.2.201
Connected to insider.example.com (192.168.2.201).
Escapte character is '^]'.
+OK Dovecot ready.
user aghori
+OK
pass Welcome01
+OK Logged in.
list
+OK 0 messages:
quit
+OK Logging out.
Connection closed by foreign host.

2. Use the *mutt* command to connect to the mailbox of user *aghori* on both 110 and 993 ports:

mutt –f pop://aghori@insider.example.com
mutt –f imaps://aghori@insider.example.com

Chapter Summary

This chapter introduced you to mail services. You examined email systems including Postfix, Dovecot, and some simple mail client programs. You learned the components of the mail system, the protocols used, and how the mail system worked. You looked at several directives defined in various configuration files. You performed exercises to set up Postfix to send out and relay incoming mail. You carried out additional exercises to configure Dovecot to receive mail using both secure and non-secure protocols. You executed exercises for verifying the functionality of both Postfix and Dovecot.

Chapter Review Questions

1. What command would you use to display or alter the current MTA?
2. What would the command *telnet localhost 25* do?
3. Which command would you use to convert the contents of Postfix configuration files into Postfix database files?
4. What is the equivalent Postfix command for *mailq*?
5. In which file can you define an entry such as "lnxadm: root"?
6. Dovecot uses the */etc/postfix/access* file for access control. True or False?
7. The *mutt* command may be used to verify the functionality of a mail server. True or False?
8. What is the default protocol and the port used by the mail system?
9. In which Postfix configuration file would you define the directive relayhost for forwarding mail?
10. What functionality does an MDA provide?
11. What would the command *postconf –e setgid_group=postdrop* do?
12. Dovecot supports POP, POP3, IMAP, and IMAPS protocols. True or False?
13. What would the entry "example.com   RELAY" in the access control file imply?
14. Dovecot is an MUA. True or False?
15. Name the default incoming and outgoing mail servers in RHEL6.
16. When should you run the *postconf check* command?
17. Which command would you run to update the aliases database?
18. Where do the mail servers log their activities?
19. What would the command *postconf –n* do?
20. What is the name of the main configuration file for Dovecot.
21. What is the name of the main daemon for the Postfix service?
22. What is the name of the Postfix access control file?

Answers to Chapter Review Questions

1. The *alternatives* command.
2. The command provifded will open a telnet connection on port 25 for the testing purpose.
3. The *postmap* command.
4. The *postqueue* command with the –p switch.
5. In the */etc/aliases* file.
6. False.
7. True.
8. The default protocol used by the mail system is SMTP on port 25.
9. In the */etc/postfix/main.cf* file.

10. An MDA is responsible for forwarding a message to a mail spool location.
11. The command provided will modify the directive's value in the *main.cf* file without having to open the file with a text editor.
12. True.
13. This entry would relay incoming mail from systems in the example.com domain.
14. False.
15. The default incoming mail server in RHEL6 is Dovecot and the default outgoing mail server is Postfix.
16. After making any changes to the *main.cf* file for syntax errors.
17. The *newaliases* command.
18. To the */var/log/maillog* file.
19. The command provided will display all the current settings for Postfix.
20. The main configuration file for Dovecot is *dovecot.conf*.
21. The main daemon for the Postfix service is called the *master* daemon.
22. The Postfix access control file is called *access*.

Labs

Lab 18-1: Establish Postfix Relay Service

Configure Postfix on *insider.example.com* in such a way that it relays all inbound mail on port 25 to *physical.example.com* via the *eth0* interface. Do not forget to set Postfix as the default MTA.

Lab 18-2: Set Up Secure POP Dovecot Service

Set up the secure POP Dovecot service on *server.example.com* so that it receives mail in the */var/spool/mail* directory using custom SSL certificates. Test the service's functionality to validate the setup.

Chapter 19

Network File System

This chapter covers the following major topics:

- ✓ Understand NFS concepts and benefits
- ✓ Overview of NFS versions, security, daemons, commands, related files, and startup scripts
- ✓ How NFS server and client interact with each other
- ✓ Configure an NFS server and an NFS client
- ✓ Configure an NFS server to export a resource for group collaboration
- ✓ Display exported and mounted resources
- ✓ Unmount and unexport a resource
- ✓ Monitor NFS read and write statistics using different tools

This chapter includes the following RHCE objectives:

27. Mount and unmount NFS network file systems
NFS: Provide network shares to specific clients
NFS: Provide network shares suitable for group collaboration

The *Network File System* (NFS) service is based on the client/server architecture whereby users on one system access files, directories, and file systems (let us collectively call them *resources*) residing on a remote system as if they exist locally on their system. The remote system that makes its resources available for shared access over the network is called an *NFS server,* and the process of making the resources accessible is referred to as *exporting.* The resources exported by the NFS server may be accessed by one or more systems. These systems are called *NFS clients,* and the process of making the resources accessible on clients is referred to as *mounting.*

Resources may be shared for group collaboration among users that have accounts on different systems. There are read and write activities that occur between the server and the client when users access shared resources.

Understanding Network File System

A system can function as both an NFS server and an NFS client at the same time. When a directory or file system resource is exported, the entire directory structure beneath it becomes available for mounting on the client. A sub-directory or the parent directory of an exported resource cannot be re-exported if it exists in the same file system. Similarly, a resource mounted by an NFS client cannot be exported further by the client. A single exported file resource is mounted on a directory mount point.

NFS is built on top of *Remote Procedure Call* (RPC) and *eXternal Data Representation* (XDR) to allow a server and a client to communicate with each other. They provide a common "language" that both the server and the client understand. This is standardized based on the fact that the NFS server and client may be running two completely different operating systems on different hardware platforms. The RPC service uses program numbers defined in the */etc/rpc* file.

Benefits

The use of NFS provides several benefits, some of which are listed below:

- ✓ Supports heterogeneous operating system platforms including Linux, UNIX, and Microsoft Windows.
- ✓ Several NFS client systems can access a single exported resource simultaneously.
- ✓ Enables sharing common application binaries and read-only information such as the *man* pages, instead of loading them on each single system. This results in reduced overall disk storage cost and administration overhead.
- ✓ Gives users access to uniform data.
- ✓ Useful when many users exist on many systems with their home directories located on every single host. In such a situation, you can create user home directories on a single system under */home* for example, and export it. Now, whichever system a user logs on to, his home directory becomes available there. This way the user will need to maintain only one home directory.

NFS Versions

RHEL6 comes with version 4 of the NFS protocol (NFSv4), which is an *Internet Engineering Task Force* (IETF) standard protocol that provides enhanced security, greater scalability, encrypted transfers, better cross-platform interoperability, improved functionality through firewalls and on the Internet, support for ACLs, better handling of system crashes, and superior efficiency compared to previous versions of NFS. NFSv4 maintains all features and benefits of NFSv3 including support for TCP and files of sizes up to 128GB (64-bit). NFSv4 uses usernames and groupnames rather than UIDs and GIDs when sharing files over the network.

Older NFS protocols – v2 and v3 – are still supported in RHEL6 for backward compatibility.

NFS Security

NFS security is paramount in v4 to ensure that NFS operates securely in a WAN environment. In older versions, authentication was performed on the NFS client side. In contrast, an exchange of information takes place in v4 between the client and server for identification, authentication, and authorization. Identification establishes identity of systems and users that will be accessing the shares, authentication confirms the identity, and authorization controls what information systems and users will have access to. Exchange of information in transit between the client and server is encrypted to prevent eavesdropping and unauthorized access to private data.

NFS Daemons, Commands, Configuration Files, and Scripts

When working with NFS, several daemons, commands, configuration files, and scripts are involved. The tables furnished below list and explain them.

Table 19-1 describes NFS daemons.

Daemon	Description
rpcbind	A server- and client-side service that converts RPC program numbers into universal addresses to facilitate RPC-based communications. This daemon uses TCP port 111 for operation.
rpc.idmapd	A server- and client-side service that controls mappings of UIDs and GIDs with their corresponding usernames and groupnames. Directives Domain, Nobody-User, and Nobody-Group need to be defined accordingly in the configuration file *idmapd.conf* of this service, which is located in the */etc* directory. Additional directives may need to be adjusted on both an NFS server and client if you intend to share */home*. You will need to start the *rpcidmapd* service, which will then take care of authentication and home directory mounts.
rpc.mountd	A server-side service that responds to client requests to mount a resource and provide status of exported and mounted resources. Access to this service may be controlled via TCP Wrappers using the */etc/hosts.allow* and */etc/hosts.deny* files. See Chapter 15 "Advanced Firewall, TCP Wrappers, Usage Reporting & Remote Logging" for details on how to control access using TCP Wrappers.
rpc.nfsd	A server-side service that responds to client requests to access files. This daemon uses TCP port 2049 for operation.
rpc.rquotad	A server- and client-side service that shows user quota information of a local resource that is mounted remotely and enables defining user quotas on NFS mounted resources, respectively.
rpc.statd	A server- and client-side service that keeps track of locks placed on files by

Daemon	Description
	an NFS client. This service releases all file locks if the NFS client is rebooted so that other clients may use them. In case of a reboot of the NFS server, the NFS client reminds the server of file locks that were in place before the server had been rebooted. This service is controlled with the */etc/init.d/nfslock* script.

Table 19-1 NFS Daemons

The running status of these daemons may be viewed by running the *rpcinfo* command, which shows all the RPC-based services along with their program number, version, protocol, and port that they are listening on.

```
# rpcinfo –p
   program vers proto  port service
    100000   4  tcp     111 portmapper
    100000   3  tcp     111 portmapper
    100000   2  tcp     111 portmapper
    100000   4  udp     111 portmapper
    100000   3  udp     111 portmapper
    100000   2  udp     111 portmapper
    100011   1  udp     875 rquotad
    100011   2  udp     875 rquotad
    100011   1  tcp     875 rquotad
    100011   2  tcp     875 rquotad
    100005   1  udp   51828 mountd
    100005   1  tcp   52453 mountd
    100005   2  udp   52544 mountd
    100005   2  tcp   57398 mountd
    100005   3  udp   59439 mountd
    100005   3  tcp   51244 mountd
    100003   2  tcp    2049 nfs
    100003   3  tcp    2049 nfs
    100003   4  tcp    2049 nfs
    . . . . . . . .
```

Table 19-2 describes NFS commands.

Command	Description
exportfs	A server-side command that exports resources listed in the */etc/exports* file and displays exported resources listed in the */var/lib/nfs/etab* file.
showmount	A server-side and client-side command that displays which resources are exported to which clients by consulting the */var/lib/nfs/etab* file on the server. It displays which clients have those resources mounted by consulting the */var/lib/nfs/rmtab* file on the client.
mount	A client-side command that mounts a resource specified at the command line or listed in the */etc/fstab* file when executed by the *root* user. It adds an entry for the mounted resource to the client's */etc/mtab* file, and server's */var/lib/nfs/rmtab* file via the *rpc.mountd* daemon. This command can be

Command	Description
	used to display mounted resources listed in the */etc/mtab* file.
nfsiostat	A client-side command that provides NFS I/O statistics by consulting the */proc/self/mountstats* file.
rpcinfo	A server-side command that checks whether NFS server daemons are registered with RPC.
nfsstat	A server-side and client-side command that displays NFS and RPC statistics by consulting the */proc/net/rpc/nfsd* and */proc/net/rpc/nfs* files, respectively.
mountstats	A client-side command that displays per-mount statistics. It consults the */proc/self/mountstats* file for information.

Table 19-2 NFS Commands

Table 19-3 describes NFS configuration and functional files.

File	Description
/etc/exports	A server-side file that contains a list of resources to be exported.
/var/lib/nfs/etab	A server-side file that contains a list of exported resources whether or not they are remotely mounted. This file is updated when a resource is exported or unexported, and is maintained by the *rpc.mountd* daemon.
/var/lib/nfs/rmtab	A server-side file that contains a list of exported resources, which are currently mounted by clients. This file is updated when a resource is remotely mounted or unmounted, and is maintained by the *rpc.mountd* daemon.
/etc/fstab	A client-side file that contains a list of resources to be mounted at system reboots or manually with the *mount* command.
/etc/mtab	A client-side file that contains a list of mounted resources. The *mount* and *umount* commands update this file.
/etc/sysconfig/nfs	A server-side and client-side configuration file that is used by NFS startup scripts.

Table 19-3 NFS Configuration and Functional Files

Table 19-4 describes NFS startup and shutdown scripts.

Script	Description
/etc/rc.d/init.d/nfs	A server-side script that starts and stops the *rpc.nfsd*, *rpc.rquotad*, *rpc.svcgssd*, *rpc.idmapd*, and *rpc.mountd* services. This script sources the */etc/sysconfig/nfs* file for configuration information.
/etc/rc.d/init.d/nfslock	A server-side and client-side script that starts and stops the *rpc.statd* server. This script sources the */etc/sysconfig/nfs* file for configuration information.
/etc/rc.d/init.d/portreserve	A server-side and client-side script that starts and stops the TCP port reserver, which prevents RPC services from picking up well-known reserved ports.
/etc/rc.d/init.d/rpcbind	A server-side and client-side script that starts and stops the *rpcbind* service, which converts RPC program numbers into universal addresses.

Script	Description
/etc/rc.d/init.d/rpcgssd	A client-side script that starts and stops the RPCSEC GSS service, which controls general RPC security services.
/etc/rc.d/init.d/rpcsvcgssd	A server-side script that starts and stops the RPCSEC GSS service, which controls general RPC security services.
/etc/rc.d/init.d/rpcidmapd	A server-side and client-side script that starts and stops the username to UID and groupname to GID mapping service.

Table 19-4 NFS Startup and Shutdown Scripts

How Does NFS Work?

NFS is a simple protocol that is used to export a resource on one system and mount it on another system. The following outlines the process of exporting and mounting a resource using the NFSv4 protocol:

- ✓ The contents of the */etc/exports* file are evaluated for syntax errors and access issues.
- ✓ Each resource listed in this file is exported and its entry is added to the */var/lib/nfs/etab* file on the server. The *showmount* command looks into this file to obtain exported resource information.
- ✓ The client issues the *mount* command on the NFS client which requests the NFS server through the *rpcbind* daemon to provide a file handle for the requested resource.
- ✓ The incoming request is evaluated by the firewall running on the server to check whether the corresponding service port (*rpcbind* on port 111) is allowed to accept the traffic.
- ✓ The request goes to the *rpc.mountd* daemon on the NFS server through the *rpcbind* daemon.
- ✓ The *rpc.mountd* daemon consults TCP Wrappers and performs an access check to validate if the client is authorized to mount the resource.
- ✓ The *rpc.mountd* daemon checks whether an SELinux boolean is applicable on the incoming request. If there is one, it checks whether the boolean is activated.
- ✓ The *rpc.mountd* daemon sends a file handle for the requested resource to the client after verifying and validating all the checks.
- ✓ The client mounts the resource if the correct *mount* command syntax is used. An entry for the resource may be added to the */etc/fstab* file to ensure automatic mounting of the resource at each client reboot.
- ✓ The *mount* command tells the *rpc.mountd* daemon on the NFS server that the resource has been mounted successfully. Upon receiving this confirmation, the daemon adds an entry to the */var/lib/nfs/rmtab* file. The *showmount* command consults this file to obtain information on remotely mounted NFS resources. When the resource is unmounted on the client, the *umount* command sends a request to the *rpc.mountd* daemon requesting it to remove the entry from this file.
- ✓ The *mount* command also adds an entry to the */etc/mtab* file for the mounted resource on the client. The *mount* and *df* commands reference this file to obtain information about the mounted resources. The *mount* and *umount* commands update this file whenever these commands are executed successfully.
- ✓ Any file access request by the client on the mounted resource is now handled by the server's *rpc.nfsd* daemon. There must be an appropriate firewall rule in place (if the firewall is active) for this daemon to provide the proper service to the client.
- ✓ The *rpc.statd* daemon is involved when the client requests the server to place a lock on a file located in the mounted resource.

Configuring NFS

This section provides an overview on how to make NFS work with SELinux activated in the enforcing mode and furnishes exercises to solidify the knowledge and understanding of the NFS functionality that has been explained in the previous section.

SELinux Requirements for NFS

As you know from Chapter 12 "Firewall & SELinux" there are two possibilities on how SELinux could affect the functionality of a service. SELinux protects systems by setting appropriate controls using file contexts and boolean values.

The SELinux security contexts need not be modified for NFS as it does not create new files or directories during its operation. The existing default file contexts are sufficient for a smooth NFS server and client operation.

However, there are several SELinux booleans available that need to be looked at carefully to see which ones might need to be enabled or disabled for NFS to function as expected. To list the booleans, you can either use the SELinux Configuration tool and filter out the NFS-related booleans or run the *getsebool* command as follows:

> # **getsebool –a | egrep 'nfs|gssd' | egrep –v 'cobbler|xen|git'**
> allow_ftpd_use_nfs --> off
> allow_gssd_read_tmp --> on
> allow_nfsd_anon_write --> off
> httpd_use_nfs --> off
> nfs_export_all_ro --> on
> nfs_export_all_rw --> on
> qemu_use_nfs --> on
> samba_share_nfs --> off
> use_nfs_home_dirs --> on
> virt_use_nfs --> off

The above output indicates that there are ten booleans that we should review and modify for our exercises. Table 19-5 describes the purpose of each of these booleans.

Boolean	Purpose
allow_ftpd_use_nfs	Enables/disables the *ftpd* daemon to access mounted NFS resources.
allow_gssd_read_tmp	Allows/disallows the *gssd* daemon to read temporary directories on clients if Kerberos authentication is active.
allow_nfsd_anon_write	Turn on/off the *nfsd* daemon to write to public directories on clients anonymously.
httpd_use_nfs	Activates/deactivates the *httpd* daemon to access mounted NFS resources.
nfs_export_all_ro	Enables/disables exporting of NFS resources with read-only permission.
nfs_export_all_rw	Turn on/off exporting of NFS resources with read and write permissions.

Boolean	Purpose
qemu_use_nfs	Specifies whether to allow the Quick Emulator to access mounted NFS resources.
samba_share_nfs	Activates/deactivates the *smbd* daemon to share mounted NFS resources on NFS clients that are configured as Samba servers.
use_nfs_home_dirs	Turn on/off NFS clients to mount user home directories.
virt_use_nfs	Specifies whether to allow NFS clients running on virtual machines to access mounted NFS resources.

Table 19-5 NFS Related SELinux Booleans

NFS Server-Side Options

When exporting a resource with NFS, there are several options available that govern the access of the resource on NFS clients. Some of the common options are described in Table 19-6. Note that the default values are shown in square brackets.

Option	Description
*	Represents all possible matches for hostnames, IP addresses, domain names, or network addresses.
all_squash (no_all_squash) [no_all_squash]	Treats all users on clients as anonymous users.
anongid=GID [65534]	Explicitly assigns this GID to anonymous groups.
anonuid=UID [65534]	Explicitly assigns this UID to anonymous users.
async (sync) [async]	Replies to client requests only before the changes have been committed.
fsid	Identifies the type of file system being exported. Options are device number, root, or UUID.
mp	Exports only if the specified resource is a file system.
root_squash (no_root_squash) [root_squash]	Prevents the *root* user on the NFS client from gaining *root* access on mounted NFS resources by mapping *root* to an unprivileged user account called *nfsnobody* with UID 65534.
rw (ro) [ro]	Allows file modifications on NFS resources.
secure / (insecure) [secure]	Allows access only on ports lower than 1024.
subtree_check (no_subtree_check) [no_subtree_check]	Checks permissions on higher level directories of an exported resource.
wdelay (no_wdelay) [wdelay]	Delays data writes to mounted NFS resources.

Table 19-6 exportfs Command Options

Exercise 19-1: Configure an NFS Server

In this exercise, you will configure *server.example.com* as an NFS server and export the */usr/local/bin* directory to a specific client *insider.example.com* in read/write mode with root squash disabled. You will also export the */home* file system to all systems in the *example.com* domain in read/write mode ensuring that the *rpc.nfsd* daemon replies to client requests only before the changes have been committed. You will ensure that the NFS server supports and advertises version 4 only. You will show the exported resources using commands and file contents.

1. Ensure that NFS packages are installed:

   # rpm –qa | egrep 'nfs|rpcbind'
   nfs–utils–lib–1.1.5–4.el6.x86_64
   nfs–utils–1.2.3–26.el6.x86_64
   rpcbind–0.2.0–8.el6.x86_64

   Install the packages if they are not already installed using one of the procedures outlined in Chapter 06 "Package Management".

2. Verify that the NFS modules are loaded in the system:

   # lsmod | grep nfsd
   nfsd   305799 11
   lockd   74270   1 nfsd
   nfs_acl   2647   1 nfsd
   auth_rpcgss   44895   1 nfsd
   exportfs   4236   1 nfsd
   sunrpc   243758 12 nfsd,lockd,nfs_acl,auth_rpcgss

   If not, run the *modprobe* command to load the modules:

   # modprobe nfsd

3. Open the */etc/sysconfig/nfs* file in a text editor and uncomment the following entries so that only NFSv4 is advertised and used. Leave other settings intact.

   MOUNTED_NFS_V2-"no"
   MOUNTED_NFS_V3-"no"
   RPCNFSDARGS-"–N 2 –N 3"

4. Open the */etc/exports* file in a text editor and add entries for */usr/local/bin* and */home*:

   /usr/local/bin   insider.example.com(no_root_squash)
   /home   *.example.com(rw,mp)

5. Activate SELinux booleans to allow NFS exports in read-only and read/write modes, and verify the activation:

   # setsebool –P nfs_export_all_ro=1 nfs_export_all_rw=1
   # getsebool –a | grep nfs_export
   nfs_export_all_ro --> on
   nfs_export_all_rw --> on

6. Configure host-based access by allowing NFS traffic on ports 111 and 2049 to pass through the firewall:

   # iptables –I INPUT –m tcp –p tcp --dport 111 –j ACCEPT
   # iptables –I INPUT –m tcp –p tcp --dport 2049 –j ACCEPT

7. Save the rule in the */etc/sysconfig/iptables* file and restart the firewall to activate the new rule:

    # **service iptables save ; service iptables restart**

8. Set the NFS service to autostart at each system reboot, and validate:

    # **chkconfig rpcbind on**
    # **chkconfig nfslock on**
    # **chkconfig nfs on**
    # **chkconfig --list | egrep 'rpcbind|nfs'**
    nfs   0:off 1:off 2:on 3:on 4:on 5:on 6:off
    nfslock   0:off 1:off 2:on 3:on 4:on 5:on 6:off
    rpcbind   0:off 1:off 2:on 3:on 4:on 5:on 6:off

9. Start the *rpcbind* process and check the running status:

    # **service rpcbind start**
    # **service rpcbind status**

10. Start the NFS file locking service and check the running status:

    # **service nfslock start**
    # **service nfslock status**

11. Start the NFS service and check the running status:

    # **service nfs start**
    # **service nfs status**

12. Run the *rpcinfo* command and verify that the required daemons for NFSv4 are running on ports 111 and 2049:

    # **rpcinfo –p | egrep '111|2049'**
    100000 4 tcp   111 portmapper
    100000 3 tcp   111 portmapper
    100000 2 tcp   111 portmapper
    100000 4 udp   111 portmapper
    100000 3 udp   111 portmapper
    100000 2 udp   111 portmapper
    100003 4 tcp   2049 nfs
    100003 4 udp   2049 nfs

    If the *rpcbind*, *nfslock*, and *nfs* services are already running, simply execute the following:

    # **exportfs –avr**
    Exporting insider.example.com:/usr/local/bin
    exporting *.example.com:/home

The –avr options instruct the command to export all resources listed in the */etc/exports* file, display details, and update the */var/lib/nfs/etab* file. The contents of the etab file are displayed below:

> # **cat /var/lib/nfs/etab**
> /usr/local/bin
> insider.example.com(ro,sync,wdelay,hide,nocrossmnt,secure,no_root_squash,no_all_squash,no_subtree_check,secure_locks,acl,anonuid=65534,anongid=65534)
> /home
> *.example.com(rw,async,wdelay,hide,nocrossmnt,secure,root_squash,no_all_squash,no_subtree_check,secure_locks,acl,mountpoint,anonuid=65534,anongid=65534)

At this point, you may run the *showmount* command also to list the exported resources:

> # **showmount –e**
> Export list for server.example.com:
> /home   *.example.com
> /usr/local/bin   insider.example.com

NFS Client-Side Options

When mounting an NFS resource, there are several options available that you may want to use. Some of the common options are described in Table 19-7. Note that the default values are shown in square brackets.

Option	Description
ac (noac) [ac]	Specifies whether to cache file attributes for better performance.
async (sync) [async]	See Table 19-6.
fg / bg [fg]	Use fg (foreground) for resources that must be available to the client to boot successfully or operate properly. If a foreground mount fails, it is retried for retry minutes in the foreground until it either succeeds or is interrupted if the intr option is also used. With bg (background), mount attempts are tried repeatedly for retry minutes in the background without hampering the system boot process or hanging the client.
hard / soft [hard]	With the hard option, the client tries repeatedly to mount a resource until either it succeeds or is interrupted if the intr option is also used. If the NFS server goes down, processes using the mounted resource hang until the server comes back up. Use soft to avoid this situation. With the soft option, if a client attempts to mount a resource for retrans times unsuccessfully, an error message is displayed.
intr / nointr [nointr]	Use intr if you want to manually interrupt a request.
retrans=n [3]	The client retransmits a read or write request for n times after the first transmission times out. If the request does not succeed after the *n* retransmissions have completed, a soft mount displays an error message and a hard mount continues to retry.

Option	Description
retry=*n* [2 minutes for fg and 10,000 minutes for bg]	For this duration the client tries to mount a resource after the first try has failed. With intr, mount attempts can be interrupted before this many retries have been attempted. If nointr is used, mount attempts must wait until this many retries have been made, mount succeeds, or the client is rebooted.
rsize=*n* [negotiated]	Specifies the size of each read request.
rw / ro [rw]	rw allows file modifications and ro prevents from doing it.
sec=mode [sys]	Specifies the type of security to be used. Some options are none, sys, and krb5.
timeo=*n* [600]	Sets timeout, in tenths of a second, for NFS read and write requests. If a request times out, this value is doubled and the request is attempted again for retrans times. When the number of retrans attempts have been made, a soft mount displays an error message while a hard mount continues to retry.
wsize=*n* [negotiated]	Specifies the size of each write request.

Table 19-7 mount Command Options

Exercise 19-2: Configure an NFS Client

In this exercise, you will configure *insider.example.com* as an NFS client and mount the */usr/local/bin* and the */home* resources. You will mount them on local directories */usr/local/bin* and */home*, respectively. Before mounting, you will ensure that both mount points are empty. You will set services and resources so that they are autostarted and automatically mounted when a system reboot occurs.

1. Enure that NFS packages are installed:

   # **rpm –qa | egrep 'nfs|rpcbind'**
   nfs–utils–1.2.3–26.el6.x86_64
   rpcbind–0.2.0–8.el6.x86_64

   Install the packages if they are not already installed using one of the procedures outlined in Chapter 06 "Package Management".

2. Execute steps 5 to 7 from Exercise 19-1:
3. Set the *rpcbind* service to autostart at each system reboot, and validate:

   # **chkconfig rpcbind on**
   # **chkconfig --list rpcbind**
   rpcbind   0:off   1:off   2:on   3:on   4:on   5:on   6:off

4. Start the *rpcbind* process and check the running status:

   # **service rpcbind start**
   # **service rpcbind status**

5. Run the *rpcinfo* command and verify that the *rpcbind* daemon is running on port 111:

```
# rpcinfo –p | grep 111
100000  4  tcp  111  portmapper
100000  3  tcp  111  portmapper
100000  2  tcp  111  portmapper
100000  4  udp  111  portmapper
100000  3  udp  111  portmapper
100000  2  udp  111  portmapper
```

6. Run the *showmount* command and check the availability of the resources from *server.example.com*:

    # showmount –e server.example.com

7. Open the */etc/fstab* file in a text editor and add the following entries for the resources so that they are mounted automatically at system reboots:

server.example.com:/usr/local/bin	/usr/local/bin	nfs	ro	0	0
server.example.com:/home	/home	nfs	rw	0	0

8. Create required mount points with the *mkdir* command if they do not already exist.
9. Execute the *mount* command with the "–at nfs" options to mount all the remote resources listed in the */etc/fstab* file:

    # mount –at nfs

    Alternatively, you can manually mount the resources in one of three ways. Repeat the first command below for each resource and specify correct options with the –o switch, run the second command that will obtain additional required information from the */etc/fstab* file and mount the resource, or execute the third command that will mount all configured NFS resources listed in the */etc/fstab* file:

    # mount –t nfs –o ro server.example.com:/usr/local/bin /usr/local/bin
    # mount /usr/local/bin
    # service netfs start
    Mouting other filesystems: [OK]

10. Issue the *mount* command again to view a list of the mounted NFS resources:

    # mount | grep nfs

A mount point should be empty when an attempt is made to mount a resource on it, otherwise, the contents of the mount point will hide. As well, the mount point must not be in use or the mount attempt will fail.

Exercise 19-3: Export a Resource for Group Collaboration

This exercise is a continuation of Exercise 4-3 that you accomplished in Chapter 04 "File Permissions, Text Editors, Text Processors & The Shell" in which you set up the setgid bit on the */sdata* directory for collaboration amongst members of the *sdatagrp* group on the local system. In

---

this exercise, you will export the */sdata* directory to *insider.example.com* for group collaboration on that system. It is assumed that Exercise 4-3 was performed on *server.example.com* and that the same group members also exist on *insider.example.com*.

1. Repeat steps 1 to 3 from Exercise 19-1.
2. Open the */etc/exports* file in a text editor and add the following entry for */sdata*:

   /sdata   insider.example.com(rw)

3. Implement steps 5 through to the end of Exercise 19-1 to complete this exercise.

Managing NFS Resources

Managing NFS resources involves exporting, mounting, viewing, unmounting, and unexporting them. Some of these tasks have been demonstrated in the previous section; others are discussed below.

Viewing Exported and Mounted Resources

To verify the functionality of both the server and the client, *cd* into the mount point of a resource and run the *ll* command. If both commands run successfully, it means the resource is exported and mounted, and that there are no issues. Several commands such as *showmount, exportfs, df,* and *mount* are available that allow you to view what resources are exported by the server, available to a particular client for mounting, and mounted on a client. Let us look at some examples.

To view exported resources, execute any of the following on the NFS server:

exportfs
showmount –e
cat /var/lib/nfs/etab

To view what resources are currently mounted by which NFS client, execute any of the following on the NFS server:

showmount –a
cat /var/lib/nfs/rmtab

To view mounted resources, execute any of the following on the NFS client:

mount | grep nfs
df –t nfs
cat /etc/mtab | grep nfs

Unmounting a Resource

Unmounting a resource detaches a mounted resource from the directory tree on an NFS client. Follow the steps below to unmount a remote resource on an NFS client:

1. Make certain that no users are accessing the resource (*/usr/local/bin* for example). If a non-critical process is using the resource or a user is accessing the mount point, list their PIDs and usernames using the *fuser* command with the −u option:

    # **fuser −cu /usr/local/bin**
    /usr/local/bin: 3412c(root)

2. Terminate any processes holding up the resource using the *fuser* command with the −k option:

    # **fuser −ck /usr/local/bin**
    /usr/local/bin: 3412c

4. Run the *umount* command to unmount the resource:

    # **umount /usr/share/man**

5. Open the */etc/fstab* file in a text editor and remove the associated entry if you wish to delete it for good.

Unexporting a Resource

After ensuring with the *showmount* command that the resource to be unexported has not been currently mounted by any clients, issue the *exportfs* command on the NFS server to unexport the resource:

exportfs −u /usr/local/bin

If you unexport a mounted resource, the next time a user on that NFS client requests access to the resource, NFS will return the "NFS stale file handle" error message.

To unexport all resources listed in the */var/lib/nfs/etab* file, execute the *exportfs* command with the −au options:

exportfs −avu

Monitoring NFS Activities

Monitoring NFS activities typically involves capturing and displaying NFS read and write statistics on the server and client. Tools such as *nfsstat*, *nfsiostat*, and *mountstats* are available and may be used for this purpose.

The *nfsstat* command provides detailed information on both NFS and RPC activities. Without any options, it captures and displays both NFS and RPC activities. This command supports several options such as −s, −c, and −r that instructs the command to capture server, client, and RPC statistics, respectively. With the −m option, it shows all activities on mounted resources.

Here is a sample output of the *nfsstat* command when it is executed without any options:

nfsstat

Server rpc stats:

calls	badcalls	badauth	badclnt	xdrcall
10	34	4	30	0

Server nfs v4:

null		compound	
10	100%	0	0%

Server nfs v4 operations:

op0-unused		op1-unused		op2-future		access		close		commit	
0	0%	0	0%	0	0%	0	0%	0	0%	0	0%

create		delegpurge		delegreturn		getattr		getfh		link	
0	0%	0	0%	0	0%	0	0%	0	0%	0	0%

lock		lockt		locku		lookup		lookup_root		nverify	
0	0%	0	0%	0	0%	0	0%	0	0%	0	0%

open		openattr		open_conf		open_dgrd		putfh		putpubfh	
0	0%	0	0%	0	0%	0	0%	0	0%	0	0%

putrootfh		read		readdir		readlink		remove		rename	
0	0%	0	0%	0	0%	0	0%	0	0%	0	0%

renew		restorefh		savefh		secinfo		setattr		setcltid	
0	0%	0	0%	0	0%	0	0%	0	0%	0	0%

setcltidconf		verify		write		rellockowner		bc_ctl		bind_conn	
0	0%	0	0%	0	0%	0	0%	0	0%	0	0%

exchange_id		create_ses		destroy_ses		free_stateid		getdirdeleg		getdevinfo	
0	0%	0	0%	0	0%	0	0%	0	0%	0	0%

getdevlist		layoutcommit		layoutget		layoutreturn		secinfononam		sequence	
0	0%	0	0%	0	0%	0	0%	0	0%	0	0%

set_ssv		test_stateid		want_deleg		destroy_clid		reclaim_comp	
0	0%	0	0%	0	0%	0	0%	0	0%

The *nfsiostat* command consults the */proc/self/mountstats* file and displays the NFS read and write statistics for each mounted resource. You need to specify a time interval and a count of iterations for the execution of this command.

Here is a sample output of the *nfsiostat* command when it is executed every 3rd second for 3 times:

nfsiostat 3 3
server:/home mounted on /home:

op/s	rpc bklog					
0.00	0.00					
read:	ops/s	kB/s	kB/op	retrans	avg RTT (ms)	avg exe (ms)
	0.000	0.000	0.000	0 (0.0%)	0.000	0.000
write:	ops/s	kB/s	kB/op	retrans	avg RTT (ms)	avg exe (ms)
	0.000	0.000	0.000	0 (0.0%)	0.000	0.000

. . . .

The *mountstats* command also consults the */proc/self/mountstats* file and displays the NFS read and write statistics for the specified mounted resource. You can specify the --nfs or the --rpc option with the command to restrict it to capture and display only NFS or RPC statistics.

Here is a sample output of the *mountstats* command when it is executed on the */home* mount point:

 # mountstats /home
Stats for server:/home mounted on /home:
 NFS mount options:
 rw,vers-4,rsize-1048576,wsize-1048576,namlen-255,acregmin-3,acregmax-60,acdirmin-30,acdirmax-60,hard
 ,proto-tcp,port-0,timeo-600,retrans-2,sec-sys,clientaddr-192.168.2.201,minorversion-0,local_lock-none
 NFS server capabilities: caps-0xffff,wtmult-512,dtsize-32768,bsize-0,namlen-255
 NFSv4 capability flags: bm0-0xfdffbfff,bm1-0xf9be3e,acl-0x3
 NFS security flavor: 1 pseudoflavor: 0

NFS byte counts:
 applications read 0 bytes via read(2)
 applications wrote 0 bytes via write(2)
 applications read 0 bytes via O_DIRECT read(2)
 applications wrote 0 bytes via O_DIRECT write(2)
 client read 0 bytes via NFS READ
 client wrote 0 bytes via NFS WRITE

RPC statistics:
 24 RPC requests sent, 24 RPC replies received (0 XIDs not found)
 average backlog queue length: 0

FSINFO:
    1 ops (4%)   0 retrans (0%) 0 major timeouts
    avg bytes sent per op: 176   avg bytes received per op: 108
    backlog wait: 0.000000 RTT: 2.000000   total execute time: 2.000000 (milliseconds)
ACCESS:
    3 ops (12%)   0 retrans (0%) 0 major timeouts
    avg bytes sent per op: 184   avg bytes received per op: 228
    backlog wait: 0.000000 RTT: 4.000000   total execute time: 4.333333 (milliseconds)
. . . .

Chapter Summary

This chapter introduced you to one of the most common system administration tasks, the administration of shared resources using the NFS service. You learned the NFS concepts, benefits, versions, and security, and understood the daemons, commands, related files, and startup scripts pertaining to NFS. You studied how NFS server and client interact with each other, and you completed exercises for setting them up. You used commands that displayed exported and mounted NFS resources, unmounted and unexported resources, and captured and displayed NFS statistical data.

Chapter Review Questions

1. Which file is referenced by the *nfsiostat* and *mountstats* commands?
2. Which file is updated when the *umount* command is executed successfully?
3. Name the three commands that are used to get NFS I/O statistical data?
4. Which command can be used to view exported resources?
5. Is the syntax of the *mount* command: *mount –t nfs –o rw <hostname>:<exported_resource>* /mnt_point correct?
6. Which option would you use with the *exportfs* command to unexport a resource?
7. Write the name of any SELinux boolean that is related to NFS.
8. Which daemon is responsible of mapping UIDs and GIDs with their corresponding usernames and groupnames?
9. Which is the main NFS configuration file?
10. Which file is updated when the *exportfs* command is executed successfully?
11. What does the *rpcinfo* command show?
12. Which two ports must be enabled in an active firewall to allow the NFS server traffic to pass through?
13. What is the default NFS version used in RHEL6?
14. Which daemon must run on both NFS server and client?
15. What is the purpose of the all_squash option used with a exported resource?
16. Which command can be used to export a resource?
17. What does the following line entry in the */etc/exports* file mean: */dir *(rw)*?
18. Which file needs to be modified if you wish to completely disable the support for older NFS versions?
19. Which daemon sends a file handle back to the NFS client?
20. One of the benefits of sharing the */home* file system is to have user home directories in one location. True or False?
21. What kind of information does the */var/lib/nfs/etab* file store?
22. Which daemon maintains the information in the *rmtab* file?

Answers to Chapter Review Questions

1. The */proc/self/mountstats* file.
2. The */var/lib/nfs/rmtab* file.
3. The *nfsstat*, *nfsiostat*, and *mountstats* commands.
4. The *showmount* command can be used to view exported resources.
5. Yes, it is correct.
6. The –u option.
7. The nfs_export_all_rw directive.
8. The *rpc.idmapd* daemon.
9. The */etc/sysconfig/nfs* file.
10. The */var/lib/nfs/etab* file.
11. The *rpcinfo* command shows the running RPC-based service daemons.
12. Ports 111 and 2049.
13. The default NFS version used in RHEL6 is version 4.
14. The *rpcbind* daemon.
15. The all_squash option treats all users on the NFS client as anonymous users.
16. The *exportfs* command.

17. This would export the */dir* resource in read/write mode to any system that needs to mount it.
18. The */etc/sysconfig/nfs* configuration file.
19. The *rpc.mountd* daemon.
20. True.
21. It stores information about all exported resources.
22. The *rpc.mountd* daemon.

Labs

Lab 19-1: Set Up an NFS Server

Set up an NFS server and export */home* and */var/log* to systems on the local network in read/write mode. Ensure that the NFS server supports all NFS versions. Mount both exported resources on another RHEL system in read/write mode. Use defaults for configuration information not provided here.

Lab 19-2: Share a Directory for Group Collaboration

Create a group and add users to it as members. Create a directory and assign it to the group. Set appropriate bits on the directory to prepare it for group collaboration. Use NFS and export the directory in read/write mode. Create a mount point on the client, create a group with the same name and assign the group to the mount point directory. Add members to the group. Mount the NFS share on the mount point. Check to ensure that users belonging to the groups on either system should be able to create files that are shareable by group members on both the NFS server and client.

Chapter 20

Samba & FTP

This chapter covers the following major topics:

- ✓ Describe Samba and its features and benefits
- ✓ Samba daemons, commands, configuration files, and scripts
- ✓ Explain Samba configuration file and software packages
- ✓ Understand SELinux requirements for Samba
- ✓ Configure a Samba server for file sharing using commands and SWAT
- ✓ Access the Samba share from another Linux system
- ✓ Share a resource for group collaboration using Samba
- ✓ Share user home directories using Samba
- ✓ Introduction to very secure File Transfer Protocol (vsFTP)
- ✓ Understand vsFTP configuration, control files, and SELinux requirements
- ✓ Configure a vsFTP server for specific user access, anonymous access, and anonymous-only download
- ✓ Test a vsFTP server using an FTP client program
- ✓ Overview of the wget utility

This chapter includes the following RHCE objectives:

27. Mount and unmount CIFS network file systems
SMB: Provide network shares to specific clients
SMB: Provide network shares suitable for group collaboration
FTP: Configure anonymous-only download

Samba is a networking protocol which allows Linux and Unix systems to share file and print resources with Windows and other Linux and Unix systems. The configuration of a Samba server is straightforward and may be accomplished by modifying the Samba configuration file using either a text editor or the web-based tool. Samba has a number of benefits associated with it, including interacting with Windows systems seamlessly.

File Transfer Protocol is a standard networking protocol for transferring files among hosts on the network and the Internet. This protocol is based on the client/server architecture and has been in use on Linux, UNIX, and other operating system platforms for decades. It may be configured to allow users to log in anonymously or with their usernames/passwords and upload or download files with or without encryption. This protocol is supported on virtually every operating system today.

The wget utility is a non-interactive utility for downloading files from the web using HTTP, HTTPS, and FTP protocols. It may be scheduled to run in the background via the Cron service to retrieve large files.

Understanding Samba

Server Message Block (SMB), now widely known as the *Common Internet File System* (CIFS), is a networking protocol developed by Microsoft, IBM, and Intel in the late 1980's to enable Windows-based PCs to share file and print resources. This protocol has been in use in Windows operating systems as the primary native protocol for file and printer sharing purposes. On the other hand, software called *Samba* was developed in the Linux world to share file and print resources with Microsoft systems using the SMB format. This enabled Linux systems to participate in Microsoft Windows workgroups and domains. As well, Samba allowed Linux systems to share file and print resources with other Linux systems, and UNIX.

In the Samba terminology, the system that shares its file and print resources is referred to as a *Samba server* and the system that accesses those resources is referred to as a *Samba client*. The resources may be shared for group collaboration among users that have accounts on different systems running a mix of Linux and Windows operating systems. Moreover, user home directories that exist on a Windows system may be shared with Linux systems, and vice versa. This would eliminate the need of having a separate home directory on each system for each user where the user logs on.

Benefits

The use of Samba provides several benefits, some of which are listed below:

- ✓ Samba shares can be accessed on Windows systems as standard drive letters, and can be navigated via Windows Explorer.
- ✓ Windows shares can be mounted as Samba resources on RHEL.
- ✓ A Samba server can be configured as a *Primary Domain Controller* (PDC) in the Windows environment.
- ✓ A Samba server can act as a *Windows Internet Name Service* (WINS) server or client for name resolution.
- ✓ A Samba server can be set up as an Active Directory member server on a Windows network.
- ✓ A Samba server can act as a print server for Windows-based CIFS clients.

- ✓ RHEL and Windows domain usernames and passwords can be used on either platform for authentication and authorization.
- ✓ Samba shares support Microsoft Access Control Lists.
- ✓ Samba shares can be accessed on other RHEL, Linux, and UNIX systems, and vice versa.

Samba Daemons, Commands, Configuration Files, and Scripts

When working with Samba, several daemons, commands, configuration files, and scripts are involved. The tables shown below list and explain them.

Table 20-1 describes Samba daemons.

Daemon	Description
smbd	The main Samba server daemon that manages services such as resource sharing, resource locking, user authentication, and user authorization. This daemon uses TCP port 445 for its operation.
nmbd	The NetBIOS server daemon that handles services such as browsing of file shares and print shares available on the network and hostname resolution. This daemon uses UDP ports 137, 138, and 139 for NetBIOS name service, NetBIOS datagram service, and NetBIOS session service, respectively.

Table 20-1 Samba Daemons

Table 20-2 describes Samba commands.

Command	Description
findsmb	Finds Samba servers on the local network.
mount	Mounts a Samba resource specified at the command line or listed in the */etc/fstab* file. It adds an entry for the mounted resource to the client's */etc/mtab* file and can be used to display mounted resources listed in this file.
nmblookup	Queries a WINS server.
pdbedit	Maintains a local user database on the server.
smbclient	Connects to Samba resources.
smbpasswd	Allows you to change Samba user passwords.
smbprint	Sends a print job to a Samba printer.
swat	A web-based tool that lets you set up a Samba server.
testparm	Tests the syntax of the *smb.conf* file.
umount	Functions opposite to that of the *mount* command.

Table 20-2 Samba Commands

Table 20-3 describes Samba configuration and functional files.

File	Description
/etc/samba/smb.conf	Samba server configuration file.
/etc/samba/smbusers	Maintains Samba and Linux user mappings.
/etc/samba/smbpasswd or /var/lib/samba/private/smbpasswd	Maintains Samba user passwords. This file is used for authentication purposes. Samba user passwords may be different from their Linux user passwords.
/var/log/samba	Directory location for Samba log files.
/etc/sysconfig/samba	Contains directives used when Samba is started.

Table 20-3 Samba Configuration and Functional Files

Table 20-4 describes Samba startup and shutdown script.

Scripts	Description
/etc/rc.d/init.d/smb	Starts, stops, restarts, and provides status of the Samba service.

Table 20-4 Samba Startup and Shutdown Script

Understanding Samba Configuration File

The *smb.conf* file is the main configuration file for setting up a Samba server. This is the file where you specify resources to be shared and several other parameters. There are two major sections in this file that cover Global Settings and Share Definitions. The former defines the directives that affect the overall Samba behavior, and includes options for networking, logging, standalone server, domain members, domain controller, browser control, name resolution, printing, and file systems. The latter defines the share-specific directives for home directories, printers, and other custom shares.

The following is an excerpt from the *smb.conf* file that show all uncommented lines and disabled directives except for the domain controller and file system portions. Each options area and its directives are explained in the Table that follows.

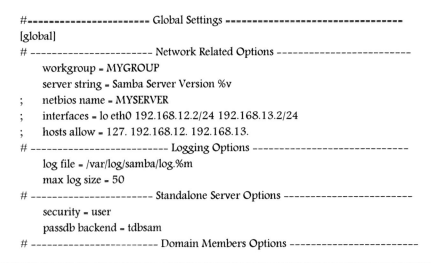

```
#------------------------ Global Settings ----------------------------------
[global]
# ---------------------- Network Related Options ------------------------
        workgroup = MYGROUP
        server string = Samba Server Version %v
;       netbios name = MYSERVER
;       interfaces = lo eth0 192.168.12.2/24 192.168.13.2/24
;       hosts allow = 127. 192.168.12. 192.168.13.
# ------------------------- Logging Options ----------------------------
        log file = /var/log/samba/log.%m
        max log size = 50
# --------------------- Standalone Server Options ----------------------
        security = user
        passdb backend = tdbsam
# --------------------- Domain Members Options ----------------------
```

Red Hat Certified System Administrator & Engineer

```
;       security = domain
;       passdb backend = tdbsam
;       realm = MY_REALM
;       password server = <NT-Server-Name>
# --------------------- Browser Control Options ---------------------------
;       local master = no
;       os level = 33
;       preferred master = yes
#------------------------ Name Resolution ----------------------------
;       wins support = yes
;       wins server = w.x.y.z
;       wins proxy = yes
;       dns proxy = yes
# -------------------------- Printing Options -------------------------
        load printers = yes
        cups options = raw
;       printcap name = /etc/printcap
;       printcap name = lpstat
;       printing = cups
#--------------------------- Share Definitions ----------------------------
[homes]
        comment = Home Directories
        browseable = no
        writable = yes
;       valid users = %S
;       valid users = MYDOMAIN\%S
[printers]
        comment = All Printers
        path = /var/spool/samba
        browseable = no
        guest ok = no
        writable = no
        printable = yes
;       [netlogon]
;       comment = Network Logon Service
;       path = /var/lib/samba/netlogon
;       guest ok = yes
;       writable = no
;       [Profiles]
;       path = /var/lib/samba/profiles
;       browseable = no
;       guest ok = yes
;       [public]
;       comment = Public Stuff
;       path = /home/samba
;       public = yes
;       writable = yes
;       printable = no
;       write list = +staff
```

Description of directives from the output above is provided in Table 20-5. Note that the ; character represents disabled directives.

Directive	Global Settings
Network Related Options	
workgroup	Name of Windows workgroup or domain.
server string	Any description that identifies this system as a Samba server.
netbios name	Name of the Samba server.
interfaces	Specifies network and host IP addresses to be served in case there are more than one network interfaces.
hosts allow / deny	Allows (denies) the specified networks to access the shares. May also be set at the individual share level.
Logging Options	
log file	Defines separate log file names in the *var/log/samba* directory for every client that connects to this Samba server.
max log size	Specifies the size in KBs for the log file to grow to before it is rotated.
Standalone Server Options	
security	Defines one of the five supported authentication options: ads = performs authentication from an Active Directory server. domain = performs authentication from a domain controller. server = performs authentication from a server. share = this system will be part of a peer to peer workgroup where the Samba users will be able to browse share contents without entering their login credentials. user = matches usernames/passwords with those on individual Windows systems. This is the default.
passdb backend	Controls which backend is to be used for user authentication. This directive is used with security = user, and it supports three options: smbpasswd = uses the *var/lib/samba/private/smbpasswd* file for authentication. This is the default. tdbsam (*Trivial Database Security Accounts Manager*) = sets up a trivial local account database in the *var/lib/samba* directory. ldapsam (*LDAP Security Accounts Manager*) = uses a remote LDAP account database.
Domain Members Options (Includes both options from Standalone Server Options plus the following)	
realm	Specifies the name of an active directory realm if the security directive under the Standalone Server Options is set to ads.
password server	Identifies the name of a server to be used for user authentication if the security directive under the Standalone Server Options is set to domain or share.
Browser Control Options	
local master	Specifies whether to use the Samba server as a local browser master.
os level	Determines the precedence of this Samba server in master browser elections.
preferred master	Specifies whether to force a local browser election providing this Samba server a better opportunity to win the election.

Directive	Global Settings
Name Resolution	
wins support	Specifies whether to enable the WINS server functionality on the local system.
wins server	Sets the IP address or hostname of a remote WINS server.
wins proxy	Specifies whether to enable non-WINS systems to use a WINS proxy.
dns proxy	Specifies whether to enable searches via DNS.
Printing Options	
load printers	Specifies whether to load the list of printers automatically instead of setting them up manually.
cups options	Specifies custom printing options.
printcap name	Specifies the file name from which to obtain the printer information.
printing	Defines "cups" as the default sub-system for printing.
Share Definitions	
comment	A short description of the share visible when a client queries the server.
browseable	Specifies whether to allow users on clients to browse each other's files and directories located on the share.
writable	Specifies whether to enable writing to the share.
valid users	Specifies a list of users and groups that are allowed access to the share (group names are prefixed with the @ sign).
path	Sets the absolute path to the share.
guest ok	Specifies whether to enable the guest account.
printable	Specifies whether to load printer configuration files.
public	Specifies whether to create public accessible shares.
write list	Specifies a list of users and groups that can write to the share (group names are prefixed with the @ sign).

Table 20-5 Directives in the smb.conf File

Samba Software Packages

There are several software packages that need to be installed on the system to configure it as a Samba server or client. Samba support is added when the Windows File Server software package group is chosen at install time. The *rpm* command below produces a list of the Samba packages that are currently installed on the system:

 # rpm –qa | egrep 'samba|smb'
samba-3.5.10-125.el6.x86_64
samba-common-3.5.10-125.el6.x86_64
samba-winbind-clients-3.5.10-125.el6.x86_64

Table 20-6 describes these packages.

Package	Description
samba	Includes the support for Samba server.
samba-common	Provides Samba man pages and general commands.
samba-winbind-clients	Includes the support for WINS clients.
samba-swat	Contains the Samba web administration tool.

Table 20-6 Samba Software Packages

Configuring Samba

This section provides an overview on how to make Samba work with SELinux activated in the enforcing mode. It also furnishes exercises to solidify the knowledge and understanding of the Samba functionality that has been explained in the previous section.

SELinux Requirements for Samba

SELinux protects systems by setting appropriate controls using file contexts and boolean values. The SELinux security context samba_share_t needs to be set on each resource before it can be shared with Samba. This should only be done on user-created and other non-system directories.

The system has several SELinux booleans that need to be looked at carefully in order to see which ones need to be enabled or disabled for Samba to function as expected. To list the booleans, you can either use the SELinux Configuration tool and filter out the Samba-related booleans or run the *getsebool* command as follows:

```
# getsebool –a | egrep 'samba|smb' | egrep –v 'cobbler|git|fusefs'
allow_ftpd_use_cifs --> off
allow_smbd_anon_write --> off
httpd_use_cifs --> off
qemu_use_cifs --> on
samba_create_home_dirs --> off
samba_domain_controller --> off
samba_enable_home_dirs --> off
samba_export_all_ro --> off
samba_export_all_rw --> off
samba_run_unconfined --> off
samba_share_nfs --> off
use_samba_home_dirs --> off
virt_use_samba --> off
```

The above output indicates that there are thirteen booleans that should be reviewed and modified for our exercises. Table 20-7 describes the purpose of each of these booleans.

Boolean	Purpose
allow_ftpd_use_cifs	Specifies whether a mounted Samba share can be used as a public file transfer location by the FTP service.
allow_smbd_anon_write	Specifies whether the Samba server is allowed to write to public directories. Such directories must have SELinux security context set to public_content_rw_t.
httpd_use_cifs	Specifies whether a mounted Samba share can be used by the Apache service.
qemu_use_cifs	Specifies whether a mounted Samba share can be used by the QEMU service.
samba_create_home_dirs	Turn on/off user ability to create home directories on Samba shares.
samba_domain_controller	Activates/deactivates Samba for use as a domain controller for authentication purposes.
samba_enable_home_dirs	Turn on/off sharing of user home directories.
samba_export_all_ro	Specifies whether to export Samba resources read-only.
samba_export_all_rw	Specifies whether to export Samba resources read/write.
samba_run_unconfined	Allows/disallows the execution of unconfined scripts located in the /var/lib/samba/scripts directory.
samba_share_nfs	Specifies whether mounted NFS resources be shared.
use_samba_home_dirs	Turn on/off Samba clients to mount user home directories.
virt_use_samba	Specifies whether Samba clients running on virtual machines be allowed to access mounted Samba shares.

Table 20-7 Samba Related SELinux Booleans

Exercise 20-1: Configure a Samba Server

In this exercise, you will configure *server.example.com* as a Samba server and share the */usr/local/bin* directory for access by *user1, user2,* and a guest user account called *guest*. It is assumed that these user accounts already exist on the system. These users will have the ability to access and browse the share from systems on the 192.168.2 network and in the *example.com* domain except for system *physical.example.com*.

1. Ensure that the required Samba packages are installed:

   # **rpm –qa | egrep 'samba|smb'**
   samba–winbind–clients–3.5.10–125.el6.x86_64
   samba–client–3.5.10–125.el6.x86_64
   samba–swat–3.5.10–125.el6.i686
   samba–common–3.5.10–125.el6.x86_64
   samba–3.5.10–125.el6.x86_64

   Install the packages if they are not already installed using one of the procedures outlined in Chapter 06 "Package Management".

2. Edit the *smb.conf* file and define the directives as follows:

```
[global]
        netbios name        = server
        workgroup           = EXAMPLE
        server string       = Samba Server
        hosts allow         = 192.168.2. .example.com
        hosts deny          = physical.example.com
        passdb backend      = smbpasswd
        security            = user
        log file            = /var/log/samba/%m.log
        max log size        = 50
[localusrbin]
        comment             = /usr/local/bin directory available to user1, user2, and guest
        browseable          = yes
        path                = /usr/local/bin
        valid users         = user1 user2
        guest ok            = yes
        guest account       = guest
        writable            = yes
        write list          = user1 user2 guest
```

3. Execute the *testparm* command to check for any syntax errors in the file. Use the –v option to display other default parameters that are not defined in the file.

```
# testparm
Load smb config files from /etc/samba/smb.conf
Processing section "[/usr/local/bin]"
Global parameter guest account found in service section!
Loaded services file OK.
[global]
      workgroup = EXAMPLE
      server string = Samba Server
      passdb backend = smbpasswd
      log file = /var/log/samba/%m.log
      max log size = 50
      hosts allow = 192.168.2., .example.com
      hosts deny = physical.example.com
[localusrbin]
      comment = /usr/local/bin directory available to user1, user2, and guest
      path = /usr/local/bin
      valid users = user1, user2
      write list = user1, user2, guest
      read only = No
      guest ok = Yes
```

4. Add users *user1* and *user2* to Samba configuration and assign them a password such as *Welcome01*. Show the contents of the *smbusers* file.

```
# smbpasswd –a user1
New SMB password.
Retype new SMB password.
Added user user1.
# smbpasswd –a user2
# cat /var/lib/samba/private/smbpasswd
user1:500:XXXXXXXXXXXXXXXXXXXXXXXXXXXXXXXX:00B2C85DDFBD8CC81602D6FC7340EB0B:[U
]:LCT-50C69C50:
user2:501:XXXXXXXXXXXXXXXXXXXXXXXXXXXXXXXX:C549EE84021E5E8372E10CEDEAFD02A8:[U
]:LCT-50C69C57:
```

5. Activate SELinux booleans to allow Samba shares in read-only and read/write modes, and verify the activation:

```
# setsebool –P samba_export_all_ro=1 samba_export_all_rw=1
# getsebool –a | grep samba_export
samba_export_all_ro --> on
samba_export_all_rw --> on
```

6. Set the correct SELinux security context on the */usr/local/bin* directory, and confirm:

```
# chcon –Rvt samba_share_t /usr/local/bin
# ll –Zd /usr/local/bin
drwxr-xr-x. root root system_u:object_r:samba_share_t:s0 /usr/local/bin
```

7. Issue the *semanage* command and modify the contexts on the directory to ensure that the new contexts survive a SELinux relabeling:

```
# semanage fcontext –a –t samba_share_t /usr/local/bin
```

8. Configure host-based access by allowing Samba traffic on UDP ports 137, 138, and 139, and TCP port 445 to pass through the firewall:

```
# iptables –I INPUT –m udp –p udp --dport 137 –j ACCEPT
# iptables –I INPUT –m udp –p udp --dport 138 –j ACCEPT
# iptables –I INPUT –m udp –p udp --dport 139 –j ACCEPT
# iptables –I INPUT –m tcp –p tcp --dport 445 –j ACCEPT
```

9. Save the rules in the */etc/sysconfig/iptables* file and restart the firewall to activate the new rules:

```
# service iptables save ; service iptables restart
```

10. Set the Samba service to autostart at each system reboot, and validate:

```
# chkconfig smb on
# chkconfig --list smb
smb       0:off  1:off  2:on  3:on  4:on  5:on  6:off
```

11. Start the Samba processes and check the running status:

> # **service smb start**
> # **service smb status**

Configuring a Samba Server Using SWAT

Samba server configuration may alternatively be accomplished using the Samba browser-based application called *Samba Web Administration Tool* (SWAT). You can install this tool using the *yum* command if it is not already there (you might have to download its package from the Internet). The following is based on the assumption that the package samba-swat-3.5.10-125.el6.x86_64.rpm has been downloaded and is located in the */tmp* directory:

> # **yum –y install /tmp/samba-swat-3.5.10-125.el6.x86_64.rpm**

Execute the following command to activate the *swat* service to autostart at each system reboot, and validate:

> # **chkconfig swat on**
> # **chkconfig --list swat**
> swat   0:off   1:off   2:on   3:on   4:on   5:on   6:off

The *swat* service is controlled by the *xinetd* daemon, which consults the */etc/xinetd.d/swat* file for necessary startup configuration information. An excerpt from this file is shown below:

```
service swat
{
        disable              = no
        port                 = 901
        socket_type          = stream
        wait                 = no
        only_from                        = 127.0.0.1
        user                 = root
        server               = /usr/sbin/swat
        log_on_failure       += USERID
}
```

The file contents show several directives. The key directives are disable, port, and only_from. The disable directive is set to no indicating that the service is enabled to autostart at system reboots, the port directive indicates the port number that SWAT uses, and the only_from directive specifies where this service can be accessed from. If you wish this service to be accessed from any system on the 192.168.2 network, specify this network address with the only_from directive.

SWAT is now ready to run. Open up a browser window on your RHEL system and type http://localhost:901. Enter username *root* and its password to connect to SWAT. The home page of this application will show up as depicted in Figure 20-1.

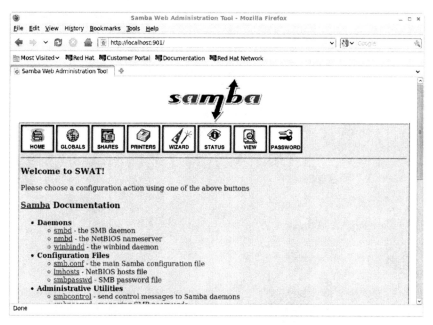

Figure 20-1 Samba Web Administration Tool (SWAT)

SWAT is flexible, feature-rich, and powerful, and provides a single interface for setting up and administering Samba shares as well as offering a wealth of documentation on Samba. There are eight buttons across the top, of which the GLOBALS button allows you to set and modify global directives affecting the overall functionality of Samba server; SHARES and PRINTERS give you the ability to configure directory and print shares; WIZARD guides you through the configuration process; STATUS displays the current status of the Samba server daemons, active connections, and shares; VIEW displays the current configuration information; and PASSWORD allows you to add, delete, enable, or disable a Samba user account and change a user password.

Create a copy of the *smb.conf* file before you make any changes with SWAT as this application overwrites any customization that you have made to the file previously.

Exercise 20-2: Access a Samba Share on Another RHEL System

In this exercise, you will configure *insider.example.com* as a Samba client and access the */usr/local/bin* directory as *user1*. You will install the package that is required to use Samba client functionality only. Likewise, you will open only UDP ports 137 and 138. You will ensure that the share is automatically mounted when the system is rebooted. You will store the username and password for *user1* in a file which is owned by the *root* user with 0400 permissions on the file.

1. Log on to *insider.example.com* and perform steps 1, 8, 9, and 10 from Exercise 20-1.
2. Run the *smbclient* command with the –L option to list available resources from *server.example.com*. Specify *user1* with the –U option and enter the password for *user1* when prompted.

```
# smbclient –L server.example.com –U user1
Enter user1's password:
Domain=[EXAMPLE] OS=[Unix] Server=[Samba 3.5.10-114.el6]

    Sharename      Type    Comment
    ---------      ----    -------
    IPC$           IPC     IPC Service (Samba Server)
    localusrbin    Disk    /usr/local/bin directory available to user1, user2, and guest
Domain=[EXAMPLE] OS=[Unix] Server=[Samba 3.5.10-114.el6]

    Server         Comment
    ------         --------


    Workgroup      Master
    ---------      ------
    EXAMPLE        server
```

3. Mount the */usr/local/bin* share and enter the password for *user1* when prompted:

    # mount –t cifs //server.example.com:/localusrbin /usr/local/bin –o username=user1

4. Execute the *df* or the *mount* command to check the mount status of the share:

    # df
    //server/localuserbin   18553596 10545208 7065900 60% /usr/local/bin
    # mount | grep localusrbin
    //server/localusrbin on /usr/local/bin type cifs (rw)

5. Edit the */etc/fstab* file and add the following entry to it:

    //server.example.com/localusrbin /usr/local/bin cifs rw,credentials=/etc/samba/smbcred 0 0

6. Create the */etc/samba/smbcred* file and add the credentials for *user1* to it:

    # vi /etc/samba/smbcred
    username=user1
    password=Welcome01

7. Ensure that the *smbcred* file is owned by *root* and the permissions on the file is 0400:

    # chown root /etc/samba/smbcred && chmod 0400 /etc/samba/smbcred

    With these entries in place, the share can be manually mounted as:

    # mount –at cifs

Exercise 20-3: Share a Resource for Group Collaboration

This exercise is a continuation of Exercise 4-3 that you accomplished in Chapter 04 "File Permissions, Text Editors, Text Processors & The Shell" in which you set up the setgid bit on the */sdata* directory for collaboration amongst members of the *sdatagrp* group on the local system. In this exercise, you will share the */sdata* directory to *server.example.com* for group collaboration. It is assumed that Exercise 4-3 was performed on *insider.example.com* and that the same group members also exist on *server.example.com*.

1. Log on to *insider.example.com* and execute step 1 from Exercise 20-1 to ensure that the required Samba packages are installed.
2. Append the following entries to the *smb.conf* file and comment out duplicates:

```
[global]
        netbios name        = insider
        workgroup           = EXAMPLE
        server string       = Samba Server
        hosts allow         = 192.168.2. .example.com
        passdb backend      = smbpasswd
        security            = user
        log file            = /var/log/samba/%m.log
        max log size        = 50
[sdata]
        comment             = /sdata directory for group collaboration
        browseable          = yes
        path                = /sdata
        valid users         = @sdatagrp
        guest ok            = no
        writable            = yes
        write list          = @sdatagrp
```

3. Execute steps 3 and 5 to 11 from Exercise 20-1. Replace */usr/local/bin* with */sdata*.

Exercise 20-4: Share User Home Directories

In this exercise, you will share user home directories located in */home* on *insider.example.com* with all systems in the *example.com* domain. You will ensure that the share is browseable and writeable, and it does not allow guest users to access it.

1. Log on to *insider.example.com* and execute step 1 from Exercise 20-1 to ensure that the required Samba packages are installed.
2. Append the following entries to the *smb.conf* file and comment out duplicates:

```
[global]
        netbios name        = insider
        workgroup           = EXAMPLE
        server string       = Samba Server
        hosts allow         = .example.com
        passdb backend      = smbpasswd
```

```
          security              - user
          log file              - /var/log/samba/%m.log
          max log size          - 50
[home_dirs]
          comment               - User home directories
          browseable            - yes
          path                  - /home
          guest ok              - no
          writable              - yes
```

3. Execute step 3 from Exercise 20-1.
4. Activate SELinux booleans to allow Samba shares in read-only and read/write modes, and verify the activation:

> **# setsebool –P samba_export_all_ro=1 samba_export_all_rw=1 **
> **samba_enable_home_dirs=1**
> **# getsebool –a | egrep 'samba_export|samba_enable_home'**
> samba_export_all_ro --> on
> samba_export_all_rw --> on
> samba_enable_home_dirs --> on

5. Execute steps 6 to 11 from Exercise 20-1. Replace */usr/local/bin* with */home*.

File Transfer Protocol (FTP)

File Transfer Protocol (FTP) is the standard protocol for transferring files on the network and the Internet, and has been used on Linux, UNIX, and other operating system platforms for decades. This protocol is based on the client/server architecture and its server component allows users to transfer files after being authenticated or anonymously. In RHEL6, an enhanced implementation of FTP called *very secure File Transfer Protocol* (vsFTP) is available and is used as the default file transfer service. This enhanced version is faster, more stable, and powerful than the standard FTP and at the same time allows you to enable, disable, and set security controls on incoming service requests. The vsFTP daemon called *vsftpd* communicates on port 21 as defined in the */etc/services* file.

Understanding the vsFTP Configuration File

The primary configuration file for the vsFTP service is the *vsftpd.conf* file located in the */etc/vsftpd* directory. There are a number of directives in this file that may be set appropriately to control the behavior of the *vsftpd* daemon. The default uncommented directives are shown below:

> **# grep –v ^# /etc/vsftpd/vsftpd.conf**
> anonymous_enable-YES
> local_enable-YES
> write_enable-YES
> local_umask-022
> dirmessage_enable-YES
> xferlog_enable-YES

```
connect_from_port_20=YES
xferlog_std_format=YES
listen=YES
pam_service_name=vsftpd
userlist_enable=YES
tcp_wrappers=YES
```

These and some other key directives are listed and explained in Table 20-8.

Directive	Description
anonymous_enable	Enables/disables anonymous FTP access. If enabled, anonymous users are able to log in to the */var/ftp* directory by default. Alternatively, you can create a directory elsewhere and define it with the anon_root directive.
local_enable	Default is YES. Enables/disables local users from logging in via FTP.
write_enable	Default is YES. Allows/disallows remote users to write. If enabled, the target directory needs to have the SELinux label public_content_rw_t set as well as either the ftp_home_dir directive or the allow_ftpd_anon_write directive enabled.
local_umask	Default is 022. If writes are enabled, remote users are able to write new files with this umask value.
anon_upload_enable	Disabled by default. Allows/disallows anonymous users from uploading files. If enabled, the target directory needs to have the SELinux lable public_content_rw_t set.
anon_mkdir_write_enable	Default is YES. Allows/disallows anonymous users from creating directories.
dirmessage_enable	Default is YES. Enables/disables displaying the contents of the *~/.message* file to logged in users.
xferlog_enable	Default is YES. Enables/disables logging file transfer activities to the */var/log/xferlog* file.
connect_from_port_20	Default is YES. Ensures that port transfer connections originate from TCP port 20.
chown_uploads	Disabled by default. Automatically changes the ownership on uploaded files.
chown_username	Disabled by default. Automatically changes the ownership on uploaded files to the specified username.
xferlog_file	Default is set to the */var/log/vsftpd.log* file. Logs file transfer activities to the specified file.
xferlog_std_format	Default is YES. Logs file transfer activities in the standard xferlog format capturing date, time, IP address, file being transferred, username, and so on.
idle_session_timeout	Default is 600. Logs out users after 600 seconds of inactivity.
chroot_local_users	Default is YES. Limits user access to a chrooted directory. May be used if local_enable is set to YES.
listen	Default is YES. Enables *vsftpd* to listen on IPv4 sockets.
pam_service_name	Default is vsftpd. Defines the service name for which PAM is to be enabled.

Directive	Description
userlist_deny	Default is YES. Allows/disallows users listed in the */etc/vsftpd/user_list* file from connecting via FTP.
userlist_enable	Default is YES. Enables/disables the *vsftpd* service to consult the */etc/vsftpd/user_list* file.
tcp_wrappers	Default is YES. Allows/disallows *vsftpd* access to incoming requests based on the contents in the */etc/hosts.allow* and */etc/hosts.deny* files.

Table 20-8 vsftpd.conf Directives

Controlling Who Can Use the vsFTP Service

There are two files that control which users can or cannot log in with an FTP client. These files are located in the */etc/vsftpd* directory and are called *ftpusers* and *user_list*. Both files contain a list of users, one line per user name and, by default, contain system users only. The *ftpusers* file lists users that are denied FTP access to the system and the *user_list* file lists users that are allowed or denied access based on the NO or YES value of the userlist_deny directive in the *vsftpd.conf* file. The *user_list* file is consulted only if the userlist_enable directive is set to YES. The *ftpusers* file is used if user authentication via PAM is enabled.

vsFTP makes use of the TCP Wrappers access control files.

SELinux Requirements for vsFTP

SELinux protects systems by setting appropriate controls using file contexts and boolean values. The key file contexts associated with the FTP service are ftpd_exec_t set on the *vsftpd* daemon file, public_content_rw_t set on public writeable directories, public_content_t set on public accessible files, and xferlog_t set on the file transfer log file */var/log/xferlog*.

The system has several SELinux booleans that need to be looked at carefully to see which ones might need to be enabled or disabled for ftp to function as expected. To list the booleans, you can either use the SELinux Configuration tool and filter out the ftp-related booleans or run the *getsebool* command as follows:

```
# getsebool –a | grep ftp | grep –v tftp
allow_ftpd_anon_write --> off
allow_ftpd_full_access --> off
allow_ftpd_use_cifs --> off
allow_ftpd_use_nfs --> off
ftp_home_dir --> off
ftpd_connect_db --> off
httpd_enable_ftp_server --> off
```

The above output indicates that there are seven booleans that we should review and modify for our exercises. Table 20-8 describes the purpose of each of these booleans.

Boolean	Purpose
allow_ftpd_anon_write	Specifies whether users are allowed to write to directories anonymously. Such directories must have SELinux security context set to public_content_rw_t.
allow_ftpd_full_access	Allows/disallows full user access to files.
allow_ftpd_use_cifs	Specifies whether a mounted Samba share be used as a public file transer location.
allow_ftpd_use_nfs	Specifies whether a mounted NFS resource be used as a public file transer location.
ftp_home_dir	Specifies whether users are allowed to read and write to their home directories.
ftpd_connect_db	Specifies whether a MySQL database is used for user authentication.
httpd_enable_ftp_server	Specifies whether the Apache web server acts as a FTP server by listening on port 21.

Table 20-9 FTP Related SELinux Booleans

Exercise 20-5: Configure a vsFTP Server for Specific User Access Only

In this exercise, you will configure *server.example.com* as an FTP server to provide full read and write access to only *user1* and *user2* and only from *physical.example.com* to the system and to their home directories. You will configure appropriate host-based access control and SELinux settings.

1. Ensure that the required vsFTP package is installed:

   # **rpm –qa | grep vsftpd**
   vsftpd-2.2.2-6.el6_0.1.x86_64

   Install the package if it is not already installed using one of the procedures outlined in Chapter 06 "Package Management".

2. Edit the *vsftpd.conf* file and define the directives as follows:

   anonymous_enable=NO
   local_enable=YES
   write_enable=YES
   local_umask=022
   dirmessage_enable=YES
   xferlog_enable=YES
   connect_from_port_20=YES
   xferlog_std_format=YES
   listen=YES
   pam_service_name=vsftpd
   userlist_enable=NO
   tcp_wrappers=YES

3. Activate SELinux booleans to provide users with read/write FTP access to the system and to their home directories, and verify the activation:

setsebool –P allow_ftpd_full_access=1 ftp_home_dir=1
getsebool –a | egrep 'ftpd_full|ftp_home'
allow_ftpd_full_access --> on
ftp_home_dir --> on

4. Configure host-based access by allowing FTP traffic on TCP port 20 and 21 to pass through the firewall:

iptables –I INPUT –m tcp –p tcp --dport 20 –j ACCEPT
iptables –I INPUT –m tcp –p tcp --dport 21 –j ACCEPT

5. Save the rules in the */etc/sysconfig/iptables* file and restart the firewall to activate the new rules:

service iptables save ; service iptables restart

6. Set the vsFTP service to autostart at each system reboot, and validate:

chkconfig vsftpd on
chkconfig --list vsftpd
vsftpd   0:off   1:off   2:on   3:on   4:on   5:on   6:off

7. Open the */etc/hosts.deny* TCP Wrappers file and enter the following to ensure that only *user1* and *user2* and only from *physical.example.com* are allowed to use the FTP service:

vsftpd : ALL EXCEPT user1@physical.example.com,user2@physical.example.com

8. Start the *vsftpd* service and check the running status:

service vsftpd start
service vsftpd status

Exercise 20-6: Configure a vsFTP Server for Anonymous Access Only

In this exercise, you will configure *server.example.com* to provide an anonymous file upload and download service using the vsFTP service. You will provide access to anonymous users from everywhere. You will ensure that they get write access into the default */var/ftp* directory. You will define appropriate directives to ensure that each uploaded file automatically gets the *bin* user ownership.

1. Log on to *server.example.com* and execute step 1 from Exercise 20-5 to ensure that the required vsFTP package is installed.
2. Edit the *vsftpd.conf* file and define the directives as follows:

```
anonymous_enable=YES
local_enable=NO
write_enable=YES
anon_upload_enable=YES
chown_uploads=YES
chown_username=bin
local_umask=022
dirmessage_enable=YES
xferlog_enable=YES
connect_from_port_20=YES
xferlog_std_format=YES
listen=YES
pam_service_name=vsftpd
userlist_enable=NO
tcp_wrappers=NO
```

3. Activate the SELinux boolean to allow anonymous user with write access and verify the activation:

    # setsebool –P allow_ftpd_anon_write=1
    # getsebool –a | grep allow_ftpd_anon_write
    allow_ftpd_anon_write --> on

4. Set the correct SELinux security context on the */var/ftp* directory to public_content_rw_t recursively, and confirm:

    # chcon –Rvt public_content_rw_t /var/ftp
    # ll –Z /var/ftp
    drwxrwxrwx. root root system_u:object_r:public_content_rw_t:s0 /var/ftp

5. Issue the *semanage* command and modify the contexts on the directory to ensure that the new contexts survive a SELinux relabeling:

    # semanage fcontext –a –t public_content_rw_t /var/ftp

6. Execute steps 4 to 6 from Exercise 20-5.
7. Open the */etc/hosts.allow* TCP Wrappers file and enter the following to ensure that all users from everywhere are allowed to use the anonymous FTP service:

    vsftpd : ALL

8. Restart the *vsftpd* daemon and check the running status:

    # service vsftpd start
    # service vsftpd status

Exercise 20-7: Configure a vsFTP Server for Anonymous-Only Download

In this exercise, you will configure *server.example.com* to allow anonymous downloads from the */var/ftp/outgoing* directory using the vsFTP service. You will provide access to anonymous users from everywhere. You will ensure that they get write access into the */var/ftp/outgoing* directory.

1. Log on to *server.example.com* and execute step 1 from Exercise 20-5 to ensure that the required vsFTP package is installed.
2. Create the */var/ftp/outgoing* directory, change ownership and group membership to root:ftp, and permissions to 755:

    # **mkdir /var/ftp/outgoing**
    # **chown root:ftp /var/ftp/outgoing**
    # **chmod 755 /var/ftp/outgoing**

3. Change the home directory for the *ftp* user to */var/ftp/outgoing*:

    # **usermod –d /var/ftp/outgoing ftp**

4. Copy some files to */var/ftp/outgoing* and change their permissions to 644.
5. Execute step 3 from Exercise 20-6.
6. Set the correct SELinux security context on the */var/ftp/outgoing* directory to public_content_t recursively, and confirm:

    # **chcon –Rvt public_content_t /var/ftp/outgoing**
    # **ll –Zd /var/ftp/outgoing**
    drwxr-xr-x. root root unconfined_u:object_r:public_content_t:s0 /var/ftp/outgoing

7. Issue the *semanage* command and modify the contexts on the directory to ensure that the new contexts survive a SELinux relabeling:

    # **semanage fcontext –a –t public_content_t /var/ftp/outgoing**

8. Execute steps 4 to 6 from Exercise 20-5.
9. Execute step 7 and 8 from Exercise 20-6.

Testing the Functionality

Once the vsFTP server is configured, use a client program such as *ftp* or *lftp* to test it. *lftp* is more powerful and includes more features than the standard *ftp* client program. Alternatively, you can use a browser and type in the IP address of the system or simply *localhost* in the URL and press the Enter key.

You need to install the package for lftp if it is not already installed on your system using the *yum* command or one of the other methods outlined in Chapter 06 "Package Management".

The procedure to use *lftp* is such that you specify the hostname or IP address of the remote system you want to connect to. For example, the following runs *lftp* from *insider* to enter *server* with IP address 192.168.2.201 as *user1*:

$ lftp 192.168.2.201 –u user1

As soon as *lftp* contacts *server* on port 21, the *vsftpd* daemon comes into action and sends the requester a login prompt to enter a username. It engages TCP Wrappers to check whether the remote system is allowed. It then checks for the validity of the user in the */etc/passwd* file and references *ftpusers* and *user_list* files whether the user is allowed. It prompts the user to enter a password and consults the */etc/shadow* file for validation. The daemon establishes a communication session with *lftp* provided both the system and the user have been validated.

To log in as an anonymous user for file transfers, run the *lftp* command as follows and type in *anonymous* as the user name and a string of characters such as an email address as password:

$ lftp server

Available commands at the *lftp:~>* prompt can be listed by typing *?*:

```
lftp:~> ?
  !<shell-command>              (commands)
  alias [<name> [<value>]]      bookmark [SUBCMD]
  cache [SUBCMD]                cat [–b] <files>
  cd <rdir>                     chmod [OPTS] mode file...
  close [–a]                    [re]cls [opts] [path/][pattern]
  debug [<level>|off] [–o <file>]du [options] <dirs>
  exit [<code>|bg]              get [OPTS] <rfile> [–o <lfile>]
  glob [OPTS] <cmd> <args>      help [<cmd>]
  history –w file|–r file|–c|–l [cnt]   jobs [–v]
  kill all|<job_no>             lcd <ldir>
  lftp [OPTS] <site>            ls [<args>]
  mget [OPTS] <files>           mirror [OPTS] [remote [local]]
  mkdir [–p] <dirs>             module name [args]
  more <files>                  mput [OPTS] <files>
  mrm <files>                   mv <file1> <file2>
  [re]nlist [<args>]            open [OPTS] <site>
  pget [OPTS] <rfile> [–o <lfile>]   put [OPTS] <lfile> [–o <rfile>]
  pwd [–p]                      queue [OPTS] [<cmd>]
  quote <cmd>                   repeat [OPTS] [delay] [command]
  rm [–r] [–f] <files>          rmdir [–f] <dirs>
  scache [<session_no>]         set [OPT] [<var> [<val>]]
  site <site_cmd>               source <file>
  torrent [–O <dir>] <file>     user <user|URL> [<pass>]
  version                       wait [<jobno>]
  zcat <files>                  zmore <files>
```

Commands such as *cd*, *ls*, *pwd*, and *quit* work the same way as they do at the Linux shell prompt. The *get* and *put* commands are used to download and upload a single file and their enhanced

---

versions *mget* and *mput* for several files. These commands support wildcard characters and tab completion. You can run the *cd*, *ls*, and *pwd* commands by preceding them with the ! sign which will execute these commands at the Linux command prompt.

Use the *quit* or *bye* command at the lftp> prompt to exit the connection and go back to the Linux command prompt.

All file transfer activities are recorded in the */var/log/xferlog* file by default.

The wget Utility

wget is a non-interactive file download utility that allows you to retrieve a single file, multiple files, or an entire directory structure from an FTP or HTTP source. Certain options are important when working with *wget*, and are listed and explained in Table 20-10. For additional options and details, refer to the command's man pages.

Option	Description
–P	Specifies the directory location in which to download the files.
–T	Specifies the number of seconds before the command times out.
–d	Turns the debug mode on.
–i	Downloads files from URLs listed in the specified file.
–nv	Hides the verbose output except for error messages and basic information.
–o	Logs messages to the specified output file.
–q	Hides the output and any errors.
–r	Turns the recursive retrieval on.
–t	Specifies the number of retries.
–v	Displays the verbose output.
--user	Specifies a username for FTP or HTTP access.
--password	Specifies a password for the --user option.

Table 20-10 wget Command Options

Let us take a look at a few examples to understand how this command works.

To download a kernel package in the current directory from *ftp.kernel.org* with verbose mode on:

```
# wget ftp.kernel.org/pub/linux/kernel/v2.6/linux-2.6.32.9.tar.gz
--2012-11-20 19:50:10-- http://ftp.kernel.org/pub/linux/kernel/v2.6/linux-2.6.32.9.tar.gz
Resolving ftp.kernel.org... 149.20.4.69
Connecting to ftp.kernel.org|149.20.4.69|:80... connected.
HTTP request sent, awaiting response... 200 OK
Length: 81922503 (78M) [application/x-gzip]
Saving to: linux-2.6.32.9.tar.gz.1
100%[------------------------------------->] 81,922,503   729K/s   in 1m 50s
2012-11-20 19:52:04 (728 KB/s) - linux-2.6.32.9.tar.gz.1
```

To download the same package and store at */var/tmp*:

wget ftp.kernel.org/pub/linux/kernel/v2.6/linux-2.6.32.9.tar.gz –P /var/tmp

To download the same package with output redirected to a file called */tmp/wget.out*:

wget ftp.kernel.org/pub/linux/kernel/v2.6/linux-2.6.32.9.tar.gz –o /tmp/wget.out

To download *index.html* file from *www.redhat.com* with the debug mode on:

wget –d www.redhat.com
DEBUG output created by Wget 1.12 on linux-gnu.

--2012-11-20 19:52:54-- http://www.redhat.com/
Resolving www.redhat.com... 23.1.167.214
Caching www.redhat.com => 23.1.167.214
Connecting to www.redhat.com|23.1.167.214|:80... connected.
Created socket 3.
Releasing 0x0000000001541830 (new refcount 1).
---request begin---
GET / HTTP/1.0
User-Agent: Wget/1.12 (linux-gnu)
Accept: */*
Host: www.redhat.com
Connection: Keep-Alive
---request end---
HTTP request sent, awaiting response...
---response begin---
HTTP/1.0 200 OK
Server: Apache
X-Powered-By: Servlet 2.5; JBoss-5.0/JBossWeb-2.1
Content-Type: text/html;charset=UTF-8
Expires: Wed, 21 Nov 2012 00:49:50 GMT
Cache-Control: max-age=0, no-cache
Pragma: no-cache
Date: Wed, 21 Nov 2012 00:49:50 GMT
Connection: close
Set-Cookie: JSESSIONID=F26BBFB139CD4FB6EF7ECF62BAE17380.d20ec1a7; Path=/
Set-Cookie: DCID=origin-www-a; expires=Wed, 21-Nov-2012 02:49:50 GMT; path=/; domain=.redhat.com
---response end---
200 OK
Stored cookie www.redhat.com -1 (ANY) / <session> <insecure> [expiry none] JSESSIONID
F26BBFB139CD4FB6EF7ECF62BAE17380.d20ec1a7
cdm: 1 2 3 4 5 6 7 8
Stored cookie redhat.com -1 (ANY) / <permanent> <insecure> [expiry 2012-11-20 21:49:50] DCID origin-
www-a
Length: unspecified [text/html]
Saving to: index.html
    [<=>   ] 78,338   --.-K/s in 0.1s
Closed fd 3
2012-11-20 19:52:56 (629 KB/s) - index.html

Chapter Summary

This chapter discussed two additional file sharing services: Samba and FTP. It provided an overview of Samba and its features and listed SELinux requirements for it. The chapter described Samba daemons, commands, configurations files, scripts, and provided basic information on Samba software packages. It explained how to configure a Samba server for sharing files using commands, the Samba Server Configuration tool and Samba Web Administration Tool. The chapter further explained how a RHEL share could be accessed on another RHEL system and a Windows system, and how a Windows share could be accessed on a RHEL system. The next major topic discussed very secure FTP service and the associated SELinux requirements and configuration file. It explained how to control user access. You studied how to configure a vsFTP server and test it with lftp client.

Chapter Review Questions

1. Both Samba and FTP are used for file and printer sharing on the network. True or False?
2. What is the default port that SWAT uses?
3. What is the name and location of the vsFTP configuration file?
4. The *wget* command may be used for uploading and downloading files. True or False?
5. Name two FTP client programs that you may use to connect to the FTP server.
6. What is the name of the main Samba server configuration file and where is it located?
7. What is the new name for the SMB protocol?
8. The *smbclient* command is used to mount Samba shares on a Linux system. True or False?
9. The vsFTP directive anonymous_enable must be set to YES for both anonymous and non-anonymous FTP access. True or False?
10. What does the directive browseable mean in Samba?
11. What would the Samba directive *write list = @dba* do?
12. Name the two files that may be used to control the user access to the FTP service?
13. What are the four network port numbers that Samba uses?
14. Where does the FTP service store its log information?
15. What is the name of the Samba web management application?
16. What is the purpose of the *testparm* command?
17. What would the directive *hosts allow = 192.168.23* in the Samba configuration file imply?
18. vsFTP does not support TCP Wrappers for user- and host-based security. True or False?
19. What SELinux file context would you set on a Samba share?
20. What is the name of the main Samba daemon?
21. Where does the Samba server log its activities?
22. Provide any three benefits of using Samba on Linux.
23. Which SELinux file context would you set on the directory to provide write access to FTP users?
24. When would you activate the Samba boolean samba_export_all_rw?
25. What is the name of the daemon for the default FTP service in RHEL?

Answers to Chapter Review Questions

1. False. FTP is not used for printer sharing.
2. Port 901 is the default port that SWAT uses.
3. The vsFTP configuration file is called *vsftpd.conf* and it is located in the */etc/vsftpd* directory.

---

4. False. The *wget* utility may only be used for file download purpose.
5. The *ftp* and *lftp* programs.
6. The main Samba server configuration file is called *smb.conf* and it is located in the */etc/samba* directory.
7. Common Internet File System (CIFS).
8. False. The *smbclient* command is used to view available Samba shares.
9. False. This directive is only activated when you want to set anonymous FTP access.
10. The browseable directive allows users to browse the contents of a Samba share.
11. This directive would allow the write access to the members of the *dba* group.
12. The *ftpusers* and *user_list* files.
13. Samba server uses ports 137, 138, 139, and 445 for its proper operation.
14. The FTP service stores its log information in the */var/log/xferlog* file.
15. The name of the Samba web management application is called Samba Web Administration Tool (SWAT).
16. The *testparm* command is used to validate the settings in the *smb.conf* file.
17. This directive would allow systems on the 192.168.23 network to access Samba shares.
18. False. vsFTP service does support TCP Wrappers for user- and host-based security.
19. The samba_share_t file context.
20. The main Samba daemon is called *smbd*.
21. Samba server logs its activities in to the */var/log/samba* directory.
22. Three benefits are: Samba shares can be accessed on Windows and vice versa, a Samba server can be configured to function as a PDC, and a Samba server can be configured to function as a WINS server.
23. The public_content_rw_t file context.
24. This Samba boolean allows a share to be exported in read/write mode.
25. The daemon name for the default FTP service in RHEL is called *vsftpd*.

Labs

Lab 20-1: Configure a Samba Server for Guest Access

Configure *physical.example.com* as a Samba server. Create a directory and share it for read/write access by a single guest account that should be able to access and browse this share from *server.example.com* only. The share should be configured on *server.example.com* to mount automatically after each system reboot. Implement necessary host-based access control rules (firewall and TCP Wrappers) and SELinux settings.

Lab 20-2: Share a Directory for Group Collaboration

Create a directory on *insider.example.com* and share it with *server.example.com* for group collaboration. Create a group account and add three members to it. Configure the access for this share on *server.example.com* as writeable and update the relevant file so that the share is automatically mounted at system reboots. Implement necessary host-based access control rules and SELinux settings.

Lab 20-3: Configure a vsFTP Server for Anonymous File Transfer

Configure *server.example.com* to provide anonymous file uploads and downloads using vsFTP. Create directories */var/ftp/uploads* for uploads and */var/ftp/downloads* for downloads. Allow anonymous users to access the service only from systems on the example.com domain and the 192.168.3 network except for systems with IP addresses 192.168.3.100 and 192.168.3.101. Give anonymous users full read and write access into the */var/ftp/uploads* and */var/ftp/downloads* directories. Define appropriate directives to ensure that each uploaded file automatically gets the ownership of the *bin* user. Implement necessary host-based access control rules (firewall and TCP Wrappers) and SELinux settings.

Chapter 21

Apache Web Server & Secure Shell Server

This chapter covers the following major topics:

- ✓ Apache web server, features, configuration and log files, and commands
- ✓ Understand SELinux, and host-based and user-based security requirements
- ✓ Set up and test the Apache web server with varying configuration
- ✓ Apache virtual host and benefits of using it
- ✓ Set up and test the Apache standard virtual host
- ✓ Set up and test the Apache secure virtual host
- ✓ Deploy a basic CGI script
- ✓ Understand the OpenSSH service and review the private/public encryption
- ✓ An overview of SSH administration commands and configuration files
- ✓ Set up the SSH server,
- ✓ Configure key-based and passwordless authentication
- ✓ Use ssh, scp, and sftp, and access the SSH server from Windows

This chapter includes the following RHCE objectives:

HTTP/HTTPS: Configure a virtual host
HTTP/HTTPS: Configure private directories
HTTP/HTTPS: Configure group-managed content
HTTP/HTTPS: Deploy a basic CGI application
SSH: Configure key-based authentication
SSH: Configure additional options described in documentation

Web servers are deployed in order to host websites for business, commercial, or personal use. Hosting websites is made possible by the software that runs on a system referred to as a *web server* and provides access to the hosted websites to users on the Internet or the corporate network. There are millions of web servers being used and they are running a variety of web server software of which Apache is the most popular. Apache provides stronger security, better stability, superior reliability, flexible scalability, and enhanced functionality compared to all other available web server software.

Secure Shell (SSH) is a network protocol that delivers a secure mechanism for data transmission between source and destination systems over insecure network channels. SSH includes a set of utilities providing remote users the ability to log in, transfer files, and execute commands securely using strong, hidden encryption and authentication. SSH tools have successfully replaced their insecure counterparts to a great extent in the corporate world.

The Apache Web Server

Apache (A PAtCHy sErver) is undoubtedly the most popular, stable, secure, efficient, and modular web server software currently being used on the Internet and corporate networks. RHEL6 includes version 2.2 of the Apache server software, which supports features such as virtual hosts, access control, dynamic module loading, and secure HTTP. Apache uses the *httpd* daemon to provide client access to web pages using the HTTP and HTTPS protocols. Since its inception back in 1995, Apache has grown to be the world's most popular and widely used web server software today.

Apache Features

Some of the key features associated with the Apache web server are:

- ✓ Support for files of sizes larger than 2GB.
- ✓ Support for encrypted authentication methods.
- ✓ Support for LDAP authentication.
- ✓ Support for virtual hosts to host several websites on a single web server with all websites sharing one IP address.
- ✓ Support for separate configuration files for separate services stored in separate locations.
- ✓ Support for proxy load balancing.
- ✓ Supported on virtually every operating system and hardware platform.

Apache Configuration Files

Apache configuration files are located in the */etc/httpd/conf* and */etc/httpd/conf.d* directories. The configuration files are listed and explained in Table 21-1.

File	Description
conf/httpd.conf	The standard Apache web server configuration file.
conf.d/ssl.conf	The secure Apache web server configuration file.
conf.d/*.conf	Configuration files for supporting services.

Table 21-1 Key Apache Files

The configuration file for setting up a standard Apache web server is the *httpd.conf* file located in the */etc/httpd/conf* directory and the configuration file for setting up a secure Apache web server is the *ssl.conf* file located in the */etc/httpd/conf.d* directory. There are many other configuration files located in the */etc/httpd/conf.d* directory with *.conf* extension and are referenced when Apache is started. An *ll* on this directory produces an output similar to the following:

ll /etc/httpd/conf.d

```
-rw-r--r--. 1 root  root   707 Sep  9   2004 auth_kerb.conf
-rw-r--r--. 1 root  root   295 May 20  2009 manual.conf
-rw-r--r--. 1 root  root  8842 Aug 1   2011 nss.conf
-rw-r--r--. 1 root  root  1796 Apr 22  2005 perl.conf
-rw-r--r--. 1 root  root   392 Oct  6  2011 README
-rw-r--r--. 1 root  root   845 Oct 31  2011 revocator.conf
-rw-r--r--. 1 root  root  9473 Dec  8  2009 ssl.conf
-rw-r--r--. 1 root  root   352 Sep  9  2004 webalizer.conf
-rw-r--r--. 1 root  root   299 May 20  2009 welcome.conf
-rw-r--r--. 1 root  root    43 Jul  14  2009 wsgi.conf
```

These files may be customized according to your requirements. Moreover, you can create as many additional configuration files in this directory for your web pages as you wish. The syntax for all of these files is almost identical to that of the *httpd.conf* file. The following sub-section provides an analysis of the *httpd.conf* file and describes the general directives used in it.

An Analysis of the httpd.conf File

The contents in *httpd.conf* are divided into three sections: Global Environment, Main Server Configuration, and Virtual Hosts, and are explained below:

Section 1: Global Environment: Directives defined in this section affect the overall operation of the Apache web server. Some of the key directives are listed and described in Table 21-2.

Directive	Description
ServerTokens	Default is OS. Restricts as to what information is to be displayed to user when the user hits a web page that does not exist on the website. This directive has six levels of restriction: Prod, Major, Minor, Min, OS, and Full, where Prod limits the information to the web server software name only and Full displays the web server software name and version, operating system name, and optional modules being used. The rest are in between.
ServerRoot	Default is */etc/httpd*. Directory location to store configuration, error, and log files.
PidFile	Default is *run/httpd.pid* with respect to ServerRoot. Stores the PID of the *httpd* process.
Timeout	Default is 60 seconds. Specifies the time before receive and send requests time out.
KeepAlive	Default is off. Enables or disables more than one requests per connection (a.k.a. a *persistent connection*).

Directive	Description
MaxKeepAliveRequests	Default is 100. Specifies the maximum number of allowable requests during a persistent connection. A value of 0 allows an unlimited number of requests.
KeepAliveTimeout	Default is 15 seconds. Specifies the time to wait for the next request from the same client on the same connection.
StartServers	Specifies the number of server processes to start.
MinSpareServers	Specifies the minimum number of spare server processes.
MaxSpareServers	Specifies the maximum number of spare server processes.
ServerLimit	Specifies the maximum number of allowable server processes for the lifetime of the server.
MaxClients	Restricts the number of concurrent requests.
MaxRequestsPerChild	Restricts requests per child server process.
MinSpareThreads	Specifies the minimum number of spare threads.
MaxSpareThreads	Specifies the maximum number of spare threads.
ThreadsPerChild	Specifies the number of threads per child server process.
Listen	Default is 80. Specifies a port number to listen for client requests. Specify an IP address and a port if you wish to limit the web server access to a specific address.
LoadModule	Specifies a module to be loaded when the web server is started.
Include	Default is *conf.d/*.conf* with respect to ServerRoot. Specifies the location of additional configuration files to be loaded when Apache starts.
ExtendedStatus	Default is off. Controls whether to generate a full (on) or partial (off) status information.
User	Default is *apache*. Specifies the owner name for the *httpd* daemon.
Group	Default is *apache*. Specifies the group name for the *httpd* daemon.

Table 21-2 Directives in Section 1 – Global Environment

Section 2: 'Main' Server Configuration: Directives defined in this section relate to the main server and also set default values for any virtual host containers that may be defined in section 3 of this file. Some of the key directives in this section are listed and explained in Table 21-3.

Directive	Description
ServerAdmin	Default is root@localhost. Specifies the email address for the webmaster.
ServerName	Specifies the server name (or IP address) and port number.
UseCanonicalName	Default is off. Uses the hostname and port supplied by the client. If "on", it will use ServerName as the official hostname.
DocumentRoot	Default is */var/www/html*. Specifies the directory location for website files.
Options	Sets features such as Indexes, FileInfo, AuthConfig, Limit, and FollowSymLinks associated with web directories.
AllowOverride	Default is None. Enables or disables overriding of directives defined in *.htaccess* files.
Order	Defines the order for evaluating allow and deny directives.
Allow	Allows access to specified hosts.
Deny	Denies access to specified hosts.

Directive	Description
UserDir	Default is disabled. Specifies the directory location for user web pages. If enabled, contents located in the ~/*public_html* directory become visible to web users.
DirectoryIndex	Default is */usr/share/doc/HTML/index.html*. Specifies the web page to be served when a user requests an index of a directory.
AccessFileName	Default is *.htaccess*. Specifies the file name in each directory which the server will use for access control information.
Satisfy	Default is All. Allows the request if both authentication directives (used for password protection) and access directives (such as allow/deny) are satisfied. The other option is Any, which allows the request if any of the two conditions are met.
TypesConfig	Default is */etc/mime.types*. Sets the file name containing default file type mappings.
DefaultType	Default is text/plain. Sets the default content type to use for a document if its MIME type is not found in */etc/mime.types*.
HostNameLookups	Default is off. Sets whether to resolve URLs for IP addresses.
ErrorLog	Default is *logs/error_log* with respect to ServerRoot. Specifies the location for the error log file.
LogLevel	Default is warn. Specifies the level of verbosity at which messages are to be logged. Other options are debug, info, notice, error, crit, alert, and emerg.
LogFormat	Sets the format for log file contents.
CustomLog	Default is combined and stored in *logs/httpd/access_log* with respect to ServerRoot. Specifies the custom log file and identifies its format.
ServerSignature	Default is on. Adds a line that includes the Apache server version, the ServerName directive, and (optionally) the ServerAdmin directive (email address) to any server-generated documents such as error messages.
Alias	Defines a directory location outside of the DocumentRoot directory.
ScriptAlias	Specifies the directory location where CGI (*Common Gateway Interface*) scripts are stored.
IndexOptions	Controls how directory listings should appear.
AddDescription	Defines a short description for a file in server-generated indexes.
ReadmeName	Default is *README.html*. Defines a README file name to be appended to the end of directory listings.
HeaderName	Default is *HEADER.html*. Defines a HEADER file name to be prepended to the start of directory listings.
IndexIgnore	Lists file extensions, partial and full file names, and wildcard expressions that are not to be included in directory listings.
AddEncoding	Allows certain browsers to uncompress information online.
AddHandler	Maps a file extension to the specified handler.
AddOutputFilter	Maps a file extension to the specified filter.
BrowserMatch	Allows the server to send custom replies to different types of browsers.

Table 21-3 Directives in Section 2 – Main Server Configuration

Section 3: Virtual Hosts: Directives defined in this section are used if you wish to run multiple websites on the same server. Some of the key directives are listed and explained in Table 21-4.

Directive	Description
NameVirtualHost	Specifies the hostname or the IP address (and port) for the virtual host.
ServerAdmin	Specifies the e-mail address for the webmaster.
DocumentRoot	Specifies the root directory where virtual host files are located.
ServerName	Specifies the URL for the virtual host.
ErrorLog	Specifies the location for an error log file with respect to ServerRoot.
CustomLog	Specifies the location for a custom log file with respect to ServerRoot.

Table 21-4 Directives in Section 3 – Virtual Hosts

A lot of information is defined within containers in the *httpd.conf* file. There are four types of containers – virtual host, module, file, and directory – that are commonly used. Each of these containers begins and ends in the way listed in Table 21-5.

Container Begins	Container Ends
<VirtualHost .... >	</VirtualHost>
<IfModule .... >	</IfModule>
<Files .... >	</Files>
<Directory .... >	</Directory>

Table 21-5 Container Begin and End Syntax

Apache Log Files

Apache log files are located in the */var/log/httpd* directory, which is symbolically linked from the */etc/httpd/logs* directory. An *ll* on the directory is shown below. The *access_log* and *error_log* files log access to the web server and error messages, respectively. The *ssl_access_log*, *ssl_error_log*, and *ssl_request_log* files are only used if SSL certificates are employed.

```
# ll /var/log/httpd/*_log
-rw-r--r--. 1 root  root     0 Nov  4 04:27 access_log
-rw-r--r--. 1 root  root  2561 Nov 19 18:01 error_log
-rw-r--r--. 1 root  root     0 Oct 31 09:38 ssl_access_log
-rw-r--r--. 1 root  root   708 Nov 19 18:01 ssl_error_log
-rw-r--r--. 1 root  root     0 Oct 31 09:38 ssl_request_log
```

The LogFormat directive as defined in the *httpd.conf* file sets the standard log file format for Apache logs. Log related directives defined in this file are grepped and displayed below:

```
# cat /etc/httpd/conf/httpd.conf | grep Log | grep –v ^#
ErrorLog logs/error_log
LogLevel warn
LogFormat "%h %l %u %t \"%r\" %>s %b \"%{Referer}i\" \"%{User-Agent}i\"" combined
LogFormat "%h %l %u %t \"%r\" %>s %b" common
LogFormat "%{Referer}i -> %U" referer
LogFormat "%{User-agent}i" agent
CustomLog logs/access_log combined
```

The ErrorLog directive sets the log file name for Apache errors. The LogLevel is set to "warn", which instructs to log all warning and higher level messages. The LogFormat directive supports four formats: combined, common, referrer, and agent. "combined" combines what the other three formats log; "common" logs the name of the requesting host, the username and user ID of the client, the date and time of the request, the response returned, and the number of bytes returned; "referrer" specifies the referrer system from where the client was redirected here; and "agent" specifies the client software. Finally, set the CustomLog directive to where you wish the log information to go. It is recommended that you choose separate log files for each website configured on the server.

Apache Administration Commands

Table 21-11 lists and explains some common Apache administration commands.

Command	Description
apachctl	Controls the *httpd* daemon.
httpd	Server program for the Apache web server. With the –t option, it checks for any syntax errors in the Apache configuration files.
htpasswd	Creates and updates files to store usernames and passwords for basic authentication of Apache users. Specify the –c option to create a file and the –m option to use MD5 encryption for passwords.

Table 21-6 Apache Administration Command

Apache Web Server Security Overview

Securing an Apache web server can be done at three levels: firewall, SELinux, and directives in the *httpd.conf* file. These levels of security are described in the following sub-sections.

Security at the Firewall Level

Ports are defined with the Listen and NameVirtualHost directives in the *httpd.conf* file for the standard and secure Apache web server, respectively. Along with the port numbers, you may also specify IP address(es) of specific clients from where you want the standard or the secure Apache server accessed.

The iptables firewall in RHEL6 allows you to secure your system at the host level. You can permit or restrict access to your web server by opening port 80, 443, or any other port that you have defined with the Listen or the NameVirtualHost directive. Additionally, you may also want to specify the IP address of a system or network, or a domain name that you plan to allow access or restrict from gaining access.

Refer to Chapter 12 "Firewall & SELinux" for details on how to use the *iptables* command to set, save, and apply new firewall rules.

Security at the SELinux Level

There are a number of booleans associated with the Apache service. Run the SELinux Configuration tool *system-config-selinux*, go to Boolean, and then filter out apache. This will list all associated booleans; from all of them only six are enabled by default.

 # getsebool –a | egrep 'http|xguest_connect'

allow_httpd_anon_write --> off
allow_httpd_mod_auth_ntlm_winbind --> off
allow_httpd_mod_auth_pam --> off
allow_httpd_sys_script_anon_write --> off
httpd_builtin_scripting --> on
httpd_can_check_spam --> off
httpd_can_network_connect --> off
httpd_can_network_connect_cobbler --> off
httpd_can_network_connect_db --> off
httpd_can_network_memcache --> off
httpd_can_network_relay --> off
httpd_can_sendmail --> off
httpd_dbus_avahi --> on
httpd_enable_cgi --> on
httpd_enable_ftp_server --> off
httpd_enable_homedirs --> off
httpd_execmem --> off
httpd_read_user_content --> off
httpd_setrlimit --> off
httpd_ssi_exec --> off
httpd_tmp_exec --> off
httpd_tty_comm --> on
httpd_unified --> on
httpd_use_cifs --> off
httpd_use_gpg --> off
httpd_use_nfs --> off
xguest_connect_network --> on

Some of the key booleans are listed and described in Table 21-7.

Boolean	Description
allow_httpd_anon_write	Allows/disallows Apache to write to public directories anonymously.
allow_httpd_mod_auth_ntlm_winbind	Allows/disallows Apache to use the mod_auth_ntlm_winbind module.
allow_httpd_mod_auth_pam	Allows/disallows Apache to use the mod_auth_pam module.
allow_httpd_sys_script_anon_write	Allows/disallows Apache to write to public directories.
httpd_builtin_scripting	Allows/disallows Apache to use built-in scripting such as PHP.
httpd_can_check_spam	Allows/disallows Apache to check for spam.
httpd_can_network_connect	Allows/disallows Apache scripts and modules to connect to the network.
httpd_can_network_connect_cobbler	Allows/disallows Apache scripts and modules to connect to cobbler over the network.
httpd_can_network_connect_db	Allows/disallows Apache scripts and modules to connect to databases over the network.

Boolean	Description
httpd_can_network_memcache	Allows/disallows Apache to connect to the memcache server.
httpd_can_network_relay	Allows/disallows Apache to act as a relay.
httpd_can_sendmail	Allows/disallows Apache to send mail.
httpd_dbus_avahi	Allows/disallows Apache to talk to the avahi service via dbus.
httpd_enable_cgi	Enables/disables CGI support.
httpd_enable_ftp_server	Allows/disallows Apache to act as a FTP server by listening on port 21.
httpd_enable_homedirs	Enables/disables Apache's access to user home directories.
httpd_execmem	Allows/disallows Apache scripts and modules execmem.
httpd_read_user_content	Allows/disallows Apache to read user content.
httpd_setrlimit	Allows/disallows Apache to set read limit.
httpd_ssi_exec	Allows/disallows Apache to run SSI executables.
httpd_tmp_exec	Allows/disallows Apache to execute tmp content.
httpd_tty_comm	Allows/disallows Apache to talk to the terminal.
httpd_unified	Allows/disallows Apache to handle all content files.
httpd_use_cifs	Allows/disallows Apache to use mounted Samba shares.
httpd_use_gpg	Allows/disallows Apache to use GPG signatures.
httpd_use_nfs	Allows/disallows Apache to use mounted NFS resources.
xguest_connect_network	Allows/disallows xguest to configure Network Manager and connect to Apache ports.

Table 21-7 SELinux Booleans for Apache

In addition to the SELinux booleans, having proper SELinux file context on Apache files and directories is mandatory for Apache to operate smoothly. There are three key directories where the Apache information is typically stored, and these are */etc/httpd*, */var/www*, and */var/log/httpd*. Run the *ll* command with the –Zd options on these directories to check the current SELinux context:

```
# ll –Zd /etc/httpd
drwxr-xr-x. root root system_u:object_r:httpd_config_t:s0 /etc/httpd
# ll –Zd /var/www
drwxr-xr-x. root root system_u:object_r:httpd_sys_content_t:s0 /var/www
# ll –Zd /var/log/httpd
drwx------. root root system_u:object_r:httpd_log_t:s0 /var/log/httpd
```

The output indicates that SELinux labels httpd_config_t, httpd_sys_content_t, and httpd_log_t are set on directories containing the Apache configuration information, web contents, and log files, respectively. Individual files beneath these directories inherit their parent directory's context. Additionally, you will need to set the httpd_sys_content_ra_t or the httpd_sys_content_rw_t label on files providing web pages for read/append or read/write operation using the *chcon* command.

Refer to Chapter 12 "Firewall & SELinux" for details on how to use the *chcon* command and other tools to alter SELinux context on files and directories.

Security in the httpd.conf File

In addition to using the firewall and SELinux for securing access to the web server, there are several directives in the *httpd.conf* file that can be set appropriately to further enhance the level of security for the web server. Some of the directives such as ServerTokens and UserDir work at the global and main levels, respectively, as explained in Table 21-2 and 21-3 earlier in this chapter. Other directives may be set at the directory container level and are explained below.

Security in the httpd.conf File (General Security Directives)

Within a directory container, you can configure several general security attributes using the directives such as Options and AllowOverride to limit the use of the directory container and hence the access to the website and web pages defined therein. Table 21-8 lists and explains both directives.

Directive	Description
Options	ExecCGI: Allows web pages to execute CGI scripts. FollowSymLinks: Allows directories external to DocumentRoot to have symlinks. Indexes: Displays a list of files on the web server if no *index.html* file is available in the stated directory. Includes: Allows server-side includes. MultiViews: Allows substitution of file extensions. All: Allows all options besides MultiViews. None: Denies unauthorized users from executing CGI scripts located in DocumentRoot.
AllowOverride	None: Denies normal users from modifying permissions on files in DocumentRoot. All: Allows all options except for MultiViews. Limit: Allows the host access control directives such as allow, deny, and order to be used. AuthConfig: Allows a directory to use the *.htaccess* file for overriding default values of authorization directives. FileInfo: Allows a directory to use the *.htaccess* file for overriding default values controlling document types.

Table 21-8 General Security Directives

A sample directory container is shown below with some selected directives from Table 21-8:

```
<Directory /var/www/example>
    Options ExecCGI FollowSymLinks Includes
    AllowOverride AuthConfig FileInfo
</Directory>
```

Security in the httpd.conf File (Security at the Host Level)

This level of security lets you allow or deny access requests to your web server from specific hosts, networks, or domains based on the settings of the allow and deny directives. Table 21-9 provides examples to explain the usage of the two directives.

Allow / Deny Directive	Effect
Allow/Deny from all	Allows/Denies access to all hosts.
Allow/Deny from outsider	Allows/Denies access to *outsider.example.net*.
Allow/Deny from 192.168.2	Allows/Denies access to systems on the 192.168.2 network.
Allow/Deny from 192.168.4.0/24	Allows/Denies access to systems on the 192.168.4 network with default class C netmask.
Allow/Deny from 192.168.2.205	Allows/Denies access to the system with IP 192.168.2.205.
Allow/Deny from .example.com	Allows/Denies access to systems on the *example.com* domain.

Table 21-9 Host Level Allow/Deny Directives

Table 21-10 describes a couple of combinations of the allow and deny directives in which they are evaluated by the Order directive, and explains the impact of the settings.

Order Directive	Effect
Order allow,deny	Allows access to hosts defined with the allow directive, and denies all other hosts.
Order deny,allow	Denies access to hosts defined with the deny directive, and allows all other hosts.

Table 21-10 Host Level Order Directive

An example directory container is shown below with added host level access control directives:

```
<Directory /var/www/example>
    Options ExecCGI FollowSymLinks Includes
    AllowOverride AuthConfig FileInfo
    Order allow,deny
    Allow from example.com
</Directory>
```

Security in the httpd.conf File (Security at the User Level)

Securing an Apache server at the user level sets controls so that only authorized users are able to access the website. These users will be assigned passwords that may or may not be different from their RHEL passwords. Several directives are available for controlling access at this level and are described in Table 21-11.

Directive	Description
AuthType	Sets the basic authentication.
AuthName	Adds general comments.
AuthUserFile	Defines the file that contains authorized user passwords.
AuthGroupFile	Defines the file that contains authorized group passwords.
Require	Specifies users or groups that are allowed. For example: Require user *user1* → limits access to user *user1* only. Require group *dba* → limits access to members of the *dba* group.

Table 21-11 User Level Security Directives

Configuring an Apache Web Server

Having gone through the material provided in this chapter so far and the knowledge and understanding developed based on that, you should now be able to perform the exercises provided in this section. You might need to adjust settings presented in these exercises to make them work on your systems.

Exercise 21-1: Configure a Web Server for Basic Operation

In this exercise, you will set up an Apache web server for basic operation using the default directive settings in the *httpd.conf* file. You will test the web server's functionality by accessing the default page from both firefox and elinks web browsers.

1. Check whether Apache software packages are installed:

   # **rpm –qa | grep httpd**
   httpd–tools–2.2.15–15.el6_2.1.x86_64
   httpd–2.2.15–15.el6_2.1.x86_64

   Check whether the graphical web browser *firefox* and the textual web browser *elinks* are installed:

   # **rpm –qa | egrep 'firefox|elinks'**
   elinks–0.12–0.20.pre5.el6.x86_64
   firefox–3.6.24–3.el6_1.x86_64

   Install missing packages if they are not already installed using one of the procedures outlined in Chapter 06 "Package Management".

2. Configure host-based access by allowing Apache traffic on TCP port 80 to pass through the firewall:

   # **iptables –I INPUT –m tcp –p tcp --dport 80 –j ACCEPT**

3. Save the rule in the */etc/sysconfig/iptables* file and restart the firewall to activate the new rule:

---

service iptables save ; service iptables restart

4. Set the Apache service to autostart at each system reboot, and validate:

    # **chkconfig httpd on**
    # **chkconfig --list httpd**
    httpd   0:off 1:off 2:on 3:on 4:on 5:on 6:off

5. Start the Apache service using any of the following and check the running status:

    # **apachectl start**
    # **service httpd start**
    # **service httpd status**

6. Open up a firefox browser window and type in the URL http://localhost to view the default web page. Alternatively, you can run the *elinks 127.0.0.1* from the command line and the default web page will show up as depicted in Figure 21-1.

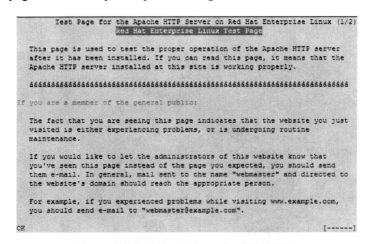

Figure 21-1 Default Apache Web Page

Run the *tail* command on the *error_log* and *access_log* files located in the */var/log/httpd* directory to view any error and access messages generated.

Exercise 21-2: Configure a Web Server with Host-Based Security

In this exercise, you will configure a web server on *insider.example.com* with DocumentRoot set to */var/www/insider*, web contents located in the */var/www/insider/webcontents* directory, and the name of the index file *index.html*. You will ensure that this website is accessible on port 80 and only from systems on 192.168.1 and 192.168.2 networks. You will ensure that users on the 192.168.3 network are refrained from accessing this website. You will configure appropriate host-based access control and SELinux settings.

---

1. Execute step 1 from Exercise 21-1.
2. Open the *httpd.conf* file and add or replace the directives as per below. Leave other settings intact.

   ```
   Listen 80
   DocumentRoot "/var/www/insider"
   <Directory /var/www/insider>
           Options Indexes FollowSymLinks
           Order allow,deny
           Allow from 192.168.1 192.168.2
           Deny from 192.168.3
   </Directory>
   DirectoryIndex index.html
   ErrorLog logs/error_log
   CustomLog logs/access_log combined
   ```

3. Create DocumentRoot directory and a sub-directory *webcontents* for storing website contents:

   # **mkdir –p /var/www/insider/webcontents**

4. Create the *index.html* file in the */var/www/example/webcontents* directory and add some text to it:

   # **vi /var/www/insider/webcontents/index.html**
   This web page is to test host-based security.

5. Check the syntax of the *httpd.conf* file for any errors:

   # **httpd –t**
   Syntax OK

6. Execute steps 2 to 4 from Exercise 21-1.
7. Modify SELinux security contexts recursively on the */var/www/insider* directory so that the *httpd* daemon is able to read them:

   # **chcon –Rvu system_u –t httpd_sys_content_t /var/www/insider**
   changing security context of `/var/www/insider/webcontents/index.html'
   changing security context of `/var/www/insider/webcontents'
   changing security context of `/var/www/insider'

8. Issue the *semanage* command and modify the contexts on the directory to ensure that the new contexts survive a SELinux relabeling:

   # **semanage fcontext –a –s system_u –t httpd_sys_content_t /var/www/insider**

9. Restart the *httpd* daemon using any of the following and check the running status:

   # **apachectl start**
   # **service httpd start**

        # service httpd status

10. Test the configuration.

    To access the web page using the *elinks* command:

        # elinks 127.0.0.1/webcontents   # on the local system
        # elinks 192.168.2.202/webcontents   # on another system

    Alternatively, open up the firefox browser and type the URLs to view the web page from the local and a remote system.

Run the *tail* command on the *error_log* and *access_log* files located in the */var/log/httpd* directory to view error and access logs.

Exercise 21-3: Configure a Web Server with User-Based Security

In this exercise, you will restrict the website created in Exercise 21-2 to users *user1*, *user2*, and members of the *webgrp* group (*user3*) only.

1. Execute step 1 from Exercise 21-1.
2. Open the *httpd.conf* file and add or replace the directives as per below. Leave other settings intact.

        Listen 80
        DocumentRoot "/var/www/insider"
        <Directory /var/www/insider>
           Options Indexes FollowSymLinks
           AllowOverride authconfig
        </Directory>
        DirectoryIndex index.html
        AccessFileName .htaccess
        ErrorLog logs/error_log
        CustomLog logs/access_log combined

3. Execute steps 3 to 6 from Exercise 21-2.
4. Create the *.htaccess* file in DocumentRoot and add appropriate directives to it:

        # cd /var/www/insider && vi .htaccess
        AuthType Basic
        AuthName "This site is password protected. Enter a valid username and password to continue."
        AuthUserFile "/etc/httpd/conf/.userdb"
        Require user user1 user2
        AuthGroupFile "/etc/httpd/conf/.groupdb"
        Require group webgrp

5. Set passwords for *user1* and *user2* using the *htpasswd* command. Enter passwords when prompted and display the file contents.

```
# cd /etc/httpd/conf && htpasswd –cm .userdb user1
New password:
Re-type new password:
Adding password for user user1
# htpasswd –m .userdb user2
New password:
Re-type new password:
Adding password for user user2
# cat .userdb
user1:$apr1$ieTL4FVA$fKMRYgoBpVSnDiUebkYsJ.
user2:$apr1$36YkSVKm$.nPhjpmrfCxK4p6iBDBL//
```

6. Create the */etc/httpd/conf/.groupdb* file and add the following line to it:

```
# vi .groupdb
webgrp: user3
```

7. Change group membership on the *.userdb* and *.groupdb* files to *apache*:

```
# chgrp apache .userdb .groupdb
```

8. Execute step 7 to 9 from Exercise 21-2.
9. Execute step 10 from Exercise 21-2 to test the configuration. Enter *user1*, *user2*, or *user3* and the associated password when prompted.

Run the *tail* command on the *error_log* and *access_log* files located in the */var/log/httpd* directory to view error and access logs.

Exercise 21-4: Configure a Web Server to Provide Access to Private Directories

In this exercise, you will configure a web server with DocumentRoot set to the */var/www/insider* directory, web contents stored in the */var/www/insider/webcontents* directory, and the name of the index file will be *index.html*. This website will provide access to user home directories from everywhere.

1. Execute step 1 from Exercise 21-1.
2. Append the following directive to the code entered in step 2 in Exercise 21-3.

```
UserDir public_html
```

3. Execute steps 4 and 5 from Exercise 21-3 for *user1* only.
4. Check the syntax of the *httpd.conf* file for any errors:

```
# httpd –t
```

5. Create a sub-directory under */home/user1*, and set appropriate ownership, group membership, and permissions:

```
# mkdir /home/user1/public_html
# chown –R user1:user1 /home/user1
# chmod 711 /home/user1
# chmod –R 711 /home/user1/public_html
# setfacl –m u:apache:x /home/user1
# setfacl –m u:apache:x /home/user1/public_html
```

6. Create *index.html* file in the */home/user1/public_html* directory and add some text to it:

    ```
    # vi /home/user1/public_html/index.html
    ```
    This web page is to test access to private directories.

7. Execute step 2 to 4 from Exercise 21-1.
8. Activate the SELinux boolean to allow access to user home directories and verify the activation:

    ```
    # setsebool –P httpd_enable_homedirs=1
    # getsebool –a | grep httpd_enable_homedirs
    ```

9. Execute steps 7 to 9 from Exercise 21-2.
10. Execute step 10 from Exercise 21-2 to test the configuration. Connect as *user1* and access the web page located in the home directory.

Run the *tail* command on the *error_log* and *access_log* files located in the */var/log/httpd* directory to view error and access logs.

Exercise 21-5: Configure a Web Server to Support Group-Managed Contents

This exercise is a continuation of Exercise 4-3 that you accomplished in Chapter 04 "File Permissions, Text Editors, Text Processors & The Shell" in which you set up the setgid bit on the */sdata* directory for collaboration amongst members of the *sdatagrp* group on the local system.

To configure a web server to allow members of the *sdatagrp* group to share their files located in the */sdata* directory, replace the UserDir directive in Exercise 21-4 with the following in the *httpd.conf* file:

```
UserDir  sdata
```

Run all the steps from Exercise 21-4 and you should have the access to the */sdata* directory available to the members of the *sdatagrp* group via the web server.

The Apache Virtual Host

Apache allows you to build and run multiple virtual hosts on a single system to provide shared web hosting for several distinct websites. This technique of sharing a single system for hosting a number of websites offers a low cost hosting solution for customers. Each hosted website can either share a common IP address or configured to have its own unique address. Virtual hosts sharing a common

IP address are referred to as *name-based* and those having their own dedicated IP addresses are referred to as *IP-based*. These two commonly-used mechanisms direct the inbound traffic to the appropriate virtual host. The primary configuration file for defining standard virtual hosts using the HTTP protocol is the *httpd.conf* file located in the */etc/httpd/conf* and that for secure virtual hosts using the HTTPS protocol is the *ssl.conf* file located in the */etc/httpd/conf.d* directory.

Section 3 in the *httpd.conf* file defines virtual hosts within separate VirtualHost containers. You can define as many virtual hosts as you wish. Modify the directives for each virtual host as necessary. Directives used within VirtualHost containers are explained in Table 21-4 earlier in this chapter.

Exercise 21-6: Configure a Standard IP-Based Virtual Host

In this exercise, you will configure an IP-based virtual host for *server.example.com* to use IP address 192.168.2.201 on port 80. You will configure appropriate host-based access control and SELinux settings.

1. Perform step 1 from Exercise 21-1.
2. Open the *httpd.conf* file and add or replace the directives as per below. Leave other settings intact.

```
NameVirtualHost 192.168.2.201:80
<VirtualHost 192.168.2.201:80>
        ServerAdmin user1@server
        DocumentRoot /var/www/server
        ServerName server
        ErrorLog logs/server-error_log
        CustomLog logs/server-access_log combined
</VirtualHost>
```

3. Execute steps 3 to 9 from Exercise 21-2. Use appropriate directive values.
4. Test the configuration by loading *http://server/index.html* on port 80 in both text and graphical browsers from the local and a remote system.

While visiting the website, run the *tail* command with the –f option on the *server-access_log* file in DocumentRoot to view what is going on.

An Analysis of the ssl.conf File

The *ssl.conf* file is part of the mod_ssl package and it is used to define SSL-enabled web servers and virtual hosts. The contents in this file are divided into two sections: SSL Global Context and SSL Virtual Host Context. There are two directives – LoadModule and Listen – that are included at the very beginning of the file and are not part of either section.

```
LoadModule ssl_module modules/mod_ssl.so
Listen 443
```

The LoadModule directive instructs the *httpd* daemon to load the SSL module at startup and the Listen directive indicates the port that the web server listens on.

The SSL Global Context section includes some directives that need not be modified as their default values are sufficient for most requirements.

The SSL Virtual Host Context section contains several directives of which the more important directives with their default values are provided below:

```
## SSL Virtual Host Context
<VirtualHost _default_:443>
#DocumentRoot "/var/www/html"
#ServerName www.example.com:443
ErrorLog logs/ssl_error_log
TransferLog logs/ssl_access_log
LogLevel warn
SSLEngine on
SSLProtocol all -SSLv2
SSLCipherSuite ALL:!ADH:!EXPORT:!SSLv2:RC4+RSA:+HIGH:+MEDIUM:+LOW
SSLCertificateFile /etc/pki/tls/certs/localhost.crt
SSLCertificateKeyFile /etc/pki/tls/private/localhost.key
<Files ~ "\.(cgi|shtml|phtml|php3?)$">
    SSLOptions +StdEnvVars
</Files>
<Directory "/var/www/cgi-bin">
    SSLOptions +StdEnvVars
</Directory>
CustomLog logs/ssl_request_log \
     "%t %h %{SSL_PROTOCOL}x %{SSL_CIPHER}x \"%r\" %b"
</VirtualHost>
```

The <VirtualHost _default_:443> directive identifies the port number with an IP or * for IP-based and name-based virtual host setups, respectively. The next five directives – DocumentRoot, ServerName, ErrorLog, TransferLog, and LogLevel – and the CustomLog directive have the same meaning that was provided under the *httpd.conf* file analysis. The SSLEngine directive must be set to on if you intend to use SSL. The next four directives – SSLProtocol, SSLCipherSuite, SSLCertificateFile, and SSLCertificateKeyFile – specify the SSL version to use, the encryption method to use, the location of the default SSL certificate file, and the location of the default SSL key file. The <Files> and <Directory> sub-containers specify the file types containing dynamic contents and their location.

Generating a Self-Signed SSL Certificate

The directives – SSLCertificateFile and SSLCertificateKeyFile – in the *ssl.conf* file indicate file names and their locations for the SSL certificate and key files. You need to generate a self-signed local certificate for each secure virtual host that you intend to use. To generate a certificate for the virtual host *testserver.example.com*, change into the */etc/pki/tls/certs* directory and run the *genkey* command on the virtual host:

 # cd /etc/pki/tls/certs

 # genkey testserver.example.com

A screen similar to the one shown in Figure 21-2 will appear.

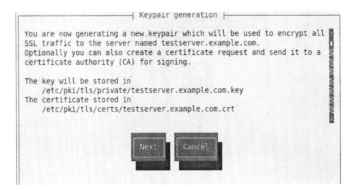

Figure 21-2 Generating a Self-Signed SSL Certificate – 1 of 6

It shows the name of the certificate files that will be created and their locations. Enter Next to continue. Choose the recommended key size, as shown in Figure 21-3, and enter Next to go to the next screen.

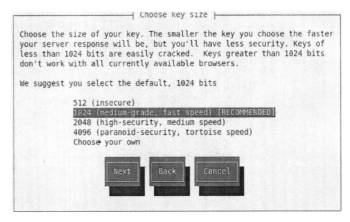

Figure 21-3 Generating a Self-Signed SSL Certificate – 2 of 6

You will be asked whether you want a Certificate Request (CSR) generated and send to a CA. Select No if you are not actually planning to purchase one.

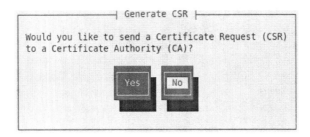

Figure 21-4 Generating a Self-Signed SSL Certificate – 3 of 6

You will be prompted to answer whether you would like the key encrypted with a passphrase as shown in Figure 21-5. Select Encrypt the private key and choose Next to continue.

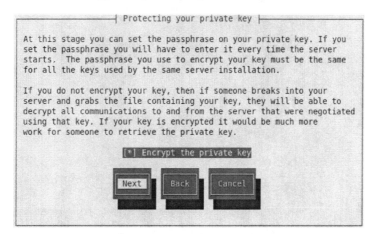

Figure 21-5 Generating a Self-Signed SSL Certificate – 4 of 6

Enter a passphrase twice and press Next to continue.

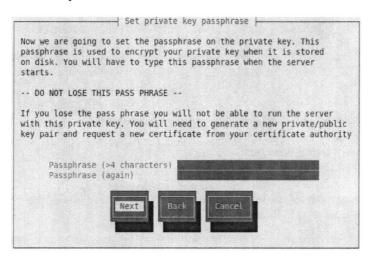

Figure 21-6 Generating a Self-Signed SSL Certificate – 5 of 6

The final screen for this setup prompts you to enter details for the certificate. Enter the information appropriate to your organization and continue by choosing the Next button.

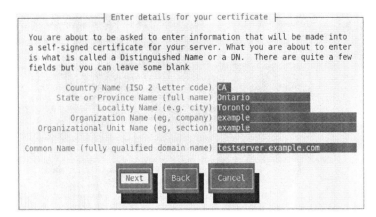

┤ Enter details for your certificate ├

You are about to be asked to enter information that will be made into
a self-signed certificate for your server. What you are about to enter
is what is called a Distinguished Name or a DN. There are quite a few
fields but you can leave some blank

        Country Name (ISO 2 letter code) CA
     State or Province Name (full name) Ontario
            Locality Name (e.g. city) Toronto
         Organization Name (eg, company) example
    Organizational Unit Name (eg, section) example

Common Name (fully qualified domain name) testserver.example.com

                Next   Back   Cancel

Figure 21-7 Generating a Self-Signed SSL Certificate – 6 of 6

Once the *genkey* command has finished, you will see certificate files created for the new virtual host:

-rw-r-----. 1 root root   903 Nov 27 10:13 /etc/pki/tls/certs/testserver.example.crt
-r--------. 1 root root 1089 Nov 27 10:13 /etc/pki/tls/private/testserver.example.com.key

Exercise 21-7: Configure a Secure Name-Based Virtual Host

In this exercise, you will configure a name-based secure virtual host for *server.example.com* to use IP address 192.168.2.202 on port 443. You will configure appropriate host-based access control and SELinux settings.

1. Check whether the mod_ssl software package is installed:

    # **rpm –qa | grep mod_ssl**
    mod_ssl-2.2.15-15.el6.x86_64

    Install the missing package if it is not already installed using one of the procedures outlined in Chapter 06 "Package Management".

2. Open the *ssl.conf* file and add or replace the directives as per below. Leave other settings intact.

    LoadModule ssl_module modules/mod_ssl.so
    Listen 443
    NameVirtualHost *:443
    <VirtualHost *:443>
            ServerAdmin user1@server
            DocumentRoot /var/www/server
            ServerName server
            ErrorLog logs/server-ssl_error_log
            CustomLog logs/server-ssl_access_log combined
    </VirtualHost>

3. Create DocumentRoot directory and a sub-directory *webcontents* for storing website contents:

> # **mkdir –p /var/www/server/webcontents**

4. Create the *index.html* file in the */var/www/server/webcontents* directory and add some text to it:

> # **vi /var/www/server/index.html**
> This is a secure virtual host website for server.

5. Execute steps 2 to 4 from Exercise 21-1 for port 443.
6. Execute step 7 and 8 from Exercise 21-2 on DocumentRoot and then step 9 from the same exercise.
7. Test the configuration.

> To access the web page using the *elinks* command:

> # **elinks https://127.0.0.1/webcontents**　　# on the local system
> # **elinks https://insider/webcontents**　　　# on another system

Alternatively, open up the firefox browser and type the URLs to view the web page from the local and a remote system.

While visiting the website, run the *tail* command with the –f option on the *server-ssl_access_log* files in DocumentRoot to view what is going on.

Deploying a Basic CGI Application

This section briefly talks about how to set up a script for execution when it is accessed via the web. The *httpd.conf* file contains the LoadModule and ScriptAlias directives. The LoadModule directive instructs the Apache server to read CGI files and the ScriptAlias directive allows web users to find CGI files through their browsers.

Exercise 21-8: Run a Basic CGI Application

In this exercise, you will configure a standard Apache web server to execute a CGI script and display its output in a browser window when the page is accessed from anywhere. You will prevent regular users from making changes into the *cgi-bin* directory. You will configure appropriate host-based access control and SELinux settings.

1. Perform steps 1 to 4 from Exercise 21-1.
2. Create a script called *systime.sh* in the */var/www/cgi-bin* directory and add the following text to it:

> # **vi /var/www/cgi-bin/systime.sh**
> #!/bin/bash
> echo "Content-type: text/html"
> echo
> echo "The current system time is `date`"

3. Add the execute permission to this script for everyone:

   # **chmod +x /var/www/cgi-bin/systime.sh**

4. Set the SELinux context on the script to httpd_sys_script_exec_t:

   # **chcon –t httpd_sys_script_exec_t /var/www/cgi-bin/systime.sh**

5. Open the *httpd.conf* file and add or replace the directives as per below. Leave other settings intact:

   ScriptAlias /cgi-bin/ "/var/www/cgi-bin/"
   AddHandler cgi-script .cgi .sh
   <Directory /var/www/cgi-bin>
           AllowOverride None
           Options ExecCGI
           Order allow,deny
           Allow from all
   </Directory>

6. Check the syntax of the *httpd.conf* file for any errors:

   # **httpd –t**
   Syntax OK

7. Execute steps 2 to 4 from Exercise 21-1.
8. Restart the HTTP service using any of the following and check the running status:

   # **apachectl start**
   # **service httpd start**
   # **service httpd status**

9. Test the configuration.

   To access the web page using the *elinks* command:

   # **elinks http://localhost/cgi-bin/systime.sh**   # on the local system
   # **elinks http://192.168.2.202/cgi-bin/systime.sh**   # on another system

Figure 21-8 CGI Script Web Page

Alternatively, open up the firefox browser and type the URLs to view the web page from the local and a remote system.

The OpenSSH Service

Secure Shell (SSH) was created in 1995 to provide a secure mechanism for data transmission between source and destination systems over IP networks. SSH includes a set of utilities providing remote users the ability to log in, transfer files, and execute commands securely using strong, hidden encryption and authentication. Due to their built-in powerful security features, SSH utilities have widely replaced the conventional, unsecured *telnet, rlogin, rsh, rcp*, and *ftp* services in computing environments.

OpenSSH is a free, open source implementation of SSH. Once applied successfully on a system, the *telnetd, rlogind, rshd*, and *ftpd* services can be disabled in the */etc/xinetd.d* directory, provided no user or application functionality is impacted. The secure command that has replaced *rlogin, rsh*, and *telnet* is called *ssh* and those that have replaced *rcp* and *ftp* are referred to as *scp* and *sftp*, respectively.

Secure shell uses TCP Wrappers for access control.

Private and Public Encryption Keys

Encryption is a way of transforming information into a scrambled form to hide the information from unauthorized users. Encryption is done at the sending end and the reverse process called *de-encryption* happens on the receiving side. The sending system transforms the information using a *private* key and the receiving system reads it by using a *public* key. The private key is like a lock that requires a public key to open. When a client sends information to a server where the keys were generated, the server decrypts the incoming public key and see if it matches with the private key. If the two matches, the server allows the connection to establish, otherwise, it simply rejects the connection request. Both private and public keys are randomly-generated long strings of alphanumeric characters attached to messages during the ssh communication.

The private key must be kept secure since it is private to that one system only. The public key is distributed to clients. The system serves incoming ssh requests and based on the matching or mismatching of the private/public key combination, it allows or denies these requests.

DSA and RSA are widely used encryption methods used with OpenSSH. DSA stands for *Digital Signature Algorithm* and RSA for *Rivest, Shamir, Adleman*.

SSH Encryption Without Private/Public Keys

SSH is based on the client/server model where a client piece (*ssh, scp*, or *sftp*) sends a connection request and the server process *sshd* responds to it. Here is how the ssh communication channel is established without the involvement of the private and public keys:

- ✓ A user issues a connection request to the specified server using one of the ssh utilities.
- ✓ TCP Wrappers on the server evaluates the request for access control.
- ✓ The *sshd* daemon shares ssh protocol versions with the client and switches to a packet-based protocol for communication.
- ✓ The server prompts the user for a password.
- ✓ The user responds by entering a password.

✓ The server process invokes a PAM authentication module and matches the password entered against the encrypted password defined in the */etc/shadow* file. Upon validation, an encrypted communication channel is established for the user.

SSH Authentication With Private/Public Keys

Here is how SSH performs authentication using private/public keys:

✓ The first three steps are the same from the above.
✓ The server prompts the user for the public key (a.k.a. *passphrase*).
✓ The user responds by entering the public key.
✓ The server process decrypts the public key and matches it with the private key stored in that user's home directory. Upon validation, the user is allowed to get in and an encrypted communication session is established for the user.

Commands and Daemons, Configuration and Log Files

Several commands, daemons, and configuration and log files are involved in the ssh communication setup process. Table 21-12 lists and explains them.

	Description
Command	
ssh-agent	Authentication Agent. Holds private keys used for the RSA/DSA authentication.
ssh-add	Adds RSA/DSA characteristics to ssh-agent.
ssh-copy-id	Copies RSA/DSA keys to other systems.
ssh-keygen	Generates private and public keys for RSA/DSA. Specify a key type using the –t option. Default is RSA.
ssh-keyscan	Gathers ssh public keys.
ssh	Secure alternative to *telnet* and *rlogin*.
Daemon	
sshd	The ssh server daemon that serves ssh client requests. It listens on port 22 by default.
sftp-server	Autostarted by *sshd* to serve incoming *sftp* requests.
Configuration and Log Files	
/etc/ssh/ssh_config	Defines system-wide ssh client configuration defaults.
/etc/ssh/sshd_config	Defines system-wide ssh server configuration defaults.
/etc/sysconfig/sshd	The startup configuration file for *sshd*.
/etc/rc.d/init.d/sshd	The startup script for *sshd*.
/var/log/secure	Logs all authentication information.

Table 21-12 OpenSSH Commands, Daemon, and Configuration and Log Files

An excerpt from the main configuration file */etc/ssh/sshd_config* showing only the key directives is displayed below:

cat /etc/ssh/sshd_config
#Port 22
#AddressFamily any
#ListenAddress 0.0.0.0
. . . .

There are numerous directives that you may be set in the *sshd_config* file. Some of the key directives are explained in Table 21-13.

Directive	Description
Port	Default is 22. Specifies the port for the ssh communication.
ListenAddress	Specifies the IP to be used for incoming ssh requests.
Protocol	Default is 2. Specifies the ssh protocol version to be used.
SyslogFacility	Default is AUTHPRIV. Defines the facility code to be used when logging messages to the */var/log/secure* file. This is based on the configuration in the */etc/rsyslog.conf* file.
LogLevel	Default is INFO. Defines the level of criticality for the messages to be logged.
LoginGraceTime	Default is 2 minutes. Defines the period after which the server disconnects the user if the user has not successfully logged in.
PermitRootLogin	Default is yes. Allows the *root* user to log in directly on the system.
MaxAuthTries	Default is 6. Specifies the number of authentication attempts per connection.
MaxSessions	Default is 10. Allows to open up to this many number of ssh sessions.
RSAAuthentication	Default is yes. Specifies whether to allow the RSA authentication.
PubKeyAuthentication	Default is yes. Specifies whether to enable public key-based authentication.
AuthorizedKeysFile	Default is ~/.ssh/authorized_keys. Specifies the name and location of the file containing the authorized keys.
PasswordAuthentication	Default is yes. Specifies whether to enable local password authentication.
PermitEmptyPasswords	Default is no. Specifies whether to allow null passwords.
ChallengeResponseAuthentication	Default is no. Disables user authentication via PAM.
UsePAM	Default is yes. Enables or disables user authentication via PAM. If enabled, only *root* will be able to run the daemon.
AllowAgentForwarding	Default is yes. Enables or disables the *ssh-agent* command to forward private keys to other systems.
AllowTCPForwarding	Default is yes. Specifies whether to allow forwarding TCP communication over an ssh channel.
AcceptEnv	Allows a client to set global variables.
X11Forwarding	Default is yes. Allows or disallows remote access to GUI applications.
TCPKeepAlive	Default is yes. Specifies whether to send TCP keepalive signals to the ssh server to check its accessibility.

Directive	Description
UseLogin	Default is no. Allows or disallows the use of the login command for interactive login sessions.
Compression	Default is delayed. Specifies whether to allow or delay compression until the user has authenticated successfully.
ClientAliveInterval	Default is 0. Defines a timeout interval in seconds after which if no data has been received from the client, the *sshd* daemon will send a message to the client for a response.
ClientAliveCountMax	Default is 3. If the ClientAliveInterval directive is set to a non-zero value, this directive specifies the maximum number of messages to be sent to the client.
Subsystem sftp	Enables encryption for sFTP file transfers.
AllowUsers	Except for the listed users, all other users will be denied ssh access into the system. For example: AllowUsers *user1 user2* AllowUsers *user3@server.example.com* AllowUsers *user4@192.168.1.33 user5@192.168.3.20*
AllowGroups	Except for the members of the listed groups, all other users and group members will be denied ssh access into the system. For example: AllowGroups *dba unixadmins* AllowGroups *dba@server.example.com*
DenyUsers	Except for the listed users, all other users will be allowed ssh access into the system. For example: DenyUsers *user1 user2* DenyUsers *user3@server.example.com* DenyUsers *user4@192.168.1.33 user5@192.168.3.20*
DenyGroups	Except for the members of the listed groups, all other users and group members will be allowed ssh access into the system. For example: DenyGroups *dba unixadmins* DenyGroups *dba@server.example.com*

Table 21-13 OpenSSH Server Configuration File

In addition to using the AllowUsers, DenyUsers, AllowGroups, and DenyGroups directives in the *sshd_config* file for setting user-based security for the ssh service, you can alternatively define such restrictions in the TCP Wrappers files *hosts.allow* and *hosts.deny* as appropriate. Furthermore, you can restrict access into the system from specific systems, domains, or networks via port 22 using the iptables firewall.

Exercise 21-9: Configure an SSH Server

In this exercise, you will configure the OpenSSH service on *server.example.com* to allow ssh access into the system for *user1* from 192.168.2.200 system and *user2* from *example.net*. You will restrict access to *user3* from 192.168.2.201. You will disallow ssh access into the system from 192.168.1 network using the iptables firewall.

1. Check whether the required OpenSSH software packages are installed:

    # rpm –qa | grep openssh
    openssh-server-5.3p1-70.el6.x86_64
    openssh-clients-5.3p1-70.el6.x86_64
    openssh-askpass-5.3p1-70.el6.x86_64

    Install the packages if they are not already installed using one of the procedures outlined in Chapter 06 "Package Management".

2. Set the secure shell service to autostart at each system reboot, and validate:

    # chkconfig sshd on
    # chkconfig --list sshd
    sshd   0:off   1:off   2:on   3:on   4:on   5:on   6:off

3. Open the */etc/hosts.allow* file and add the following entry to it:

    sshd : user1@192.168.2.200 user2@example.net

4. Open the */etc/hosts.deny* file and add the following entry to it:

    sshd : user3@192.168.2.201

5. Configure host-based access by allowing ssh traffic on port 22 to pass through the firewall except from 192.168.1 network:

    # iptables –I INPUT –p tcp –m tcp –dport 22 –s !192.168.1 –j ACCEPT

6. Save the rule in the */etc/sysconfig/iptables* file and restart the firewall to activate the new rule:

    # service iptables save ; service iptables restart

7. Restart the *sshd* daemon and check the running status:

    # service sshd start
    # service sshd status

Accessing the SSH Server Using User Password

To access the SSH server configured in Exercise 21-10, run the *ssh* command from insider.

 # ssh server.example.com
The authenticity of host 'server.example.com (192.168.2.200)' can't be established.
RSA key fingerprint is b8:28:24:60:b1:20:cb:99:99:3e:a2:1b:f2:5d:f9:d0.
Are you sure you want to continue connecting (yes/no)? **yes**
Warning: Permanently added 'server.example.com' (RSA) to the list of known hosts.
root@server.example.com's password:
Last login: Sun Nov 25 06:13:28 2012 from 192.168.2.16

As displayed in the output above, you will need to enter yes when prompted and then the user's password on *server.example.com* to log in. This command assumes that you wish to log in to the remote ssh server as the local user. If you want to log on as a different user such as *user1*, specify the user's name with the –l option as demonstrated below:

ssh server.example.com –l user1

An alternative method to the above is to specify both the username and the hostname as follows:

ssh user1@server.example.com

On each subsequent login attempts from *insider* to *server*, *user1* will only have to enter his password to access *server*. He will not be prompted for confirming the authenticity of the server.

The following would open an X session on *server.example.com* as the *root* user:

ssh –X server.example.com

Exercise 21-10: Configure Key-Based and Passwordless Authentication

In this exercise, you will generate RSA encryption keys for *user1* and use them to allow this user from *insider.example.com* to connect to *server.example.com* using the *ssh*, *scp*, and *sftp* commands without entering a password. You will configure SELinux settings as required.

1. Log on to *server* as *user1*.
2. Modify the SELinux file type on */home/user1* to user_home_dir_t:

   $ **chcon –t user_home_dir_t /home/user1**

3. Generate RSA encryption keys and save them in files in the *~/.ssh* sub-directory. Press the Enter key to accept the default values and when prompted for a passphrase (Enter a passphrase if you want *user1* to be prompted at each login attempt).

   $ **ssh–keygen**
   Generating public/private rsa key pair.
   Enter file in which to save the key (/home/user1/.ssh/id_rsa):
   Created directory '/home/user1/.ssh'.
   Enter passphrase (empty for no passphrase):
   Enter same passphrase again:
   Your identification has been saved in /home/user1/.ssh/id_rsa.
   Your public key has been saved in /home/user1/.ssh/id_rsa.pub.
   The key fingerprint is:
   50:a3:34:a8:f4:ab:75:23:bd:39:d3:8f:62:61:42:00 user1@server.example.com
   The key's randomart image is:
   +--[RSA 2048]----+
   |E. .o o |

```
|  o .. + .       |
|  . + o          |
|  . o .          |
|  . o S          |
|   - *           |
|  o - *          |
|  . B ..         |
|    . +...        |
+-----------------+
```

The contents of *id_rsa* and *id_rsa.pub* files are shown below:

$ cat ~/.ssh/id_rsa

-----BEGIN RSA PRIVATE KEY-----

MIIEoQIBAAKCAQEAm3qDePb0W7VefuyuoVSrL/9UwKkC5uJjq0Hn4IIQIZx7/rYiqQgcpn99XxaZW/wjv46
0F37NxSdEetyG294Dtiil57Cp78yE3QeWIv6+qJfkmo9TUbpkCpEW+4PZQiYWs1MzZ8gSZdZXEXc8YU7kU
0f0+mkCj9EHdHLF7DlAHKgsjUIFKsC6IuoihmFQR2BVxEb/psYAYpz5VBBbobrarvKLKO00mbwTVxfgH+Qm
V9NRNxTMbDWfxrYWzpw4bfZVBQdz96u9wfVIkzN6GCoHLhPXR8PRU7pLlgDuRYqdr6D2JDU4krvPvs+
bhPIB3TPXKall4fyUiNAKWt4EvhFUPwIBIwKCAQA1TpN57kUmwddBdbg3UDqxXts6vZqYTZ6D3BT8hF1N
WjkkITB0d89PByOrkrg8y3n4iLLGOhqp/tWmd36x
yHZNFUAyLfEfBErPcFC7ioqC8j+/9qA5Rzg+Ix3Sk5ryG7AC9/ReGLXZxdS1h/dx0efiCgrZgxbSOQneyEPb9l8f
xBjBbaRWzPTfTIa4Wyca8k3anXX+O/RL5FiXIykSPcNZ5GocSEWL/g5vk39r+vG/SXZK7XNLjh8hqJG8QGTk
OwV9ia9L855zxzu4xcgrZVXBHX5jS8YKCqo66NeMnZzE/i40tfQNGRTC+VrUqjT2uW/pz3v+B0OX5rtHxc0
IqxRLAoGBAMhSvgB2TfKUpJHJhPDIaHq9RnAiebNbQ1m9u+Im4+z+9wYezUWP+x9AxKT1OFJUmCnRf/TK
0HftgI4Vs4a5aBDmaz3F5MKX4DWP5PvoEzPn1+ryNM9xK9UNRFIz8+Ae3Qe71/13SXDRmr8vLSuHghzxo
66+pGQ7TkXfsa4pElKHAoGBAMaxBDq3
R6gC+7J4fS1wX81JX/deyBnUqkfdg0GwTXPpy/o3onmGxp9Q5QeFD+g/+jkLCvlFBqJhtIpT14twN3QbH/bplc
JYZU99uEuxhJkoUJEJR1WamgYHlXy3dbRSM+ej9TNKk+JfMui+1GaGFN54veLTwuTHy3eUgMCRDAaJAoG
AFuTig7W/w/O7CVjb/kLKHKfqyvygMcFJhplXTQvCRviRQoctZwkkA5JfndLh3YYCwvNfFKl26SJ1F41zml5V
CT7nr0nRAEvfHBBx8OdSptiq9keRAcPKfr+wCWUF7bpiZ0iq+GVnepucqCKmE5p1RSLuIplx35G/zXiml4+q
UpMCgYEAtakocDKKqD095QBycq/ij84OmQYzStEQtrvluGatNscg5MUm1YKJtjtVDjCDkoOi8lM9NQvozvo
+qltIugAys0v/97D984QE2vaZ3s4w
FwBJqS0OBR+iw68plpkbKIWrzHilqzze+tqyM+GsQHqWvMYp8/TPchXBV2Mz40LJKosCgYA+E/IqMf8T82FP
aP7YNqda19gXM/eDf6U2z7+SlrvmZYano+LrEZPUI7sI
1f+/rgnmexP787zi+LjKnD+NZ3AfV1pzi125vCjXpwHwL4uoZ9qYs7S/afo/USqrjVkeTSCx1imUggo2IVqrNEx
ceJoglBcqcbjnjTiUoH5ivulnqw==

-----END RSA PRIVATE KEY-----

$ cat ~/.ssh/id_rsa.pub

ssh-rsa

AAAAB3NzaC1yc2EAAAABIwAAAQEAm3qDePb0W7VefuyuoVSrL/9UwKkC5uJjq0Hn4IIQIZx7/rYiqQgcpn
99XxaZW/wjv460F37NxSdEetyG294Dtiil57Cp78yE3QeWIv6+qJfkmo9TUbpkCpEW+4PZQiYWs1MzZ8gS
ZdZXEXc8YU7kU0f0+mkCj9EHdHLF7DlAHKgsjUIFKsC6IuoihmFQR2BVxEb/psYAYpz5VBBbobrarvKLKO0
0mbwTVxfgH+QmV9NRNxTMbDWfxrYWzpw4bfZVBQdz96u9wfVIkzN6GCoHLhPXR8PRU7pLlgDuRYqd
r6D2JDU4krvPvs+bhPIB3TPXKall4fyUiNAKWt4EvhFUPw== user1@server.example.com

4. Copy the *id_rsa.pub* file to insider under the */home/user1/.ssh* sub-directory as *authorized_keys*. Enter yes when prompted and enter the password for *user1* on insider.

```
$ ssh-copy-id –i ~/.ssh/id_rsa.pub insider.example.com
The authenticity of host '192.168.2.201 (192.168.2.201)' can't be established.
RSA key fingerprint is f6:2b:57:e2:c8:b7:bd:f1:0f:6d:1b:6f:bb:cd:56:fa.
Are you sure you want to continue connecting (yes/no)? yes
Warning: Permanently added '192.168.2.201' (RSA) to the list of known hosts.
user1@192.168.2.201's password:
Now try logging into the machine, with "ssh '192.168.2.201'", and check in:
  .ssh/authorized_keys
to make sure we haven't added extra keys that you weren't expecting.
```

At the same time this command also creates a file called *known_hosts* on server and stores insider's hostname (or IP address) and a unique key for identifying it:

```
$ cat ~/.ssh/known_hosts
192.168.2.201 ssh-rsa
AAAAB3NzaC1yc2EAAAABIwAAAQEA4/RfYZJbso/0jIF2wSjuYy5ZutBUUunfmgFWuGewMrKORxwdxV3D
mjT90IqgJsYPjUQmsZf2fJ6f1w5uzRPOU/9bRB6KOl189cVYDK5gfcXnT+juRGSFU+X9DECKSnu7EZhqgrgci/
W4Z2CPpxMaIYnkcDWR0ihdK3k+ZJMtexK+KpsIT+vjmEmvuw1o6QPElQyFrprsAZCvCD1lq2tI1U/P4xhEm
7u5iTVbN8/YKccMpvYX0H3WBmQd2xzZl86kv88jh/gAXY9q1VIwIiKXAK0XYH/piMLbxKdu2P82xDMaK
7lZGt6d0n8VBjNYsE0eptWXQBtW+KmUYt0YWsQPnw==
```

5. Log in as *user1* on *insider.example.com* from *server.example.com* with the *ssh* command. Answer yes when prompted and supply *user1*'s password:

```
$ ssh insider.example.com
The authenticity of host 'insider.example.com (192.168.2.201)' can't be established.
RSA key fingerprint is f6:2b:57:e2:c8:b7:bd:f1:0f:6d:1b:6f:bb:cd:56:fa.
Are you sure you want to continue connecting (yes/no)? yes
Warning: Permanently added 'insider.example.com' (RSA) to the list of known hosts.
Last login: Tue Nov 20 11:59:09 2012 from 192.168.2.16
```

On each subsequent login attempt, *user1* will be allowed a passwordless access into insider provided a passphrase was not entered in step 3. If a passphrase was supplied, then each time *user1* tries to access insider, he will be prompted to enter the passphrase.

View the login attempt activities for *user1* in the */var/log/secure* file on insider.

Executing a Command Remotely Using ssh

With the ssh keyless/passwordless configuration in place, you can execute the *ssh* command on *server.example.com* to invoke the *ls* command remotely on *insider.example.com* to list the files located in *user1*'s home directory:

```
$ ssh insider /bin/ls –la
total 32
drwxr-xr-x. 4 user1 user1 4096 Nov 28 08:58 .
drwxr-xr-x. 6 user1 user1 4096 Nov 24 18:19 ..
-rw-------. 1 user1 user1 1207 Nov 28 09:17 .bash_history
-rw-r--r--. 1 user1 user1   18 Jan  27 2011 .bash_logout
```

```
-rw-r--r--.  1 user1  user1  176 Jan   27 2011 .bash_profile
-rw-r--r--.  1 user1  user1  124 Jan   27 2011 .bashrc
drwxr-xr-x.  2 user1  user1 4096 Nov  20 13:09 .lftp
drwx------.  2 user1  user1 4096 Nov  28 08:58 .ssh
```

Copying Files Using scp

Similarly, you can execute the *scp* command on *server.example.com* to copy a file called *file1* from *user1*'s home directory to the */tmp* directory on *insider.example.com*:

$ scp file1 insider:/tmp
```
file1                         100%   0    0.0KB/s  00:00
```

Transferring Files Using sftp

In the same manner, you can execute the *sftp* command on *server.example.com* to connect to *insider.example.com* and transfer files:

$ sftp insider.example.com
```
Connecting to insider...
sftp>
```

Type ? at the prompt to list available commands. A short description of what each command does is also displayed.

```
sftp> ?
Available commands:
Bye                          Quit sftp
cd path                      Change remote directory to 'path'
chgrp grp path               Change group of file 'path' to 'grp'
chmod mode path              Change permissions of file 'path' to 'mode'
chown own path               Change owner of file 'path' to 'own'
df [-hi] [path]              Display statistics for current directory or filesystem containing 'path'
exit                         Quit sftp
get [-P] remote-path [local-path] Download file
help                         Display this help text
lcd path                     Change local directory to 'path'
lls [ls-options [path]]      Display local directory listing
lmkdir path                  Create local directory
ln oldpath newpath           Symlink remote file
lpwd                         Print local working directory
ls [-1aflnrSt] [path]        Display remote directory listing
lumask umask                 Set local umask to 'umask'
mkdir path                   Create remote directory
progress                     Toggle display of progress meter
put [-P] local-path [remote-path] Upload file
pwd                          Display remote working directory
quit                         Quit sftp
rename oldpath newpath       Rename remote file
```

rm path	Delete remote file
rmdir path	Remove remote directory
symlink oldpath newpath	Symlink remote file
version	Show SFTP version
!command	Execute 'command' in local shell
!	Escape to local shell
?	Synonym for help

Some of the common commands at the sftp> prompt are *cd/lcd*, *get/put*, *ls/lls*, *pwd/lpwd*, *mkdir/rmdir*, *quit/exit*, *rename*, and *rm*.

Accessing the SSH Server from a Windows System

On the Windows side, several ssh client programs such as PuTTY, are available. PuTTY can be downloaded free of charge from the Internet. Figure 21-9 shows the PuTTY interface.

Figure 21-9 PuTTY Interface

Supply the hostname or IP address of the remote system to log in to and check SSH under Connection Type. Make certain to use port 22. Assign a name to this session and click Save to save the data to avoid retyping this information in the future.

Chapter Summary

This chapter discussed the Apache web server and the secure shell server. It started off with an overview of the Apache web server and its features. The chapter provided an analysis of the Apache web server's main configuration file and touched upon its log files and administration commands. It explained the security requirements from firewall and SELinux perspectives, and covered several directives related to security controls within the web server's configuration file. After getting a grasp on the basics, the chapter presented a series of exercises providing you with an opportunity to configure Apache web server based on varying requirements. The focus of this chapter then switched to the Apache virtual host which supported shared hosting of many secure and non-secure websites on a single system, presented exercises for solidifying the concepts, demonstrated how to generate a self-signed SSL certificate, and explained and demonstrated how to deploy a basic CGI script.

The next major topic in this chapter was on secure shell. The chapter provided an introduction to private and public encryption keys, ssh commands, daemons, and its configuration and log files. It presented exercises to configure an ssh server and a key-based/passwordless authentication setup. This chapter ended with demonstrating how to use the secure shell utilities to execute commands on remote systems, how to transfer files, and how to connect to the ssh server from a Windows system.

Chapter Review Questions

1. Why would you use the AutoUserFile and Require directives in the Apache configuration file?
2. What is the name and location of Apache's main configuration file?
3. What is the secure equivalent for the *rcp* command?
4. Provide the difference between the name-based and IP-based web servers.
5. Which command would you run to generate a self-signed SSL certificate?
6. The secure shell may be defined to use TCP Wrappers for access control. True or False?
7. What are the ServerAdmin, ServerName, and Options directives used for?
8. The primary secure shell server configuration file is *ssh_config*. True or False?
9. How does a virtual host container begin and end? Provide the syntax.
10. What are the two standard encryption methods used in RHEL6?
11. List any three features of the Apache web server software.
12. What would the SELinux boolean httpd_enable_homedirs imply?
13. What is the secure shell equivalent for the *telnet* command?
14. Provide the names of the four directives that may be used to set user-level access control for the secure shell server.
15. What are the two host-level security directives in the *httpd.conf* file?
16. Can the *apachectl* command be used to start, restart, or stop the Apache web server daemon? Yes or no.
17. Where are the user ssh keys stored?
18. Write the difference between a regular web server and a virtual host.
19. What is the location of the Apache log files?
20. What is the use of the *htpasswd* command?
21. What would the command *ssh-copy-id* do?
22. Does the directive *VirtualHost *:80* define a named-based virtual host or an IP-based?
23. What is the purpose for running the *httpd* command with the –t switch?
24. Where would you define the user-level security for Apache?
25. Describe the use of ServerRoot, DocumentRoot, and LoadModule directives.
26. What is the significance of the ScriptAlias directive?
27. What is the use of the *ssh-keygen* command?
28. What is the name and location of the secure web server's configuration file?
29. Where would you define the host-level security for Apache?
30. What are the default ports used by the Apache web server and secure shell server daemons?

Answers to Chapter Review Questions

1. The AutoUserFile directive points to the file that contains authorized user passwords and the Require directive specifies users or groups that are allowed.
2. The Apache's main configuration file is *httpd.conf* and it is located in the */etc/httpd/conf* directory.
3. The *scp* command is the secure equivalent for the *rcp* command.
4. Name-based virtual servers share a common IP address whereas IP-based virtual servers have their own dedicated IP addresses.
5. The *ssh-keygen* command.
6. True.
7. The ServerAdmin directive defines the administrator's email address, the ServerName directive specifies the hostname of the web server, and the Options directive specifies options to be used.
8. False. The primary secure shell configuration file is *sshd_config*.
9. The begin and end of a virtual host container is <VirtualHost .... VirtualHost/>
10. The two standard encryption methods used in RHEL6 are RSA and DSA.
11. Three features of the Apache web server include the ability to share a single IP address amongst multiple virtual hosts, support for LDAP authentication, and support to run on virtually every hardware platform and operating system software.
12. The httpd_enable_homedirs directive is used to enable or disable Apache's access to user home directories.
13. The secure equivalent for *telnet* is the *ssh* command.
14. The four directives that can be used to set user-level access control for the ssh server are AllowUsers, DenyUsers, AllowGroups, and DenyGroups.
15. The Allow and Deny are two host-level directives in the *httpd.conf* file.
16. Yes, the *apachectl* command can be used to start, restart, or stop the Apache web server daemon.
17. In the *.ssh* sub-directory under the user's home directory.
18. A regular Apache web server has the ability to run a single website whereas you can define as many virtual hosts on a single system as the hardware capacity of the system supports.
19. The Apache log files are located in the */var/log/httpd* directory.
20. The *htpasswd* command is used to create and update files where Apache usernames and passwords are stored.
21. The *ssh-copy-id* command is used to install public keys on remote systems.
22. The directive VirtualHost *:80 would define a name-based virtual host.
23. The *httpd* command with the –t option checks for any syntax errors in the *httpd.conf* file.
24. User-level access control for the Apache web server may be defined in the *httpd.conf* file.
25. The ServerRoot directive specifies the directory location where the configuration, error, and log files are stored; the DocumentRoot directive defines the location where the documents are stored; and the LoadModule directive specifies the module to be loaded when the web server is started.
26. The ScriptAlias directive is used for specifying the directory location for CGI scripts.
27. The *ssh-keygen* command is used to generate authentication keys for the ssh use.
28. The secure Apache web server's configuration file is *ssl.conf* and it is located in the */etc/httpd/conf.d* directory.
29. Apache's host-level security can be set in the iptables firewall and/or in the *httpd.conf* file.

30. The default port used by the Apache web server is 80 and that for the secure shell server is 22.

Labs

Lab 21-1: Configure a Web Server with User- and Host-Based Security

Establish the web service on port 80 on *server.example.com* to provide both user-based and host-based security. Create DocumentRoot in */var/server.example.com* and store web contents in the */var/server.example.com/webcnts* directory. Create a custom *index.html* file and store it in an appropriate directory location. Restrict access to this website from *example.com* only for *user4* and all the members of group *dba*.

Lab 21-2: Configure Apache to Host Two Websites

Configure the Apache service on *insider.example.com* to host two name-based websites *www.website1.com* and *www.website2.com*. Configure the first website to use port 1001 and the second one to use port 1002. Use your own ideas for missing information.

Lab 21-3: Configure an SSL-Enabled, IP-Based Virtual Host

Configure an IP-based SSL-enabled virtual host on *physical.example.com* on port 443. Use your own ideas for missing information.

Lab 21-4: Deploy a CGI Script

Configure a web server on *server.example.com* using the default port to execute the *for_do_done.sh* script created in Chapter 14 "Shell Scripting, Building an RPM Package, & iSCSI". Use your own ideas for missing information.

Lab 21-5: Configure the OpenSSH Service

Configure the OpenSSH service on *insider.example.com* to permit ssh access into the system for all users in the *example.com* domain and prevent all users from *example.net* domain. Use your own ideas for missing information.

Lab 21-6: Establish Passwordless/Keyless Entry for a User

Establish ssh access for *user2* so that he is not prompted for either a password or a passphrase when he attempts to log in from *insider.example.com* to *physical.example.com*, and vice versa.

Appendix A: Sample RHCSA and RHCE Exams

Time Duration: 2.5 hours
Passing Score: 70%

Instruction 01: Exercises furnished in this appendix are in addition to the exercises and labs provided in the RHCSA section of this book. No solutions are provided as it is expected that the installation and configuration tasks covered in this book have been performed successfully. The exercises are not in any sequence.

Instruction 02: RHCSA exam is presented in the electronic format. You will be provided with a list of tasks and expected to complete them within the 2.5 hour time window.

Instruction 03: For each exercise provided, ensure that the settings and modifications made are available across system reboots. Also ensure that configured services work properly with SELinux activated in the enforcing mode and the iptables firewall enabled.

Instruction 04: To perform the exercises presented here, ensure that you have a physical system called *physical1* on the 192.168.2 network with RHEL6 (or Scientific Linux or CentOS) software installed. This system should also have the KVM virtualization, desktop, and the X Window packages loaded. See hardware requirements for RHEL6 installation on a physical system in Chapter 01.

Instruction 05: Do not refer to the material provided in this book or browse the Internet to perform the sample exam presented here. However, you can use the manual pages or the documentation located in the /usr/share/doc directory for successfully accomplishing the exercises.

Instruction 06: Use any method or tool that you feel comfortable with to complete these exercises. You may use graphical or text tools, or the commands. It is the result that matters.

Instruction 07: The following tasks should be performed on *server1* created in one of the exercises unless otherwise stated.

Exercise 01: Create a virtual machine on *physical1* and load RHEL6 in it. Do not configure a hostname or an IP address yet.

Exercise 02: Assuming that you have forgotten the *root* user password on the VM that you have just built and no *root* user session is currently open. Change the *root* password to Welcome01.

Exercise 03: Configure the first primary network interface on the VM with hostname *server1.example.com* (alias *server1*) and IP address 192.168.2.220/24 with gateway 192.168.2.1 (you may use a different IP address).

Exercise 04: Create another virtual machine on *physical1* and load RHEL6 in it. Configure *client1.example.com* (alias *client1*) as the hostname and 192.168.2.221 as the IP address.

Exercise 05: Modify the configuration on both *server1* and *client1* so that the default run level is set to 5.

Exercise 06: Configure both *server1* and *client1* to autostart when the virtualization services are brought up on *physical1*.

Exercise 07: Configure a virtual storage pool on *physical1* with 50gb of storage.

Exercise 08: Modify the SELinux mode on both *server1* and *client1* and set it to enforcing.

Exercise 09: Activate the iptables firewall on both VMs.

Exercise 10: Create *user1, user2, user3,* and *user4* accounts on *client1* with home directories in */home*. Set their passwords to Temp123$ and make *user2* and *user3* accounts to expire on December 31, 2013.

Exercise 11: Create a file called *testfile* in *user2*'s home directory on *client1* and configure ACLs on it to allow *user3* and *user4* to be able to read and execute it.

Exercise 12: Create a directory called *testdir1* in *user2*'s home directory on *client1* and configure it for group collaboration among members of the *admins* and the *dba* groups. Create the groups and add members as needed.

Exercise 13: Create a directory called *testdir* in *user3*'s home directory on *client1* and configure default ACLs on it for *user1* and *user4* for read and write permissions.

Exercise 14: Add *users1* and *user4* to the *linuxadm* group on *client1*. Give them full access to the */linuxadm* directory to read, write, and execute each other's files without modifying any permissions or ownership/group membership.

Exercise 15: Configure an FTP-based yum repository at *server1.example.com/pub*.

Exercise 16: Configure an NFS-based yum repository at *server1.example.com/var/ftp/pub*.

Exercise 17: Set up an HTTP-based installation server on *server1*.

Exercise 18: On *client1*, create a logical volume called *linuxadm* of size 1.5GB in the *vgtest* volume group with mount point */linuxadm* and ext4 file system structures. Define a group called *linuxadmin* and change group membership on */linuxadm* to *linuxadmin*. Set read/write/execute permissions on */linuxadm* for the owner and group members, and no permissions for others. Use the UUID of the file system and set it to automatically mount at each system reboot.

Exercise 19: On *client1*, create a logical volume *lvol1* of size 300MB in the *vg02* volume group with mount point */mntlvol1* and ext4 file system structures. Also create a swap logical volume called *swapvol1* of size 320MB in the *vg02* volume group. Use the UUID to automatically mount the file system and activate the swap at system reboots.

Exercise 20: On *client1*, create a logical volume *luksvol* of size 150MB in the *vg02* volume group with mount point */luksmnt* and ext4 file system structures. Configure this file system with LUKS encryption and set it to get automatically mounted at system reboots.

Exercise 21: On *client1*, create a standard partition of size 200MB on any available disk and format it with the ext4 file system structures. Create a mount point called */stdmnt* and mount the file system on it manually. Use the UUID of the file system and set it to automatically mount at each system reboot.

Exercise 22: On *client1*, create a standard partition of size 100MB on any available disk and set it to activate at system reboots.

Exercise 23: On *client1*, extend *linuxadm* in *vgtest* online by 300MB without losing any data.

Exercise 24: On *client1*, enable the cron access for *root* and *user3* users only and deny to everyone else.

Exercise 25: On *client1*, configure a cron job for the *root* user to search for core files in the */usr* directory and delete them the first day of each week at 11:49pm system time.

Exercise 26: On *client1*, set permissions on */linuxadm* so that all files created underneath get the parent group membership.

Exercise 27: On *client1*, configure */mntlvol1* so that users can delete their own files only and not other users'.

Exercise 28: Upgrade *client1* to a higher kernel version and make the new kernel the default. Do not overwrite the existing kernel.

Exercise 29: On *client1*, set the primary command prompt for user *root* to display the hostname, username, and the current working directory information.

Exercise 30: On *client1*, change the default base home directory for new users to */usr*.

Reboot the system and validate the configuration performed above.

Sample RHCE Exam

Time Duration: 2 hours
Passing Score: 70%

Instruction 01: Exercises furnished in this appendix are in addition to the exercises and labs provided in the RHCE section of this book. No solutions are provided as it is expected that the installation and configuration tasks covered in this book have been performed successfully.

Instruction 02: RHCE exam is presented in the electronic format. You will be provided with a list of tasks and expected to complete them within the 2 hour time window.

Instruction 03: For each configuration task that you perform, install appropriate packages, configure the SELinux support, configure the service to autostart at reboots, and ensure that user-based and host-based security is taken into consideration. This instruction may not apply to all the configuration tasks presented here.

Refer to Appendix A for additional instructions.

Exercise 01: On *server1*, configure the caching DNS service for the local network.

Exercise 02: On *server1*, add the forwarding DNS service capability to the caching DNS server.

Exercise 03: On *client1*, configure the DNS client functionality to use the DNS server configured on *server1*.

Exercise 04: On *client1*, configure the LDAP client functionality with search base dc=example and dc=com with locally signed TLS SSL encryption enabled. This exercise is based on the assumption that *server1* is configured as an LDAP server to service the systems on the *example.com* domain.

Exercise 05: On *client1*, enable the ssh service and allow access from *server1* and deny from *example.net* and *example.org*.

Exercise 06: On *client1*, disable direct *root* user logins via *ssh*.

Exercise 07: On *client1*, create a custom rpm package to include the */etc/yum.conf* file.

Exercise 08: Allow the *telnet* access into *client1* from 192.168.2 and 192.168.3 networks only.

Exercise 09: Configure *server1* to act as an IP forwarder.

Exercise 10: Configure *client1* to forward syslog messages to *server1*.

Exercise 11: Configure *server1* to receive syslog messages from *client1*, and record them locally.

Exercise 12: On *client1*, write an interactive shell script in */home/user2* that displays the directory contents and prompts to enter a file name for deletion.

Exercise 13: On *server1*, configure anonymous FTP access for both file downloads and uploads. Allow the access to systems on the 192.168.2 and 192.168.3 networks and disallow from everywhere else.

Exercise 14: Configure the FTP service on *client1* with no access from *example.org* and *example.net* domains. Enable access for *user1* and *user2* and deny for *user3* and *user4*.

Exercise 15: Configure Samba on *server1* to share the */linuxadm* directory accessible to *user1* and *user3* from the local network only with write and browse capabilities. Access the share on *client1* and make it auto-mountable at system reboots.

Exercise 16: Configure NFS on *server1* to share */mntlvol1* in read-only and */mntpart01* in read/write mode to users on the local network only. Use the standard mount method to mount */mntlvol1* and AutoFS to automount */mntpart01* on *client1*. Create file systems/mount points if they do not already exist. Set timeout for the AutoFS file system to 8 minutes.

Exercise 17: Configure web services on *server1* with access wide open to everyone to view the contents of the *index.html* file in a web browser window.

Exercise 18: Configure web services on *client1* to listen on port 3000 with DocumentRoot */var/www/client1*, web contents in */var/www/client1/webcnts*, index file *index.html*, and accessible to *user1* and *user2* from the local network only. Protect this web server from systems in the *example.net* and *example.org* domains.

Exercise 19: Configure a non-secure name-based virtual host on *server1* and *client1* to share IP address 192.168.2.222 on port 8000.

Exercise 20: Configure a 2-way ssh single sign-on for *user1* between *client1* and *server1*.

Exercise 21: On *server1*, configure a secure name-based virtual host for *server2.example.com* and *server3.example.com* on port 3020.

Exercise 22: Set up Postfix on *server1* for the local network only and have the mail for *user2* forwarded to *user4*, *example.com* to *user3*, and any mail for user *admin* forwarded to *user1*.

Exercise 23: Assuming that an iSCSI target is accessible to *server1*, configure an LVM-based file system with ext4 structures and make it automatically mounted at system reboots.

Exercise 24: Enable pop3s and imaps protocols on *server1* and allow access to local network users with mail from *example.net* and *example.org* blocked.

Exercise 25: Configure *client1* so that *user2* is able to access mail using imaps.

Exercise 26: Configure Postfix so that all mail go out via *server1* and appear to be originating from *example.com*. Have all mail for *info@example.com* received by *admin@example.com*.

Exercise 27: Add a rule to the iptables firewall on *client1* to allow incoming ftp and ssh traffic on *example.com*, drop traffic on *example.net*, and reject from *example.org*.

Exercise 28: Add a new chain called nat_chain to the iptables firewall on *server1* to allow the ftp and ssh traffic on *eth0* and *eth1* interfaces, respectively.

Exercise 29: Add a new route to *server1* so that it is able to access systems on the 192.168.3 network via the *eth0* interface.

Exercise 30: Configure the NTP client functionality on *client1* to get time updates from *server1*. Configure *server1* to provide the time.

Reboot the system and validate the configuration performed above.

Bibliography

The following websites, forums, and guides were referenced in writing this book:

1. www.redhat.com
2. docs.redhat.com/docs/en-US
3. www.centos.org
4. www.scientificlinux.org
5. www.hp.com
6. www.ibm.com
7. www.unix.org
8. www.linux.org
9. www.linuxhq.com
10. www.isc.org
11. www.wikipedia.org
12. www.gnome.org
13. www.pathname.com/fhs
14. www.ntp.org
15. www.samba.org
16. www.sendmail.org
17. www.postfix.org
18. www.dovecot.org
19. www.apache.org
20. www.openssh.org
21. www.netfilter.org
22. www.nsa.gov/selinux
23. www.opensource.org
24. RedHat installation guide
25. RedHat deployment guide
26. RedHat release notes
27. RedHat security guide
28. RedHat Security Enhanced Linux guide
29. RedHat storage administration guide
30. RedHat virtualization administration guide
31. RedHat virtualization getting started guide
32. RedHat virtualization host configuration and guest installation guide
33. RedHat logical volume manager administration guide
34. HP-UX 11i v3 book by Asghar Ghori (ISBN: 978-1-606-436547)

Glossary

. (single dot)	Represents current directory.
.. (double dots)	Represents parent directory of the current directory.
Absolute mode	A method of giving permissions to a file or directory.
Absolute path	A pathname that begins with a /.
Access Control List	A method of allocating file permissions to a specific user or group.
Access mode	See File permissions.
ACL	See Access Control List.
Address Resolution Protocol	A protocol used to determine a system's Ethernet address when its IP address is known.
Address space	Memory location that a process can refer.
Anaconda	RHEL's installation program.
Apache	A famous HTTP web server software.
Archive	A file that contains one or more compressed files.
Argument	A value passed to a command or program.
ARP	See Address Resolution Protocol.
ASCII	An acronym for American Standard Code for Information Interchange.
Auditing	System and user activity record and analysis.
Authentication	The process of identifying a user to a system.
AutoFS	The NFS client-side service that automatically mounts and unmounts an NFS resource on an as-needed basis.
Automounter	See AutoFS.
Background process	A process that runs in the background.
Backup	Process of saving data on an alternative media such as a tape or another disk.
Bash shell	A feature-rich default shell available in Red Hat Enterprise Linux.
Berkeley Internet Name Domain	A UC Berkeley implementation of DNS. See also DNS.
BIND	See Berkeley Internet Name Domain.
BIOS	Basic I/O System. Software code that sits in the computer's non-volatile memory and is executed when the system is booted.
Block	A collection of bytes of data transmitted as a single unit.
Block device file	A device file associated with devices that transfer data in blocks. For example, disk, CD and DVD devices.
Boot	The process of starting up a system.
Bootloader	A program that loads the operating system.
Broadcast client	An NTP client that listens to time broadcasts over the network.
Broadcast server	An NTP server that broadcasts time over the network.
Bus	Data communication path among devices in a computer system.
Cache	A temporary storage area on the system where frequently accessed information is duplicated for quick future access.
Caching DNS	A system that obtains zone information from a primary or slave DNS server and caches in its memory to respond quickly to client queries.

CentOS	A 100% unsponsored rebuild of Red Hat Enterprise Linux OS available for free.
CGI	See Common Gateway Interface.
Character special file	A device file associated with devices that transfer data serially. For example, disk, tape, serial and other such devices.
Child process	A sub-process started by a process.
CIFS	Common Internet File System. Allows resources to be shared among Linux, UNIX and non-UNIX systems. Also may be referred to as Samba.
Common Gateway Interface	A method for web server software to delegate the generation of web content to executable files.
Command	An instruction given to the system to perform a task.
Command aliasing	Allows creating command shortcuts.
Command history	A feature that maintains a log of all commands executed at the command line.
Command interpreter	See Shell.
Command line editing	Allows editing at the command line.
Command prompt	The OS prompt where you type commands.
Compression	The process of compressing information.
Core	A core is a processor that shares the chip with another core. Dual-core and quad-core processor chips are common.
Crash	An abnormal system shutdown caused by electrical outage or kernel malfunction, etc.
Current directory	The present working directory.
Daemon	A server process that runs in the background and responds to client requests.
De-encapsulation	The reverse of encapsulation. See Encapsulation.
Default	Pre-defined values or settings that are automatically accepted by commands or programs.
Defunct process	See Zombie process.
Desktop manager	Software such as GNOME that provides graphical environment for users to interact with the system.
Device	A peripheral such as a printer, disk drive, or a CD/DVD device.
Device driver	The software that controls a device.
Device file	See Special file.
Direct Memory Access	Allows hardware devices to access system memory without processor intervention.
Directory structure	Inverted tree-like Linux/UNIX directory structure.
Disk partitioning	Creating multiple partitions on a given hard drive so as to access them as separate logical containers for data storage.
Distinguished name	A fully qualified object path in LDAP DIT.
DIT	Directory Information Tree. An LDAP directory hierarchy.
DMA	See Direct Memory Access.
DNS	Domain Name System. A widely used name resolution method on the Internet.
Domain	A group of computers configured to use a service such as DNS or NIS.
Dovecot	A mail transfer agent used for receiving email.
Driver	See Device driver.
Encapsulation	The process of forming a packet through the seven OSI layers.
Encryption	A method of scrambling information for hiding it from unauthorized access.
Encryption keys	A private and public key combination that is used to encrypt and decrypt data.

EOF	Marks the End OF File.
EOL	Marks the End Of Line.
Ethernet	A family of networking technologies designed for LANs.
Export	Making a file, directory or a file system available over the network as a share.
Extent	The smallest unit of space allocation in LVM. It is always contiguous. See Logical extent and Physical extent.
Fedora	Red Hat sponsored community project for collaborative enhancement of Red Hat Entprise Linux OS.
Fibre channel	A family of networking technologies designed for storage networking.
File descriptor	A unique, per-process integer value used to refer to an open file.
File permissions	Read, write, execute or no permission assigned to a file or directory at the user, group or public level.
File system	A grouping of files stored in special data structures.
Filter	A command that performs data transformation on the given input.
Firewall	A software or a dedicated hardware device used for blocking unauthorized access into a computer or network.
FireWire	A bus interface standard designed for very fast communication.
Firstboot process	Program that starts at first system reboot after a system has been installed to customize authentication, firewall, network, timezone and other services.
Forwarding DNS	A system that forwards client requests to other DNS servers.
Full path	See Absolute path.
Gateway	A device that links two networks that run completely different protocols.
GID	See Group ID.
Globbing	See Regular expression.
GNOME	GNU Object Model Environment. An intuitive graphical user environment.
GNU	GNU Not Unix. A project initiated to develop a completely free Unix-like operating system.
GPL	General Public License that allows the use of software developed under GNU project to be available for free to the general public.
Group	A collection of users that requires same permissions on a set of files.
Group collaboration	A collection of users belonging to more than one groups to share files.
Group ID	A unique identification number assigned to a group.
GRUB	GRand Unified Bootloader is a GNU bootloader program that supports multiboot functionality.
GUI	Graphical User Interface.
Hangup signals	Signals that are meant to terminate a running process.
HAL	See Hardware Abstraction Layer.
Hardware Abstraction Layer	A piece of software implemented in between the kernel and the underlying system hardware to hide differences in hardware from the kernel to enable it to run on a variety of hardware platforms.
Hardware Compatibility List	A list of computer devices that have been tested by Red Hat to support the Red Hat Enterprise Linux version 6.
Home directory	A directory where a user lands when he logs into the system.
Host-based security	Security controls put in place for allowing or disallowing hosts, networks, or domains into the system.
Hostname	A unique name assigned to a node on a network.
Hypervisor	Software loaded directly on the physical computer to virtualize its hardware.
IMAP	See Internet Message Access Protocol.

Inode	An index node number holds a file's properties including permissions, size and creation/modification time as well as contains a pointer to the data blocks that actually store the file data.
Interface card	A card that allows a system to communicate to external devices.
Internet	A complex network of computers and routers.
Internet Message Access Protocol	A network protocol that is used to retrieve email messages.
Internet Small Computer System Interface	An IP-based storage networking protocol for sharing mass storage devices.
Interrupt request	A signal sent by a device to the processor to request processing time.
I/O redirection	A shell feature that gets input from a non-default location and sends output and error messages to non-default locations.
IP address	A unique 32- or 128-bit software address assigned to a node on a network.
IPTables	A host-based packet-filtering firewall software to control the flow of data packets.
IRQ	See Interrupt request.
iSCSI	See Internet Small Computer System Interface.
Job control	A shell feature that allows a process to be taken to background, brought to foreground, and to suspend its execution.
Job scheduling	Execution of commands, programs or scripts at a later time in future.
Kerberos	A network protocol for authenticating users over unsecure networks.
Kernel	Software piece that controls an entire system including all hardware and software.
Kernel-based Virtual Machine	Hypervisor software that supports virtualization of a physical computer.
Kickstart	A technique to perform a hands-off, fully-customized and automated installation of RHEL.
KVM	See Kernel-based Virtual Machine.
LAN	See Local Area Network.
LDAP	Lightweight Directory Access Protocol.
LDIF	LDAP Data Interchange Format. A special format used by LDAP for importing and exporting LDAP data among LDAP servers.
Link	An object that associates a file name to any type of file.
Link count	Number of links that refers to a file.
Linux	An open source version of the UNIX operating system.
Linux Unified Key Setup	A disk-encryption specification.
Load balancing	A technique whereby more than one servers serve client requests to share the load.
Local Area Network	A campus-wide network of computing devices.
Logical construct	If and case statements used in shell scripting for conditional execution of a piece of script.
Logical extent	A unit of space allocation for logical volumes in LVM.
Logical volume	A logical container that holds a file system or swap.
Login	A process that begins when a user enters a username and password at the login prompt.
Login directory	See Home directory.

Looping construct	Looping statements used in shell scripting for conditional repetitive execution of a piece of script.
LUKS	See Linux Unified Key Setup.
LVM	Logical Volume Manager. A disk partitioning solution.
MAC address	A unique 48-bit hardware address of a network interface. Also called physical address, ethernet address and hardware address.
Machine	A computer, system, RHEL workstation, RHEL desktop or a RHEL server.
Mail Delivery Agent	A program that is used to deliver email to recipient inboxes.
Mail Submission Agent	A program that is used to submit email messages to a mail transfer agent.
Mail Transfer Agent	A program that is used to transfer email messages from one system to another.
Mail User Agent	A mail client application that a user interacts with.
Major number	A number that points to a device driver.
Masquerading	A technique whereby systems on an internal network access external networks or the Internet using an intermediary device (router, firewall or RHEL system) so that requests originated from any of the internal systems appear to the outside world as being originated from the intermediary device.
MDA	See Mail Delivery Agent.
Metacharacters	Characters that have special meaning to the shell.
Minor number	A unique number that points to an individual device controlled by a specific device driver.
Module	Device drivers used to control hardware devices and software components.
Mounting	Attaching a device (a file system, a CD/DVD) to the directory structure.
MSA	See Mail Submission Agent.
MTA	See Mail Transfer Agent.
MUA	See Mail User Agent.
Name resolution	A technique to determine IP address by providing hostname.
Name space	A hierarchical organization of all DNS domains on the Internet.
NAT	See Network Address Translation.
Netfilter	A framework that provides a set of hooks within the kernel to enable it to intercept and manipulate data packets.
Netmask	See Subnet mask.
Network	Two or more computers joined together to share resources.
Network Address Translation	Allows a system on an internal network to access external networks or the Internet using a single, registered IP address configured on an intermediary device (router, firewall or RHEL system).
Network Time Protocol	A protocol used to synchronize system clock.
NFS	Network File System. Allows Linux and UNIX systems to share files, directories and file systems.
NFS client	A system that mounts an exported Linux or UNIX resource.
NFS server	A system that exports a resource for mounting by an NFS client.
Node	A device connected directly to a network port and has a hostname and an IP address associated with it. A node could be a computer, an X terminal, a printer, a router, a hub, a switch, and so on.
Node name	A unique name assigned to a node.
NTP	See Network Time Protocol.
Octal mode	A method for setting permissions on a file or directory using octal numbering system.

Octal numbering system	A 3 digit numbering system that represents values from 0 to 7.
Open source	Any software whose source code is published and is accessible at no cost to the general public under GNU GPL for copy, modification and redistribution.
OpenSSH	Free implementation of secure shell services and utilities.
Open Systems Interconnection	A layered networking model that provides guidelines to networking equipment manufacturers to develop their products for multi-vendor interoperability.
Orphan process	An alive child process of a terminated parent process.
OSI	See Open Systems Interconnection.
Owner	A user who creates a file or starts a process.
PackageKit	A group of graphical package management tools.
PAM	See Pluggable Authentication Module.
Parent directory	A directory one level above the current directory in the file system hierarchy.
Parent process ID	The ID of a process that starts a child process.
Password aging	A mechanism that provides enhanced control on user passwords.
Pattern matching	See Regular expression.
Performance monitoring	The process of acquiring data from system components for analysis and decision-making purposes.
Permission	Right to read, write or execute.
Physical extent	A unit of space allocation on physical volumes in LVM.
Physical volume	A hard drive or a partition logically brought under LVM control.
PID	See Process ID.
Pipe	Sends output from one command as input to the second command.
Plug and play	Ability of the operating system to add a removable device to the system without user intervention.
Pluggable Authentication Module	A set of library routines that allows using any authentication service available on a system for user authentication, password modification and user account validation purposes.
POP	See Post Office Protocol.
Port	A number appended to an IP address. This number could be associated with a well-known service or is randomly generated.
POST	Power On Self Test runs at system boot to test hardware.
Post Office Protocol	A network protocol that is used to retrieve email messages.
Postfix	A mail transfer agent used for sending out email.
PPID	See Parent process ID.
Primary DNS	A system that acts as the primary provider of DNS zones.
Primary prompt	The symbol where commands and programs are typed for execution.
Process	Any command, program or daemon that runs on a system.
Process ID	An identification number assigned by kernel to each process spawned.
Processor	A CPU. It may contain more than one cores.
Prompt	See Primary prompt and Secondary prompt.
Protocol	A common language that two nodes understand to communicate.
Proxy	A system that acts on behalf of other systems to access network services.
RAID	Redundant Array of Independent Disks. A disk arrangement technique that allows for enhanced performance and fault tolerance.
RAID array	A disk storage subsystem that uses hardware RAID.
Realm	A term commonly used to identify a domain in LDAP.
Recovery	A function that recovers a crashed system to its previous normal state. It may require restoring lost data files.

Redirection	Getting input from and sending output to non-default destinations.
Referral	An entity defined on an LDAP server to forward a client request to some other LDAP server that contains the client requested information.
Regular expression	A string of characters commonly used for pattern matching and globbing.
Relative path	A path to a file relative to the current user location in the file system hierarchy.
Remote logging	A configuration that allows one or several servers to send their alerts to a centralized remote loghost.
Replica	A slave LDAP server that shares master LDAP server's load and provides high availability.
Repository	A directory location to store software packages for downloading and installing.
Rescue mode	A special boot mode for fixing and recovering an unbootable system.
Resolver	The client-side of DNS.
RHEL	Red Hat Enterprise Linux.
Root	See Superuser.
Router	A device that routes data packets from one network to another.
Routing	The process of choosing a path over which to send a data packet.
RPM	Red Hat Package Manager. A file format used for software packaging.
Run control levels	Different levels of RHEL operation.
Samba	A client/server network protocol for sharing file and print resources.
Samba client	A system that accesses a Samba-shared resource.
Samba server	A system that shares a resource for Samba clients.
SATA	Serial Advanced Technology Attachment. This disk technology is a successor to the PATA drives.
Schema	A set of attributes and object classes.
Scientific Linux	A 100% unsponsored rebuild of Red Hat Enterprise Linux OS available for free.
Script	A text program written to perform a series of tasks.
SCSI	Small Computer System Interface. A parallel interface used to connect peripheral devices to the system.
Search path	A list of directories where the system looks for the specified command.
Secondary prompt	A prompt indicating that the entered command needs more input.
Secure shell	A set of tools that gives secure access to a system.
Security context	SELinux security attributes set on users, processes, and files.
Security Enhanced Linux	An implementation of Mandatory Access Control architecture for enhanced and granular control.
SELinux	See Security Enhanced Linux.
Server (hardware)	Typically a larger and more powerful system that offers services to network users.
Server (software)	A process or daemon that runs on the system to serve client requests.
Set Group ID	Sets real and effective group IDs.
Set User ID	Sets real and effective user IDs.
Setgid	See Set group ID.
Setuid	See Set user ID.
Shadow password	A mechanism to store passwords and password aging data in a secure file.
Shared memory	A portion in physical memory created by a process to share it with other processes that communicate with that process.
Shell	The Linux/UNIX command interpreter that sits between a user and kernel.
Shell program	See Script.
Shell script	See Script.

Shell scripting	Programming in a Linux/UNIX shell to automate a given task.
Signal	A software interrupt sent to a process.
Single user mode	An operating system state in which the system cannot be accessed over the network.
Slave DNS	A system that acts as an alternate provider of DNS zones.
Special characters	See Metacharacters.
Special file	A file that points to a specific device.
Standard error	A location to send error messages generated by a command.
Standard input	A location to obtain input from. The default is the keyboard.
Standard output	A location to send output, other than error messages, generated by a command.
Stderr	See Standard error.
Stdin	See Standard input.
Stdout	See Standard output.
Sticky bit	Disallows non-onwers to delete files located in a directory.
Stratum level	The categorization of NTP time sources based on reliability and accuracy.
String	A series of characters.
Subnet	One of the smaller networks formed by dividing an IP address.
Subnet mask	Segregates the network bits from the node bits.
Subnetting	The process of dividing an IP address into several smaller subnetworks.
Superblock	A small portion in a file system that holds the file system's critical information.
Superserver	A term used for the xinetd daemon.
Superuser	A user with unlimited powers on a RHEL system.
Swap	Alternative disk or file system location for demand paging.
Switch	A network device that looks at the MAC address and switches the packet to the correct destination port based on the MAC address.
Symbolic link	A shortcut created to point to a file located somewhere in the file system tree.
Symbolic mode	A method of setting permissions on a file using non-decimal values.
Symlink	See Symbolic link.
System	A computer or partition in a computer that runs RHEL.
System Administrator	Person responsible for installing, configuring and managing a RHEL system.
System call	A mechanism that applications use to request service from the kernel.
System console	A display terminal that acts as the system console.
System recovery	The process of recovering an unbootable system.
Tab completion	Allows completing a file or command name by typing a partial name at the command line and then hitting the Tab key twice.
TCP/IP	Transmission Control Protocol / Internet Protocol. A stacked, standard suite of protocols for computer communication.
TCP Wrappers	A security software for limiting access into a system at user and host levels.
Terminal	A window where commands are executed.
Test conditions	Conditions used in logical and looping constructs for decision making.
Text processor	Tools to manipulate rows and columns of text.
Tilde substitution	Using tilde character as a shortcut to move around in the directory tree.
Tty	Refers to a terminal.
UID	See User ID.
Unmounting	Detaching a mounted file system or a CD/DVD from the directory structure.
Upstart	An event-based replacement for the legacy init program.
USB	Universal Serial Bus. A bus standard to connect peripheral devices.

User-based security	Security controls put in place for allowing or disallowing users from hosts, networks, or domains into the system.
User ID	A unique identification number assigned to a user.
User Private Group	Referred to the GID that matches with the user's UID for the purpose of safeguarding the user's private data from other users on the system.
Variable	A temporary storage of data in memory.
Virtual console	One of several console screens available for system access.
Virtual file system	A file system that is created in memory at system boot and gets destroyed when the system is shut down.
Virtual host	An approach to host more than one websites on a single system using unique or shared IP addresses.
Virtualization	A technology that allows a single physical computer to run several independent logical computers (called virtual machines) with complete isolation from one another.
Virtual machine	A logical computer running on a physical computer.
Virtual Network Computing	A platform-independent client/server graphical desktop sharing application.
Volume group	A logical container that holds physical volumes, logical volumes, file systems and swap.
Web	A system of interlinked hypertext documents accessed over a network or the Internet via a web browser.
Web server	A system or service that provides web clients access to website pages.
Wide Area Network	A network with systems located geographically apart.
Wildcard characters	See Metacharacters.
X Window System	A protocol that provides users with a graphical interface for system interaction.
Zombie process	A child process that terminated abnormally and whose parent process still waits for it.
Zone	A delegated portion of a DNS name space.

Index

CPSIA information can be obtained at www.ICGtesting.com
Printed in the USA
BVOW06s0206211014

371664BV00006B/21/P